Fundamentals of Network Analysis and Flow Optimization

Alberto Garcia-Diaz
Don T. Phillips

"Fundamentals of Network Analysis and Flow Optimization," by Alberto Garcia-Diaz and Don T. Phillips. ISBN 978-1-63868-048-2 (casebound).

Published 2022 by Virtualbookworm.com Publishing Inc., P.O. Box 9949, College Station, TX 77842, US.

PREFACE

Our motivation for writing this textbook comes from our assessment and conviction that the field of network analysis and network flow optimization currently needs a single bibliographical source with an effective and comprehensive mix of theory and applications in industrial engineering, with emphasis on business analysis and operations research. The topical coverage includes the development, application and integration of both specialized linear programming network flow algorithms and other non-linear programming network procedures with the goal to provide a comprehensive treatment of network analysis and optimization. Although the topics included are well-known in the field of network analysis and have been covered by several authors, this book represents a unique resource that facilitates the study of a wide range of specialized network representations of industrial problems, illustrates the application of network methodologies to a wide range of relevant problems, and provides the solution of these problems using a specialized non-commercial *n\Network Optimization Program* (NOP) fully described in Chapter 1.

The two authors wrote a book titled *Fundamentals of Network Analysis* in 1981. Although the book went out of print over three decades ago, college professors continued using it as a textbook in graduate/undergraduate classes on network analysis at Texas A&M University, the University of Tennessee and other university Industrial and Systems Analysis departments for a number of years. Through these years we became aware of additions, deletions, modifications, corrections and reorganizations that would significantly improve the content of the book as a resource to advanced undergraduate students, master's students and beginning PhD students, in both Industrial Engineering and quantitative Business classes. Additionally, we kept in mind that the book should be a valuable resource to practitioners in the field of industrial engineering and operations research. After examining each topic in the original book and including new developments, our collection of network flow optimization topics resulted in this new book.

Three major accomplishments that make this book especially valuable for instructors, students and practitioners are:

1. A wider well-balanced and comprehensive topical coverage that includes traditional network flow optimization procedures and network analysis techniques including some that are outside the boundaries of traditional linear network flow representations, such as the Graphical Evaluation and Review Technique (GERT) for stochastic network applications, multi-traveling salesman problems, network reliability, and some project management methodologies that allow the consideration on nonlinear costs.
2. The development of a non-commercial Network Optimization Computer Program (NOP) which can be used to solve all of the Applications presented in this textbook, with the exception of network reliability analysis.
3. The following teaching aids to enhance the pedagogical process:
   - A *Reading Guide* containing a summary of important definitions, concepts, procedures, and formulas.
   - A *Power Point Presentation* for each topic (chapter) especially designed to support and enhance both the teaching and learning of all materials covered in class.
   - The *Network Optimization* Computer *Program* (NOP)

These teaching adds can be accessed through the link

https://drive.google.com/drive/folders/1HyBTiBlp6nZLZnORD9PSvK3a0GaGqhIA?usp=sharing

This book was written for two key audiences. It can serve as a textbook on the theory and application of network models for advanced undergraduate students and graduate students, especially Master's students and introductory level PhD students in Industrial Engineering, and Quantitative Graduate Business Programs. The textbook is also intended to be used as a modeling and solution resource for industrial practitioners in these areas.

The book is divided into nine chapters. Chapter 1 is an introduction to Network Analysis with brief sections on its history: literature bibliographical references outlining the development of the field, application areas, definitions, fundamental concepts, matrix representation of network graphs and parameters, and a description of the Network Optimization Program (NOP). Chapter 2 focuses on a wide variety of algorithmic procedures to solve problems in engineering and business. Efficient and optimal solutions to a wide range of problems which can be represented as a network. It is this representation and solution of problems which can be solved by NOP that makes Network Analysis a valuable addition to the Operations Research field. Chapters 3 and 4 discuss specialized dual and primal network-based methodologies for an efficient solution of larger network models. Chapters 5, 6, and 7 develop network-based methodologies for special application areas including the Traveling Salesman Problem for single and multiple salesmen a, Project Management applications and GERT (Graphical Evaluation and Review Technique) for a special class of stochastic networks. Chapter 8 introduces the students to the fundamentals of Multicommodity Networks. Finally, Chapter 9 focuses on fundamental principles and procedures for Network Reliability.

In closing, we want to express our gratitude to a number of students and instructors that have shared with us valuable suggestions on topical coverage, level of treatment, and organization of the material included in our first book on network analysis. Special recognition is due to many of our former students who have helped us with the network optimization computer code. In particular we extend our appreciation to Sandeep Gaudana for developing the first version of the VB code which through the years was modified and reorganized to result in the current version of NOP.

Dr. Alberto Garcia-Diaz
    Professor Emeritus, Department of Industrial and Systems Engineering
    University of Tennessee, Knoxville, Tennessee

Dr. Don T. Phillips
    Distinguished Research Professor, College of Engineering
    Department of Industrial and Systems Engineering
    Lamar University, Beaumont, Texas

TABLE OF CONTENTS

CHAPTER 1: INTRODUCTION TO NETWORK ANALYSIS 1

    1.1 Scope of Network Analysis 1

    1.2 Brief Historical Perspective 2

    1.3 Survey Articles, Books, Journals and Applications 4

    1.4 Definitions, Concepts and Notation 8

    1.5 Matrix Representation of Networks...................................... 12

    1.6 Conservation of Flow 15

    1.7 Maximum-Flow and Minimum-Cut 16

        1.7.1 Example for Network with Arc Capacities 18

        1.7.2 Example for Network with Node and Arc Capacities 19

    1.8 Network Optimization Program (NOP)..................................... 20

        1.8.1 Overview of NOP 20

        1.8.2 NOP Computer Screens 23

        1.8.3 Algorithms Available in NOP 24

        1.8.4 Data File Format for NOP Procedures........................ 26

        1.8.5 Sample Run (Dijkstra's Shortest Route Algorithm) 28

CHAPTER 2: NETWORK-FLOW MODELS AND APPLICATIONS 36

    2.1 Maximum Flow and Minimum Cost Flow Models 36

        2.1.1 Maximum Flow Model 37

        2.1.2 Maximum Flow Model 37

    2.2 Selected Applications of Network Models 37

        2.2.1 Equipment Replacement...................................... 38

        2.2.2 Project Planning 39

        2.2.3 Scheduling Tanker Operations 40

        2.2.4 The Transportation Model 41

        2.2.5 The Caterer Problem 42

2.2.6   Employment Scheduling 44

2.2.7   Fleet Scheduling Problem... 48

2.2.8   Production and Inventory Planning...................................... 50

2.2.9   Aircraft Fuel Allocation Problem.. 52

2.2.10 Conclusions ....

2.3 Linear Programming and Networks.. 54

2.3.1   The Transportation Problem.. 55

2.3.2   The Assignment Problem 55

2.3.3   The Maximum Flow Problem 56

2.3.4   The Shortest Path Problem 56

2.4 The Total Unimodularity Property 57

2.4.1   Theorem 1.. 57

2.4.2   Theorem 2.. 58

2.4.3   Theorem 3.. 59

2.4.4   Fundamental Result 59

2.5 Shortest Route Problem 62

2.5.1   Dijkstra's Algorithm 63

2.5.2   Example of Dijkstra's Algorithm 63

2.5.3   Shortest Route Problem: A Car Replacement Application 65

2.5.4   Ford's Algorithm 66

2.5.5   Example of Ford's Algorithm... 67

2.6 Shortest Path Model with Turn Penalties.................................. 68

2.6.1   Theorem 69

2.6.2   Algorithm 70

2.6.3   An Illustrative Example 71

2.7 Multi-Terminal and Multi-Source Shortest-Path Problem......... 73

2.7.1   Floyd's Algorithm 75

2.7.2   Example of Floyd's Algorithm 75

2.8 Special cases 79

    2.8.1 Ford's Algorithm to Find the Longest Route 80

    2.8.2 Most Reliable Route 80

    2.8.3 Most Reliable Route Between Each Pair of Nodes 80

2.9 The *K*-Shortest Path Problem 81

    2.9.1 The Double-Sweep Method 80

    2.9.2 Sample Problem for the Double-Sweep Method 80

2.10   Computational Complexity of Selected Shortest-Path Algorithms... 91

    2.10.1 Computational Complexity of Dijkstra's Method 91

    2.10.2 Computational Complexity of Floyd's Algorithm 92

    2.10.3 Computational Complexity of the Double-Sweep Method 92

2.11   Minimal Spanning Tree Problem 93

    2.11.1 Algorithm 93

    2.11.2 Example 93

2.12   The Maximum Flow Problem 95

    2.12.1 Labeling Procedure 96

    2.12.2 Example of the Labeling Procedure 97

2.13   Multi-Terminal Maximum Flow Problem 100

    2.13.1 The Gomory-Hu Algorithm 100

    2.13.2 Example 103

2.14   Maximal Capacity Path Problem 106

    2.14.1 Max-Capacity Chain for Networks with Single Source and Terminal 107

    2.14.2 Max-Capacity Chains for Networks with Multiple Sources and Terminals 107

    2.14.3 Example 107

CHAPTER 3: THE OUT-OF-KILTER ALGORITHM 125

PART I: NETWORK-FLOW OPTIMIZATION WITH THE OKA: FUNDAMENTAL
CONCEPTS 125

    3.1 Relevant Terminology 125

3.2 Primal and Dual Model Formulation 127

3.3 Optimality Conditions 129

3.4 Corrective Actions for Out-of-Kilter Arcs 131

3.5 Labeling Procedure for Flow Changes 131

3.6 Changes in Node Dual-Variable Values 132

3.7 Description of the Out-of-Kilter Algorithm 133

3.8 Numerical Example 133

PART II: NETWORK-FLOW OPTIMIZATION WITH THE OKA: MODELING CONCEPTS 140

3.9 Special Cases 141

    3.9.1 Transportation Problem.. 141

    3.9.2 Assignment Problem 142

    3.9.3 Maximum Flow Problem.. 142

    3.9.4 Shortest Path Problem.. 142

    3.9.5 The Shortest-Path Tree Problem 143

    3.9.6 The Transshipment Problem 143

3.10 Nonlinear Arcs Flow Costs 144

3.11 A Production-Distribution Problem 145

3.12 Summary and Conclusions 151

PART III. NETWORK-FLOW OPTIMIZATION WITH THE OKA: SELECTED APPLICATIONS 152

3.13 Bottleneck Assignment Problem 152

3.14 Scheduling Workers to Time-Dependent Tasks 154

3.15 A Wholesale Storage and Marketing Problem 157

CHAPTER 4: PRIMAL SIMPLEX PROCEDURES FOR NETWORK OPTIMIZATION 164

4.1 External Flows 164

4.2 Network Transformation Procedure 166

4.3 Pure Network Flow Cost Minimization Problem 167

    4.3.1 Overview of the Network Specialization of the Simplex Method.................... 169

    4.3.2 Basic Feasible Solution.. 169

    4.3.3 Dual Variables Computation... 171

4.3.4   Optimality Condition and Selection of Entering Variable............................171

4.3.5   Selection of the Leaving Variable...172

4.3.6   Changing the Basis 173

4.3.7   Example 173

4.3.8   Final Remarks..179

4.4   Generalized Network Flow Cost Minimization Problem 179

4.4.1   The Inverse of the Basis Matrix...181

4.4.2   Basic Feasible Solution 181

4.4.3   Dual Variables Computation 182

4.4.4   Optimality Condition 181

4.4.5   Selection of the Leaving Variable...181

4.4.6   Changing the Basis 182

4.4.7   Example 182

4.5   Finding an Initial Feasible Basis 188

4.5.1   Creating Basic Arcs Connected to the Slack Node...............................188

4.5.2   Creating Artificial Basic Arcs Connected to the Slack Node 188

CHAPTER 5: THE TRAVELING SALESMAN PROBLEM 194

5.1   The Traveling Salesman Problem 194

5.1.1   Initial Lower Bound...195

5.1.2   Branching..195

5.1.3   Penalties..196

5.1.4   Distance Matrix Modifications..196

5.2   Little's Algorithm 196

5.2.1   Steps of the Algorithm...197

5.2.2   Example 197

5.3   The Multi-Traveling Salesman Problem 205

5.3.1   Transformation Technique...205

5.3.2   A Heuristic Procedure..207

5.3.3   Example with $M = 2$ 209

5.3.4   Example with $1 \leq M \leq 2$ 211

5.3.5   Comparison of Procedures...213

CHAPTER 6: NETWORK-BASED PROJECT MANAGEMENT PROCEDURES............219

PART I: PROJECT MANAGEMENT WITH CPM AND PERT.................................219
  6.1 Origin and Use of PERT...219
  6.2 Network Construction...221
  6.3 Event Earliest and Latest Realization Times.................................224
  6.4 Event Slacks, Earliest/Latest Activity Completion Times and Activity
      Floats...225
  6.5 Illustrative CPM Example 228
  6.6 PERT Methodology 230
  6.7 Illustrative PERT Example 219

PART II: RESOURCE ALLOCATION IN PROJECT NETWORKS....................234
  6.8   Time Vs. Cost: Dollar Allocations 235
  6.9   Resource Loading and Resource Leveling 238

PART III: OPTIMIZATION OF TIME/COST TRADE-OFFS IN PROJECT NETWORKS...241
  6.10 Time/Cost Trade-Offs in CPM Networks 241
      6.10.1 A Network Flow Algorithm 244
      6.10.2 Description of the Algorithm 247
      6.10.3 Illustrative Example 250
  6.11 Nonlinear Activity Cost Functions...256
      6.11.1 Mathematical Model Formulation..257
      6.11.2 Solution Approach 258
      6.11.3 Illustrative Example 261

CHAPTER 7: GRAPHICAL EVALUATION AND REVIEW TECHNIQUE (GERT)..........274
  7.1 Network Representation 274
  7.2 GERT Basic Procedures 276
  7.3 Basic Concepts of Linear Flowgraphs 274
  7.4 Definitions...281
  7.5 Mason's Rule for Closed Flowgraphs...283
  7.6 GERT Procedural Steps 283
  7.7 GERT Applications...285
      7.7.1 Production of a High-Risk Item.......................................286
      7.7.2 Material Processing 286
      7.7.3 Determination of Time Standards 289

CHAPTER 8: AN INTRODUCTION TO MULTICOMMODITY NETWORK
             FLOW OPTIMIZATION 299
PART I: MULTICOMMODITY NETWORK FLOWS 299

8.1 Linear Programming Formulations..301

8.2 A Special Class of Integer Multicommodity Networks 302

8.3 Approximate Solutions of Multicommodity Transportation Problems...307

    8.3.1 A Fruit Distribution Problem 309

    8.3.2 Error Bounds for Aggregation...310

8.4 Maximal Flows in Multicommodity Networks 311

8.5 Multicommodity Flows in Undirected Networks............................314

    8.5.1 A Two-Commodity Flow Problem 316

    8.5.2 Algorithmic Steps 316

8.6 Maximal Flows and Funnel Nodes 318

8.7 Applications of Multicommodity Networks...................................319

    8.7.1 Tanker Scheduling 319

    8.7.2 Urban Transportation Planning...321

    8.7.3 Computer Communication Models 321

8.8 Notes and Remarks 322

PART II: MULTICOMMODITY NETWORK FLOW COST MINIMIZATION PROCEDURES
....323

8.9   Mathematical Formulation..323

8.10 Solution Procedures 326

    8.10.1 Price-Directive Dantzig-Wolfe Decomposition Method 327

    8.10.2 An Example of the Dantzig-Wolfe Decomposition Method 329

    8.10.3 Lagrangian Relaxation Method 335

    8.10.4 An Example of the Lagrangian Relaxation Method 336

    8.10.5 Resource-Directive Decomposition Method 340

CHAPTER 9: AN INTRODUCTION TO NETWORK RELIABILITY 348

9.1 Introduction 348

9.2 Node and Arc Failures in Deterministic Networks 349

    9.2.1 Minimum Number of Arcs to Disconnect a Network.............350

    9.2.2 An Illustrative Example..350

    9.2.3 Minimum Number of Nodes to Break all Paths 351

    9.2.4 An Illustrative Example..351

    9.2.5 Theorem..353

    9.2.6 An Illustrative Example..354

9.3 Deterministic Network Reliability..356

    9.3.1 Proof 356

    9.3.2 Examples 357

    9.3.3   Other Reliability Measures...357

9.4 Probabilistic Network Reliability 358

    9.4.1   State-Space Enumeration Method 358

    9.4.2   Example of State-Space Enumeration Method.....................359

    9.4.3   Inclusion-Exclusion Method...360

    9.4.4   Example of Inclusion-Exclusion Method............................361

    9.4.5   Disjoint Product Method...361

    9.4.6   Example of Disjoint Product Method...................................362

    9.4.7   Factoring Method 363

    9.4.8   Example of Factoring Method 364

9.5 Introduction to Algebraic Methods 365

9.6 Closing Paragraphs..367

BIBLIOGRAPHY...367

Chapter 1

INTRODUCTION TO NETWORK ANALYSIS

*"Cheshire-Puss", she began, rather timidly,
"Would you tell me, please, which way I
ought to go from here"?
"That depends a great deal on where you
want to get to", said the cat.
"I don't much care where-," said Alice.
"Then it doesn't matter which way you go,"
said the cat.
"-so long as I get somewhere," Alice added.*

From *Alice in Wonderland*
Lewis Carroll

As the Cheshire cat so astutely observed in Wonderland, progress can often be made if one wanders around long enough. There are often better ways to search for a solution than the aimless wandering of Alice. Network modeling techniques often provide the framework and computational structure to greatly improve many traditional approaches to systems analysis. The purpose of this chapter is to present the necessary machinery to both understand and apply the fundamental algorithms presented in this textbook. The application of network analysis techniques often requires not only *"where you want to get to"* but also *"which way you ought to go."* Hopefully, this chapter will aid in defining both strategies.

1.1 SCOPE OF NETWORK ANALYSIS

A contemporary society can be viewed, in part, as a system of interacting networks for transportation, communication, and distribution of energy, goods, and services. The complex structure and cost of these networks requires that existing networks be efficiently used and that new ones be rationally designed. Network analysis techniques can be of great value in the design, improvement, and rationalization of a wide range of complex large-scale systems.

Network-flow models and solution techniques provide a rich and powerful framework from which many engineering problems can be formulated and solved. The visual and logical structure of network-flow analysis often provides a fresh and natural approach from which further engineering analysis can proceed. Once only a small segment in the field of operations research, network analysis techniques have recently emerged as a viable and computationally tractable approach to solving significant problems faced by modern engineering analysts.

Network models and analysis are widely used in operations research for diverse applications, such as the analysis and design of large-scale irrigation systems, computer networks, cable television networks, transportation systems, and ground and satellite communication networks. Efficient network methodologies have been implemented to solve industrial problems, such as the warehousing and distribution of goods, project scheduling, equipment replacement, cost control, traffic studies, queueing analysis, assembly-line balancing, inventory control, and manpower allocation, to name a few. Pritsker [27] provides some insight as to the popularity and usefulness of network analysis techniques:

1

Networks and network analyses are playing an increasingly important role in the description and improvement of operational systems primarily because of the ease with which systems can be modeled in network form. This growth in the use of networks can be attributed to:

1. The ability to model complex systems by compounding simple systems.
2. The mechanistic procedure for obtaining system figure-of-merits from networks.
3. The communication mechanism to discuss the operational system in terms of its significant features.
4. A means for specifying the data requirements for analysis of the system.
5. A starting point for analysis and scheduling of the operational system.

Item 5 was the original reason for network construction and use. The advantages that accrued outside of the analysis procedure soon justified the expanded use of a network approach. Considerable work is motivated by the need for extending present analysis procedures to keep pace with applications of networks.

Network analysis is not a discipline confined to only one branch of academia or industry. Indeed, the real strength of the network approach lies in the fact that it can be successfully applied to almost any problem when the modeler has enough knowledge and insight to construct the proper network representation. The advantages of using network models can be stated as follows:

1. Network models accurately represent many real-world systems.
2. Network models seem to be more readily accepted by non-analysts than perhaps any other type of models used in operations research. This phenomenon appears to stem from the notion that "a picture is worth a thousand words." Managers seem to accept a network diagram more easily than they do abstract symbols. Additionally, since network models are often related to physical problems, they can be easily explained to people with little quantitative background.
3. Network algorithms facilitate extremely efficient solutions to some large-scale models.
4. Network algorithms can often solve problems with significantly more variables and constraints than can be solved by other optimization techniques. This phenomenon is due to the fact that a network approach often allows the exploitation of particular structures in a model.

1.2   BRIEF HISTORICAL PERSPECTIVE

In order to follow the historical development of network analysis we can start at the kingdom of Prussia. Prussia was a kingdom and the largest and most important of the German States. Berlin was its capital. Prussia remained a kingdom in the German empire until Germany became a republic in 1918. It was abolished as a state in 1947 and divided among West Germany, East Germany, USSR, and Poland. One of the cities in Prussia was Konigsberg. In this city there is an island called Kneiphof, surrounded by the river Pregel, as shown in Figure 1.1. In this figure, A represents an island and B, C and D represent land areas. The branches of the river are crossed by seven bridges *a, b, c, d, e, f,* and *g*. Is it possible to arrange a route from any place to any other place such that each bridge is crossed exactly once?

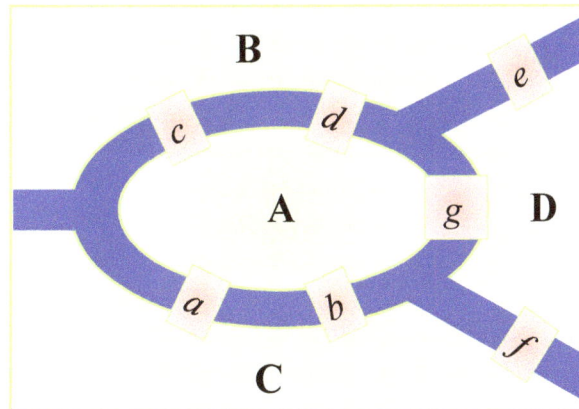

Figure 1.1. Konigsberg's bridges.

In 1736, at the age of 29, Euler modeled this problem using nodes to represent land areas and arcs to represent bridges. In his famous paper "*The solution of a problem relating to the geometry of position*" (translated from German) [11], Euler demonstrated that this problem had no solution and indicated the required conditions for a solution to exist. Euler's procedure for solving the Konigsberg bridge problem can be found in several bibliographical references. Examples of these include the references by Biggs, Lloyd, and Wilson [5], Newman [26], and Scientific American [29].

Network analysis relies heavily on graph theory, a branch of mathematics that evolved with Leonhard Euler's formulation and solution of the famous *Konigsberg bridge problem* in 1736 [5]. More than a century later, James Clerk Maxwell and Gustav Robert Kirchhoff discovered certain basic structures in network formulations that could be computationally exploited in their analysis of electric circuits. Since then, network analysis has become an important tool in the design and optimization of electrical systems. For example, early in the twentieth century, telephone engineers in Europe and the United States devised network computational algorithms to determine the best capacity of telephone trunk lines and switching centers in order to guarantee specified levels of customer serviceability. In the 1940s, during the period of World War II, the development of pioneering work in modern network analysis was conducted by Hitchcock [17] in 1941 and Koopmans [22] in 1947. Since then, network analysis has been a very active and productive research area with well over 1000 published papers. The emphasis of research in the 1950s and early 1960s was on the formulation of new models and development of new algorithms. Later, emphasis shifted to the extension, computer implementation, and analysis of previously developed models and algorithms.

In 1975 the Royal Swedish Academy of Science awarded the Nobel Prize in Economic Science to Professor Leonid Kantorovich of the USSR and Professor Tjalling C. Koopmans of the U.S. for their contributions to theory of optimum allocation of resources. They are associated with some of the first papers describing network flow problems as we know them today.

In 1956 Alex Orden proposed a generalization of the *transportation problem* known as the *transshipment problem*. At approximately the same time, the maximal flow problem and the minimal-cost flow problem were formulated and implemented by Lester Ford and Delbert Fulkerson. From 1950 through 1965 much activity was directed toward developing algorithms for linear network flow models. These algorithms can be generally classified as *primal simplex methods* and *primal-dual simplex methods*.

The primal simplex *network specialization* theory began with the work of George Dantzig at Stanford University and culminated with a paper by Ellis Johnson. This work has been published in books by Dantzig and by Charnes and Cooper. The primal-dual methods originated with Harold Kuhn's *Hungarian algorithm* for the assignment problem and culminated with the *out-of-kilter algorithm* by Fulkerson.

Most of the work since the mid-1960s has involved the efficient computational implementation of these basic techniques and their extension to (1) linear networks with arc flow gain/loss factors; (2) linear multi-commodity networks; (3) linear networks with side constraints; and (4) network problems with nonlinear convex cost functions. New methods invented in the eighties challenged old ones, both in terms of practical efficiency and theoretical worst-case performance. Two of these methods, originally proposed by Dimitri P. Bertsekas are called *relaxation and auction*.

1.3 SURVEY ARTICLES, BOOKS, JOURNALS AND APPLICATIONS

This section is intended to provide the reader with a list of significant bibliographical references that address the scope of network models and help to understand and appreciate their impact in solving relevant problems in a modern society. The section briefly presents several survey articles, network analysis and optimization books, technical journals, and typical applications of the network methodology.

Survey Articles

The purpose of a literature review is to gain an understanding of the existing research and discussions relevant to a particular area of study with the aim of building knowledge in a field, such as network analysis and optimization. There are many survey articles available in the published literature on historical developments and reviews of network procedures for specific applications. A comprehensive list with relevant discussions of these articles is beyond the scope of this book. Our main goal is to provide the reader with a foundation of knowledge and identify areas of prior scholarship relevant to selected topics of network analysis and optimization which will be presented in this book. With this purpose in mind the following articles have been selected among many excellent publications.

"A Survey of Deterministic Networks" [6]

The state-of-the-art of deterministic networks is surveyed with a discussion of shortest path, transportation, assignment, transshipment, maximum flow, minimum spanning tree, Chinese postman, Euler-path and multi-commodity flow models. The computational complexity of these network models is also discussed. This survey paper concentrates on models and algorithms that can be used to solve large-scale problems.

"Flow Networks and Combinatorial Operations Research" [15]

The initial part of the paper proves the following results: (1) For any network the maximum flow amount from source to sink is equal to the minimum cut capacity relative to the source and the sink. (2) A flow from source to sink is maximal if and only if there is no flow-augmenting path. (3) If all arc capacities are integers, the maximal flow and all arc flows are integer. Later, some feasibility theorems are proved. Then minimal cost flows, maximal dynamic flow, multi-terminal maximal flows are discussed. The second part of the paper discusses combinatorial problems like network potentials and shortest chains, optimal chains in a cyclic network, and the well-known knapsack problem. This combinatorial problem is viewed as finding the longest chain in a suitable acyclic network. Finally, the following problems are discussed: equipment replacement, project planning, assignment, production and inventory planning, optimal capacity scheduling, minimal spanning tree, traveling salesman, and minimal K-connected network problems.

"Some Network Flow Models in Management Science" [10]

This reference gives an expository treatment of four important network models: shortest path problems, maximal flow models, signal flow graphs and activity networks. This paper can be considered an attempt to answer the following questions that are often asked by practitioners and students in the field of Management Science and Operation Research: What is the theory of network model formulations, which operational systems have been modeled by networks, and how useful are such models?

"Transportation Planning: Network models and their Implementation" [23]

Transportation planning plays an essential role in shaping regional and urban lifestyle. Complex decisions regarding policy alternatives for railroads, shipping, airline, and roadway traffic can be formulated and analyzed using network optimization techniques. In this paper the authors survey applications of network algorithms to transportation planning, and efficient computer implementation. Contributions and use of shortest paths, minimum cost networks flows, traffic equilibrium, vehicle routing, and network design are listed.

On the History of Combinatorial Optimization [30]

A well-known reference on the history of combinatorial optimization until 1960 is an article by Alexander Schrijver. In it we find historical perspectives and developments of a number of network flow optimization procedures including the assignment problem, the transportation problem, the maximum flow problem, the minimal spanning tree, the shortest path problem and the traveling salesman problem.

A Survey of Linear Cost Multicommodity Network Flows [21]

Multicommodity network flow models with linear costs are uniquely structured linear programming formulations that lend themselves to be solved with network specializations of the simplex method. This article presents several models with specific properties identified for each type of formulation presented. Three basic approaches known as price-directive decomposition, resource-directive decomposition, and partitioning methods are reviewed. Price-directive decomposition methods coordinates a master program and several subprograms by changing the objective function of the subprograms. Resource-directive decomposition methods iteratively distribute arc capacities among individual commodities. At each iteration, a minimal flow cost optimization problem is solved for each commodity. Partitioning methods are specialized formulations of linear programs where the current basis is partitioned to exploit the structure of the model formulations. Basic algorithms for each type are developed and discussed in great detail.

Network Books

The following list shows authors and bibliographical references of well-known books published between 1961 and 1999 devoted to *network analysis* or *linear programming* with substantial portions addressing the topic of *network analysis and optimization*: These books have been and continue to be used by students, instructors, practitioners and researchers in the area of networks despite the fact that some of them are out of print. Although we have tried to include all significant references, we cannot claim that our list is comprehensive. In particular, we have not included books in very specialized areas since this is not the main focus of this list.

Charnes and Cooper, 1961 [8]
Ford and Fulkerson, 1962 [13]
Dantzig, 1963 [9]
Busacker and Saaty, 1965 [7]
Hu, 1969 [18]
Frank and Frisch, 1971 [14]
Whitehouse, 1973 [31]
Bazaraa and Jarvis, 1977 [2]
Minieka, 1978 [25]
Jensen and Barnes, 1980 [20]
Phillips and Garcia-Diaz, 1981 [27]
Lawler, Lenstra, Rinnoy Kan and Shmoys, 1985 [23]
Bertsekas, 1991 [3]
Evans and Minieka, 1992 [12]
Glover, Klingman and Phillips, 1992 [16]
Ahuja, Magnanti and Orlin, 1993 [1]
Bertsekas, 1998 [4]

Technical Journals

The list of publication outlets for network analysis and applications can be very extensive. To provide guidance to the students of network flow models and related topics, we will list three very well-known journals addressing network models and applications at varying degrees of both theoretical approach and application scope. The succinct statements provided for the journals are based on the journal descriptions available online.

Networks: An International Journal. The goal of this journal is to provide a central forum for the distribution of timely information about network problems, their design and mathematical analysis, as well as efficient algorithms for carrying out optimization on networks. The nonstandard modeling of diverse processes using networks and network concepts is also of interest. Consequently, the disciplines that are useful in studying networks are varied, including applied mathematics, operations research, computer science, discrete mathematics, and economics. *Networks* publishes material on the modeling of problems using networks, the analysis of network problems, the design of computationally efficient network algorithms, and innovative case studies of successful network applications. Since the audience for this journal is then necessarily broad, articles that impact multiple application areas or that creatively use new or existing methodologies are especially appropriate.

Operations Research (INFORMS, formerly Operations Research and Management Science). The main purpose of the journal is to publish high-quality articles that represent the true breadth of the methodologies and applications that define O.R. A number of excellent articles published in this journal address the theory and applications of network models. INFORMS publishes a collection of journals covering several areas including decision analysis, information systems research, applied analytics, computing, data science, optimization, management science, manufacturing and service operations, marketing science, operations research, stochastic systems transportation science, and a few others.

IISE Transactions (formerly IIE Transactions for the years 1983-2016 and AIIE Transactions for the years 1969-1982). As the flagship journal of the Institute of Industrial and Systems Engineering, IISE Transactions publishes original high-quality papers on a wide range of topics of interest to industrial engineers who want to remain current with the state-of-the-art technologies. Specifically, the journal publishes research on *design manufacturing, operations engineering and analytics, quality and reliability engineering,* and *scheduling and logistics.*

Typical Network Applications

Applications of network formulations cover a very wide range of disciplines and theoretical content of the analysis. An attempt has been made to identify the areas of application where the network models are more frequently used or where there is a significant potential to enhance the contribution of the network approach to solving problems that are relevant to a modern society. A representative, but not exhaustive, list of typical applications of network models include those listed below.

Warehousing and distribution
Project scheduling
Equipment replacement
Cost control
Traffic studies
Transportation systems
Queueing analysis
Assembly line balancing
Inventory control
Manpower allocation
Homeland security
Telecommunications
Energy management
Health care system design
Computer system integration
Reliability of integrated systems

Although a comprehensive treatment of all of these applications is beyond the scope of this book, nevertheless a number of the applications will be considered at varying levels of detail in this textbook.

1.4 DEFINITIONS, CONCEPTS AND NOTATION

Nodes and Arcs

A *node* is a point where a commodity flow is originated, relayed, or terminated. If the flow is originated at the point, the node is called a *source node*; if the flow is relayed through the point, the node is called an intermediate node; if the flow is terminated at the point, the node is called a *terminal node*. Formally, if the net flow (defined as flow out minus flow in) is strictly positive the node is a source, if it is equal to zero the node is an intermediate node, and if it is strictly negative the node is a terminal.

The nodes of a network can represent highway intersections, power stations, telephone exchanges, railroad yards, airline terminals, water reservoirs, computers, well-defined occurrences in time, or simply project milestones. To preserve the individuality of the nodes and keep track of them in the process of analysis, they are ordinarily assigned numerical labels. Although it sometimes helps the mental process to number the nodes in an orderly fashion, this obviously has no significance whatsoever and, in general, the numbering is assumed to be arbitrary.

An *arc* is a generic channel for transmission of flow from a given node to another node directly connected to it. The arcs of a network can represent roads, power lines, telephone lines, airline routes, water mains, or generalized channels through which entities flow. In some instances, the arcs have no physical meaning but serve to direct flow in a logical sequence, or to maintain a specified precedence relationship.

A directed arc has an *orientation* indicated by an arrowhead. There are three types of arcs (also called *branches*): *undirected*, *directed*, and *bidirected*, depending on the specification of the direction along which flow can move through the arcs. An undirected arc has no direction, while a directed arc has a unique direction of possible flow specified in advance. A bidirected arc has two opposite directions for flow. To represent networks graphically, we use circles to indicate nodes, lines to indicate arcs, and arrowheads to indicate net flow orientation. We also use the notation (i, j) for a directed arc leading from node i to node j. Figure 1.2(a) shows an undirected arc, Figure 1.2(b) a directed arc, and Figure 1.2(c) a bidirected arc.

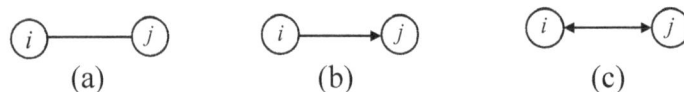

Figure 1.2. Node-arc representations: (a) undirected arc; (b) directed arc; (c) bidirected arc.

If an arc is *traversed* by flow in the same direction as the orientation, the arc is said to be a *forward* arc. If it is traversed in direction opposite to its orientation, the arc is said to be a *reverse* arc. For most practical purposes there is no necessity to distinguish between undirected and bidirected arcs. Further, it is evident that a bidirected arc between two arbitrary nodes i and j can be represented by two directed arcs (i, j) and (j, i), as shown in Figure 1.3.

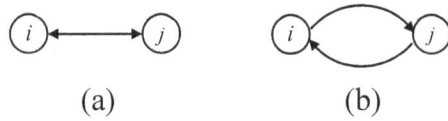

(a)   (b)

Figure 1.3. Representation of a bidirected arc by two directed arcs with opposite orientations: (a) bidirected arc; (b) equivalent representation.

When the flow across an arc is constant, the arc is referred to as a *pure arc*; otherwise, it is referred to as a *generalized arc*. Generalized arcs allow *gains* and *losses* on arc flows. In these arcs the flow at the end-node is equal to the flow at the start-node multiplied by a positive value, referred to as the arc *multiplier*. If the multiplier is less than 1 there is a loss, if it is 1 there is no loss or gain, and if it is larger than 1 there is a gain. If not specifically indicated, it will be assumed that arc loss/gain multipliers are equal to 1 and then the flow at the beginning node is equal to the flow at the end node of each arc.

Networks and Graphs

A *network* is an interconnected collection of nodes and arcs, with one or more quantitative elements associated with each arc or node. Networks can be classified as *pure networks* if all arc multipliers are equal to 1, or *generalized networks*, otherwise. As previously mentioned, the arcs of a network can be considered as channels through which a generic commodity can *flow*. The net amount of commodity at a given node is equal to the difference between the amount of flow leaving the node and the amount of flow arriving at the node. If the net amount of commodity is strictly positive, the node is called a source node. If it is strictly negative, the node is called a sink or terminal node.

A network may contain more than one sink or source node. For convenience, we will frequently refer to a *source node* as any node that only originates flow, and a *sink node* as any node that only terminates flow. For example, in the network of Figure 1.4, nodes 1 and 2 are source nodes and nodes 6 and 7 are sink nodes.

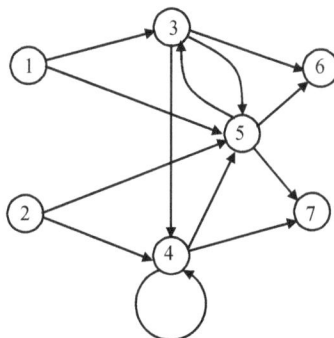

Figure 1.4. Illustrative network.

A graph G = $(\mathbf{N}, \tau)$ is the pair consisting of a set $\mathbf{N}$ and function τ. Whenever possible the elements of set $\mathbf{N}$ will be represented by points in the plane and if x and y are two points such that

9

$y \in \tau x$, then they will be joined by a continuous line with an arrow head pointing from x to y. Hence, an element of **N** is called a point or vertex or node of the graph while the pair (x,y) with $y \in \tau x$ is called an arc of the graph. The set of arcs of the graph is denoted by **A** and instead of G $= (\mathbf{N}, \tau)$ we can write G = (**N, A**). As in the case of a network, a *connected* graph is one such that at least one path exists between every pair of nodes. As an illustration of this definition, the graph shown in Figure 1.4 will be considered again. Here, **N** = {1,2,3,4,5,6,7}. Also, $\tau_1 = \{3,5\}$, $\tau_2 = \{4,5\}$, $\tau_3 = \{4,5,6\}$, $\tau_4 = \{4,5,7\}$, $\tau_5 = \{3,6,7\}$, $\tau_6 = \tau_7 = \varnothing$. Therefore, **A** = {(1,3), (1,5), (2,4), (2,5), ..., (5,7)}.

Paths, Chains, Cycles, Circuits and Self-loops

Two important terms used in network flow analysis are path and chain. These two terms are defined as sequences of nodes connected by arcs. The difference between the two terms is caused by the orientation of the arcs. In this textbook a *path* is defined as a connected sequence of arcs, not necessarily all having the same orientation. If all arcs have the same orientation, the path is referred to as a *chain*. A *cycle is* a finite path that begins and terminates at the same node. A special class of cycle is a circuit. A *circuit is* defined as a finite chain with the first and last nodes being identical. Categorically, cycles are *closed paths* and circuits are *closed chains*. A *self-loop is* a special case of a circuit or cycle, consisting of a single node and arc. In the network shown in Figure 1.4 the sequence of arcs (4, 7), (4, 5), and (5, 7) forms a cycle. In the same network, the sequence of arcs (3, 4), (4, 5), and (5, 3) forms a circuit. Additionally, arc (4, 4) is a loop. It is noted that several articles and books switch the definitions of paths and chains.

A network is said to be *circuitless* if it does not contain circuits. Circuitless networks possess certain desirable characteristics that are exploited in some applications of specialized shortest-route and longest route problems. Such characteristics form the central logic in the derivation of CPM/PERT algorithms. As illustrations, the network in Figure 1.5(a) has circuits, but the network in Figure 1.5(b) is circuitless. Figure 1.5(c) represents an *acyclic* network, that is, one which contains no cycles. A network is said to be *connected* if there is at least one path or chain joining any pair of arbitrary distinct nodes.

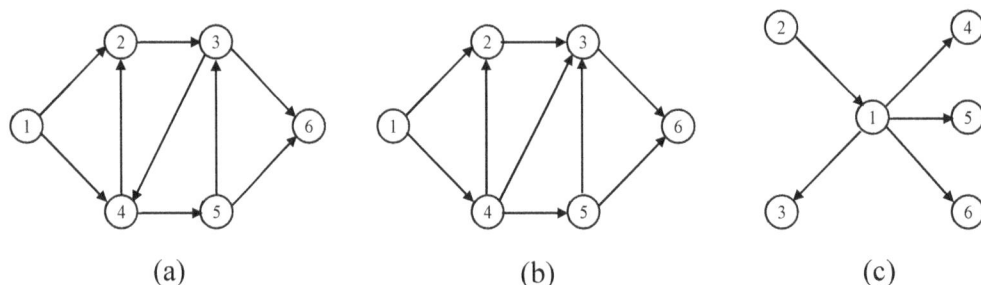

(a)   (b)   (c)

Figure 1.5. (a) network with circuits; (b) network without a circuit; (c) acyclic network.

Now that some basic definitions are understood, it is possible to reconsider the Konigsberg problem. Euler's treatment of this problem can be summarized as follows. The map of the city was replaced by a graph in which nodes represent land areas and arcs represent bridges, as shown in Figure 1.6. Using this figure, the object of the Konigsberg bridge problem can be restated as finding a path that contains each arc of the graph exactly once. A path of this kind is now known as a *Eulerian path*.

A specialized term known as the *valency* (or *degree*) of a node is defined as the number of arcs either entering or leaving the node. For example, in Figure 1.6, the valency of node A is equal to 5. Using this terminology, Euler's main result can be stated as follows: *if a connected graph has more than two nodes of odd valency, then it cannot contain a Eulerian path. If the graph has no nodes of odd valency, or two such nodes, then it contains a Eulerian path.* Using this result, it can be verified that the Konigsberg bridge problem has no Eulerian path, since all nodes in the graph have odd valency.

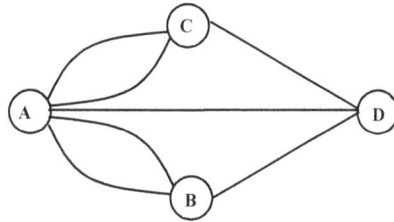

Figure 1.6. Konigsberg graph.

Trees and Arborescences

In a directed graph each arc $a = (i,j)$ is an *ordered pair* of two directly connected nodes i and j with specified orientation from node i to node j. Alternatively, in an undirected graph, arc $a = (i,j)$ or equivalently $a = (j,i)$, is an *unordered pair* of two directly connected nodes i and j with no specified orientation. Although some authors use the notation $a = \{i,j\}$ to indicate an undirected arc, we will continue using the standard notation with the understanding that undirected arcs have no orientation.

A *tree* is a finite connected undirected or directed graph with no cycles and possessing at least two nodes. Therefore, between any two nodes in a tree there is a unique path. A particular class of tree known as an *arborescence* or an *arborescent tree* (**X, U**) with root x is defined for a *directed* graph as a tree having the following properties:

(a) Every node different from x is the terminal node of a single arc
(b) Node x is not the terminal node of any arc
(c) (**X, U**) has no circuits

For a network containing n nodes, any two of the following three conditions will serve to define a subgraph of k nodes, $k \leq n$, as a tree.

(a) The subgraph is connected
(b) The subgraph has no cycles
(c) The number of arcs in the subgraph is k-1

A *spanning tree is* a tree that contains every node in the graph of a network. Hence, if a network contains n nodes, a tree with n nodes and n-1 arcs is a spanning tree. Trees and spanning trees are defined for either directed or undirected networks. However, an arborescence can be defined only on a directed network. An arborescence can also be called an *arborescent tree,* and a spanning tree that is also an arborescence will be called an *arborescent spanning tree.*

Figure 1.7 illustrates the definition of spanning tree using the *undirected* network shown in Figure 1.7 (a). It also illustrates the definition of spanning arborescent tree using the *directed* network of Figure 1.7 (d). In this figure the root node is the node shown with gray background.

11

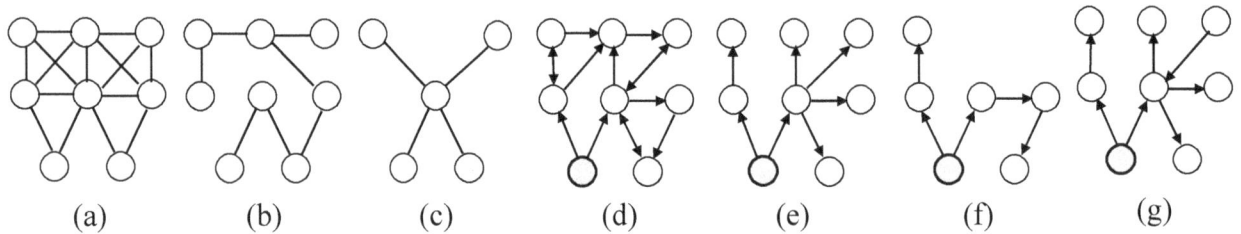

Figure 1.7. Illustrations of trees and arborescences.

The undirected network of Figure 1.7 (a) has 8 nodes and 15 arcs. A spanning tree for this network is shown in Figure 1.7 (b). Note that this tree has 8 nodes and 7 arcs. The tree shown in Figure 1.7 (c) has 5 nodes and 4 arcs; thus, it is not a spanning tree. The directed network of Figure 1.7 (d) has 8 nodes, 9 directed arcs and 3 bidirected arcs, or 15 arcs in total. Figure 1.7 (e) shows an arborescent spanning tree for this network. Note that the tree has 8 nodes and 7 arcs. The directed tree shown in Figure 1.7 (f) is an arborescence but not a spanning arborescent tree because it has fewer than 8 nodes. Finally, Figure 1.7 (g) shows a spanning tree that is not an arborescent tree for the specified root. Clearly, a network may have more than one spanning tree or arborescent tree for a given root.

If we associate with each arc a value such as distance, cost, time, or any other *path-wise additive* parameter to indicate the natural limitations or capabilities of the arcs, the resulting graph is known as a network. It is assumed that the arc parameters can be added along either an *undirected* path, or a *directed* path with all arcs having the *same orientation* along the path. Under this assumption, the *weight* of a tree in an undirected network, or of an arborescent tree in a directed network, is defined as the sum of the parameters of the arcs in the tree. Finally, a *minimum spanning tree* is defined as a tree having minimum weight, and a *maximum spanning tree* as a tree having maximum weight.

An important class of problems, including the interconnection of several buildings with a telephone line, building roads to connect towns, and several other applications, can be solved by finding minimum spanning trees. An algorithm for finding a minimal spanning tree in an undirected network will be presented along with a numerical example in Section 2.11. Section 2.13 shows a computational application of a maximal spanning tree in an undirected network in the development of an algorithm to find maximal flows with multiple sources and terminals.

1.5 MATRIX REPRESENTATION OF NETWORKS

Adjacency Matrix

The *adjacency matrix* describes the connectivity of a directed network by specifying the nodes that are directly connected by each arc of the network. It is defined as $\mathbf{X} = [x_{ij}]$, where $x_{ij} = 1$ if (i,j) exists, and 0 otherwise. This matrix is square but not necessarily symmetric. Its main limitation is that in a computer representation of a sparse network a significant portion of the matrix has zero-entries, which practically means that computer memory is inefficiently used. As an illustration, consider the network shown in Figure 1.8. For this network, $\mathbf{N} = \{1,2,3,4,5,6\}$ and $\mathbf{A} = \{a_1, a_2, a_3, a_4, a_5, a_6, a_7, a_8, a_9\}$.

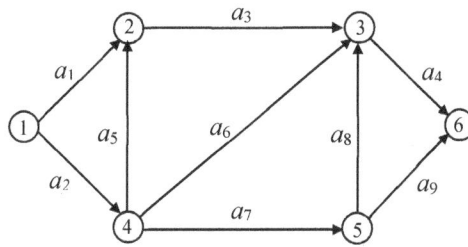

Figure 1.8. Directed network consisting of 6 nodes and 9 arcs.

The adjacency matrix for the network of Figure 1.8 is shown below:

$$X = \begin{pmatrix} 0 & 1 & 0 & 1 & 0 & 0 \\ 0 & 0 & 1 & 0 & 0 & 0 \\ 0 & 0 & 0 & 0 & 0 & 1 \\ 0 & 1 & 1 & 0 & 1 & 0 \\ 0 & 0 & 1 & 0 & 0 & 1 \\ 0 & 0 & 0 & 0 & 0 & 0 \end{pmatrix}$$

Node-Arc Incidence Matrix

The *node-arc incidence matrix* summarizes the connectivity of a directed network by specifying the orientation of each arc incident to each node. It is defined as $Z = [z_{ik}]$, where $z_{ik} = +1$ if node i is the starting node of arc a_k; $z_{ik} = -1$ if node i is the ending node of arc a_k; otherwise, $z_{ik} = 0$. For the network of Figure 1.8, the node-arc incidence matrix is shown below:

$$Z = \begin{pmatrix} +1 & +1 & 0 & 0 & 0 & 0 & 0 & 0 & 0 \\ -1 & 0 & +1 & 0 & -1 & 0 & 0 & 0 & 0 \\ 0 & 0 & -1 & +1 & 0 & -1 & 0 & -1 & 0 \\ 0 & -1 & 0 & 0 & +1 & +1 & +1 & 0 & 0 \\ 0 & 0 & 0 & 0 & 0 & 0 & -1 & +1 & +1 \\ 0 & 0 & 0 & -1 & 0 & 0 & 0 & 0 & -1 \end{pmatrix}$$

Connectivity Lists

Connectivity lists are the most efficient representation of a directed network in a computer. It is assumed that the nodes are *labeled* using the numbers 1, 2, …, n. The representation is done by means of two lists:

List 1. An ordered list of *starting nodes* for all *directed arcs terminating* at nodes 1, 2, 3, …, n. Arcs ending at node 1 are listed first, followed by arcs ending at node 2, and so on.

List 2. An ordered list of the number of *directed arcs ending* at nodes 1, 2, 3, …, n. The number of arcs ending at node 1 is listed first, followed by the number of arcs ending at node 2, and so on.

As an illustration, the directed network given in Figure 1.8 is considered again. The two ordered lists for this network are shown below:

List 1:   -   1   4   2   4   5   1   4   3   5

13

Note that the first entry in list 1 is not defined because no arc is ending at node 1. List 1 has nine *numerical* entries representing the labels for the *end nodes* of the nine arcs of the network. List 2 indicates that no arc ends at node 1, two arcs end at node 2, three arcs end at node 3, and so on. Based on the information provided by list 2, list 1 can be partitioned into five consecutive sections (one for each non-zero entry of list 2) as shown below:

List 1: | 1   4 | 2   4   5 | 1 | 4 | 3   5 |

The two arcs corresponding to the first section end at node 2; the first arc starts at node 1 and the second arc starts at node 4. Furthermore, the three arcs corresponding to the second section end at node 3; the first one starts at nodes 2, the second arc starts at node 4, and the third arc starts at node 5; and so on. As a result, the two lists represent arcs (1,2), (4,2), (2,3), (4,3), (5,3), (1,4), (4,5), (3,6) and (5,6) ordered according to their end nodes.

Distance Matrix

The distance matrix $\mathbf{C} = [c_{ij}]$ summarizes the natural limitations and capabilities of the arcs of a directed network $G = (\mathbf{N},\mathbf{A})$. The parameter c_{ij} represents the *length* of arc (i,j) in $\mathbf{A}$. This parameter can actually represent other values, such as cost, time or any other path-wise additive value associated with the arc. It is noted that $c_{ij} = \infty$ when a single directed arc does not connect node i to node j. As an illustration, the directed network of Figure 1.9 is considered. The network has four nodes and eight arcs. It can be seen in the figure that single directed arcs do not exist from nodes 2 and 3 to node 1; additionally, single directed arcs do not exist from node 4 to each of the nodes 2 and 3. For this reason the corresponding entries in the matrix are set equal to ∞.

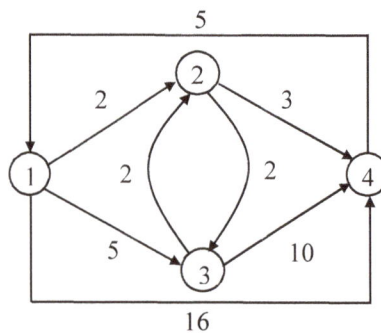

Figure 1.9. Network with arc lengths.

The distance matrix for the network is shown below:

$$\mathbf{C} = \begin{pmatrix} 0 & 2 & 5 & 16 \\ \infty & 0 & 2 & 3 \\ \infty & 2 & 0 & 10 \\ 5 & \infty & \infty & 0 \end{pmatrix}$$

Additionally, the connectivity lists for the network are

List 1: 4 1 3 1 2 1 2 3

List 2: 1 2 2 3

The entries of the distance matrix can be more efficiently stored in a third list that takes into consideration the configuration of list 1 and list 2. These lists represent arcs (4,1), (1,2), (3,2), (1,3), (2,3) , (1,4), (2,4) and (3,4). The corresponding lengths of these arcs are shown in the following ordered list:

List 3: 5 2 2 5 2 16 3 10

In conclusion the connectivity and arc lengths for the network can be more efficiently represented in a computer utilizing lists 1, list 2 and list 3, instead of the distance matrix.

1.6 CONSERVATION OF FLOW

Let s and t be the source and terminal nodes of a given network. Let U_{ij} be the capacity of arc (i,j). Furthermore, let β_i be the set of nodes directly connected into node i, and let α_i be the set of nodes directly connected from node i. For any node, the *net flow* is defined as the difference between the total flow out of the node and the total flow into the node; furthermore, *flow conservation* at the node means that the *net flow* is equal to zero. For a *pure* network, the formulation of this condition is given in Eq. (1-1):

$$\sum_{j\in\alpha_i} f_{ij} - \sum_{j\in\beta_i} f_{ji} = 0, \quad i \neq s,t \tag{1-1}$$

where it is assumed that the arc flows are within specified lower and upper bounds, as indicated in Eq. (1-2):

$$0 \leq f_{ij} \leq U_{ij}, \text{ for all } (i,j)\in \mathbf{A} \tag{1-2}$$

For a *generalized* network one or more arcs have gain/loss factors different from one. In this case the flow conservation condition given in Eq. (1-2) is reformulated as shown in Eq. (1-3) where A_{ji} is the gain/loss factor of arc (j,i).

$$\sum_{j\in\alpha_i} f_{ij} - \sum_{j\in\beta_i} A_{ji} f_{ji} = 0, \quad i \neq s,t \tag{1-3}$$

It can be verified that the coefficient matrix corresponding to Constraints (1-1) is the $\mathbf{Z}$ matrix (node-arc incidence matrix) defined in Section 1.5. To illustrate this, the *pure* network shown in Figure 1.10(a) will be considered. The numbers associated with the arcs of the network are the arc multipliers or *gain/loss* factors, which by definition are all equal to one. Figure 1.10(b) shows a *generalized* network with gain/loss factors given next to each arc.

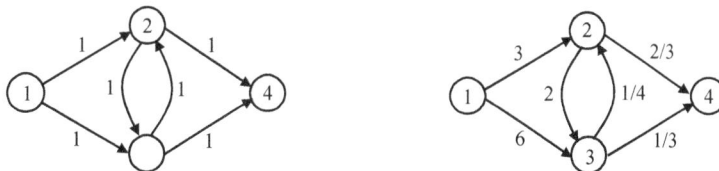

Figure 1.10. (a) Pure network; (b) Generalized network.

The coefficient matrix corresponding to Constraints (1-1) is shown below for the *pure* network given in Figure 1.10(a):

$$\begin{pmatrix} 1 & 1 & 0 & 0 & 0 & 0 \\ -1 & 0 & 1 & 1 & -1 & 0 \\ 0 & -1 & -1 & 0 & 1 & 1 \\ 0 & 0 & 0 & -1 & 0 & -1 \end{pmatrix}$$

Similarly, the coefficient matrix corresponding to Constraints (1-3) is shown below for the *generalized* network shown in Figure 1.10(b):

$$\begin{pmatrix} 1 & 1 & 0 & 0 & 0 & 0 \\ -3 & 0 & 1 & 1 & -1/4 & 0 \\ 0 & -6 & -2 & 0 & 1 & 1 \\ 0 & 0 & 0 & -2/3 & 0 & -1/3 \end{pmatrix}$$

If we now assume that node 1 has a supply of 5 units of flow and node 2 has a demand of one unit in the network shown in Figure 1.10(b), the flow at the end of path 1-3-2-4 is equal to $[5(6)(1/4) -1](2/3) = 13/3$. Moreover, the flow conservation condition for node 3 is $f_{32}+f_{34}-2f_{23}-6f_{13}= 0$.

1.7 MAXIMUM-FLOW AND MINIMUM-CUT

Let V be the value of the flow that can be shipped from the source node s to the terminal node t through the arcs of a directed network $G = (\mathbf{N}, \mathbf{A})$. The maximal value of V is limited by arc capacities under the assumption that the arcs in set $\mathbf{A}$ have finite capacity. The value of the maximum flow is determined by a fundamental structural property of the network referred to as a *cut* or *cut set*. A cut can be defined for specified source and terminal (sink) nodes as a set of arcs whose removal would not allow flow to move from the source node to the terminal node. As an illustration, in the network shown in Figure 1.11, a cut consisting of arcs (2, 4) and (3, 4) disconnects the terminal node 4 from the group of nodes 1, 2, 3.   Thus, its removal will preclude any flow from moving from source node 1 to terminal node 4. The capacity of a cut, or *cut value,* in a directed network with lower bounds on arc flows equal to *zero*, is defined as the sum of the flow capacities of the arcs of the cut. In the example network of Figure 1.11, the capacity of the cut shown is equal to $c_{24} + c_{34}$, where c_{ij} is the capacity of arc (i,j).

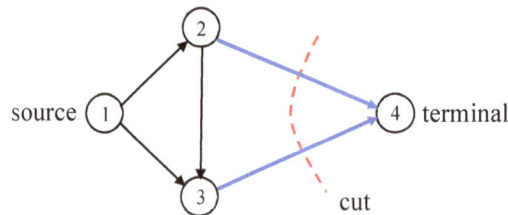

Figure 1.11. Illustration of a cut for specified source and terminal nodes.

One of the most important results in the theory of network flows, the *maximum flow/minimum cut theorem,* proved by Ford and Fulkerson [8] can now be stated: The maximum feasible flow from a source node to a terminal node is equal to the value of the minimum-cut among the cuts of the network that would not allow flow to move from the source to the sink nodes. A proof of this theorem will be given in Chapter 2. As an illustration of this result, it can be verified that the maximal flow that can be shipped from node s to node t in the network of Figure 1.12 is equal to 3. In this figure, the numbers shown on the arcs are the corresponding flow capacities in the specified directions. The minimum cut consists of arcs $(1,2)$ and $(3,4)$, and has a capacity equal to 3 units.

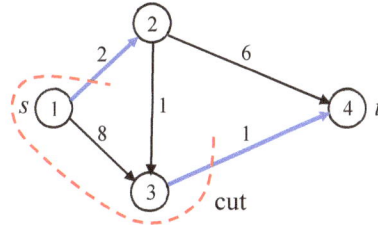

Figure 1.12. Minimum cut for a directed network.

Another cut disconnecting node 1 from node 4 in the network of Figure 1.12 consists of arcs $(1,3)$, $(2,3)$, and $(2,4)$, as shown in Figure 1.13. This cut has a capacity equal to 15 and is not, therefore, a minimum cut.

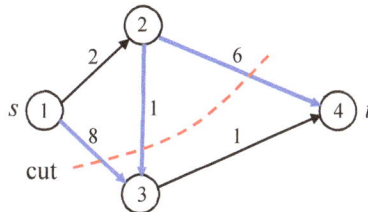

Figure 1.13. Non-minimum cut.

Finally, if the network is *undirected*, the maximal flow from node s to node t is equal to 4, since the minimum cut in this case consists of arcs $(1,2)$, $(2,3)$, and $(3,4)$, as shown in Figure 1.14.

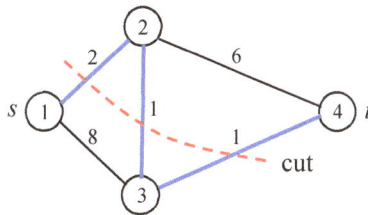

Figure 1.14. Minimum cut for an undirected network.

More generally, the value of a cut between two nodes s and t on a directed pure network with *positive* lower bound L_{ij} and upper bound U_{ij} on arc flow f_{ij} is defined in Eq. (1-4):

$$C = \sum_{i \in \mathbf{N}_1} \sum_{j \in \mathbf{N}_2} U_{ij} - \sum_{i \in \mathbf{N}_1} \sum_{j \in \mathbf{N}_2} L_{ji} \qquad (1\text{-}4)$$

Using the definition given in Eq. (1-4), the set of nodes can be partitioned into two sets $\mathbf{N}_1$ and $\mathbf{N}_2$, such that $s \in \mathbf{N}_1$ and $t \in \mathbf{N}_2$, as illustrated in Figure 1.15.

17

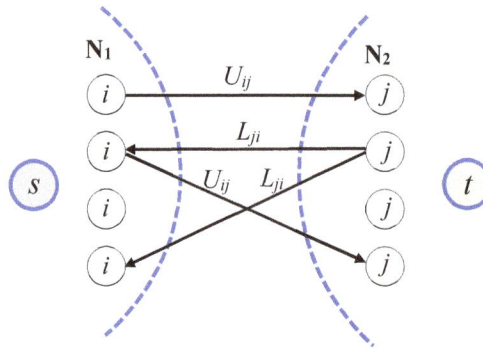

Figure 1.15. Illustration of cut-set value.

1.7.1 Example for Network with Arc Capacities

The definition of cut-set value given in Eq. (1-4) will be illustrated using the network of Figure 1.16, where each arc has two parameters $[L_{ij}, U_{ij}]$ such that $L_{ij} \leq f_{ij} \leq U_{ij}$. It is desired to verify that the cut shown is a minimal cut whose value is equal to the maximal flow from node $s=1$ to node $t=5$. As shown in the figure, the cut shown *partitions* the set of nodes into two sets $N_1 = \{1,2\}$ and $N_2 = \{3,4,5\}$.

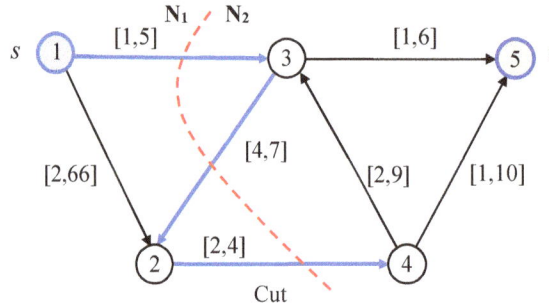

Figure 1. 16. Minimum cut illustration.

For the given cut, according to Eq. (1-4), the cut value is equal to

$$C = \sum_{i \in N_1} \sum_{j \in N_2} U_{ij} - \sum_{i \in N_1} \sum_{j \in N_2} L_{ji} = 5 + 4 - 4 = 5$$

By inspection it is possible to determine that the maximal flow is equal to 5. Therefore, the cut shown in Figure 1.16 is a minimum cut. The corresponding arc flows resulting in a maximal flow of 5 being delivered to node 5 are shown in Figure 1.17(a).

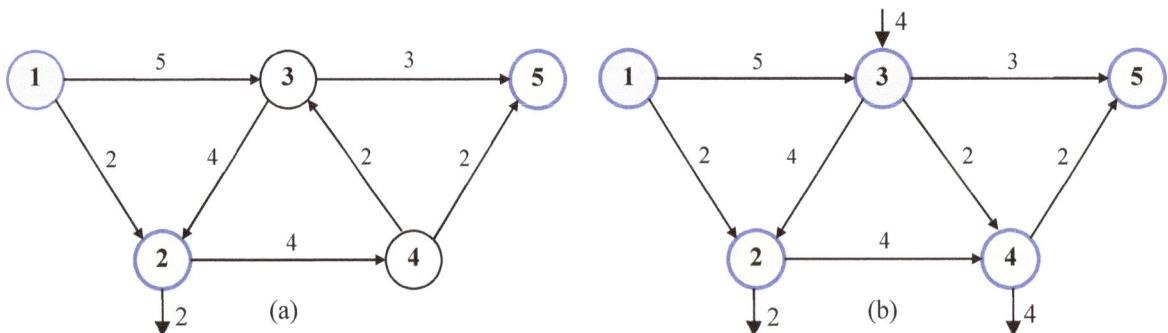

Figure 1.17. Maximum flow illustrations.

18

Note that for the feasibility of the maximal flow, node 2 can be a terminal node with a demand equal to 2 units. If the demand is less than 2 units, the amount of flow arriving at node 2 is insufficient to satisfy the condition that the net flow must be larger than or equal to zero at a terminal node. Node 2 can actually be a terminal node with a flow demand equal to at least 2 and at most 66, which is the value of the upper bound on flow in arc (1,2).

To further illustrate the concept of cuts, let us assume that instead of arc (4,3) in Figure 1.16 we have arc (3,4) with parameters [2,9]. In this case, the minimum cut is still the same as in Figure 1.15, and the arc flows are as in Figure 1.17(a). However, for the feasibility of the maximal flow it is necessary to make further assumptions. Specifically, node 3 becomes a source with supply equal to 4 and node 4 becomes a terminal with demand equal to 4, as shown in Figure 1.17(b).

1.7.2 Example for Network with Node and Arc Capacities

In the network of Figure 1.18, the numbers associated with arcs represent *arc* capacities. Additionally, each node $i = 1,2,3,4$ has a *capacity* equal to the value k_i shown in brackets. (a) Transform this network into an equivalent *directed* network having *only* arc capacities. (b) By inspection, find the value of the maximum flow allowed from node s to node t as well as the corresponding arc flows. (c) Identify the minimum cut.

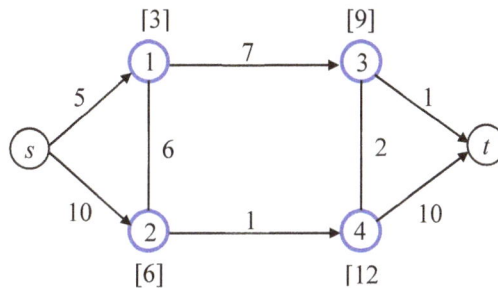

Figure 1.18. Network with arc and node capacities.

(a) First, the undirected arcs are replaced by two arcs having opposite directions and capacity equal to original arc capacity. Second, each original node i is represented as an arc (i_a, i_b) with capacity equal to k_i. Third, each original arc (i,j) is replaced by arc (i_b, j_a). The transformed network is shown in Figure 1.19 (a).

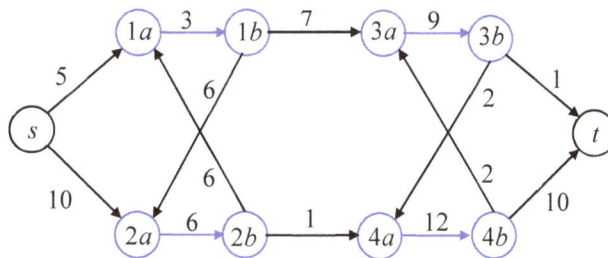

Figure 1.19 (a). Directed network with only arc capacities.

(b) The maximal flow is 4. The corresponding arc flows are shown in Figure 1.19 (b). In this figure, bold-faced arc flows correspond to *saturated* arcs $(1a,1b)$, $(2b,4a)$, $(3b,4a)$, and $(3b,t)$ since the arc flows are equal to the arc capacities.

19

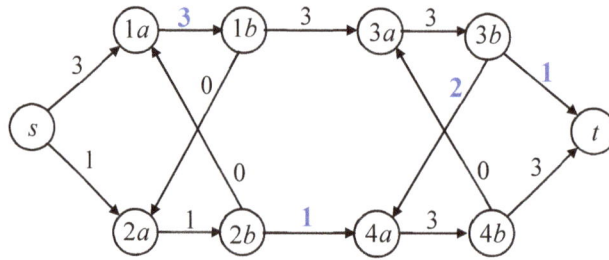

Figure 1.19 (b). Arc flows resulting in maximal flow at the terminal node.

There are two minimum-cuts. The first cut is shown in Figure 1.19(c) and the second cut is shown in Figure 1.19(d). The value of the minimum cut is equal to 4. The first cut consists of node 1 and arc (2,4) in the original network. The second cut consists of arcs $(3, t)$, (3,4) and (2,4) in the original network.

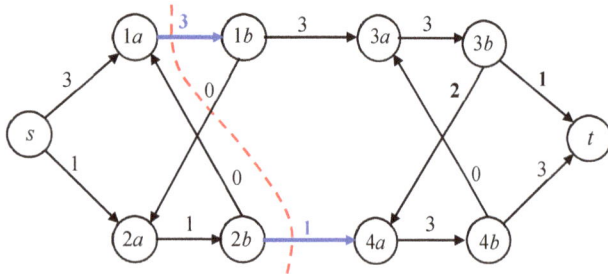

Figure 1.19 (c). First minimum cut.

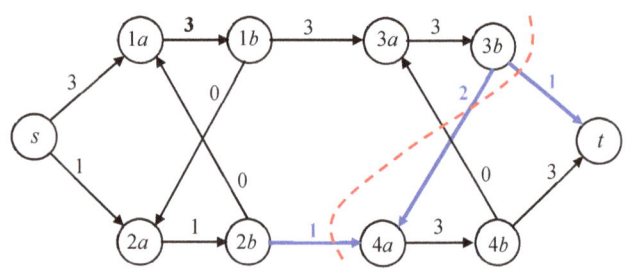

Figure 1.19 (d). Second minimum cut.

1.8 NETWORK OPTIMIZATION PROGRAM (NOP)

NOP is a non-commercial Visual Basic program designed to facilitate the learning and application of the network algorithms discussed in this book. It is easy to use and not difficult to learn or understand. The program makes all common actions very accessible to promote efficient navigation through all the procedures (algorithms) included in the program. Opening and saving input files can easily be done and data can be transferred between applications using the clipboard. This helps the user to view all the input files available before choosing a desired one. Most of the forms used in NOP have scroll bars that allow the user to scroll up and down, in case the file is large. NOP consists of six menu bar options. They are the File menu, Edit menu, Options menu, Procedures menu, Window menu and the Help menu. In addition to NOP, a Visual Basic program that creates a LINDO environment for linear programming models has been designed to solve any LP model. This is particularly useful for solving the special class of linear multi-commodity network flow problems studied in Chapter 8. This chapter shows a sample numerical example.

1.8.1 Overview of NOP

A simplified flowchart of the Network Optimization Program (NOP) is shown in Figure 1.20. It shows the options available to the user. Briefly, it is possible to create a new file or retrieve an existing one. Also, the program allows access to a file directory that can be created with samples of files for each type of problem solvable by NOP. Once a file is created or made available, it can be edited to either change, delete, or add data.   After this, the file can be stored for future use.

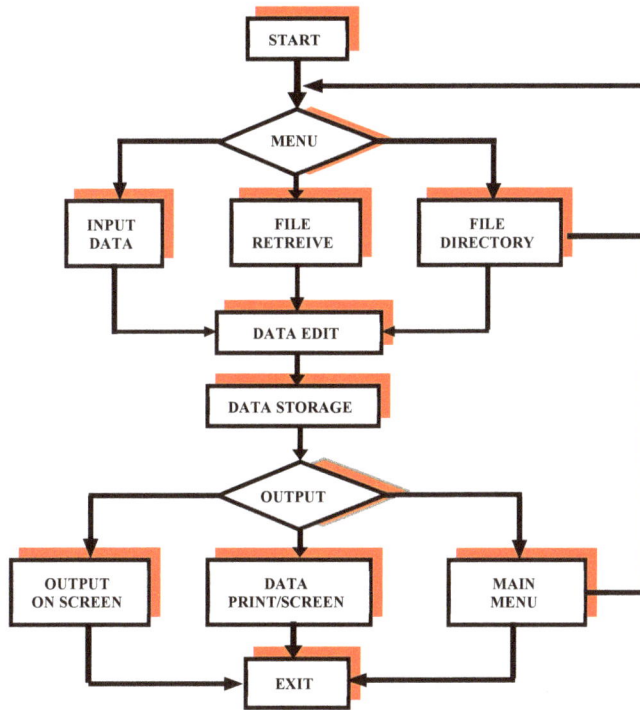

Figure 1.20. Network Optimization Program.

Table 1.1 shows the list of the seventeen algorithmic procedures available in NOP. The corresponding data files required to run each procedure are recognized by the extensions listed in the table. For each procedure NOP shows the computer output on a screen and allows the option of printing it if desired by the user.

Table 1.1. PROCEDURES AND FILE EXTENSIONS

Name of the algorithm	File extension
1. Dijkstra's Algorithm	.shr
2. Ford's Algorithm	.frd
3. Floyd's Algorithm	.msc
4. Double-Sweep Algorithm	.ksp
5. Minimal Spanning Tree	.mst
6. Maximum Flow	.mxf
7. Multi-terminal Maximum Capacity	.mmc
8. Shortest path with turn penalties	.sht
9. Gomory-Hu	.mmf
10. Traveling Salesman algorithm	.tsp
11. Multi-Traveling Salesman algorithm	.mts
12. Out of Kilter algorithm	.oka
13. Critical Path method (CP)	.cpm
14. Project Evaluation and Review Technique (PERT)	.prt
15. Graphical Evaluation and Review Technique (GERT)	.grt
16. Generalized network	.gen
17. Time Cost Trade Offs Analysis	.cos

As can be seen in Figure 1.21, NOP starts by presenting to the user a number of choices for selecting file options, network procedures for a specific application, and help regarding the procedures. The *file menu* handles file input and output.

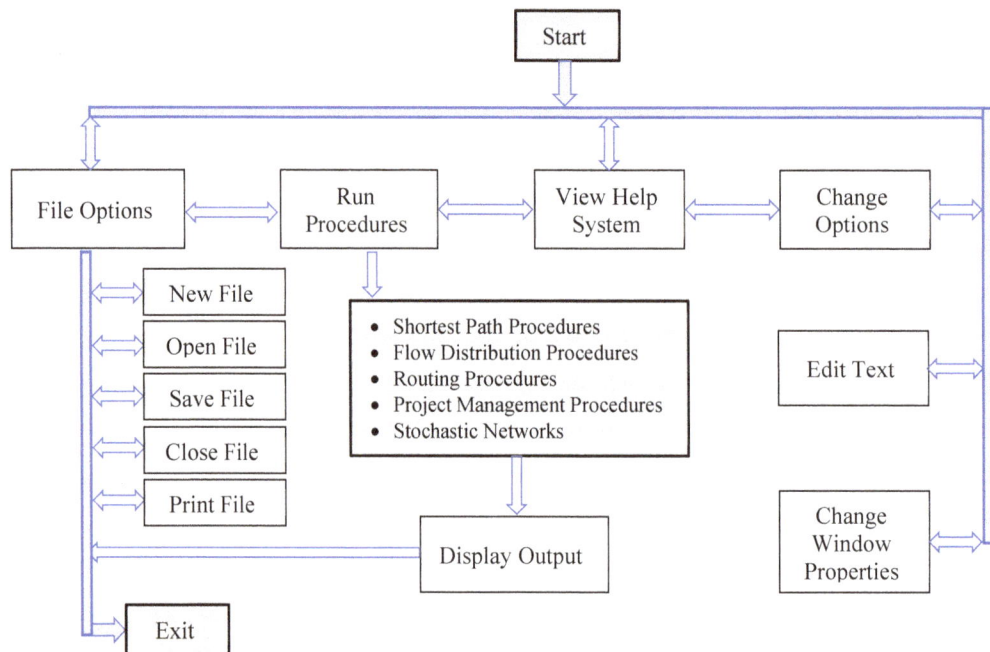

Figure 1.21. Choices Available in the Network Optimization Program.

The **File | New option** allows the user to create a new network model input file. The help system gives the user a detailed description to write a procedure-specific input file.

A typical *open dialog box* appears when the **File | Open** option is chosen. The **File | Close** option is used to close input files that are no longer being used. Data file extensions for the network procedures of NOP are automatically assigned. The **File | Save** option saves the file into the desired folder. This option prompts the user to specify a file name when it is saved for the first time. The network model can be saved in a different file by using the **File | Save As** option. The program also saves the name of the four most recently used files. The file names can be found in the **File menu** and keeps on changing as the user opens and closes files. The **File | Print** option that is found on most Windows applications is used to print the contents of windows. The **File | Exit** option enables the user to exit the program by the click of a mouse button.

The **Edit** menu contains buttons that enable data transfer between applications. This menu can also be accessed using the shortcut key ALT + E. Highlighted text from any instance of the SDI form can be placed onto the clipboard using the CUT or COPY keys. Text from the clipboard can be pasted into any of these windows using the PASTE option. The delete button is used to delete highlighted text. The select all button is used to select all the contents of a window.

The **options** menu is used to change the display options in NOP. The toolbar containing commands to open new or and existing file and the edit keys can be turned on or off. The type of font used can also be changed to suit the needs of the user. The procedure option contains all the algorithms available in NOP. Seventeen procedures are classified into five main types of models:

1. Path finding procedures
2. Flow distribution procedures
3. Routing procedures
4. Project management procedures
5. Stochastic network procedures (GERT).

The **procedure selection menu** is the heart of the network optimization program. As previously indicated, the sixteen procedures contained in NOP are listed in Table 1.1 along with the corresponding data file extensions. The algorithms are selected either by clicking on the procedures item or using keyboard shortcuts. For example, ALT + P + G can be used to access the Gomory-Hu algorithm and ALT + P + Z to access the Generalized networks procedure.

The **Window** option is used to view all the active windows. This includes all active input files, the log window and the output window. These windows can be minimized, maximized and closed using the buttons on the top right-hand corner of the windows. These commands apply to all the four SDI windows and the MDI window. The last option is the Help option. This pop-up menu bar item contains two buttons. One of them displays an 'about' box with copyright information and the other one brings up a standard **window help system**. The help system was developed with a standard Helpfile compiler.

1.8.2   NOP Computer Screens

Figure 1.22 is based on the general flow-chart described in Figure 1.21. It shows the *sequence* of NOP *computer screens* for an application of the software. The sequence starts with an MCI screen showing the name of the software, names of the authors and title of the book. This screen is followed by the formulation of the model screen. This screen consists of two sections; the bottom section is for the formulation of the model and the top section is for confirmation of the model and the indication that it has been read.

Initially the two sections of the screen are empty. Subsequent screens show specific selections from the File-Edit-Options-Procedures-Window Menu. After selecting either the choice of generating a new model or opening an available model file, the control of the program returns to the bottom section of the model formulation screen where the corresponding formulation is shown. After this the Procedures screen is chosen. A procedure is selected from the list and then the computer output is shown.

Figure 1.23 in Section 1.8.5 illustrates the sequence of screens for a sample run of the NOP procedure to find the shortest path from the source node of a network (node 1) to a specified terminal node.

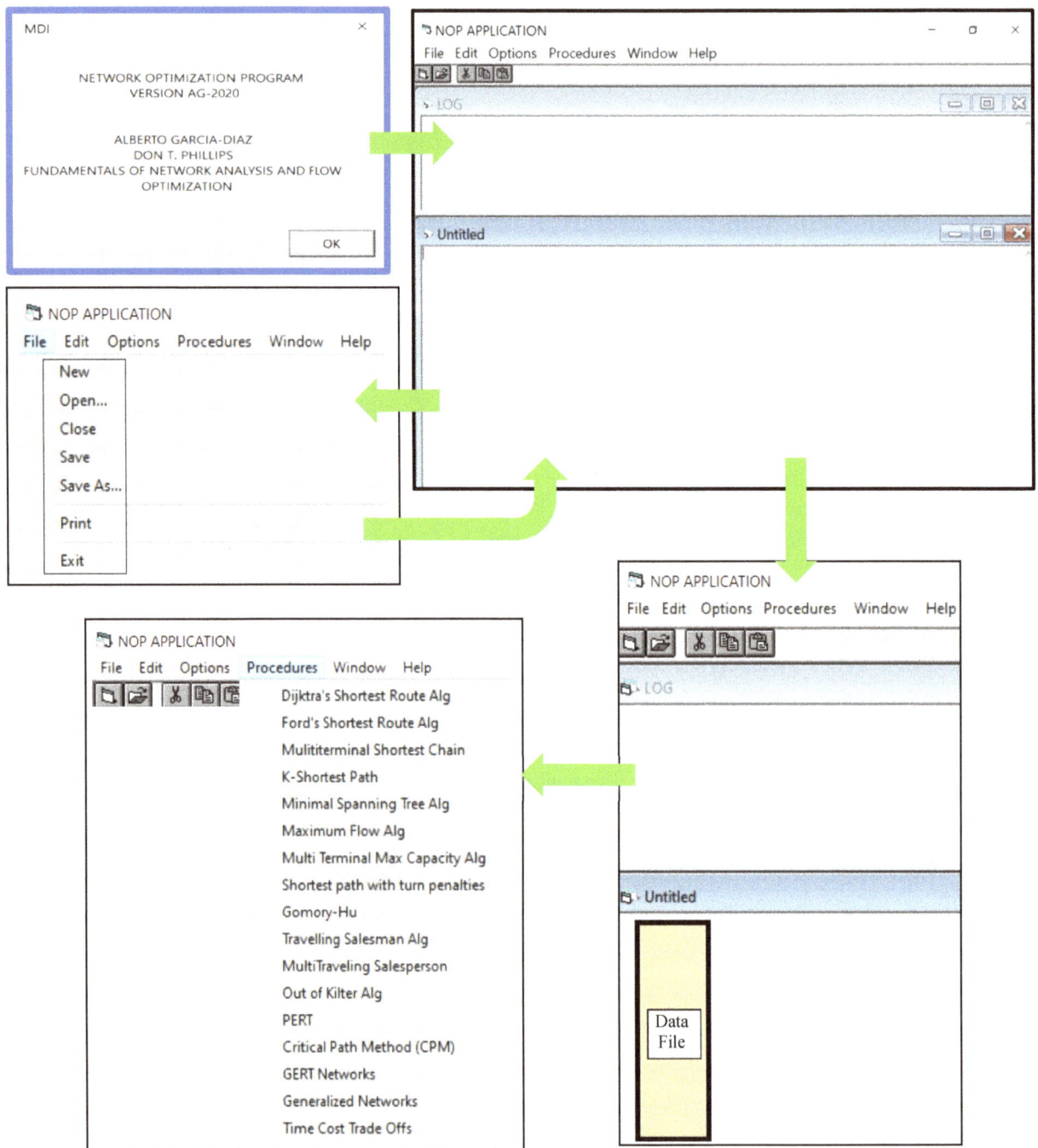

Figure 1.22. Sequence of NOP Computer Screens.

1.8.3 Algorithms Available in NOP

The NOP software executes seventeen algorithms. A brief description of each algorithm is given below.

1. **Dijkstra's Shortest Route Algorithm.** This program finds the shortest route between the source (node 1) and any other node in the network. The program runs until the terminal node gets a permanent label. Negative arc lengths are not allowed.

24

2. **Ford's Shortest Route Algorithm.** This program always finds the shortest route between the source (node 1) and any other node in the network. The program runs until the terminal node gets a permanent label. It is similar to Dijkstra's Algorithm with the only difference that it can accept negative arc lengths.

3. **Multi Terminal Shortest Chain Problem.** This problem is solved using Floyd's Algorithm. Floyd's algorithm is used to find the shortest distance between any two nodes in a network.

4. **K–Shortest Path Problem.** This problem is solved using the Double-Sweep method. The program computes K shortest path lengths from any node to all the other nodes in the network.

5. **Minimal Spanning Tree Problem.** This problem is solved using a Greedy algorithm. The program finds the minimal spanning tree in an undirected network. Because of this, only the upper triangular portion of the arc length matrix is entered.

6. **Maximum Flow Problem.** This problem is solved using a labeling procedure. The program computes the maximum flow along arcs from a source to a sink node, subject to flow conservation constraints.

7. **Multi Terminal Maximum Capacity Problem.** This problem is solved using a modified version of Floyd's method. The program computes the route with maximum possible capacity between pairs of nodes in a directed network.

8. **Shortest Path Problem with Turn Penalties.** This problem is solved using a modified version of Dijkstra's method. The program computes the shortest paths between the source and sink nodes while accounting for turn penalties. Turn penalties are entered via an input form. In this form the value of a penalty is specified for each node that can be turned. This requires as additional input the three nodes involved in each turn.

9. **Multi Terminal Maximal Flow Problem.** For a given an undirected capacitated network, the maximal flow associated with each source/terminal pair is found using Gomory-Hu's method. Maximum flows are determined assuming that all arc capacities are exclusively used for each choice of source and terminal.

10. **Traveling Salesman Problem.** Each city must be visited exactly once by the salesman in such a way that the tour covering all cities has minimal total length. The tour starts and ends at the base city (headquarter of company). This problem is solved using Little's Branch and Bound algorithm.

11. **Multi Traveling Salesperson Problem.** This problem is a generalization of the TSP problem. Each city is visited by exactly one salesman and all must depart from the base city (headquarter of company). The problem is solved using Garcia-Diaz's method. The program computes the route after transforming the given network and applying the Out-of-Kilter algorithm to the modified network. Negative flow limits are not allowed. When you run the algorithm, it will ask you to enter the number of salesmen and their corresponding fixed costs.

12. **Out of Kilter Algorithm.** The dual linear programing algorithm is used to solve a variety of minimal-cost problems. Flow conservation conditions are assumed for all intermediate nodes and flows are within upper and lower bounds specified for each arc. Negative flow limits are not allowed.

13. **Project Evaluation and Review Technique.** This problem is used to find the critical or longest path in an activity network (representing precedence relationships for all activities)

assuming that all activities of the project have pessimistic, realistic, and optimistic duration estimates.

14. **Critical Path Method.** This procedure finds the critical or longest path in an activity network (representing precedence relationships for all activities) assuming that each activity has a unique constant duration.

15. **Graphical Evaluation and Review Technique.** This program uses Pritsker's GERT procedure to find the mean time and the probability of reaching a desired node in a network, knowing the probability of choosing one arc out of each node, and knowing the statistical distributions for the duration of all arcs.

16. **Minimal Cost Flows in Generalized Networks.** This program uses Primal Simplex procedures to satisfy flow requirements at minimum cost subject to flow conservation at each node and gains or losses along arcs.

17. **Cost-Time Trade-Offs** This program uses an algorithm developed by Fulkerson to determine optimal project and activity durations assuming that activity cost-duration functions are linear.

1.8.4 Data File Format for NOP Procedures

The input data format for each of the sixteen algorithms is described below.

1. **Dijkstra's Shortest Route Algorithm** (Source Node is Node 1)
   The user will be asked to enter the terminal node of the shortest path from Node 1.
   NODES n; (n denotes the number of nodes)
   ST;
   START_NODE   END_NODE   ARC_LENGTH; (one line for each arc)
   END;

2. **Ford's Shortest Route Algorithm** (Source Node is Node 1)
   The user will be asked to enter the terminal node of the shortest path from Node 1.
   NODES n; (n denotes the number of nodes)
   ST;
   START_NODE   END_NODE   ARC_LENGTH; (one line for each arc)
   END;

3. **Multi Terminal Shortest Chain Algorithm**
   NODES n; (n denotes the number of nodes)
   ST;
   START_NODE   END_NODE   ARC_LENGTH; (for each arc)
   END;

4. **K–Shortest Path Algorithm**
   NODES n; (n denotes the number of nodes)
   ST;
   START_NODE   END_NODE   ARC_LENGTH; (for each arc)
   END;
   The user will be asked to enter the following additional data using input boxes: K, PMAX, and origin (X) and destination (Y) pairs for paths to be traced. Several X,Y pairs can be entered. PMAX is the maximum number of paths to be traced for each specified origin (X) and destination (Y) pair and a given value of K. The origin and destination pairs are entered as shown below:
   ST;
    X Y;
    ⋮
    X Y:
    END;

5. Minimal Spanning Tree Algorithm

Enter the upper triangular matrix only.
NODES n; (n denotes the number of nodes)
ST;
START_NODE   END_NODE   ARC_LENGTH; (for each arc)
END;

6. Maximum Flow Labeling Algorithm (Source Node is Node 1, Terminal Node is Node n)

NODES n; (n denotes the number of nodes)
ST;
START_NODE   END_NODE   ARC_CAPACITY; (for each arc)
END;

7. Multi Terminal Maximum Capacity Algorithm

NODES n; (n denotes the number of nodes)
ST;
START_NODE   END_NODE   ARC_CAPACITY; (for each arc)
END;

8. Shortest Path Problem with Turn Penalties

NODES n; (n denotes the number of nodes)
ST;
START_NODE   END_NODE   ARC_LENGTH; (for each arc)
END;
The user will then be asked to enter turn penalties via an input form. Remember that three nodes characterize a turn. Enter the corresponding penalty and the three nodes after running the algorithm. Penalties and lengths must be specified in the same unit.

9. Multi Terminal Maximal Flow Algorithm

NODES n; (n denotes the number of nodes)
ST;
START_NODE   END_NODE   ARC_CAPACITY; (for each arc)
END;

10. Traveling Salesman Algorithm

NODES n; (n denotes the number of nodes)
ST;
START_NODE   END_NODE   ARC_LENGTH; (for each arc)
END;

11. Multi Traveling Salesperson Heuristic Algorithm

NODES n; (n denotes the number of nodes)
ST;
START_NODE   END_NODE   ARC_LENGTH; (for each arc)
END;
When you run the algorithm, you will be asked to enter the number of salesmen and their corresponding fixed costs.

12. Out of Kilter Algorithm

Negative flows are not allowed. If upper bounds are INFINITE type 9999.
NODES n; (n denotes the number of nodes)
ST;
START_NODE END_NODE   UPPER_LIMIT   LOWER_LIMIT COST_PER_UNIT; (for each arc)
END;

13. Program Evaluation and Review Technique (PERT)

The START NODE label must be less than the label of the END NODE for each arc (no circuits allowed). If an arc duration is INFINITE type 9999.

NODES *n*; (*n* denotes the number of nodes)
ST;
START_NODE   END_NODE   MOST_LIKELY   PESSIMISTI COPTIMISTIC; (for each arc)
END;

14. **Critical Path Method** (CPM)

START_NODE value must be less than END_NODE value for each arc (no circuits).
NODES *n*; (*n* denotes the number of nodes)
ST;
START_NODE   END_NODE   ACTIVITY_DURATION; (for each arc)
END;

15. **Graphical Evaluation and Review Technique** (GERT)

Negative arc lengths are not allowed.
NODES *n*; (*n* denotes the number of nodes)
ST;
START_NODE END_NODE PROBABILITY_DISTRIBUTION PARAMETER1 PARAMETER2;
END;
Codes for distributions supported by GERT
BIN   Binomial
CON   Constant
EXP   Exponent
GAM   Gamma
GEO   Geometric
NOR   Normal
POI   Poisson
UNI   Uniform
NB   Negative

16. **Generalized Network Simplex Algorithm**

The data for fixed and variable external flows starts with STN; and it is finished with ENDN; It is not necessary to enter zero external node flows. The arc data starts with STA; and it is finished with ENDA; Type 9999 for infinite values.
NODES *n*; (*n* denotes the number of nodes)
STN;
NODE_NUMBER   FIXED_FLOW   VARIABLE_FLOW
UNIT_COST_FOR_VARIABLE_FLOW; (for each node)
ENDN;
STA;
ST_NODE   END_NODE   UPPER_BOUND   LOWER_BOUND   UNIT_COST
ARCMULTIPLIER; (for each arc)
ENDA

17. **Time Cost Trade Offs Algorithm**

The START_NODE value must be less than the END_NODE value for each arc. Each activity has a constant cost to reduce its duration. If this cost is ∞ type 9999.
NODES *n*; (*n* denotes the number of nodes)
ST;
START_NODE   END_NODE   UPPER_LIMIT   LOWER_LIMIT   CONSTANT_COST; (for each arc)
END;

1.8.5   **Sample Run (Dijkstra's Shortest Route Algorithm)**

For the network shown in Figure 1.9 it is desired to find the shortest path from node 1 to node 4. Figure 1.23 shows the sequence of ten NOP computer screens. These screens include those screens described in Figure 1.22 plus four additional ones.

28

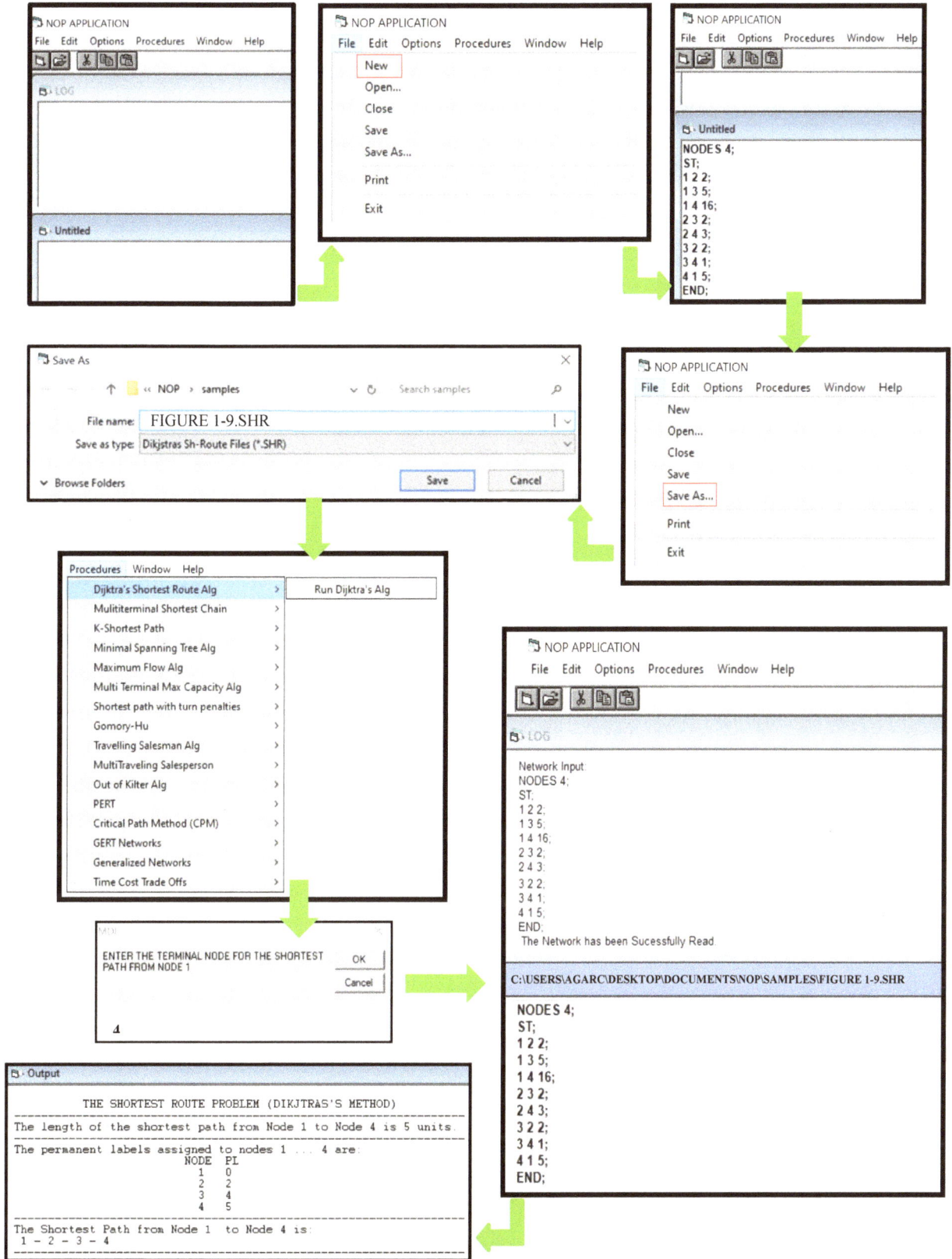

Figure1.23. Sequence of NOP Computer Screens for Sample Application.

This section provides the sequence of NOP screens starting with the execution of NOP.exe, following with the creation of the data file and, after a few steps, terminating with the shortest path.

The first screen shows the name of the software, names of the authors, and the title of the textbook. The second is the formulation of the model screen. It is noted that both top and bottom sections are empty; furthermore, the file is shown as untitled. The third screen shows the file options. After an option is selected (either generating new data or using available data) the program flow returns to the model formulation screen and shows the data file at the bottom portion of the screen. Finally, the procedure selection menu is chosen and the appropriate algorithm is selected. After this the output results are shown.

EXERCISES

1. List four reasons for the preference of network analysis techniques over other operations research techniques when the network representation is possible.

2. Define the following terms:

   (a) Arc
   (b) Node
   (c) Directed arc
   (d) Bidirected arc
   (e) Undirected arc
   (f) Forward and reverse arcs
   (g) Source node
   (h) Sink or terminal node
   (i) Intermediate node
   (j) Pure and generalized networks

3. Define a chain and a path. What is the difference between the two? Illustrate both definitions using the following network.

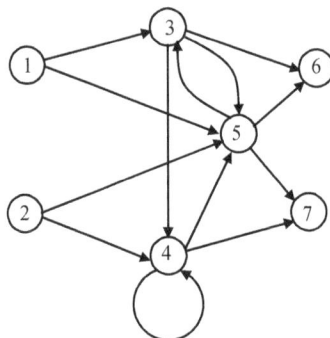

   Define a cycle and a circuit. What is the difference between the two? Illustrate both definitions using the graph shown in Exercise 3.

4. Define a graph and illustrate the mathematical definition using the networks shown in Figure 1.5 (a), (b), (c).

5. In minimizing network-flow costs, why might cyclic networks with negative arc costs cause computational problems? Illustrate your answer using a shortest-path example.

6. Define a tree, a spanning tree, and a minimum spanning tree. What is the condition for an arc not to be in a minimum spanning tree? Provide examples.

7. Define an arborescence. What is the difference between this and a spanning tree? Provide examples.

8. Using the network shown, draw (a) a tree; (b) a spanning tree.

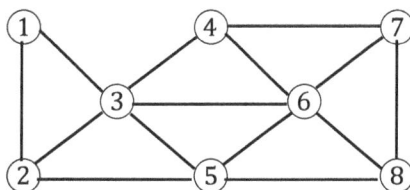

9. What facts can you deduce from the node-arc incidence matrix of the network given in Figure 1.8?

10. Formulate the flow-conservation condition for node i of (a) a pure network; (b) a generalized network.

11. What is (a) a cut; (b) cut value; (c) a minimum cut?

12. Find the node-arc incidence matrix and the adjacency matrix of the undirected network shown.

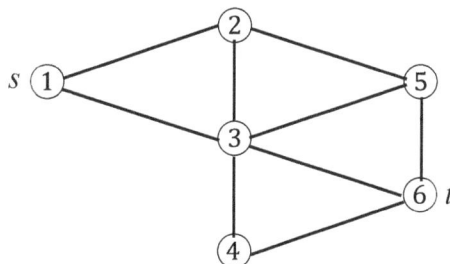

13. Find the node-arc incidence matrix, adjacency matrix, and connectivity lists for the directed network shown.

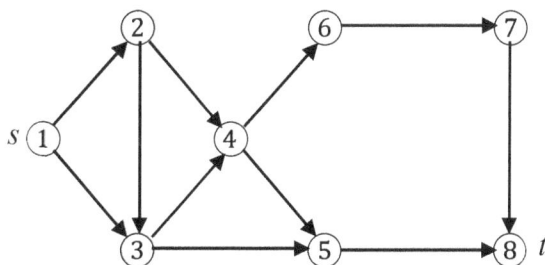

14. For the network given in Exercise 13, find a cycle, a circuit, a path from s to t, and a chain from s to t. Repeat this for the network given in Exercise 14.

31

15. Consider an undirected network with node-arc incidence matrix as shown. (a) Draw the network. (b) Find the adjacency matrix.

	a_1	a_2	a_3	a_4	a_5	a_6	a_7	a_8	a_9	a_{10}	a_{11}	a_{12}
1	1	1	0	0	0	0	0	0	0	0	0	0
2	0	1	0	0	1	0	0	0	0	0	0	0
3	1	0	1	1	0	0	0	0	0	0	0	0
4	0	0	0	1	0	1	1	0	0	0	0	0
5	0	0	0	0	1	0	1	0	0	0	1	0
6	0	0	0	0	0	0	0	0	0	1	1	1
7	0	0	0	0	0	1	0	0	1	1	0	0
8	0	0	1	0	0	0	0	1	0	0	0	0
9	0	0	0	0	0	0	0	1	1	0	0	1

16. Consider a directed network G = (**N**, **A**) with **N** = {1, 2, 3, 4, 5} and **A** = {(1,2), (1,3), (3,1), (2,3), (2,4), (2,5), (3,5), (4,5)}. (a) Draw the network. (b) Find the adjacency matrix. (c) Find the node-arc incidence matrix. (d) Find the connectivity lists.

17. In the network shown, what is the minimal value allowed for parameter a if node 4 is wanted in the shortest acyclic path from node 1 to node 7? Under what condition on a can the length of the shortest path be reduced as much as desired?

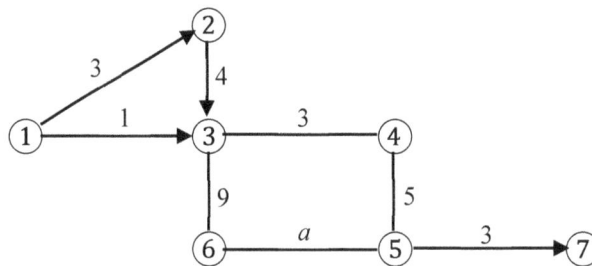

18. What is the value of the maximum flow allowed from node s to node t in the network show? Identify the minimum cut. Each number represents the capacity of the corresponding arcs.

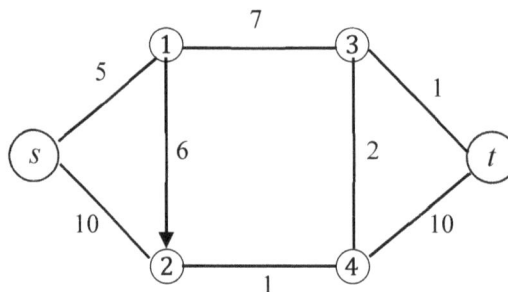

19. In Figure 1.16, assume that all arcs have lower bounds equal to zero. By inspection, determine both the maximum flow and a minimum cut.

20. Represent the network of Figure 1.8 using the notation that the first list shows nodes at the end of arcs coming from nodes 1, 2, 3, ..., n, and the second list indicates the number of arcs from each node.

21. For the network of Exercise 19, write the flow conservation constraints for nodes *s*, *3*, and *t*.

22. In the network shown in Figure 1.16, determine the value of the cut that separates nodes 1, 2, and 3 from nodes 4 and 5.

23. Design a procedure to transform a network with both node and arc capacities into an equivalent network with only arc capacities. Illustrate your procedure on the network of Figure 1.11, assuming that each node has a capacity equal to 3.

24. In the following figure each arc of the directed network is labeled with two numbers representing lower and upper bounds on flow, respectively. Node *s* is the source node, and node *t* is the terminal node. (a) Compute the values of the four cuts shown. (b) Find the minimum cut and show arc flows resulting in maximum flow. (c) Verify that the value of the minimum cut is equal to the value of the maximum flow.

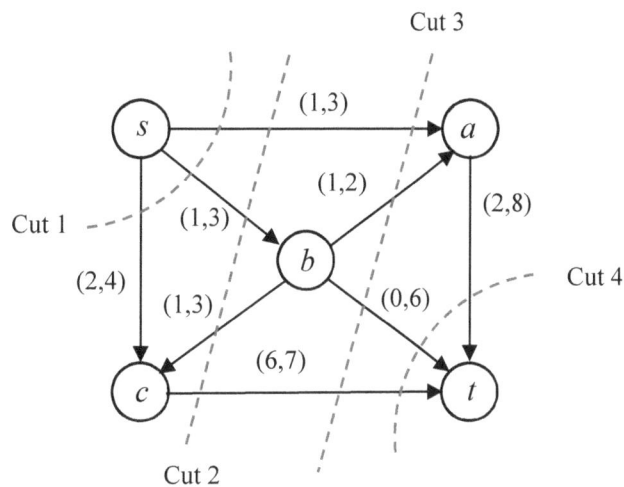

25. In Exercise 25 assume that nodes *a* and *c* have capacities equal to 2. (a) Transform the network into an equivalent network with capacities only on arcs. (b) Calculate the values of the four cuts shown in the graph. (b) By inspection find the maximal flow and the corresponding minimal cut. (c) For cut 4 find the corresponding feasible flows.

REFERENCES

[1] AHUJA R.K., T.L. MAGNANTI, AND J.B. ORLIN, *Network Flows: Theory, Algorithms, and Applications*, Prentice-Hall, Englewood Cliffs, New Jersey, 1993.

[2] BAZARRA, M., AND J. J. JARVIS, *Linear Programming and Network Flows*. John Wiley & Sons, Inc., New York, 1978.

[3] BERTSEKAS, D.P., *Linear Network Optimization*, The MIT Press, Cambridge, Massachusetts, 1991.

[4] BERTSEKAS, D. P., *Network Optimization: Continuous and Discrete Models*. Athena Scientific, 1998.

[5] BIGGS, N.L, E.K. LLOYD, AND R.J. WILSON, *Graph Theory: 1736-1936,* Clarendon Press, Oxford, 1976.

[6] BRADLEY, G. H., "A Survey of Deterministic Networks," *AIIE Transactions, 7,*222-234 (1975).

[7] BUSACKER, R. G., AND T. SAATY, *Finite Graphs and Networks.* McGraw-Hill Book Company, New York, 1965.

[8] CHARNES, A., AND W. W. COOPER, *Management Models and Industrial Applications of Linear Programming, Vols. 1* and 2. New York: John Wiley & Sons, Inc, 1961.

[9] DANTZIG, G. L., *Linear Programming and Extensions.* Princeton, N.J.: Princeton University Press, 1963.

[10] ELMAGHRABY, *S., Some Network Models in Management Science.* New York: Springer-Verlag, Inc., 1970.

[11] EULER, L., "Solutio Problematis Ad Geometriam Situs Pertinentis [The solution of a problem relating to the geometry of position], Translated into English: BIGGS, N.L, E.K. LLOYD, AND R.J. WILSON, *Graph Theory: 1736-1936,* Clarendon Press, Oxford, 3-11, 1976.

[12] EVANS, J.R. AND E. MINIEKA, *Optimization Algorithms for Networks and Graphs,* 2nd Edition, New York: Marcel Dekker, Inc., 1992.

[13] FORD, L. R., AND D. R. FULKERSON, *Flows in Networks.* Princeton, N.J.: Princeton University Press, 1962.

[14] FRANK, H., AND I. T. FRISCH, *Communication, Transmission, and Transportation Networks.* Reading, Mass.: Addison-Wesley Publishing Co., Inc., 1971.

[15] FULKERSON, D. R., "Flow Networks and Combinatorial Operations Research," *American Mathematical Monthly, 73,* 115-138 (1966).

[16] GLOVER F., KLINGMAN, D., AND N.V. PHILLIPS, *Network Models in Optimization and Their Applications in Practice,* New York: John Wiley & Sons, Inc., 1992

[17] HITCHCOCK, F. L., "The Distribution of a Product from Several Sources to Numerous Localities," *Journal of Mathematics and Physics,* 20, 224-230 (1941).

[18] HU, T. *C., Integer Programming and Network Flows.* Reading, Mass.: Addison Wesley Publishing Co., Inc., 1969.

[19] IRI, M., *Network Flow, Transportation and Scheduling.* New York: Academic Press, Inc., 1969.

[20] JENSEN, P. A., AND W. BARNES, *Network Flow Programming.* New York: John Wiley & Sons, Inc., 1980.

[21] KENNINGTON, J. F., "A Survey of Linear Multicommodity Network Flows," *Operations Research,* 26(2):209-236, 1978.

[22] KOOPMANS, T. C., "Optimum Utilization of the Transportation System," *Proceedings of the International Statistical Conference,* Washington, D.C., 1947.

[23] LAWLER, LENSTRA, RINNOYKAN AND SHMOYS, 1985.

[24] MAGNANTI, T. L., AND B. L. GOLDEN, "Transportation Planning: Network Models and Their Implementation," Working Paper 77-008, University of Maryland, General Research Board, Faculty Research Award, 1977.

[25] MINIEKA, *E., Optimization Algorithms for Networks and Graphs.* New York: Marcel Dekker, Inc., 1978.

[26] NEWMAN, J.R., *The World of Mathematics,* Simon and Schuster, New York, 1956.

[27] PHILLIPS, D.T. AND A. GARCIA-DIAZ, *Fundamentals of Network Analysis,* Englewood Cliffs, New Jersey: Prentice-Hall, Inc., 1981.

[28] PRITSKER, A.A.B., AND W.W. HAPP, "GERT: PART I – Fundamentals," Journal of Industrial Engineering, 17(5), 267 (1966).

[29] SCIENTIFIC AMERICAN, *The Konigsberg Bridges*, vol. 189, pp. 66-70, 1953.

[30] SCHRIJVER, A., "On the History of Combinatorial Optimization (Till 1960)," *Handbooks in Operations Research and Management Science*, Vol. 12, pp. 2-68, ELSEVIER, 2005.

[31] WHITEHOUSE, G.E., *Systems Analysis and Design Using Network Techniques.* Englewood Cliffs, New Jersey: Prentice-Hall, 1973.

Chapter 2

NETWORK-FLOW MODELS AND APPLICATIONS

Peter Rabbit made a face at Jimmy Skunk.
"I don't like being preached to."
"I'm not preaching; I'm just telling you
what you ought to know without being told,"
replied Jimmy Skunk.

From *The Adventures of Peter Cottontail*
T. W. Burgess

The material presented in this chapter is concerned with the study of certain deterministic network-flow models of frequent use in the representation and solution of significant problems in both business and engineering. For each problem, an efficient algorithm is developed and illustrated by means of a numerical example. Our discussion will include algorithms for finding the shortest-route between two specified nodes, the shortest path between each pair of nodes in a network, shortest paths with turn penalties, K shortest paths from a source node to each node in a network, and minimal spanning trees.

The approach used to present the material throughout this chapter is one of indoctrination by example. Each topic is introduced, mathematically defined, and then dealt with through an algorithmic framework. One or more examples are then formulated and solved in detail to illustrate the fundamental concepts and computational procedures inherent in the algorithmic process.

The applications presented in this chapter can be classified into two broad categories. The first category includes applications where the solution to a problem is achieved by selecting a chain or cycle that optimizes a suitable objective function. In these problems each arc is interpreted as a directed link that allows the movement of one unit of flow from the start node to the end node of the arc. The arc parameter or *length* is a generic cost of using the arc. Usually, these models are referred to as *distance networks*. In the second category, a more general interpretation of an arc is used, namely, that of a device or channel used for shipping multiple units of flow. In this case, the arc parameter represents the maximum amount of flow per unit time that can be shipped along the arc at any time. This parameter is called the *capacity* of the arc, and the network models are referred to as *capacitated flow networks*.

2.1 MAXIMUM FLOW AND MINIMUM COST FLOW MODELS

To provide a unifying approach for the applications studied in this chapter two mathematical models are introduced in this section. Let $G = (\mathbf{N}, \mathbf{A})$ be a directed capacitated network with $\mathbf{N} = \{1, 2, ..., n\}$, arcs in set $\mathbf{A}$, and let U_{ij} be the capacity of arc (i, j). For notational convenience, it will be assumed that node 1 is the source and node n the terminal. The two important flow distribution models are now considered.

2.1.1   Maximum Flow Model

Assuming infinite availability of flow at the source node, it is desired to find the maximum flow shipped from 1 to n through the arcs in $\mathbf{A}$. Let f_{ij} be the flow from i to j along (i, j), and let V be the amount of flow into the terminal (or out of the source). The following linear programming model is a formulation for this problem:

$$Maximize\ V \tag{2-1}$$

$$subject\ to$$

$$\sum_j f_{ij} - \sum_j f_{ji} = \begin{cases} V, & i=1 \\ 0, & i \neq 1, n \\ -V, & i=n \end{cases} \tag{2-2}$$

$$0 \leq f_{ij} \leq U_{ij}, \quad \text{for all } (i, j) \in \mathbf{A} \tag{2-3}$$

2.1.2   Minimum Cost Flow Model

Let $G = (\mathbf{N}, \mathbf{A})$ be directed and capacitated. Let b_i be the flow supply at node i. Let c_{ij} be the per unit cost of flow along (i,j). Also, let L_{ij} be a lower bound on the flow on arc (i,j). The model is shown below:

$$Minimize\ \sum_i \sum_j c_{ij} f_{ij} \tag{2-4}$$

$$subject\ to$$

$$\sum_j f_{ij} - \sum_j f_{ji} = b_i, i \in \mathbf{N} \tag{2-5}$$

$$L_{ij} \leq f_{ij} \leq U_{ij}, (i, j) \in \mathbf{A} \tag{2-6}$$

In Eq. (2-5) $b_i > 0$ for source nodes, $b_i = 0$ for intermediate nodes, and $b_i < 0$ for terminal nodes.

2.2   SELECTED APPLICATIONS OF NETWORK MODELS[1]

In order to acquaint the reader with network modeling and provide a general framework through which the network algorithms can be effectively used, this section will discuss and model several selected problems considered to be representative of recurrent applications in the field. More specifically, the following applications will be studied:

1. Equipment replacement
2. Project planning
3. Tanker operations scheduling
4. Transportation and distribution
5. Employment scheduling
6. Fleet routing and scheduling
7. Aircraft fuel allocation
8. Production planning

---

[1] Portions of this section are reproduced with permission of the *Institute of Industrial Engineers* [2].

For each application a concise definition, the corresponding network structure (meaning of nodes, arcs, and parameters), as well as the network representation are provided

2.2.1 Equipment Replacement

In the field of engineering design and analysis, there are many applied problems that can be formulated as *shortest route* models. The shortest route problem can be simply stated in the following manner. Given a collection of nodes and arcs, with a set of arc parameters c_{ij}, measuring the *length* of each arc (i,j) in a directed network without negative circuits, it is desired to find a path from the source node s to the sink node t that has minimum total length. Usually, c_{ij} is viewed as the *cost* of sending one unit of flow along the arc, and in this case it is desired to find a route that minimizes the total cost of sending one unit of flow from the source node to the sink node.

As an illustration, we consider a simple model for equipment that must be replaced periodically. Specifically, it is desired to find an optimal machine replacement policy for a given multi-period planning horizon under the assumption that the machine can be replaced instead of repaired at the beginning of each period.

Network Representation

Nodes represent beginning of periods (years, months, weeks, etc.); arcs represent periods of useful service life; for example, arc (i,j) represents the alternative of buying a machine at the beginning of period i and using it for $j-i$ periods. The *length* or arc parameter is defined as the cost of the alternative, namely, the replacement cost plus the operating cost minus the salvage value. For a planning horizon consisting of n periods, the network consists of $n+1$ nodes. Any chain from $s = 1$ to $t = n+1$ represents a replacement policy for the planning horizon. The replacement policy having minimal cost corresponds to the shortest chain from node s to node t.

As an illustration, we consider a planning period of $n = 3$ years. Here $s = 1$ and $t = 4$. The optimal solution is given by the shortest *chain* from node 1 to node 4. The set of all replacement plans can be represented by the chains in the network in Figure 2.1. For example, the chain (1,3)-(3,4)represents the policy of buying a machine at the beginning of period 1, keeping it for 2 years and selling it at the end of period 2 (beginning of period 3). A new machine is then purchased at the beginning of period 3 and sold at the end of the same period. The total cost of this policy is $c_{13} + c_{34}$. A numerical exercise will be considered in Section 2.4.3.

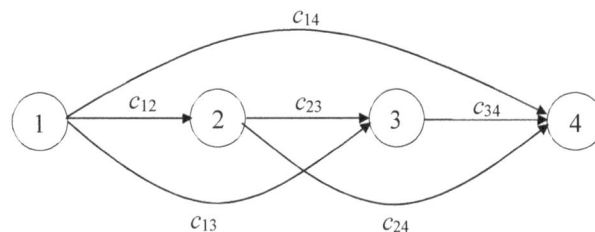

Figure 2.1. Equipment replacement network.

38

2.2.2 Project Planning

A project is defined as a collection of *activities* and *events*. An activity is any undertaking that consumes time and other resources. An event is a well-defined occurrence in time. An Activity Network is a representation of two particular aspects of a project: (a) precedence relationships among the activities; (b) duration of each activity. The precedence relationships come about from technological and other considerations. The precedence relationships are: (a) transitive, (b) non-reflexive, and (c) non-symmetric.

For a *node* to be *realized* all *arcs* ending at the node must be *completed*. Additionally, for an arc to be *started* its *beginning node* must be realized. Under these rules, it can be verified that the *critical* activities and events (those that would delay the project if they are delayed themselves) correspond to the *longest* path from the source to the terminal of the network.

An important class of network problems centers about the planning and scheduling of large projects, such as missile countdowns, research and development projects, and construction management projects. For purposes of this model, a project is envisioned to be a set of tasks and precedence relationships among the tasks. Task i has a known duration d_i. In addition, the precedence relationships require that a set of tasks must be completed before some other job can be started. For example, suppose that there are five tasks and that 1 precedes 3; 1 and 2 precede 4; and 1, 2, 3, and 4 precede 5.

Network Representation

This problem can be represented by the activity network in Figure 2.2. A dummy node s is introduced to represent the start of the project, and another node t is used to represent the project completion. The other nodes represent the beginning of each of the five tasks. Arc (i,j) with length d_i is included in the network if task i must *immediately* precede task j. Since 1 immediately precedes 3, and 3 immediately precedes 5 the arc (1,5) is implied and is not included. The longest chain from the start node to any other node represents the earliest start time for the task beginning at the given node. Therefore, the *longest chain* from the start node to the completion node represents the minimum project duration.

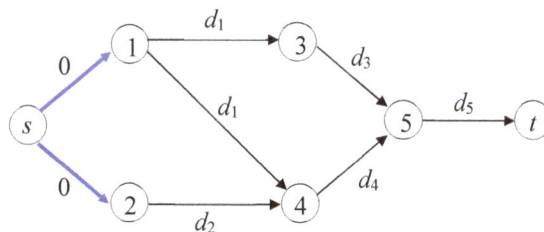

Figure 2.2. Activity-on-node representation of an activity project

In general, an activity network G = (**N,A**) satisfies three conditions: (a) it has one source, (b) it has one terminal, and (c) it has at most one arc to directly connect two nodes. When the network has circuits with positive lengths, the problem of finding the longest acyclic chain is by no means simple. Fortunately, the project network under consideration is circuitless, and the determination of its longest chain is not difficult.

The previous representation is called an *activity-on-node model since* the beginning of each task is uniquely identified with a node. An alternative network representation, called an *activity-on-arc model,* represents each task by an arc. This representation is illustrated in Figure 2.3 for the example of Figure 2.2. The number adjacent to each arc is a label used for task identification.

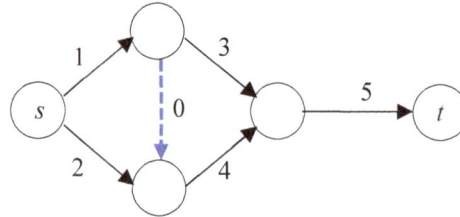

Figure 2.3. Activity-on-arc representation of an activity project network.

The dashed arc represents a dummy task of zero duration and is included to ensure that task 1 precedes task 4. The nodes in this network can be interpreted as *events*, such as the completion of the basement in a building project. As before, the longest chain from start to completion in the project network is the earliest project completion time.

The method of project planning using activity network representations is called the *critical path method (CPM)*, originally developed by Du Pont and Remington Rand to control the maintenance of chemical plants. The longest chain is called a critical path since the delay for any task on that path will result in a delay in the project completion. The same approach was developed independently for the Navy Polaris program as a stochastic duration model for the control of the research and development project. This model is called the *program evaluation and review technique* (PERT) and provides an estimate of the mean project duration and the variance of this estimate. CPM/PERT solution procedures are discussed in Chapter 6.

2.2.3 Scheduling Tanker Operations

The application described in this section is known as the *Minimal Cost-to-Time Ratio Cycle Problem*. It is desired to identify a directed cycle for which the sum of the costs divided by the sum of the transit times is minimum. In this application a ship is represented by one unit of flow traveling from port i to port j at a cost c_{ij} and taking a time equal to t_{ij} (hours). The transit time includes the time necessary to load the ship at node i, travel from node i to node j, and unload at node j. If profits are given instead of costs, they can be treated as negative costs. It can be proved that the directed cycle with a minimum value of $(\Sigma c_{ij})/(\Sigma t_{ij})$ will minimize the average profit per unit time [4,25].

As an illustration, suppose that you have a tramp steamship, and you wish to determine the best schedule of ports to visit. Specifically, it is desired to select the ports and the order in which the selected ports are visited. In the network of Figure 2.4 nodes represent ports and arcs represent connections between ports. The numbers adjacent to arc (i, j) are the cost and time c_{ij}, t_{ij} of traversing the arc. If the t_{ij}'s are all nonnegative, this problem is solved by methods similar to those employed in the shortest-route algorithms. The optimal cycle which represents the desired shipping schedule is (1,3), (3,4), (4,1), with a ratio of $(3+2-4)/(1+2+3) = 1/6$. Note that the vessel will visit ports 1, 3 and 4. Port 2 is not selected.

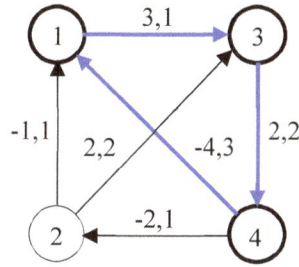

Figure 2.4. Steamship port scheduling network.

2.2.4 The Transportation Model

The *transportation problem* was one of the first network flow problems to be investigated in the field of operations research. It was first solved by Hitchcock [20] in 1941, and since then has been applied to many types of shipping and distribution problems. It is undoubtedly one of the most widely used network models. The model and its solution methodology are discussed in Section 2.10. The problem can be best described in terms of shipments from plants to warehouses. Suppose that there are m plants and n warehouses. Each plant has a supply s_i, $i = 1, 2, \ldots, m$; and each warehouse creates a demand d_j, $j = 1, 2, \ldots, n$. The problem of minimizing total shipping costs is equivalent to finding x_{ij} values such that the following model will be optimized:

$$Minimize \sum_i \sum_j c_{ij} x_{ij} \tag{2-7}$$

subject to

$$\sum_j x_{ij} = s_i \quad i = 1, 2, \ldots, m \tag{2-8}$$

$$\sum_i x_{ij} = d_j \quad j = 1, 2, \ldots, n \tag{2-9}$$

$$x_{ij} \geq 0 \quad i = 1, 2, \ldots, m;\ j = 1, 2, \ldots, n \tag{2-10}$$

In Eqs. (2-8) and (2-9) it is assumed that $\Sigma_i s_i = \Sigma_j d_j$ in order for the constraints to be consistent. This model can be described in terms of a network by letting the plants and warehouses be the nodes and the feasible shipping routes be the arcs. Under this interpretation, a route linking plant i with warehouse j is represented by a directed arc (i, j).

A sample transportation network is depicted in Figure 2.5 (a) for $m = 2$ and $n = 3$. Note that the transportation model is a special case of the minimum-cost flow model formulated in Section 2.1.2. In this case, $b_j = -d_j$, $b_i = s_i$, $L_{ij} = 0$, and $U_{ij} = \infty$. A special case of the transportation problem is the *assignment problem*, where $m = n$, $s_i = 1$, and $d_j = 1$, for all values of i and j. The assignment and transportation models will be furthered considered in Section 2.3. A network solution methodology will be developed in Chapter 3.

A *circulation network* is defined as a network containing only intermediate nodes (no sources and no terminals). Typically, when there are multiple source nodes and multiple terminal

41

nodes, as in the network of Figure 2.5 (a), a circulation network includes a *super source* node and arcs joining it to each original source, as well as a *super terminal* node and arcs joining each original terminal to it. Once this is done, a *return arc* is included to join the super terminal to the super source. The resulting circulation network is shown in Figure 2.5 (b).

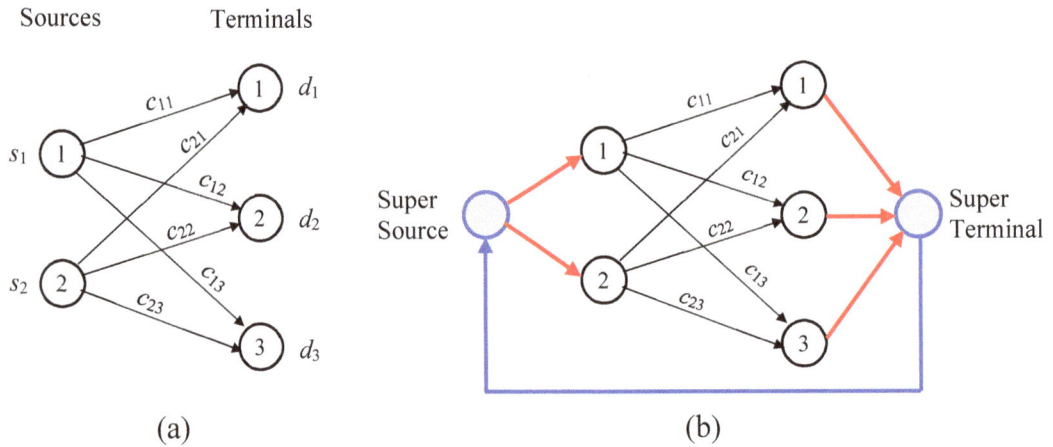

Figure 2.5. Network representations of a transportation problem.

2.2.5   The Caterer Problem

An example of a minimal cost flow problem is the *caterer problem* [24]. Assume that a caterer knows that he will require d_j clean napkins on each of n successive days, $j = 1, 2, ..., n$. He can meet his needs by purchasing new napkins and by using napkins previously laundered. Additionally, the laundry has two kinds of service, quick and slow. A napkin sent for slow service is available 2 days later, whereas a napkin sent to the fast service is available 1 day later. On day j new napkins from the store costs a_j cents each, quick laundry service is c_j cents per napkin, and slow service b_j cents per napkin, where $b_j < c_j$. How does the caterer meet his requirements at minimal cost?

Linear Programming Model (Case 1). Let p_j, s_j, q_j, and h_j be the number of napkins purchased, sent to slow service, sent to quick service, and held over on day j, respectively. The L.P. model for $n=3$ is shown below:

$$Minimize \sum_j a_j p_j + c_j q_j + b_j s_j$$

subject to

$$p_j + s_{j-2} + q_{j-1} \geq d_j \quad j = 1,2,3 \quad \text{(demand for clean napkins)}$$

$$s_j + q_j + h_j - h_{j-1} \leq d_j \quad j = 1,2,3 \quad \text{(supply of soiled napkins)}$$

$$h_j, p_j, s_j, q_j \geq 0 \quad j = 1,2,3$$

Revised Linear Programming Model (Case 2). In this case, we allow the possibility that because of lower costs at a particular time, an excess of new napkins will be purchased to satisfy future demand. Let E_j be the excess number of napkins purchased on day j. The revised model can be formulated as shown below:

$$Minimize \sum_j a_j p_j + c_j q_j + b_j s_j$$

subject to

$$p_j + s_{j-2} + q_{j-1} + E_{j-1} - E_j \geq d_j \quad j = 1,2,3 \quad \text{(demand for clean napkins)}$$
$$s_j + q_j + h_j - h_{j-1} \leq d_j \quad j = 1,2,3 \quad \text{(supply of soiled napkins)}$$
$$E_j, h_j, p_j, s_j, q_j \geq 0 \quad j = 1,2,3$$

Network Formulation

A network for Case 1 with $n = 3$ is given in Figure 2.6, where the variables are indicated adjacent to their respective arcs and the supplies (positive values) and demands (negative values) are given adjacent to the nodes. The unique aspect of this problem is that the demand for clean napkins d_j also works as a supply of dirty napkins at the end of each period.

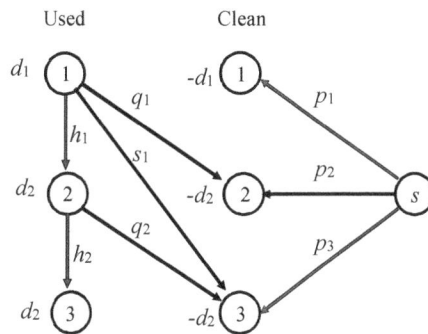

Figure 2.6. Network for caterer problem.

Let us consider Case 1 again, assuming that $a_j = a$, $c_j = c$, and $b_j = b$, for the appropriate values of the index j. Furthermore, let $d_1 = 10$, $d_2 = 20$, and $d_3 = 5$. In the model formulated below, the first three constraints are for clean napkins. The next three constraints are for dirty napkins. In these constraints the net-flow is defined as flow out minus flow in. Moreover, the net-flow is negative for terminal nodes and positive for source nodes. The seventh constraint is for network circulation, and the last one represents the non-negativity condition for arc flows.

$$Minimize \ a(p_1+p_2+p_3) + c(q_1+q_2) + bs_1$$
subject to
$$-p_1 = -10$$
$$-p_2 - q_1 = -20$$
$$-p_3 - q_2 - s_1 = -5$$
$$s_1 + q_1 + h_1 = 10$$
$$q_2 + h_2 - h_1 = 20$$
$$h_3 - h_2 = 5$$
$$p_1 + p_2 + p_3 - h_3 = 0$$
$$p_1, p_2, p_3, q_1, q_2, s_1, h_1, h_2, h_3 \geq 0$$

The *circulation networks* for Case 1 and Case 2 are shown in Figure 2.7 (a) and (b).

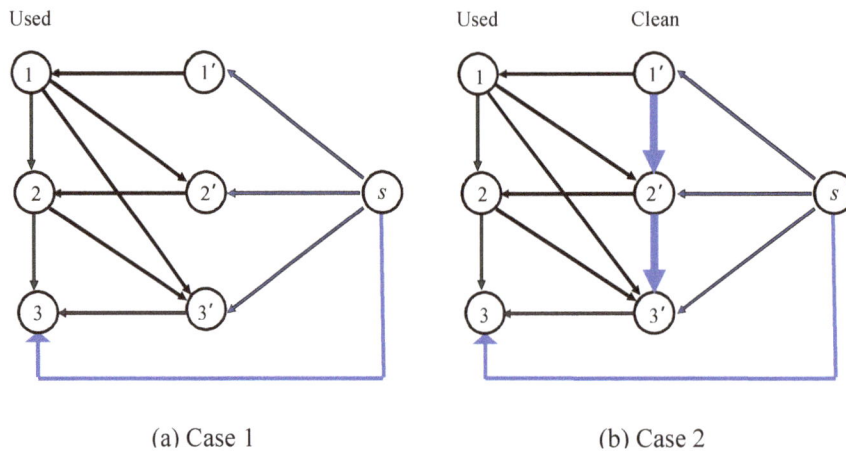

	(a) Case 1	(b) Case 2

Figure 2.7. Circulation networks for caterer problem.

In Figure 2.7, each arc has a triplet of values $[U_{ij}, L_{ij}, c_{ij}]$. These symbols represent the upper bound on flow, the lower bound on flow, and the per-unit flow cost, respectively. The return arc $(3,s)$ is used to make sure that the total number of soiled napkins at the end of the planning period is equal to the total number of purchased napkins. All arcs not yet considered, should have triplets of values $[\infty,0,0]$ assigned to them. The optimal solution corresponds to the flow resulting in minimal cost.

As indicated before, the demand for clean napkins works as a supply of dirty napkins at the end of each period. To accomplish this, each arc in Figure 2.7(a) directed from node j in the column of clean napkins to node j in the column of used napkins has lower and upper bounds set equal to d_j. Additionally, the per-unit flow cost is set equal to zero. Note that the given network does not allow holding new napkins for future days. Case 2 is a reformulation that allows this to happen by including the arcs in Figure 2.7(b) corresponding to vertical arrows in the column of new napkins. The optimal solution corresponds to the set of flows resulting in *minimal total flow cost*.

2.2.6 Employment Scheduling

The following model can be used to determine an employment policy that effectively balances the costs of hiring and terminating personnel, with the expenses of having idle employees on the payroll for a short period of time. It is assumed that the demand for labor is deterministic, although it does not need to be uniform throughout the known planning horizon. More specifically, a contract maintenance firm provides and supervises semiskilled manpower for major overhauls of chemical processing equipment. A standard job frequently requires a thousand or more men and may extend from one week to a few months. Usually, the plant site is not located in a major metropolitan area, so workers must be transported for several miles. The total labor expenses can be classified as follows: (a) operating expenses = transportation + food + housing + wages; (b) cost not depending on how long the crew remains on site = recruiting + briefing + transportation.

Consider a planning horizon consisting of periods 1, 2, ..., n-1; let us use the following notation: x_{ij} = number of employees beginning work at the start of period i and terminating at the start of period j $(1 \leq i \leq j \leq n)$, c_{ij} = total cost per employee associated with x_{ij}, D_j = demand for labor in period j, s_j = excess not used in period j. The available workforce in period j is

$$\sum_{r \leq j} \sum_{t > j} x_{rt}$$

or

$$\sum_{r=1}^{j} \sum_{t=j+1}^{n} x_{rt}$$

This is consistent with the assumption that employees must work at least during one period. Using the notation previously introduced, in the above relationships x_{rt} is the number of employees beginning work at the start of period r and terminating at the start of period t. The workforce scheduling problem for the n-1 periods of the planning horizon is given by the optimal solution to the following minimum-cost network flow model:

$$Minimize \sum_{r=1}^{n-1} \sum_{t=r+1}^{n} c_{rt} x_{rt} \tag{2-11}$$

subject to

$$\sum_{r=1}^{j} \sum_{t=j+1}^{n} x_{rt} - s_j = D_j \quad j = 1, 2, ..., n-1 \tag{2-12}$$

$$s_j \geq 0 \quad j = 1, 2, ..., n-1 \tag{2-13}$$

$$x_{rt} \geq 0 \quad r = 1, 2, ..., n-1; t = 1, 2, ..., n \tag{2-14}$$

Network Representation

This problem does not have an obvious network interpretation and it is only after some manipulation of the mathematical model that the network representation becomes evident. The procedure will be illustrated with the following example for n=4. The corresponding constraints are given in Eqs. (2-15) to (2-17):

$$x_{12} + x_{13} + x_{14} \qquad\qquad - s_1 \qquad\qquad = D_1 \tag{2-15}$$
$$x_{13} + x_{14} + x_{23} + x_{24} \qquad\qquad - s_2 \qquad = D_2 \tag{2-16}$$
$$x_{14} \qquad + x_{24} + x_{34} \qquad - s_3 = D_3 \tag{2-17}$$

Subtracting Eq. (2-15) from Eq. (2-16), Eq. (2-16) from Eq. (2-17), and writing Eq. (2-17) as a redundant equation after multiplying by -1, we obtain the following system of equations:

$$x_{12} + x_{13} + x_{14} \qquad\qquad - s_1 \qquad\qquad = D_1$$
$$- x_{12} \qquad\qquad + x_{23} + x_{24} \qquad + s_1 - s_2 \qquad = D_2 - D_1$$
$$- x_{13} \qquad - x_{23} \qquad + x_{34} \qquad + s_2 - s_3 = D_3 - D_2$$
$$- x_{14} \qquad - x_{24} - x_{34} \qquad + s_3 = -D_3$$

An examination of this set of equations reveals that each variable appears exactly twice in the equations, once with coefficient equal to +1 and once with coefficient equal to -1. This pattern of

45

constraint coefficients has special significance, and the implications of this structure are discussed fully in Section 2.3. The network for this example is given in Figure 2.8 (a), where the variables are shown next to the arcs and the supplies are shown next to the nodes. This structure is similar to the equipment replacement problem, with the exception of the arcs s_j.

In Figure 2.8 (a), nodes represent the beginning of each period. Arcs correspond to all possible employment strategies. Any arc (i,j) such that $j>i$ is used to model the number of workers hired at the beginning of period i and terminated at the beginning of period j. Additionally, any arc (j,i) such that $i<j$ is used to model the excess personnel at the beginning of period i. The parameters for the complete formulation of the problem are node demands and arc costs. The optimal solution corresponds to the *minimal-cost flow* on the network. An illustration of the model being discussed is considered now. Assume that $D_1 = 10$, $D_2 = 12$, $D_3 = 5$. A network with *feasible* flows for the above demand conditions is shown in Figure 2.8 (b). The corresponding interpretation of flows is summarized in Table 2.1.

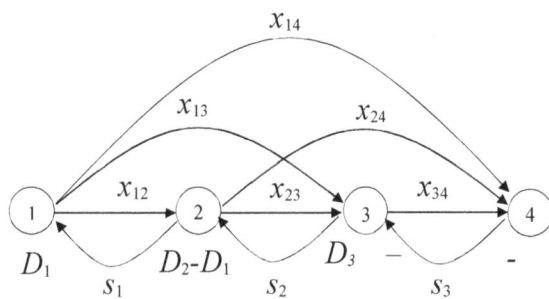

Figure 2.8 (a). Employment scheduling

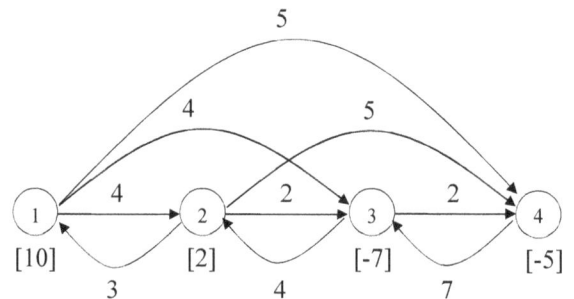

Figure 2.8 (b). Illustration of employment scheduling network.

Table 2.1. INTERPRETATION OF NETWORK FLOWS IN FIGURE 2.8 (b).

Start of Period	Hired	Terminated	Excess	Working 1 period	Working 2 periods	Working 3 periods	Available
1	13	0	3	4	4	5	13
2	7	4	4	2	5	0	16
3	2	6	7	2	0	0	12
4	0	12	0	0	0	0	0

Example

In the network shown in Figure 2.8(a), assume that the cost per employee and per period is constant and equal to \$12. Moreover, let the *net-flow requirements* be equal to 10, 2, -7 and –5, for nodes 1, 2, 3, and 4, respectively. (A) Formulate the cost-minimization linear programming model. (B) Transform the linear programming formulation into a *circulation network* model.

(A) LP Model

$$Minimize \ Z = 12x_{12} + 24x_{13} + 36x_{14} + 12x_{23} + 24x_{24} + 12x_{34}$$
Subject to

46

$$x_{12} + x_{13} + x_{14} - s_1 = 10$$
$$x_{13} + x_{14} + x_{23} + x_{24} - s_2 = 12$$
$$x_{14} + x_{24} + x_{34} - s_3 = 5$$
$$x_{ij} \geq 0 \text{ for all } i,j$$
$$s_j \geq 0 \text{ for all } j$$

It can be verified that the optimal solution for the above model is $x_{12} = 10$, $x_{23} = 7$, and $x_{24} = 5$. The minimum total cost is $Z = 324$. This solution can be interpreted as follows: hire 10 people in period 1 to work one period. Hire 7 people in period 2 to work 1 period, and 5 people to work 2 periods. It is noted that this solution and the following network solution shown in **(B)** are both optimal (*multiple* optimal solutions).

(B) Circulation Network

Nodes 1 and 2 become source nodes, and nodes 3 and 4 become terminal nodes. Figure 2.8(c) shows the circulation network. The data set corresponding to this network is given in Table 2.2. The optimal flows are shown in Table 2.3.

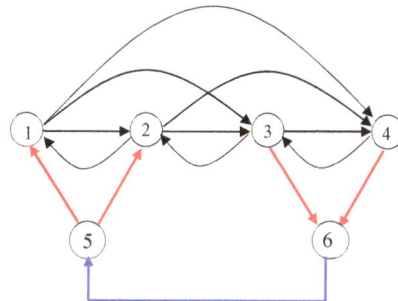

Figure 2.8(c). Circulation network for employment scheduling example.

Table 2.2. NETWORK INPUT DATA

Start Node	End Node	Upper Bound	Lower Bound	Unit Cost
1	2	12	0	12
1	3	12	0	24
1	4	12	0	36
2	1	12	0	0
2	3	12	0	12
2	4	12	0	24
3	2	12	0	0
3	4	12	0	12
3	6	7	7	0
4	3	12	0	0
4	6	5	5	0
5	1	10	10	0
5	2	2	2	0
6	5	12	12	0

Table 2.3. OPTIMAL FLOWS

Start Node	End Node	Upper Bound	Lower Bound	Unit Cost	Optimal Flow
1	2	12	0	12	0
1	3	12	0	24	5
1	4	12	0	36	5
2	1	12	0	0	0
2	3	12	0	12	2
2	4	12	0	24	0
3	2	12	0	0	0
3	4	12	0	12	0
3	6	7	7	0	7
4	3	12	0	0	0
4	6	5	5	0	5
5	1	10	10	0	10
5	2	2	2	0	2
6	5	12	12	0	12

The flows summarized in Table 2.3 can be interpreted as follows. Hire 5 people at the beginning of Period 1 to work two periods, and 5 more to work three periods. Also, hire 2 people at the beginning of Period 2 to work one period. The total cost is equal 324, as in the case of the non-network model.

2.2.7 Fleet Scheduling Problem

A problem that often arises is the scheduling and routing of vehicles to accomplish the shipment of some commodity from points of supply to points of demand [1,5]. There are two important versions of this problem. In Case 1 it is desired to maximize total profit. Case 2 focuses on the minimization of the fleet size. This can be accomplished by minimizing the number of idle ship-days. In both cases it is desired to accomplish the stated objective under the constraints that shipments must be delivered by their due dates. The optimal solution corresponds to the minimum-cost flow distribution for the given network. The mathematical model was shown in Section 2.1.2.

Example

Four shipments (each being a full shipload) are required. The corresponding information about port of origin, port of delivery, delivery date, and profit per shipment is summarized in Table 2.4. This should be interpreted to mean that there is a shipload available at port A destined for port C that must be delivered (unloaded) on day 3. The profit associated with each shipment is determined from the revenue and the operation cost directly attributed to that shipment. Additionally, there is a fixed charge of 5 to bring a ship into service. This fixed charge might reflect the company's overhead or the cost of hiring a new crew. It is desired to find the schedule for the fleet in order to maximize profits.

Table 2.4. DATA FOR FLEET SCHEDULING PROBLEM (CASE 1)

Shipment	From	To	Delivery Date	Profit
1	A	C	3	10
2	A	C	8	10
3	B	D	3	3
4	B	C	6	4

The travel time in any direction between the ports of interest for both loaded and unloaded ships is summarized below. Allowance for loading and unloading the ships has been included.

For Loaded Ships

	C	D
A	3	2
B	2	3

For Unloaded Ships

	C	D
A	2	1
B	1	2

Network Representation

There are several questions of interest for this problem. Which shipments should be made and how many ships should be used? How many trips are necessary to make all the deliveries by the specified due dates? These questions can be answered by constructing a network that represents all possible shipping schedules.

Figure 2.9 shows a network formulation of this problem. The network consists of ten nodes and fifteen arcs. The solution corresponds to the minimal-cost set of arc flows. Details on the network formulation shown in Figure 2.9 are given as follows.

A time scale (measured in days) has been placed below the network for convenience in constructing the network. In addition to a source and a terminal, nodes represent port/date combinations. A node is defined for each date and place a shipment is delivered, and for each time and place a ship starts loading. Arcs indicate all possible trips with and without cargo. All arcs emanating from the source correspond to the first use of a ship. Additionally, all arcs ending at the terminal represent the removing of ships from service. The numbers shown next to each arc are costs per unit of flow (a unit of flow represents one ship). It is noted that profits are considered here as negative costs. All arcs have infinite flow capacities except those arcs representing shipments. These arcs, having unit capacities, are (A,C) with A at time zero, (A,C) with A at time 5, (B,D) with B at time zero, and (B,C) with B at time 4.

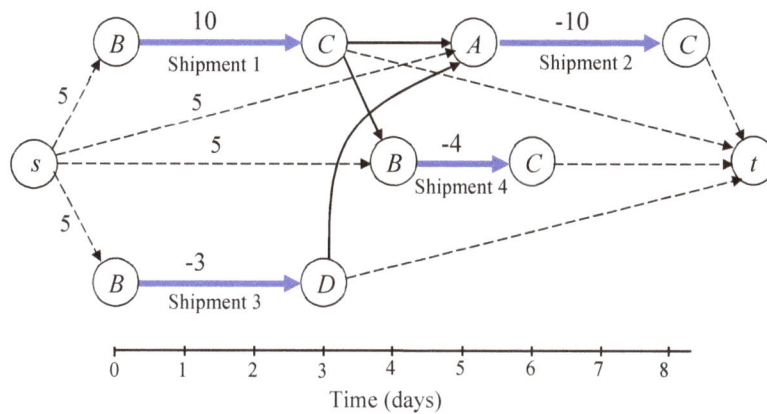

Figure 2.9. Network representation of fleet scheduling problem.

Each chain from the source node to the terminal node represents a feasible schedule for a ship. As already indicated, the optimal solution corresponds to the distribution of flow (number of ships used) having minimal cost. If a fleet of size M is used, the minimum-cost flow with a total of M units of flow will give the optimal ship schedules and also determine which shipments should be carried. If the fleet size is to be selected to maximize profit (Case 1), the problem can be solved parametrically to determine the profit profile as the number of ships is increased. It can be verified that if the fleet consists of one ship, the optimal solution is to use it for shipments 1 and 2, resulting in a total cost of -15 or a profit of 15. If the fleet consists of two ships, the optimal solution is to use one ship for shipments 1 and 4, with a cost equal to -9, and the other ship for shipments 2 and 3, with a cost equal to -8, resulting in a total profit equal to 17.

As an exercise, the student is asked to formulate the network model for minimizing the size of the fleet (Case 2) using the following notation:

c_{ij} number of *idle ship-days* associated with arc (i,j)
f_{ij} flow (ships) on arc (i,j)
L_{ij} minimum number of ships on arc (i,j)
U_{ij} maximum number of ships on arc (i,j)

2.2.8 Production and Inventory Planning

Consider a manufacturer faced with the task of meeting the demand for a product over T consecutive time periods [31]. The expected demand during period j is denoted by d_j, $j = 1, 2, \ldots, T$. The manufacturer can obtain this item from L different production or procurement sources. Let B_{ij} be the maximum number of units that can be obtained from source i in period j. Let the decision variables x_{ij}, $i = 1, 2, \ldots, L; j = 1, 2. \ldots, T$, be defined as the number of units to be obtained from source i in period j. Let c_{ij} be the cost of obtaining one unit of product from source i in period j and let c_s denote the cost of carrying 1 unit of product in inventory from one period to the next. It is assumed that the capacity of the warehouse poses no restrictions on the number of items that can be held in inventory at any time. Finally, we define I_0 as the initial inventory at the start of the planning horizon.

In formulating the model, it will be helpful to have an expression for the inventory at the end of any period t. Denoting this inventory by I_t we have the following:

$$I_t = I_0 + \sum_{j=1}^{t}\sum_{i=1}^{L} x_{ij} - \sum_{j=1}^{t} d_j \quad t = 1, 2, \cdots, T$$

At this point we might note that if we choose the values of each x_{ij} such that $I_t \geq 0$ for all t, we ensure that the demand is met in each period.

The objective function to be minimized is defined as indicated below:

$$Z = \sum_{j=1}^{T}\sum_{i=1}^{L} c_{ij} x_{ij} + c_s \sum_{j=1}^{T}\left(I_0 + \sum_{j=1}^{t}\sum_{i=1}^{L} x_{ij} - \sum_{j=1}^{t} d_j \right) \qquad (2\text{-}18)$$

The first term of the objective function (2-18) represents the cost of obtaining the product, and the second term denotes the inventory carrying costs $c_s \Sigma_t I_t$ which are incurred during the planning horizon of T periods.

The constraints of the model are of two types. The first type ensures that the available supplies are not exceeded. There are LT constraints of this type:

$$x_{ij} \leq B \quad i = 1, 2, \cdots, L; \; j = 1, 2, \cdots, T \qquad (2\text{-}19)$$

The second type of constraint ensures that the demand in each period will be met. There are T constraints of this type:

$$I_0 + \sum_{j=1}^{t}\sum_{t=1}^{L} x_{ij} - \sum_{j=1}^{t} d_j \geq 0 \quad t = 1, 2, \cdots, T \qquad (2\text{-}20)$$

As a numerical illustration, suppose that three periods are under consideration and that two production alternatives, regular time and overtime, are available in each period. The data are provided in Table 2.5.

Table 2.5. PRODUCTION PLANNING NUMERICAL EXAMPLE

Period	Capacities (units)		Unit Production Cost ($)		Demand
	Regular Time	Overtime	Regular Time	Overtime	
1	100	20	14	18	60
2	100	10	17	22	80
3	60	20	17	22	140

In order to represent this problem as a flow network, the network components must be first identified. Assume that the nodes of a network represent the sources of the product, the period in which the product is needed, and the demand for each product. There are seven possible sources: (a) initial inventory, (b) three regular time production runs, and (c) three overtime production runs. Any arc of the network will represent four characteristics: (1) the direction of allowable flow, (2) the minimum amount of flow required through the arc, (3) the maximum amount of flow possible through the arc, and (4) the cost per unit flow through the arc. Each arc (i,j) is assigned a triplet (L_{ij}, U_{ij}, c_{ij}) to indicate lower bound on flow, upper bound on flow, and per-unit flow cost, respectively. The network optimization problem is that of determining the minimum-cost flow distribution that will guarantee that all demands are met from the limited supply sources.

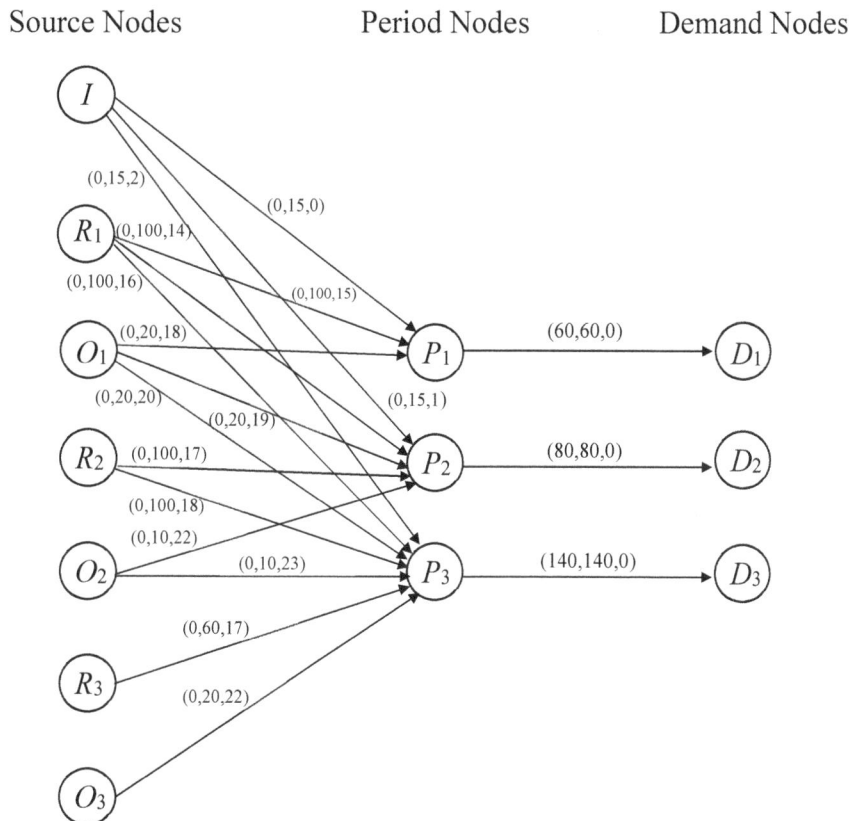

Figure 2.10. Production planning model.

The network representation of this problem is shown in Figure 2.10. Consider arc (I, P_1). The arc parameters (0, 15, 0) indicate that no use of the initial inventory in period 1 is required ($L = 0$), but 15 units are available ($U = 15$) at zero cost ($c = 0$). Similarly, the parameters of arc (I, P_2) indicate that the initial inventory of 15 units need not be used, but it will cost one dollar ($c = 1$) per unit if it is used in period 2. All arcs directed from the source nodes I, R_1, O_1, R_2, O_2, R_3, and O_3 to the period nodes P_1, P_2, and P_3 have similar interpretations. Additionally, arcs (P_1, D_1), (P_2, D_2), and (P_3, D_3), force the required demands to be satisfied. For example, arc (P_2, D_2) specifies that a minimum flow of 80 units is required ($L = 80$), a maximum flow of 80 units is allowed ($U = 80$), and the unit cost is zero ($c = 0$). Hence, arc (P_2, D_2) forces 80 units to be supplied at node D_2. Arcs (P_1, D_1) and (P_3, D_3) place similar requirements on production in periods 1 and 3, respectively.

2.2.9   Aircraft Fuel Allocation Problem

Given the route of a single aircraft, it is desired to find the minimal-cost fuel purchasing policy under the existence of price differentials among the cities in the route, as well as burn-off penalties associated with excess (beyond minimal requirements) fuel carried on the aircraft. It is also assumed that fuel availability conditions for the cities, and fuel requirements for the flights are known. This problem can be formulated as a *generalized minimum-cost network problem*. In the case of a generalized network with losses, flow arriving at the end of an arc is less than the flow entering at the beginning of an arc. This class of network lends itself to a simplified and powerful representation of the "burn-off" penalties associated with excess fuel carried on an aircraft.

Network Representation

Figure 2.11 shows a network for a four-city route. Node 1 is the source node. The intermediate nodes represent cities and flights between cities. Nodes *2*, *3*, *4*, and *5* represent the four cities *A*, *B*, *C*, and *D*, respectively. Node 6 represents the flight from city *A* to city *B*. Node 7 represents the flight from city *B* to city *C*. Node 8 represents the flight from city *C* to city *D*. Node 9 represents the flight from city *D* to *A*. Node 10 is the terminal node. Furthermore, the arcs are classified into four categories: (a) arcs coming out of the source represent buying fuel in each city; (b) arcs (2,6), (3,7), (4,8), and (5,9) represent carrying fuel in the aircraft; (c) arcs (6,3), (7,4), and (8,5) represent carrying excess fuel after flights; (d) arcs into the terminal represent the use of fuel for flight requirements.

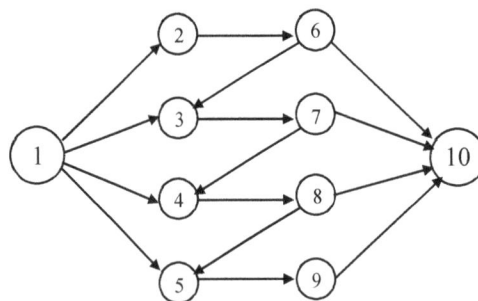

Figure 2.11. Fuel Allocation Network.

52

The arc parameters for any arc are minimum flow, maximum flow, per-unit flow cost, and the burn-off penalty factor. These parameters can be listed for arc (i,j) as $[L_{ij}, U_{ij}, c_{ij}, A_{ij}]$. The arcs associated with excess fuel left at the end of a flight are *generalized arcs* because of the burn-off losses. The arcs representing the fuel carried in the aircraft tank can be eliminated, if desired, since the fuel in the tank can be calculated by subtracting the amount used from the total available before a flight.

An illustration of the model shown in Figure 2.11 follows with arc data given in Table 2.6. The optimal flows are shown in the network of Figure 2.12. The total cost of the fuel purchasing policy shown in the network is $17,048. More details on this application are available in the article by Garcia-Diaz [15].

Table 2.6. FUEL ALLOCATION EXAMPLE

Purchasing Strategies	Fuel Carrying Strategies	Excess Fuel Strategies	Requirements Strategies
(1,2) [0,1500,4.80,1.00]	(2,6) [0,2000,0.00,1.00]	(6,3) [0,2000,0.00,0.93]	(6,10) [1134,1134,0.00,1.00]
(1,3) [0,1000,4.90,1.00]	(3,7) [0,2000,0.00,1.00]	(7,4) [0,2000,0.00,0.96]	(7,10) [642,642,0.00,1.00]
(1,4) [0,2000,4.60,1.00]	(4,8) [0,2000,0.00,1.00]	(8,5) [0,2000,0.00,0.95]	(8,10) [881,881,0.00,1.00]
(1,5) [0,1000,0.75,1.00]	(5,9) [0,2000,0.00,1.00]		(9,10) [910,910,0.00,1.00]

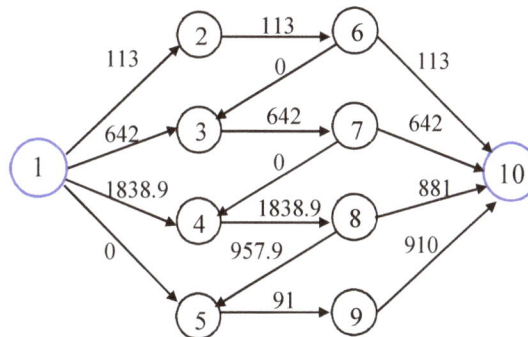

Figure 2.12. Optimal flows for fuel allocation network.

2.2.10 Conclusions

The formulations previously described were selected to illustrate the diversity of applications that can be modeled using network representations. Primary emphasis has been on network models that are deterministic, or stochastic models that have a deterministic equivalent. In this case, static models were formulated with dynamic characteristics generally included by representing the time periods as nodes, or by letting each time-location combination be a node. Models of this type are very common and can be efficiently used instead of conventional linear programming methods.

The shortest-chain models and single-commodity flow models are easily solved. These models have a very special structure that reduces the problem to finding chains in a network. This has undoubtedly contributed to their popularity.

The problems described here were intended to introduce network models. They are admittedly vast simplifications of the real problems. However, they are useful because they are easily understood and solved. In addition, they provide a visual aid to the solution of each problem. Additional network applications can be found in publications by Bennington [2], Bradley [3], Evans [9], Frank and Frisch [13], and Glover and Klingman [17].

In the sections that follow, we concern ourselves with techniques and algorithmic procedures that can be used to solve these and other problems.

2.3 LINEAR PROGRAMMING AND NETWORKS

Let $G = (\mathbf{N}, \mathbf{A})$ be a directed and capacitated network. Set $\mathbf{N}$ can be partitioned into three mutually exclusive subsets: $\mathbf{N_1}$ is the set of *sources*, $\mathbf{N_2}$ is the set of *transshipment* (or intermediate) nodes, and $\mathbf{N_3}$ is the set of *terminals*. The *pure minimum-cost problem* introduced in Section 2.1.2 can be reformulated as

$$Minimize \sum_i \sum_j c_{ij} f_{ij} \tag{2-21}$$

subject to

$$\sum_j f_{ij} - \sum_j k_{ji} f_{ji} \le a_i \quad i \in \mathbf{N_1} \tag{2-22}$$

$$\sum_j f_{ij} - \sum_j k_{ji} f_{ji} = 0 \quad i \in \mathbf{N_2} \tag{2-23}$$

$$\sum_j f_{ij} - \sum_j k_{ji} f_{ji} \ge b_i \quad i \in \mathbf{N_3} \tag{2-24}$$

$$L_{ij} \le f_{ij} \le U_{ij} \qquad (i,j) \in \mathbf{A} \tag{2-25}$$

In Eqs. (2-22) and (2-24), a_i and b_i are positive. The generalized minimum cost problem can be formulated as indicated below, where k_{ji} is the gain/loss factor for arc (j,i) and f_{ji} is the flow exiting node j in the direction of node i, along arc (j,i):

$$Minimize \sum_i \sum_j c_{ij} f_{ij} \tag{2-26}$$

subject to

$$\sum_j f_{ij} - \sum_j k_{ji} f_{ji} \le a_i \quad i \in \mathbf{N_1} \tag{2-27}$$

$$\sum_j f_{ij} - \sum_j k_{ji} f_{ji} = 0 \quad i \in \mathbf{N_2} \tag{2-28}$$

$$\sum_j f_{ij} - \sum_j k_{ji} f_{ji} \ge b_i \quad i \in \mathbf{N_3} \tag{2-29}$$

$$L_{ij} \le f_{ij} \le U_{ij} \qquad (i,j) \in \mathbf{A} \tag{2-30}$$

The following four problems illustrate the general applicability of the minimum-cost network flow models formulated in Eqs. (2-22)-(2-25) and Eqs. (2-26)-(2-30). The pure (non-generalized) network formulation will be considered since all arc multipliers are assumed to be equal to 1.0. These problems are well-known as the transportation, assignment, maximum flow and shortest path problems.

2.3.1 The Transportation Problem

In the special case where the commodity is shipped directly from source nodes to sink nodes, there are no intermediate nodes. Hence, $\mathbf{N_2} = \varnothing$. The resulting problem is a classical capacitated *transportation problem* formulated as follows:

$$Minimize \sum_i \sum_j c_{ij} f_{ij} \tag{2-31}$$

subject to

$$\sum_j f_{ij} \leq a_i \qquad i \in \mathbf{N_1} \tag{2-32}$$

$$\sum_j f_{ji} \geq b_i \qquad i \in \mathbf{N_3} \tag{2-33}$$

$$0 \leq f_{ij} \leq U_{ij} \quad (i,j) \in \mathbf{A} \tag{2-34}$$

If shipments are allowed from any source to other sources and from any terminal to other terminals, the model in Eqs. (2-31)-(2-34) can be reformulated as the well-known *transshipment problem*.

2.3.2 The Assignment Problem

In the special case where $U_{ij} = \infty$ for each arc $(i,j) \in \mathbf{A}$, $a_i = 1$ for each node $i \in \mathbf{N_1}$, and $b_i = 1$ for each node $i \in \mathbf{N_3}$, the formulation of the transportation problem becomes an *assignment problem*:

$$Minimize \sum_i \sum_j c_{ij} f_{ij} \tag{2-35}$$

subject to

$$\sum_j f_{ij} \leq 1 \quad i \in \mathbf{N_1} \tag{2-36}$$

$$\sum_i f_{ij} \geq 1 \quad i \in \mathbf{N_3} \tag{2-37}$$

$$f_{ij} \geq 0 \quad (i,j) \in \mathbf{A} \tag{2-38}$$

In general, a necessary condition for a feasible solution is that the number of sources must be equal to or greater than the number of sinks in the formulation given in Eqs. (2-35)-(2-38). Note that the constraints will be satisfied as strict equalities if there are as many sink nodes in set $\mathbf{N_3}$ as source nodes in set $\mathbf{N_1}$.

2.3.3 The Maximum Flow Problem

In the model formulated in Eqs. (2-21)-(2-25) consider the case where there is a single source node s and a single terminal t node. Therefore, $\mathbf{N_1} = \{s\}$, $\mathbf{N_2} = \mathbf{N} - \{s,t\}$, and $\mathbf{N_3} = \{t\}$. Furthermore, assume the following conditions: (a) there is an unlimited supply at the source node, $a_s = \infty$; (b) for each arc $(i,j) \in \mathbf{A}$ the lower bound is $L_{ij} = 0$; (c) for each arc connecting a node i to the terminal node the per-unit flow cost is equal to $c_{it} = -1$; (d) all remaining arc costs are assumed to be equal to zero. The *pure minimum-cost problem* becomes that of maximizing the total amount of flow into the terminal node t:

$$Maximize \ \sum_i f_{it} \tag{2-39}$$

subject to

$$\sum_j f_{ij} - \sum_j f_{ji} = 0 \quad all \ i \neq s, i \neq t \tag{2-40}$$

$$0 \leq f_{ij} \leq U_{ij} \quad (i,j) \in \mathbf{A} \tag{2-41}$$

2.3.4 The Shortest Path Problem

In the shortest path problem, it is assumed that the parameter c_{ij} for arc (i,j) represents its *length* and that there are no negative cycles in the network. In several applications the parameter c_{ij} actually represents the cost or time required to send one unit of flow from node i to node j.

If we define $\mathbf{N_1} = \{1\}$, $\mathbf{N_2} = \{2, 3, ..., n-2\}$, $\mathbf{N_3} = \{n\}$, $a_1 = 1$ and $b_n = 1$, the problem formulated in Eqs. (2-21)-(2-25) becomes one of determining the shortest chain from node 1 (source) to node n (sink). Although the flow on each arc (i,j) is such that $0 \leq f_{ij} \leq 1$, the unit of flow remains undivided as it moves through the arcs of the network. This property will be studied in Section 2.4. Mathematically, the shortest path model is formulated as follows:

$$Minimize \ \sum_i \sum_j c_{ij} f_{ij} \tag{2-42}$$

subject to

$$\sum_{j \in \mathbf{N}} f_{1j} \leq 1 \tag{2-43}$$

$$\sum_{j \in \mathbf{N}} f_{jn} \geq 1 \tag{2-44}$$

$$\sum_j f_{ij} - \sum_j f_{ji} = 0 \quad all \ i \neq 1, i \neq n \tag{2-45}$$

$$f_{ij} \geq 0 \quad (i,j) \in \mathbf{A} \tag{2-46}$$

It should be noted that the inequality constraints (2-43) and (2-44) will hold as strict equalities in the optimal solution.

2.4 THE TOTAL UNIMODULARITY PROPERTY

The previous discussion has shown that the formulation of Eqs. (2-21) to (2-25) encompasses a wide variety of important network-flow problems. Algorithms to solve these and other relevant problems will be developed later in this chapter and subsequent chapters. It is now our purpose to show that any basic, feasible solution to the general formulation must be integer, given that a_i, b_i and U_{ij} are integer. Some *definitions* are now presented before proving this result.

A *unimodular* matrix is a square, integer matrix with determinant equal to 0, +1, or -1. An integer matrix (not necessarily square) is *totally unimodular* if every square submatrix of it is unimodular. The following statements are equivalent:

1. **A** is totally unimodular
2. $\mathbf{A}^T$ is totally unimodular
3. $\left[\frac{\mathbf{A}}{\mathbf{I}}\right]$ is totally unimodular
4. Every matrix obtained by deleting a row or column of **A** is totally unimodular
5. Every matrix obtained by multiplying a row or column of **A** by -1 is totally unimodular
6. Every matrix obtained by interchanging two rows or two columns of **A** is totally unimodular
7. Every matrix obtained by duplicating rows or columns of **A** is totally unimodular
8. Every matrix obtained by a *pivot operation* on **A** is totally unimodular. A pivot operation is exactly the same operation in the Gaussian elimination technique resulting in a *column* with an entry equal to 1 in the position of the pivot and all other entries being equal to 0.

As an exercise, the student is asked to prove statement 8 in the above proposition. See for example, *Integer and Combinatorial Optimization* [28].

Now we turn our attention to the following system of linear inequality constraints (2-47) and (2-48) which defines the feasible region for the elements of a nonnegative vector **F**. It is assumed that the matrix **D** and the vector **b** are integer:

$$\mathbf{DF} \leq \mathbf{b} \qquad\qquad (2\text{-}47)$$
$$\mathbf{F} \geq \mathbf{0} \qquad\qquad (2\text{-}48)$$

2.4.1 Theorem 1

If the matrix **D** is totally unimodular, *every* basic solution to $\mathbf{DF} = \mathbf{b}$ and $\mathbf{F} \geq \mathbf{0}$ is an integer solution.

This theorem can be proved as follows. Let **D** and **F** be partitioned as $\mathbf{D} = [\mathbf{D_B D_N}]$ and $\mathbf{F} = [\mathbf{F_B} \ \mathbf{F_N}]$. It is assumed here that $\mathbf{D_B}$ is not singular. Therefore, $\mathbf{D_B F_B} + \mathbf{D_N F_N} = \mathbf{b}$. A basic solution is given by $\mathbf{F_B} = \mathbf{D_B^{-1} b}$ and $\mathbf{F_N} = \mathbf{0}$. The total unimodularity of **D** implies that $\mathbf{D_B}$ is unimodular. Recall that $\mathbf{D_B^{-1}} = \mathbf{D_B^*}/|\mathbf{D_B}|$, where $\mathbf{D_B^*}$ (the adjoint matrix) is integer. Since $|\mathbf{D_B}| = \pm 1$, we conclude that $\mathbf{D_B^{-1}}$ is an integer matrix. Since, by assumption, **b** is integer, we conclude that $\mathbf{F_B}$ is integer.

57

The total unimodularity of **D** is not necessary for integer basic solutions. For example, the system of equations $5x_1 + 2x_2 = b_1$, $2x_1 + x_2 = b_2$, where b_1 and b_2 are integer has a unique integer solution, although the matrix of coefficients is not totally unimodular.

2.4.2 Theorem 2

Consider an arbitrary integer matrix **D**. The following two conditions are equivalent with respect to matrix **D**: (a) **D** is totally unimodular. (b) All basic solutions of the set of constraints **DF** ≤ **b**, **F** ≥ **0** are integer.

(a) We first prove the sufficiency of total unimodularity. It suffices to show that the total unimodularity of **D** implies that **(D,I)** is totally unimodular. Given that **(D,I)** is totally unimodular, Theorem 1 can be applied to the constraints **DF** + **IY** = **b**, **F** ≥ **0**, **Y** ≥ **0**. To show that **(D,I)** is totally unimodular, consider an arbitrary square and nonsingular submatrix **P** of **(D,I)**. There are three mutually exclusive possibilities. If **P** does not contain columns of **I**, then $|\mathbf{P}| = \pm 1$ since **D** is totally unimodular. If **P** does not contain columns of **D**, then clearly $|\mathbf{P}| = \pm 1$. If **P** contains columns of **D** and columns of **I**, then it is possible to permute its rows to obtain

$$\mathbf{P'} = \begin{bmatrix} \overline{\mathbf{D}}_1 & \mathbf{0} \\ \overline{\mathbf{D}}_2 & \hat{\mathbf{I}} \end{bmatrix}$$

where $\overline{\mathbf{D}}_1$ and $\hat{\mathbf{I}}$ are square submatrices (not necessarily of the same dimension) and **0** is a submatrix (not necessarily square) with all entries equal to zero. Note that $|\mathbf{P}| = |\mathbf{P'}|$, and that $|\mathbf{P'}| = |\overline{\mathbf{D}}_1||\hat{\mathbf{I}}| = \pm 1$. Since **D** is totally unimodular and **P** is assumed to be non-singular, then $|\mathbf{P}| = \pm 1$.

(b) Now we prove the necessity of total unimodularity. Let **B** be a basis matrix for **DF** + **IY** = **b**. Furthermore, let $\mathbf{e_j}$ be the jth column of **B**, and let $\hat{\mathbf{b}}_j$ be the jth column of $\mathbf{B}^{-1}$. Note that $\hat{\mathbf{b}}_j = \mathbf{B}^{-1}\mathbf{e}_j$. Now, let **t** be any *integer* vector such that $\mathbf{t} + \hat{\mathbf{b}}_j \geq \mathbf{0}$. Additionally, let $\mathbf{b(t)} = \mathbf{Bt} + \mathbf{e_j}$. Now, consider the set **DF** ≤ **b(t)**. Note that $\mathbf{B}^{-1}\mathbf{b(t)} = \mathbf{B}^{-1}(\mathbf{Bt} + \mathbf{e_j}) = \mathbf{t} + \mathbf{B}^{-1}\mathbf{e_j} = \mathbf{t} + \hat{\mathbf{b}}_j \geq \mathbf{0}$. This implies that $(\mathbf{t} + \hat{\mathbf{b}}_j, \mathbf{0})$ is an extreme point of $\{\mathbf{F}: \mathbf{DF} \leq \mathbf{b(t)}, \mathbf{F} \geq \mathbf{0}\}$. Thus, $\mathbf{t} + \hat{\mathbf{b}}_j$ is an integer vector, and then $\hat{\mathbf{b}}_j$ must be integer, since **t** was chosen to be integer. This in turn implies that $\mathbf{B}^{-1}$ is integer. Now consider **A**, an arbitrary square non-singular submatrix of **D**. Using **A** and appropriate columns of **I** the following basis matrix can be constructed:

$$\mathbf{B} = \begin{bmatrix} \mathbf{A} & \mathbf{0} \\ \mathbf{A}_1 & \mathbf{I} \end{bmatrix}$$

Furthermore,

$$\mathbf{B}^{-1} = \begin{bmatrix} \mathbf{A}^{-1} & \mathbf{0} \\ -\mathbf{A}_1\mathbf{A}^{-1} & \mathbf{I} \end{bmatrix}$$

Since $\mathbf{B}^{-1}$ is integer, $\mathbf{A}^{-1}$ is integer. Also, $|\mathbf{A}|$ and $|\mathbf{A}^{-1}|$ are integer. Now, $|\mathbf{A}|.|\mathbf{A}^{-1}| = |\mathbf{A}\mathbf{A}^{-1}| = \pm 1$. This implies that $|\mathbf{A}| = |\mathbf{A}^{-1}| = \pm 1$.

2.4.3 Theorem 3 [19]

An integer matrix $\mathbf{A} = [a_{ij}]$ with $a_{ij} = 0, 1, -1$, for all i and all j, is totally unimodular if: (1) No more than two nonzero elements appear in each column. (2) The rows can be partitioned into two sets $\mathbf{E}_1$ and $\mathbf{E}_2$ such that (a) if a column contains two nonzero values with the same sign, one value is in each of the sets; (b) if a column contains two nonzero elements of opposite sign, both elements are in the same set.

This theorem is true for a submatrix $\mathbf{A}_1$ of order 1, since each entry of matrix $\mathbf{A}$ is equal to 0, 1 or -1. Now, assume that the theorem is true for all submatrices $\mathbf{A}_{k-1}$ of order k-1, and let $\mathbf{A}_k$ be an arbitrary square submatrix of order k. If $\mathbf{A}_k$ has a null vector, its determinant $|\mathbf{A}_k| = 0$. If $\mathbf{A}_k$ has some column with exactly one non-zero entry, it is possible to expand its determinant along that column and conclude that $|\mathbf{A}_k|$ is equal to either 0, 1 or -1, using the induction hypothesis. If $\mathbf{A}_k$ has *every* column with exactly two non-zero entries, the following relation between the rows $\mathbf{A}_{ki}$ of $\mathbf{A}_k$ is true from conditions 2(a) and 2(b) for $j = 1, ..., k$:

$$\sum_{\mathbf{A}_{ki} \in \mathbf{E}_1} a_{ij} = \sum_{\mathbf{A}_{ki} \in \mathbf{E}_2} a_{ij}$$

The above result implies that $|\mathbf{A}_k| = 0$, and the proof of the theorem is thus completed.

In general, the conditions of Theorem 3 are not necessary for total unimodularity. However, it can be proved that for the class of matrices satisfying condition (1), condition (2) is necessary.

2.4.4 Fundamental Result

The arc-node incidence matrix $\mathbf{Z}$ contains exactly two nonzero entries in each column. These entries are +1 and –1. Thus conditions (a) and (b) of Theorem 3 are satisfied. Condition (c) is satisfied by defining set $\mathbf{E}_1$ as containing all rows and set $\mathbf{E}_2$ as empty. Hence, matrix $\mathbf{Z}$ is totally unimodular. It follows from statement 3 of the proposition that the matrix $\left[\frac{Z}{I}\right]$ is also totally unimodular. The pure minimum-cost flow model can be formulated as minimizing $\mathbf{CF}$ subject to $\mathbf{DF} \leq \mathbf{b}$ and $\mathbf{F} \geq \mathbf{0}$, where $\mathbf{b}$ is an integer vector and $\mathbf{D} = \left[\frac{Z}{I}\right]$. Since $\mathbf{Z}$ is totally unimodular, then $\mathbf{D}$ is totally unimodular. In the case of a generalized network, the arc costs, lower bounds, and upper bounds correspond to flows *leaving* the beginning node of each arc. Furthermore, the entry for arc (i,j) equal to -1 in the $\mathbf{Z}$ matrix is changed to $-k_{ij}$. Because of this the flows cannot be guaranteed to be integer even if the right-hand constants of the model are integer.

For convenience and clarity of exposition, consider an arbitrary non-generalized network representation of Eqs. (2-21) to (2-25) with node 1 being a single source, node 6 being a sink, and nodes 2, 3, 4, 5, 7 and 8 being intermediate nodes. Hence, $\mathbf{N}_1 = \{1\}$, $\mathbf{N}_\beta = \{6\}$ and $\mathbf{N}_\phi = \{2,3,4,5,7,8\}$. The graphical representation of the problem to be considered is given in Figure 2.13.

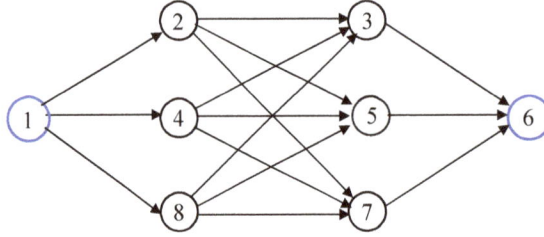

Figure 2.13. Flow network graph.

The linear programming model for this problem is formulated as follows:

$$Minimize \sum_{i} \sum_{j} c_{ij} f_{ij}$$

subject to

$$f_{12} + f_{14} + f_{18} \leq a_1$$
$$f_{23} + f_{25} + f_{27} - f_{12} = 0$$
$$f_{43} + f_{45} + f_{47} - f_{14} = 0$$
$$f_{83} + f_{85} + f_{87} - f_{18} = 0$$
$$f_{36} - f_{23} - f_{43} - f_{83} = 0$$
$$f_{56} - f_{25} - f_{45} - f_{85} = 0$$
$$f_{76} - f_{27} - f_{47} - f_{87} = 0$$
$$f_{76} - f_{27} - f_{47} - f_{87} = 0$$
$$f_{36} + f_{56} + f_{76} \geq b_6$$
$$0 \leq f_{ij} \leq U_{ij} \quad (i, j) \in \mathbf{A}$$

To prove that any basic, feasible solution to this problem is all-integer, it is sufficient to show that the coefficient matrix for the general problem is totally unimodular. The constraint matrix can be represented by the following partition of the constraint coefficients:

$$\mathbf{D} = \left[\frac{\mathbf{Z}}{\mathbf{I}} \right] \tag{2-49}$$

The **Z** matrix is generated from the coefficients of the conservation of flow constraints and the supply-demand constraints. The **I** matrix is an identity matrix formed from the coefficients of the single-term bounding constraints. The objective is to prove the total unimodularity of the matrix **D**. The submatrices **Z** and **I** of the coefficient matrix **D** are shown below. It is noted that the rows of submatrix **Z** are listed in the order 1-2-4-8-3-5-7-6.

$$\mathbf{Z} = \begin{array}{c} \\ 1 \\ 2 \\ 4 \\ 8 \\ 3 \\ 5 \\ 7 \\ 6 \end{array}
\begin{array}{ccccccccccccccc}
f_{12} & f_{14} & f_{18} & f_{23} & f_{25} & f_{27} & f_{43} & f_{45} & f_{47} & f_{83} & f_{85} & f_{87} & f_{36} & f_{56} & f_{76} \\
\end{array}$$

$$\mathbf{Z} = \begin{pmatrix}
1 & 1 & 1 & 0 & 0 & 0 & 0 & 0 & 0 & 0 & 0 & 0 & 0 & 0 & 0 \\
-1 & 0 & 0 & 1 & 1 & 1 & 0 & 0 & 0 & 0 & 0 & 0 & 0 & 0 & 0 \\
0 & -1 & 0 & 0 & 0 & 0 & 1 & 1 & 1 & 0 & 0 & 0 & 0 & 0 & 0 \\
0 & 0 & -1 & 0 & 0 & 0 & 0 & 0 & 0 & 1 & 1 & 1 & 0 & 0 & 0 \\
0 & 0 & 0 & -1 & 0 & 0 & -1 & 0 & 0 & -1 & 0 & 0 & 1 & 0 & 0 \\
0 & 0 & 0 & 0 & -1 & 0 & 0 & -1 & 0 & 0 & -1 & 0 & 0 & 1 & 0 \\
0 & 0 & 0 & 0 & 0 & -1 & 0 & 0 & -1 & 0 & 0 & -1 & 0 & 0 & 1 \\
0 & 0 & 0 & 0 & 0 & 0 & 0 & 0 & 0 & 0 & 0 & 0 & -1 & -1 & -1 \\
\end{pmatrix}$$

$$\mathbf{I} = \begin{pmatrix}
1 & 0 & 0 & 0 & 0 & 0 & 0 & 0 & 0 & 0 & 0 & 0 & 0 & 0 & 0 \\
0 & 1 & 0 & 0 & 0 & 0 & 0 & 0 & 0 & 0 & 0 & 0 & 0 & 0 & 0 \\
0 & 0 & 1 & 0 & 0 & 0 & 0 & 0 & 0 & 0 & 0 & 0 & 0 & 0 & 0 \\
0 & 0 & 0 & 1 & 0 & 0 & 0 & 0 & 0 & 0 & 0 & 0 & 0 & 0 & 0 \\
0 & 0 & 0 & 0 & 1 & 0 & 0 & 0 & 0 & 0 & 0 & 0 & 0 & 0 & 0 \\
0 & 0 & 0 & 0 & 0 & 1 & 0 & 0 & 0 & 0 & 0 & 0 & 0 & 0 & 0 \\
0 & 0 & 0 & 0 & 0 & 0 & 1 & 0 & 0 & 0 & 0 & 0 & 0 & 0 & 0 \\
0 & 0 & 0 & 0 & 0 & 0 & 0 & 1 & 0 & 0 & 0 & 0 & 0 & 0 & 0 \\
0 & 0 & 0 & 0 & 0 & 0 & 0 & 0 & 1 & 0 & 0 & 0 & 0 & 0 & 0 \\
0 & 0 & 0 & 0 & 0 & 0 & 0 & 0 & 0 & 1 & 0 & 0 & 0 & 0 & 0 \\
0 & 0 & 0 & 0 & 0 & 0 & 0 & 0 & 0 & 0 & 1 & 0 & 0 & 0 & 0 \\
0 & 0 & 0 & 0 & 0 & 0 & 0 & 0 & 0 & 0 & 0 & 1 & 0 & 0 & 0 \\
0 & 0 & 0 & 0 & 0 & 0 & 0 & 0 & 0 & 0 & 0 & 0 & 1 & 0 & 0 \\
0 & 0 & 0 & 0 & 0 & 0 & 0 & 0 & 0 & 0 & 0 & 0 & 0 & 1 & 0 \\
0 & 0 & 0 & 0 & 0 & 0 & 0 & 0 & 0 & 0 & 0 & 0 & 0 & 0 & 1 \\
\end{pmatrix}$$

Example

Show that the following matrix is totally unimodular:

$$\begin{pmatrix}
1 & 1 & 0 & 0 & 0 & 0 & 0 \\
-1 & 0 & 1 & 1 & -1 & 0 & -1 \\
0 & -1 & -1 & 0 & 1 & 1 & 1 \\
0 & 0 & 0 & -1 & 0 & -1 & 0 \\
1 & 1 & 0 & 0 & 0 & 0 & 0 \\
\end{pmatrix}$$

Note that the above matrix consists of a node-arc incidence matrix Z and one additional row and one additional column:

$$\begin{pmatrix}
1 & 1 & 0 & 0 & 0 & 0 & \mathbf{0} \\
-1 & 0 & 1 & 1 & -1 & 0 & \mathbf{-1} \\
0 & -1 & -1 & 0 & 1 & 1 & \mathbf{1} \\
0 & 0 & 0 & -1 & 0 & -1 & \mathbf{0} \\
\mathbf{1} & \mathbf{1} & \mathbf{0} & \mathbf{0} & \mathbf{0} & \mathbf{0} & \mathbf{0} \\
\end{pmatrix}$$

61

Let matrix **A** be a matrix containing **Z** and column 7 resulting from the duplication of column 5 of **Z**. Now, let **D** be a matrix containing matrix **A** and row 5 as the duplication of row 1 of **A**. Therefore, by statement 7 of the proposition given above, **D** is totally unimodular.

Example

Consider the same matrix of the previous example. After performing a pivot operation using the element at the second row and seventh column as pivot, the following matrix is obtained:

$$
\begin{pmatrix}
1 & 1 & 0 & 0 & 0 & 0 & 0 \\
1 & 0 & -1 & -1 & 1 & 0 & 1 \\
-1 & -1 & 0 & 1 & 0 & 1 & 0 \\
0 & 0 & 0 & -1 & 0 & -1 & 0 \\
1 & 1 & 0 & 0 & 0 & 0 & 0
\end{pmatrix}
$$

By statement 8, of the proposition previously given, the matrix shown above is totally unimodular.

2.5 SHORTEST ROUTE PROBLEM

Given a directed network $G = (N, A)$ with c_{ij} defined as the *length* of arc (i,j) or the *cost* of shipping *one* unit of flow along arc (i,j), for all arcs (i,j) in **A**, it is desired to find the minimum-cost route for *one* unit of flow which is sent from the source s to the terminal t through the arcs of the network. Let $b_s = 1$ be the flow supply at node s, and let $b_t = -1$ be the flow demand at node t. The linear programming model for this problem is formulated in Eqs. (2-50) to (2-54).

Eq. (2-51) guarantees that one unit of flow leaves the source. Eq. (2-53) ensures that one unit of flow arrives at the terminal node. The remaining constraints force the unit of flow to move through a subset of the intermediate nodes. After solving this model, the shortest path (or cheapest route) can be identified by connecting all arcs with $f_{ij} = 1$. It is assumed that there are no negative circuits in the shortest path.

$$Minimize \quad \sum_i \sum_j c_{ij} f_{ij} \tag{2-50}$$

subject to

$$\sum_j f_{sj} - \sum_j f_{js} = +1 \tag{2-51}$$

$$\sum_j f_{ij} - \sum_j f_{ji} = 0, i \neq s, i \neq t \tag{2-52}$$

$$\sum_j f_{tj} - \sum_j f_{jt} = -1 \tag{2-53}$$

$$f_{ij} \geq 0, (i,j) \in A \tag{2-54}$$

2.5.1 Dijkstra's Algorithm

Let us define a *label* for node j as the estimate (temporary or permanent) of the length of the shortest path from the source node s to node j. If the node label is temporary, it will be represented by δ_j, and if it is permanent, by $[\delta_j]$. Permanent labels represent lengths of actual shortest paths from the source node. Assuming that $c_{ij} \geq 0$, for each arc (i,j), the following iterative procedure, known as Dijkstra's method [6], can be used to effectively find the length of the shortest path:

Step 0: $[\delta_s] = 0$; $\delta_i = c_{si}$

Node s is the source node and c_{si} is the length of arc (s,i). If there is no arc connecting node s to node i, then $\delta_i = \infty$.

Step 1: $[\delta_j] = min_{i \in T} \{\delta_i\}$. Here j is the last node to get a permanent label, and $\mathbf{T}$ is the set of nodes with temporary label.

Step 2: If $[\delta_t]$ found, stop. Otherwise, go to Step 3.

Step 3: *new* $\delta_i = min \{old\ \delta_i; [\delta_j] + c_{ji}\}$ for $i \in \mathbf{L}$, where $\mathbf{L}$ is the set of nodes with temporary labels that can be reached from the last permanently labeled node.

Step 4: Go to Step 1.

2.5.2 Example of Dijkstra's Algorithm

The purpose of this example is to illustrate the application of Dijkstra's algorithm on the network given in Figure 2.14. In this figure, $s = 1$ is the source, $t = 5$ is the sink, and c_{ij} is the length or *cost* of each arc (i,j).

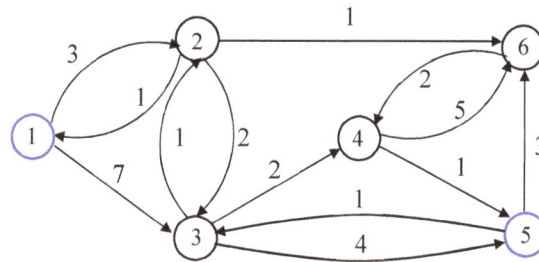

Figure 2.14. Sample network for Dijkstra's method.

A. Algorithmic Procedure

The algorithm is started by assigning the permanent label [0] to the source node $s=1$, and temporary labels $\delta_j = c_{sj}$ to nodes $j = 2,3,4,5,6$. That is, $\delta_2 = 3$, $\delta_3 = 7$, and $\delta_j = \infty$ for $j = 4,5,6$. Since $\delta_2 = 3$ is the minimum of all temporary labels, node 2 receives a permanent label, $\delta_2 = [3]$.

Nodes 3 and 6 are directly connected to node 2. Also, $\delta_2 + c_{23} = 3 + 2 = 5 < 7$ and $\delta_2 + c_{26} = 3 + 1 = 4 < \infty$. Therefore, the new temporary labels for nodes 3 and 6 are $\delta_3 = 5$ and $\delta_6 = 4$, respectively. Node 6 becomes permanently labeled since $\delta_6 < \delta_3$. That is, $\delta_6 = [4]$. Since node 6 is not the terminal node, the execution of the algorithm cannot be terminated yet.

At this point, the temporary labels are $\delta_3 = 5$ and $\delta_4 = \delta_5 = \infty$. The last node given a permanent label was node 6, which is directly connected to only node 4. Note that $\delta_6 + c_{64} = 4 + 2 = 6 < \infty$. Hence, the new temporary label for node 4 is $\delta_4 = 6$. Since $\delta_3 = \min \{\delta_3, \delta_4, \delta_5\}$, node 3 receives a permanent label equal to [5]. Continuing in this fashion, the problem is terminated when node $t = 5$ is permanently labeled. The results for this example are summarized in Table 2.7. In this table a step is an improvement in at least one temporary label or the identification of a permanent label. As can be seen in Table 2.7, the permanent label of node $t = 5$ is $\delta_t = [7]$. Therefore, the length of the shortest path from node s to node t is equal to 7.

Table 2.7. RESULTS FROM DIJKSTRA'S METHOD

Step	1	2	3	4	5	6
1	[0]	3	7	∞	∞	∞
2	[0]	[3]	7	∞	∞	∞
3	[0]	[3]	5	∞	∞	4
4	[0]	[3]	5	∞	∞	[4]
5	[0]	[3]	5	6	∞	[4]
6	[0]	[3]	[5]	6	∞	[4]
7	[0]	[3]	[5]	6	9	[4]
8	[0]	[3]	[5]	[6]	9	[4]
9	[0]	[3]	[5]	[6]	7	[4]
10	[0]	[3]	[5]	[6]	[7]	[4]

The shortest path is found by identifying arcs for which the difference in the permanent labels of two directly connected nodes is exactly equal to the length of the arc connecting the nodes. In symbols, if $[\delta_i]$ and $[\delta_j]$ are permanent labels for nodes i and j, respectively, the condition for these nodes to be on the shortest path is

$$\delta_j = \delta_i + c_{ij} \qquad (2\text{-}55)$$

The relationship given in Eq. (2-55) can be used recursively in a backward fashion starting with node t. Once a node i connected to node t is identified, we repeat the procedure to identify the next node on the shortest path. This tracing procedure is terminated when node s is reached. It can be verified that the shortest path for the example under consideration corresponds to the sequence of nodes 1-2-6-4-5.

B. Solution by NOP

Figure 2.15 shows (1) the input file, (2) the selection of the terminal node, and (3) the NOP output results. The input file has the number of nodes and the length of each arc. NOP assumes that the source node is node 1. For this reason, when a given network has the source as a node different from node 1, the nodes must be renumbered so that node 1 becomes the source. However,

64

the terminal node can be any other node of the network. In the numerical example the chosen terminal is node 5. The NOP output results provide the length of the shortest path, the permanent labels for all nodes, and the nodes for in the shortest path.

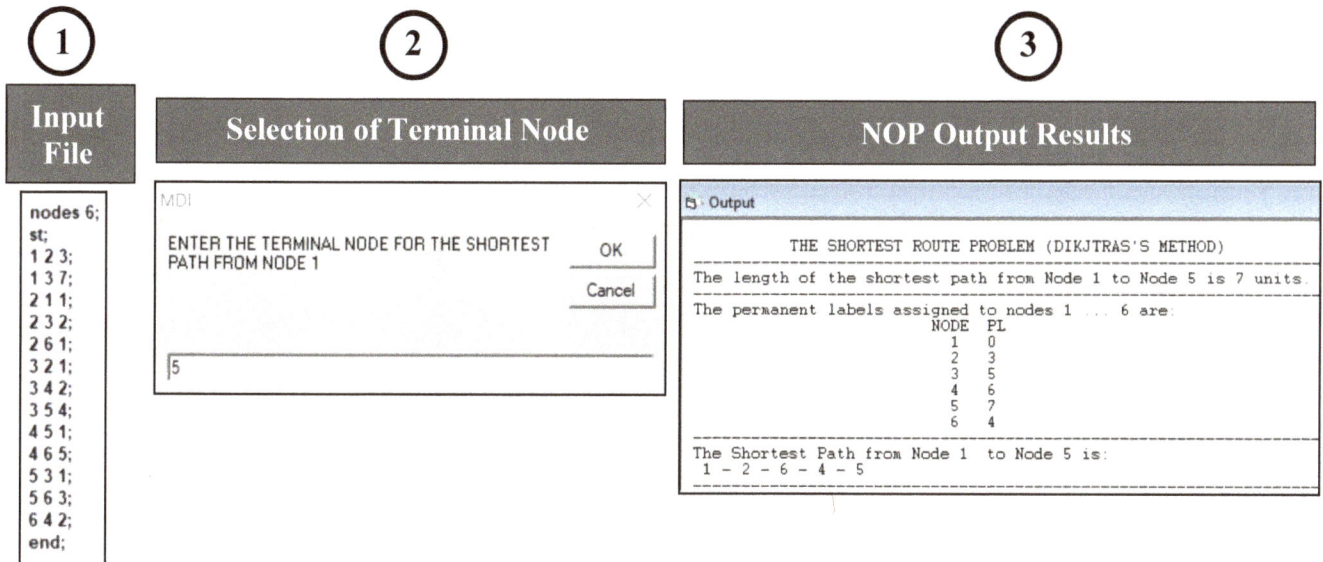

Figure 2.15. Sequence of Computer Screens for Shortest Path Example.

2.5.3   Shortest Route Problem: A Car Replacement Application

In deciding when to buy a new car, capital costs plus increasing maintenance costs should be considered. In this section we use the shortest route algorithm to decide how often to replace a car over a period of 8 years to minimize total costs. It is assumed that a decision is made at the beginning of each year on the basis of the purchase cost of a new car, the maintenance cost for the period the car would be kept, and the salvage value of the car when it is replaced. Assume that a person starts without a car and wishes to buy a new car at least every 4 years.

The network representation for the system under consideration is given in Figure 2.16. The beginning of each year is represented by a node. If a car is bought at the beginning of year i and replaced by a new one at the beginning of year j, this replacement alternative is represented by arc (i,j). Let c_{ij} be the total cost associated with this alternative. Hence, the *length* of arc (i,j) is

$$c_{ij} = P_i + \sum_{k=i}^{j-1} m_k - S_j$$

where

   P_i = purchase price at the beginning of year i
   m_k = maintenance cost during year k
   S_j = salvage value at the beginning of year j

Projected costs c_{ij} are shown on each arc (i, j) of the network given in Figure 2.16. Total costs over planning periods of the same length are not constant but vary due to inflationary effects and increasing purchase and maintenance costs. The optimal solution to this problem is given by the shortest path from node $s = 1$ to node $t = 9$. The path can be interpreted as the route followed

65

by 1 unit of *flow* going from node 1 to node 9. A summary of the computations performed to obtain the shortest chain using Dijkstra's algorithm is shown in Table 2.8. The minimum-cost policy is to buy a car in periods 1 and 5 and keep it until the end of the planning horizon. The corresponding total cost is $11,250.

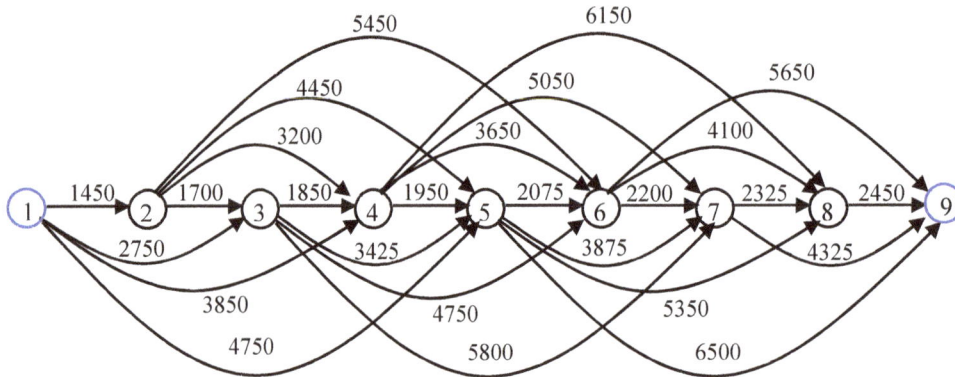

Figure 2.16. Network for car replacement application.

Table 2.8. CALCULATIONS FOR CAR REPLACEMENT APPLICATION

Step	1	2	3	4	5	6	7	8	9
1	[0]	1450	2750	3850	4750	∞	∞	∞	∞
2	[0]	[1450]	2750	3850	4750	∞	∞	∞	∞
3	[0]	[1450]	2750	3850	4750	6900	∞	∞	∞
4	[0]	[1450]	[2750]	3850	4750	6900	∞	∞	∞
5	[0]	[1450]	[2750]	3850	4750	6900	8550	∞	∞
6	[0]	[1450]	[2750]	[3850]	4750	6900	8550	∞	∞
7	[0]	[1450]	[2750]	[3850]	4750	6900	8550	10,000	∞
8	[0]	[1450]	[2750]	[3850]	[4750]	6900	8550	10,000	∞
9	[0]	[1450]	[2750]	[3850]	[4750]	6825	8550	10,000	11,250
10	[0]	[1450]	[2750]	[3850]	[4750]	[6825]	8550	10,000	11,250
11	[0]	[1450]	[2750]	[3850]	[4750]	[6825]	[8550]	10,000	11,250
12	[0]	[1450]	[2750]	[3850]	[4750]	[6825]	[8550]	10,000	11,250
13	[0]	[1450]	[2750]	[3850]	[4750]	[6825]	[8550]	[10,000]	11,250
14	[0]	[1450]	[2750]	[3850]	[4750]	[6825]	[8550]	[10,000]	11,250
15	[0]	[1450]	[2750]	[3850]	[4750]	[6825]	[8550]	[10,000]	[11,250]

2.5.4 Ford's Algorithm

This algorithm is a modified version of Dijkstra's method to find the shortest path in a network with arcs having negative lengths. Circuits with total negative length are not allowed. In order to describe the difference between this and Dijkstra's procedure, a change in terminology will be made. Labels will be considered to be either *boxed* or *non-boxed* instead of *permanent* or *temporary*.

Step 0: Assign a boxed label $[\delta_i] = 0$ to the source node s. Considering all other nodes $i = 1, 2, \dots, m$, if there is a directed arc connecting node s to node i the initial non-boxed label for this node is $\delta_i = c_{si}$; otherwise, $\delta_i = \infty$.

Step 1: Assign a boxed label to node j if $[\delta_j] = min_{i \in T} \{\delta_i\}$. Here j is the *last node* to get a boxed label, and **T** is the set of nodes with non-boxed labels.

Step 2: For each node $i \in N$ compare its old label δ_i to the value $[\delta_j] + c_{ji}$, and assign the lowest value to node i, that is, *new $\delta_i = min \{old \; \delta_i ; [\delta_j] + c_{ji}\}$*. If a node with a boxed label can be reached from the last node that received a boxed label, its label is declared to be non-boxed.

Step 3: Terminate when all nodes have boxed labels and Step 2 fails to alter any label. Otherwise, go to Step 1.

2.5.5   Example of Ford's Algorithm

Consider the network shown in Figure 2.17. It is assumed that the source node is node 1 and the terminal node is node 4. Since arc (2,3) has negative length, Ford's algorithm is used instead of Dijkstra's method to find the shortest path and its length.

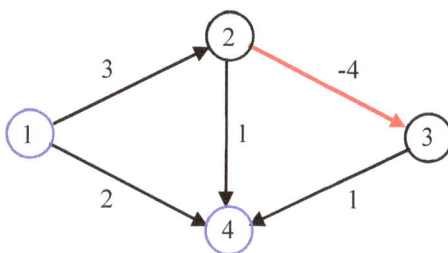

Figure 2.17. Network for Ford's algorithm.

A. Algorithmic Procedure

A summary of steps is given in Table 2.9. The shortest path can be traced as done in Dijkstra's method. The resulting path is the connected sequence of arcs (1,2), (2,3), (3,4) with length equal to zero. It is noted that in Step 4 the label for node 4 becomes boxed to be unboxed in Step 5. Furthermore. The label is reduced to 0 I Step 6 and becomes boxed in Step 7.

Table 2.9. RESULTS FOR FORD'S ALGORITHM

Step	1	2	3	4
1	[0]	3	∞	2
2	[0]	3	∞	[2]
3	[0]	[3]	∞	[2]
4	[0]	[3]	-1	[2]
5	[0]	[3]	[-1]	2
6	[0]	[3]	[-1]	0
7	[0]	[3]	[-1]	[0]

B. Solution by NOP

The three computer screens for the NOP software are shown in Figure 2.18. Screen 1 shows input data, Screen 2 shows the selection of the terminal node, and Screen 3 displays the NOP output results.

67

```
  ┌─┐  ┌──────────┐        ┌─┐  ┌───────────────────────────────────────┐
  │1│  │  Input   │        │2│  │      Selection of Terminal Node       │
  └─┘  │  File    │        └─┘  └───────────────────────────────────────┘
       └──────────┘             ┌───────────────────────────────────────┐
                                │MDI                                  ✕ │
       ┌──────────┐             │                                       │
       │NODES 4;  │             │ ENTER THE TERMINAL NODE OF THE SHORTEST│  OK  │
       │ST;       │             │ PATH FROM NODE 1                      │      │
       │1 2 3;    │             │                                       │ Cancel│
       │1 4 2;    │             │                                       │      │
       │2 3 -4;   │             │ ┌───────────────────────────────────┐ │
       │2 4 1;    │   ┌─┐       │ │4                                  │ │
       │3 4 1;    │   │3│       └───────────────────────────────────────┘
       │END;      │   └─┘  ┌───────────────────────────────────────┐
       └──────────┘        │            NOP Output Results          │
                           └───────────────────────────────────────┘
```

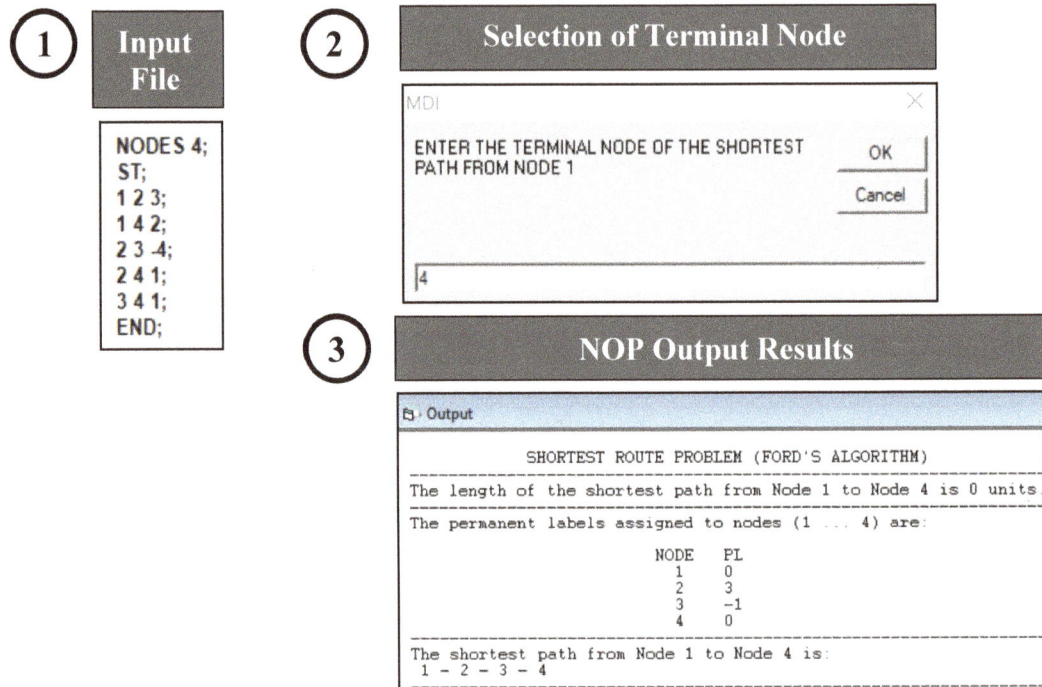

Figure 2.18. Sequence of Computer Screens for Shortest-Path Example

2.6 SHORTEST PATH MODEL WITH TURN PENALTIES

In previous sections several algorithms have been developed to determine shortest or minimum-cost routes. These algorithms, and indeed all shortest-route algorithms, are based upon the following fundamental observation: If the shortest route from node s to node t passes through node k, then the segment of the route from node s to node k is also the shortest route from s to k. Similarly, the remaining segment is the shortest route from node k to node t.

We will now study a particular class of problems for which this fundamental observation will not be true. This class of problems is known as the shortest-route model with *fixed charges*. As an illustration of this type of problem, let us consider the shipping of commodities from known storage locations to several centers of consumption, under the assumption that there is a fixed cost for the initial use of the transshipment points of a distribution network. Another example would be the routing of freighters on sea voyages between ports of call, under the consideration of port entry charges. These charges vary from port to port, depending upon the size of the freighters, the nature of the shipment, and government control policies. In general, such problems occur when transshipments facilities must be built, bought, or rented for interim usage.

Several iterative procedures based upon Lagrangian multipliers and/or duality theory have been proposed to solve fixed-charge network models of the type previously described [17]. Although the analysis of these approaches is beyond the scope of this book, we would like to address a particular class of transportation problems that can be readily solved using the methodologies presented in earlier sections of this book.

68

Many transportation planning problems are based upon the optimal routing of traffic through a network of one-way and two-way streets. In the analysis of traffic-flow problems, it is common to include delays at intersections associated with *turn penalties* that can be expressed as *fixed costs*. The penalties are usually dependent upon the direction of entry and the direction of exit at an intersection. A *prohibitive turn* can be treated as one with an infinite cost or penalty.

To illustrate the computational aspects of this problem, let us consider the rectangular grid network of Figure 2.19. This network represents a segment of city streets with five intersections leading from an entry point at node 1 to an exit point at node 8. Nodes 2, 3, 4, 5 and 7 represent the intersections. The cost of traversing each arc is given beside each arc. Additionally, it is assumed that any turn will incur a penalty or cost equal to 3. It is desired to identify the minimum-cost route to travel from node 1 to node 8. It should be noted that the *length* of the arcs and the *turn penalties* must be expressed in the same units, such as cost, time, or any other measure that is additive along any path.

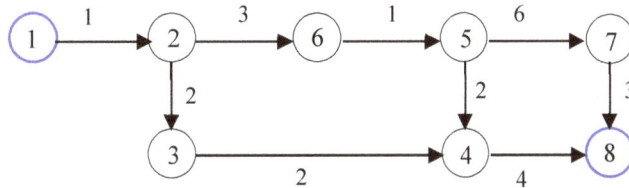

Figure 2.19. Sample network with turn penalties.

2.6.1 Theorem [13]

In a network with turn penalties, the shortest route from node s to node t through an intermediate node k may not include the shortest route from node s to node k, or from node k to node s.

We will omit the proof of this result. However, the theorem can be verified by examining the network given in Figure 2.19. It is easy to find the minimum-cost route by enumerating all possible routes from mode 1 to node 8 and proceeding to identify the route with the lowest total cost. The corresponding results are summarized in Table 2.10.

From the results given in Table 2.10 it can be seen that the shortest path from node 1 to node 8 is P_3 with a total cost equal to 15. Note that the cost of the subpath of P_3 which connects node 1 to node 4 is equal to 11. From the inspection of the arcs of P_1 and P_2, it is concluded that the other alternative route from node 1 to node 4 is a subpath of P_2 consisting of arcs (1, 2), (2, 6), (6, 5), and (5, 4). The total cost of this subpath is 10. Therefore, this subpath is the shortest path from node 1 to node 4, and it does not lie on the shortest path from node 1 to node 8. Hence, the shortest path problem with turn penalties cannot be directly solved using the conventional algorithms discussed in this textbook up to this point.

Table 2.10. PATHS FOR EXAMPLE WITH TURN PENALTIES

Path	Arc Sequence in the Path	Travel Cost	Turn Cost	Total Cost	
P_1	(1, 2)	1	0	1	
	(2, 6)	3	0	3	
	(6, 5)	1	0	1	
	(5, 7)	6	0	6	
	(7, 8)	3	3	6	17
P_2	(1, 2)	1	0	1	
	(2, 6)	3	0	3	
	(6, 5)	1	0	1	
	(5, 4)	2	3	5	
	(4, 8)	4	3	7	17
P_3	(1, 2)	1	0	1	
	(2, 3)	2	3	5	
	(3, 4)	2	3	5	
	(4, 8)	4	0	4	15

2.6.2 Algorithm

A modification of the basic problem structure can be made to allow the application of conventional techniques [13]. In more specific terms, it is desired to find the shortest (or least-cost) chain from a source node s to a terminal node t given: (a) the length (or cost) of each arc (i,j); and (b) turn penalties for each node i, depending on the two arcs used to enter and exit the node. Source and terminal nodes are not turned. This problem can be solved as a standard shortest-path model by transforming the original network and then using Dijkstra's algorithm. The steps of the algorithm are shown below:

Step 1: Modify the original network, which consists of n nodes and m arcs, by adding a *pseudo source* $\bar{s}$ and a *pseudo terminal* $\bar{t}$ and arcs $(\bar{s}, s)$ and $(t, \bar{t})$. Thus, the modified network consists of $n+2$ nodes and $m+2$ arcs. The pseudo source is not actually needed in networks with exactly one arc coming out of the source. Furthermore, the pseudo terminal is not needed when the network has exactly one arc connected to the terminal node.

Step 2: Assign labels 1, 2, ..., $m+2$ to all $m+2$ arcs of the modified network. It is assumed that label 1 is associated with arc $(\bar{s}, s)$ and label $m+2$ is associated with arc $(t, \bar{t})$.

Step 3: Find the *activity-on-node* representation of the network of Step 2. In this representation we will assume that the nodes are represented by squares instead of circles. Thus, they are represented as ▢1, ▢2, and ▢m+2. For each arc (L_i, L_j) define its weight as $d(L_i, L_j) = c(L_i) + p(L_i, L_j)$ where $c(L_i)$ is the length (cost) of original arc L_i and $p(L_i, L_j)$ is the cost penalty associated with turning the original node common to L_i and L_j using L_i to enter and L_j to exit. It is emphasized that both the length of the arcs and the turn penalties must be measured in the same units.

Step 4: Find the shortest path from the source to the terminal in the *activity-on-node* network and identify the corresponding solution in the original *activity-on-arc* network.

70

2.6.3 An Illustrative Example

To illustrate the steps of the procedure to find the shortest path with fixed charges we will find the shortest path from node 1 to node 8 in the network of Figure 2.19 assuming that each turn has a penalty equal to 3.

A. Algorithmic Procedure

The results for steps 1, 2, and 3 are shown Figure 2.19(a), 2.19(b), and 2.19(c), respectively. The network of Figure 2.19(c) also shows the result of Step 4 of the algorithm. Although the pseudo source $\bar{s}$ is not actually needed in the network of Figure 2.19, it will be used to show the steps of a complete application of the algorithm. Table 2.11 shows all the arc costs for the network of Step 3.

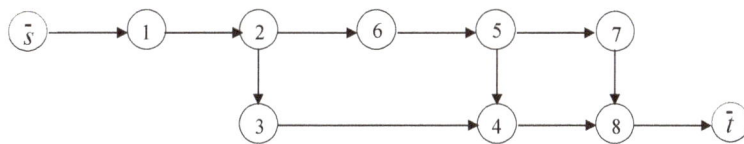

(a) Modified Original Network.

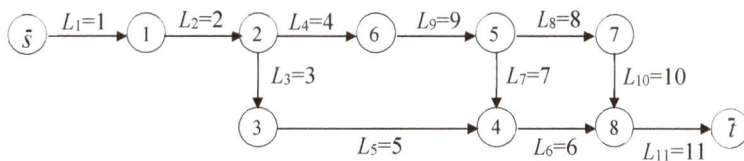

(b) Modified Network with Arc Labels.

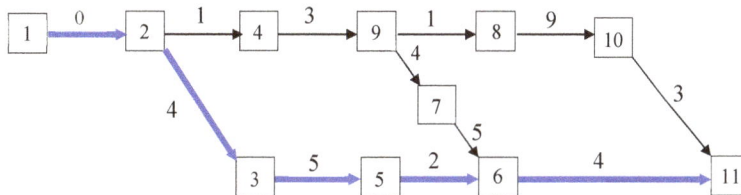

(c) Shortest Path on Activity-on-Node Representation of Modified Network.

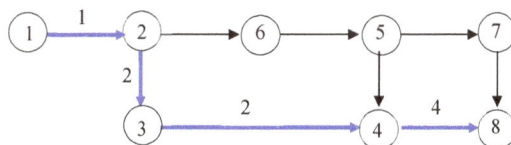

(d) Shortest path in original network.

Figure 2.19. Steps for turn penalty algorithm.

Translating the solution shown in the Figure 2.19 (c) back into the original network formulation, we obtain the results given in Table 2.13 and represented in Figure 2.19 (d). The least-cost route in the original network corresponds to the sequence of nodes 1, 2, 3, 4, 8, and has a total cost of 1 + 5 + 5 + 4 =15. The total cost corresponds to a cost of 9 for traversing the arcs, plus a cost of 6 for turning nodes.

Table 2.11. ARC COSTS FOR MODIFIED NETWORK

From j	To i	Cost c_{ij}
1	2	0
2	3	4
2	4	1
3	5	5
4	9	3
5	6	2
6	11	4
7	6	5
8	10	9
9	7	4
9	8	1
10	11	3

Table 2.12. OPTIMAL ROUTE IN MODIFIED NETWORK

From Node	To node	Cost
1	2	0
2	3	4
3	5	5
5	6	2
6	11	4

Table 2.13. OPTIMAL ROUTE IN ORIGINAL NETWORK

Arc	Original Node	Cost
(L_1, L_2)	1	-
(L_2, L_3)	2	1
(L_3, L_5)	3	5
(L_5, L_6)	4	5
(L_6, L_{11})	8	4

B. Solution by NOP

Figure 2.20 shows the computer screens for the NOP solution: (1) Input data file and turn penalties, (2) Selection of terminal node, and (3) NOP output results. As shown in Figure 2.20, the shortest path consists of arcs (1,2), (2,3), (3,4) and (4,8). The total *length* of this path is 15, including a cost of 9 for traversing the arcs plus a penalty cost of 6 for two turns.

72

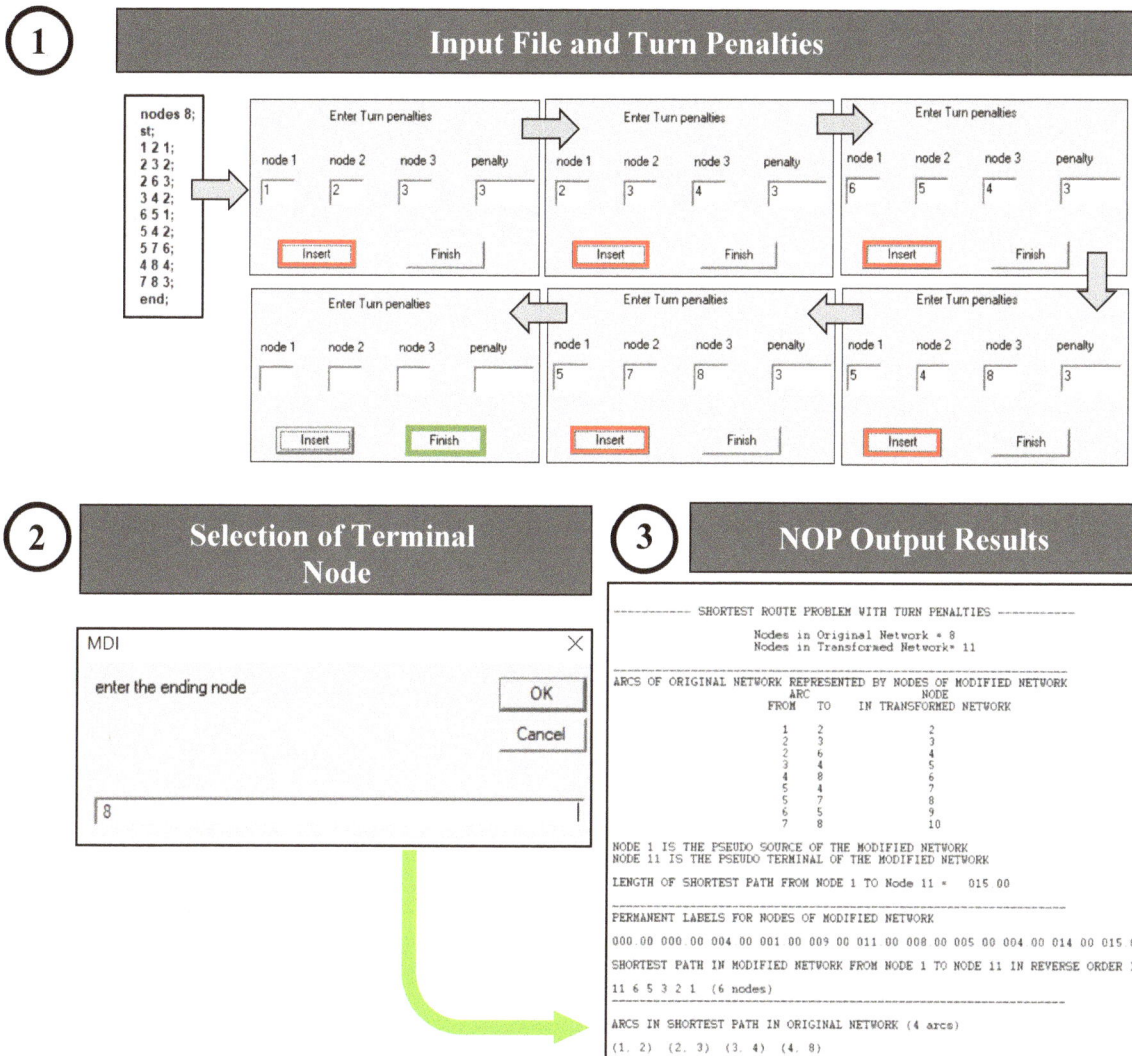

Figure 2.20. Sequence of Computer Screens for Shortest Path with Turn Penalties Example.

2.7 MULTI-TERMINAL AND MULTI-SOURCE SHORTEST-PATH PROBLEM

In this section we are concerned with finding shortest chains between all pairs of nodes of a network G = (**N, A**) which is directed and without negative circuits. Let us consider the following notation:

$\mathbf{N} = \{1, 2, ..., n\}$

c_{ij} = length of arc (i,j); $-\infty < c_{ij} < +\infty$

d_{ik}^{j} = length of shortest chain from i to k with *intermediate* nodes from the set $\{1, 2, ..., j\}$,
   $i \neq j \neq k$.

The algorithm to be discussed was developed by Floyd [10], and our exposition follows the presentations by Hu [22] and Dreyfus [7].

73

The procedure starts with $j = 0$ and $d_{ik}^0 = c_{ik}$ for each entry corresponding to arc (i,k). If this arc does not exist, $d_{ik}^0 = \infty$.

The fundamental idea behind Floyd's algorithm is to enlarge the set of *intermediate nodes* by including one node at each iteration and verifying if there are any reductions in the length of the shortest paths between the nodes of the network. More specifically, at the *j-th* iteration ($j = 1$, $2, ..., n$) we proceed to enlarge the set of intermediate nodes $\{1, 2, ..., j-1\}$ by inserting node j. The new set of intermediate nodes now becomes $\{1, 2, ..., j\}$. Choosing intermediate nodes from this set, the length of the shortest path from node i to node k is calculated and represented by d_{ik}^j. The lengths of the shortest paths are sequentially updated as the set of intermediate nodes is enlarged, and the optimal lengths are obtained when this set becomes $\{1, 2, ..., n\}$. The optimal paths have length equal to d_{ik}^n.

The above procedure can be effectively and efficiently implemented by sequentially executing the following operation at iterations $j = 1, 2, ..., n$. This operation is referred to as a *triple operation* because it includes a summation, a minimization, and a substitution:

$$d_{ik}^j = min\left\{d_{ik}^{j-1}; d_{ij}^{j-1} + d_{jk}^{j-1}\right\}, \quad i \neq j \neq k \qquad (2\text{-}56)$$

If $d_{ik}^j < d_{ij}^{j-1} + d_{jk}^{j-1}$ then bringing node j into the set of intermediate nodes results in a shorter path from node i to node k because the new length of the path is less than the existing length d_{ij}^{j-1}.

We use a matrix-based procedure to keep track of the lengths of the shortest chains and of the arcs in those chains. At each iteration of the algorithm, we construct two matrices. The first matrix, referred to as the *shortest-path matrix*, contains the current lengths of the shortest chains. That is, at the *j*th iteration the matrix is defined as $\mathbf{D}^j = \left[d_{ik}^j\right]$. The algorithm starts with $\mathbf{D}^0 = \left[d_{ik}^0\right]$, where $d_{ik}^0 = c_{ik}$. Then $\mathbf{D}^1$ is obtained by performing the triple operation (2-56) on all entries of $\mathbf{D}^0$, and so on, until matrix $\mathbf{D}^n$ is obtained.

The purpose of the second matrix, referred to as the *route matrix*, is to identify the intermediate nodes (if any) of the shortest chains. In other words, it is to keep track of the node insertions that result in improved shortest path-lengths. The route matrix at the *j*th iteration is defined as $\mathbf{R}^j = \left[r_{ik}^j\right]$ where r_{ik}^j is an intermediate node of the shortest chain from node i to node k, using intermediate nodes from the set $\{1, 2, ..., j\}$ and such that $i \neq j \neq k$. The algorithm can be initialized with $r_{ik}^0 = k$. At the *j*th iteration, the following relationship is used:

$$r_{ik}^j = \begin{cases} j & \text{if } d_{ij}^{j-1} + d_{jk}^{j-1} < d_{ik}^j \\ r_{ik}^{j-1} & otherwise \end{cases} \qquad (2\text{-}57)$$

74

2.7.1 Floyd's Algorithm

After initialization, the following procedure is repeated successively for $j = 1, 2, ..., n$ using elements of the $(j-1)$-*th* matrix for all calculations. At the *j-th* iteration, node *j* is referred to as the *pivot node*.

Step 1: (a) delete pivot row; (b) delete pivot column; (c) delete all rows with ∞-entries in the pivot column, and all columns with ∞-entries in the pivot row.

Step 2: After all entries are deleted as indicated in Step 1, the triple operation of Eq. (2-56) is performed for each non-deleted entry of the matrix. This can be performed as follows. For the non-deleted entry at the intersection of the *i*th row and the *k*th column, compare the value of d_{ik}^{j-1} against the sum $d_{ij}^{j-1} + d_{jk}^{j-1}$. These two values correspond to the *projections* of the entry under consideration on the pivot column and pivot row, respectively. In other words, each non-deleted entry is compared to the sum of its projections on the pivot column and pivot row. If the entry is larger than the sum of its projections, it is replaced by this sum. Otherwise, it is left unchanged.

Step 3: Update the route matrix using Eq. (2-57).

2.7.2 Example of Floyd's Algorithm

To illustrate the application of Floyd's algorithm, let us find the shortest path between every pair of nodes in the network of Figure 2.21. Since $n = 8$, the number of iterations for the algorithm is equal to 8. Furthermore, every undirected arc is replaced by two directed arcs with the same length but opposite orientations.

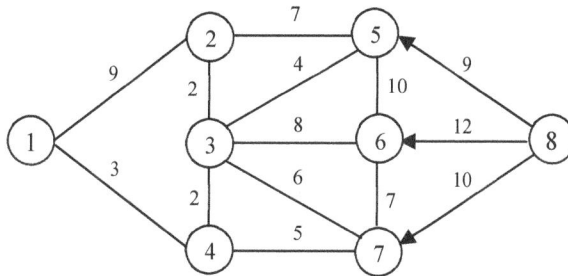

Figure 2.21. Multi-terminal network.

The iteration counter is set equal to zero with the corresponding shortest-path matrix $\mathbf{D}^0$ and the route matrix $\mathbf{R}^0$ initialized as indicated below. The algorithmic procedure shows how the triple operation is used in each of the eight iterations to update the two matrices until there are no improvements in the lengths of the multi-terminal paths.

$$
\mathbf{D}^0 = \begin{array}{c} \\ 1 \\ 2 \\ 3 \\ 4 \\ 5 \\ 6 \\ 7 \\ 8 \end{array}
\begin{array}{cccccccc}
1 & 2 & 3 & 4 & 5 & 6 & 7 & 8 \\
\left(0 \right. & 9 & \infty & 3 & \infty & \infty & \infty & \infty \\
9 & 0 & 2 & \infty & 7 & \infty & \infty & \infty \\
\infty & 2 & 0 & 2 & 4 & 8 & 6 & \infty \\
3 & \infty & 2 & 0 & \infty & \infty & 5 & \infty \\
\infty & 7 & 4 & \infty & 0 & 10 & \infty & \infty \\
\infty & \infty & 8 & \infty & 10 & 0 & 7 & \infty \\
\infty & \infty & 6 & 5 & \infty & 7 & 0 & \infty \\
\infty & \infty & \infty & \infty & 9 & 12 & 10 & \left. 0 \right)
\end{array}
\qquad
\mathbf{R}^0 = \begin{array}{c} \\ 1 \\ 2 \\ 3 \\ 4 \\ 5 \\ 6 \\ 7 \\ 8 \end{array}
\begin{array}{cccccccc}
1 & 2 & 3 & 4 & 5 & 6 & 7 & 8 \\
\left(1 \right. & 2 & 3 & 4 & 5 & 6 & 7 & 8 \\
1 & 2 & 3 & 4 & 5 & 6 & 7 & 8 \\
1 & 2 & 3 & 4 & 5 & 6 & 7 & 8 \\
1 & 2 & 3 & 4 & 5 & 6 & 7 & 8 \\
1 & 2 & 3 & 4 & 5 & 6 & 7 & 8 \\
1 & 2 & 3 & 4 & 5 & 6 & 7 & 8 \\
1 & 2 & 3 & 4 & 5 & 6 & 7 & 8 \\
1 & 2 & 3 & 4 & 5 & 6 & 7 & \left. 8 \right)
\end{array}
$$

A. Algorithmic Procedure

Iteration 1: Here $j = 1$ is defined as the pivot node. Hence, we delete the first row and the first column of matrix $\mathbf{D}^0$. Additionally, columns 3, 5, 6, 7, and 8 have ∞ entries in the pivot row; and rows 3, 5, 6, 7, and 8 have ∞ entries in the pivot column. Therefore, only the entries of $\mathbf{D}^0$ which are shown in Figure 2.22 need to be analyzed with the triple operation.

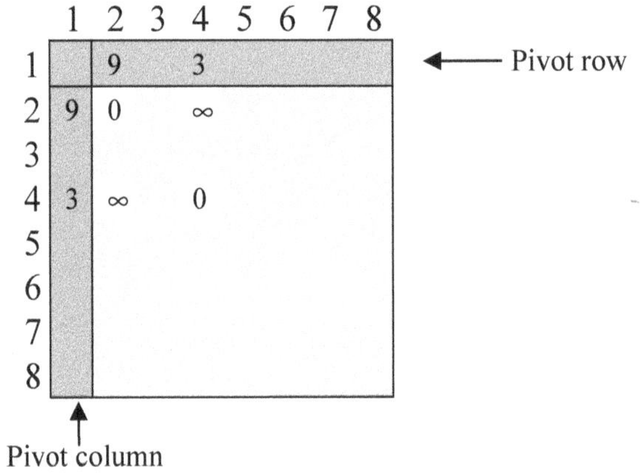

Figure 2.22. Entries to be checked in Iteration 1.

Ignoring the diagonal elements of $\mathbf{D}^0$, the estimates that must be investigated are only d_{24}^0 and d_{42}^0. The corresponding triple operations yield the following results:

$$d_{24}^1 = min\left\{d_{24}^0; d_{21}^0 + d_{14}^0\right\} = min\left\{\infty; 9+3\right\} = 12$$

$$d_{42}^1 = min\left\{d_{42}^0; d_{41}^0 + d_{12}^0\right\} = min\left\{\infty; 3+9\right\} = 12$$

Note that $d_{24}^1 = 12$ and $d_{42}^1 = 12$ are better estimates than $d_{24}^0 = \infty$ and $d_{42}^0 = \infty$, respectively. Since the use of the pivot node 1 has resulted in shorter chains from node 2 to node 4, and from node 4 to node 2, we set $r_{24}^1 = 1$ and $r_{42}^1 = 1$. All other entries of $\mathbf{D}^0$ and $\mathbf{R}^0$ remain unchanged, and then we can write for $\mathbf{D}^1$ and $\mathbf{R}^1$:

$$\mathbf{D}^1 = \begin{array}{c} \\ 1 \\ 2 \\ 3 \\ 4 \\ 5 \\ 6 \\ 7 \\ 8 \end{array}\begin{pmatrix} 0 & 9 & \infty & 3 & \infty & \infty & \infty & \infty \\ 9 & 0 & 2 & 12 & 7 & \infty & \infty & \infty \\ \infty & 2 & 0 & 2 & 4 & 8 & 6 & \infty \\ 3 & 12 & 2 & 0 & \infty & \infty & 5 & \infty \\ \infty & 7 & 4 & \infty & 0 & 10 & \infty & \infty \\ \infty & \infty & 8 & \infty & 10 & 0 & 7 & \infty \\ \infty & \infty & 6 & 5 & \infty & 7 & 0 & \infty \\ \infty & \infty & \infty & \infty & 9 & 12 & 10 & 0 \end{pmatrix} \qquad \mathbf{R}^1 = \begin{array}{c} \\ 1 \\ 2 \\ 3 \\ 4 \\ 5 \\ 6 \\ 7 \\ 8 \end{array}\begin{pmatrix} 1 & 2 & 3 & 4 & 5 & 6 & 7 & 8 \\ 1 & 2 & 3 & 1 & 5 & 6 & 7 & 8 \\ 1 & 2 & 3 & 4 & 5 & 6 & 7 & 8 \\ 1 & 1 & 3 & 4 & 5 & 6 & 7 & 8 \\ 1 & 2 & 3 & 4 & 5 & 6 & 7 & 8 \\ 1 & 2 & 3 & 4 & 5 & 6 & 7 & 8 \\ 1 & 2 & 3 & 4 & 5 & 6 & 7 & 8 \\ 1 & 2 & 3 & 4 & 5 & 6 & 7 & 8 \end{pmatrix}$$

Iteration 2: Now $j = 2$ is defined as the pivot node. Hence, we delete the second row and the second column of matrix $\mathbf{D}^1$. Additionally, columns 6, 7, and 8 have ∞ entries in the pivot row; and rows 6, 7, and 8 have ∞ entries in the pivot column. Therefore, only the entries of $\mathbf{D}^1$ shown in Figure 2.23 need to be analyzed with the triple operation.

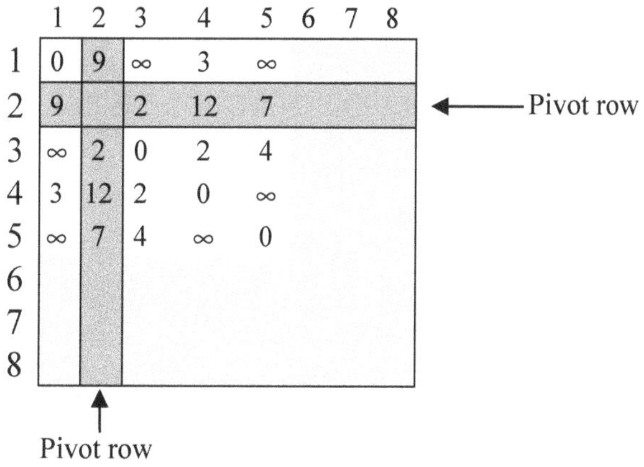

Figure 2.23. Entries to be checked in iteration 2.

Ignoring the diagonal elements of $\mathbf{D}^1$, the elements that must be investigated are d_{13}^1, d_{14}^1, d_{15}^1, d_{31}^1, d_{34}^1, d_{35}^1, d_{41}^1, d_{43}^1, d_{45}^1, d_{51}^1, d_{53}^1, and d_{54}^1. It can be verified that only the following estimates can be improved: d_{13}^1, d_{15}^1, d_{31}^1, d_{45}^1, d_{51}^1, and d_{54}^1. The new estimates are shown below.

$$d_{13}^2 = min\left\{d_{13}^1; d_{12}^1 + d_{23}^1\right\} = min\left\{\infty; 9 + 2\right\} = 11$$

$$d_{15}^2 = min\left\{d_{15}^1; d_{12}^1 + d_{25}^1\right\} = min\left\{\infty; 9 + 7\right\} = 16$$

$$d_{31}^2 = min\left\{d_{31}^1; d_{32}^1 + d_{21}^1\right\} = min\left\{\infty; 9 + 2\right\} = 11$$

$$d_{45}^2 = min\left\{d_{45}^1; d_{42}^1 + d_{25}^1\right\} = min\left\{\infty; 12 + 7\right\} = 19$$

$$d_{51}^2 = min\left\{d_{51}^1; d_{52}^1 + d_{21}^1\right\} = min\left\{\infty; 7 + 9\right\} = 16$$

$$d_{54}^2 = min\left\{d_{54}^1; d_{52}^1 + d_{24}^1\right\} = min\left\{\infty; 7 + 12\right\} = 19$$

Accordingly, $r_{13}^2 = r_{15}^2 = r_{31}^2 = r_{45}^2 = r_{51}^2 = r_{54}^2 = 2$ and $r_{ik}^2 = r_{ik}^1$ for all other entries such that $d_{ik}^2 = d_{ik}^1$. The new matrices $\mathbf{D}^2$ and $\mathbf{R}^2$ can be written as follows:

$$
\mathbf{D}^2 = \begin{array}{c} \\ 1 \\ 2 \\ 3 \\ 4 \\ 5 \\ 6 \\ 7 \\ 8 \end{array}
\begin{pmatrix}
0 & 9 & 11 & 3 & 16 & \infty & \infty & \infty \\
9 & 0 & 2 & 12 & 7 & \infty & \infty & \infty \\
11 & 2 & 0 & 2 & 4 & 8 & 6 & \infty \\
3 & 12 & 2 & 0 & 19 & \infty & 5 & \infty \\
16 & 7 & 4 & 19 & 0 & 10 & \infty & \infty \\
\infty & \infty & 8 & \infty & 10 & 0 & 7 & \infty \\
\infty & \infty & 6 & 5 & \infty & 7 & 0 & \infty \\
\infty & \infty & \infty & \infty & 9 & 12 & 10 & 0
\end{pmatrix}
\qquad
\mathbf{R}^2 = \begin{pmatrix}
1 & 2 & 2 & 4 & 2 & 6 & 7 & 8 \\
1 & 2 & 3 & 1 & 5 & 6 & 7 & 8 \\
2 & 2 & 3 & 4 & 5 & 6 & 7 & 8 \\
1 & 1 & 3 & 4 & 2 & 6 & 7 & 8 \\
2 & 2 & 3 & 2 & 5 & 6 & 7 & 8 \\
1 & 2 & 3 & 4 & 5 & 6 & 7 & 8 \\
1 & 2 & 3 & 4 & 5 & 6 & 7 & 8 \\
1 & 2 & 3 & 4 & 5 & 6 & 7 & 8
\end{pmatrix}
$$

Continuing in this fashion, we obtain the results shown in Table 2.14 for iterations $j = 3$, 4, 5. As can be verified, no improvements are possible in iterations 6, 7, and 8. Hence, the optimal solution corresponds to matrices $\mathbf{D}^5$ and $\mathbf{R}^5$.

Table 2.14. SUMMARY OF RESULTS FROM FLOYD'S ALGORITHM

Iteration j	$\mathbf{D}^j$	$\mathbf{R}^j$
3	$\begin{pmatrix} 0 & 9 & 11 & 3 & 15 & 19 & 17 & \infty \\ 9 & 0 & 2 & 4 & 6 & 10 & 8 & \infty \\ 11 & 2 & 0 & 2 & 4 & 8 & 6 & \infty \\ 3 & 4 & 2 & 0 & 6 & 10 & 5 & \infty \\ 15 & 6 & 4 & 6 & 0 & 10 & 10 & \infty \\ 19 & 10 & 8 & 10 & 10 & 0 & 7 & \infty \\ 17 & 8 & 6 & 5 & 10 & 7 & 0 & \infty \\ \infty & \infty & \infty & \infty & 9 & 12 & 10 & 0 \end{pmatrix}$	$\begin{pmatrix} 1 & 2 & 2 & 4 & 3 & 3 & 3 & 8 \\ 1 & 2 & 3 & 3 & 3 & 3 & 3 & 8 \\ 2 & 2 & 3 & 4 & 5 & 6 & 7 & 8 \\ 1 & 3 & 3 & 4 & 3 & 3 & 7 & 8 \\ 3 & 3 & 3 & 3 & 5 & 6 & 3 & 8 \\ 3 & 3 & 3 & 3 & 5 & 6 & 7 & 8 \\ 3 & 3 & 3 & 4 & 3 & 6 & 7 & 8 \\ 1 & 2 & 3 & 4 & 5 & 6 & 7 & 8 \end{pmatrix}$
4	$\begin{pmatrix} 0 & 7 & 5 & 3 & 9 & 13 & 8 & \infty \\ 7 & 0 & 2 & 4 & 6 & 10 & 8 & \infty \\ 5 & 2 & 0 & 2 & 4 & 8 & 6 & \infty \\ 3 & 4 & 2 & 0 & 6 & 10 & 5 & \infty \\ 9 & 6 & 4 & 6 & 0 & 10 & 10 & \infty \\ 13 & 10 & 8 & 10 & 10 & 0 & 7 & \infty \\ 8 & 8 & 6 & 5 & 10 & 7 & 0 & \infty \\ \infty & \infty & \infty & \infty & 9 & 12 & 10 & 0 \end{pmatrix}$	$\begin{pmatrix} 1 & 4 & 4 & 4 & 4 & 4 & 4 & 8 \\ 4 & 2 & 3 & 3 & 3 & 3 & 3 & 8 \\ 4 & 2 & 3 & 4 & 5 & 6 & 7 & 8 \\ 1 & 3 & 3 & 4 & 3 & 3 & 7 & 8 \\ 4 & 3 & 3 & 3 & 5 & 6 & 3 & 8 \\ 4 & 3 & 3 & 3 & 5 & 6 & 7 & 8 \\ 4 & 3 & 3 & 4 & 3 & 6 & 7 & 8 \\ 1 & 2 & 3 & 4 & 5 & 6 & 7 & 8 \end{pmatrix}$
5	$\begin{pmatrix} 0 & 7 & 5 & 3 & 9 & 13 & 8 & \infty \\ 7 & 0 & 2 & 4 & 6 & 10 & 8 & \infty \\ 5 & 2 & 0 & 2 & 4 & 8 & 6 & \infty \\ 3 & 4 & 2 & 0 & 6 & 10 & 5 & \infty \\ 9 & 6 & 4 & 6 & 0 & 10 & 10 & \infty \\ 13 & 10 & 8 & 10 & 10 & 0 & 7 & \infty \\ 8 & 8 & 6 & 5 & 10 & 7 & 0 & \infty \\ 18 & 15 & 13 & 15 & 9 & 12 & 10 & 0 \end{pmatrix}$	$\begin{pmatrix} 1 & 4 & 4 & 4 & 4 & 4 & 4 & 8 \\ 4 & 2 & 3 & 3 & 3 & 3 & 3 & 8 \\ 4 & 2 & 3 & 4 & 5 & 6 & 7 & 8 \\ 1 & 3 & 3 & 4 & 3 & 3 & 7 & 8 \\ 4 & 3 & 3 & 3 & 5 & 6 & 3 & 8 \\ 4 & 3 & 3 & 3 & 5 & 6 & 7 & 8 \\ 4 & 3 & 3 & 4 & 3 & 6 & 7 & 8 \\ 5 & 5 & 5 & 5 & 5 & 6 & 7 & 8 \end{pmatrix}$
6	Unchanged	Unchanged
7	Unchanged	Unchanged
8	Unchanged	Unchanged

As an illustration of how to use the results shown in Table 2.14, let us analyze the shortest chain from node 1 to node 5. The length of this chain is equal to $d_{15}^5 = 9.$ In order to find the corresponding sequence of nodes, refer to matrix $\mathbf{R}^5$ and proceed as follows. The entry $r_{15}^5 = 4$ indicates that node 4 is an intermediate node in the shortest path from node 1 to node 5. Now we obtain $r_{45}^5 = 3$ which identifies node 3 as an intermediate node between nodes 4 and 5. Furthermore, $r_{35}^5 = 5$ indicates that there is no intermediate node between nodes 3 and 5. Therefore, the shortest path from node 1 to node 5 corresponds to the sequence of nodes 1-4-3-5.

B. Solution by NOP

Figure 2.24 shows the NOP screens for this example: (1) Input data file, (2) NOP Output results. Non-existing paths corresponding to ∞-entries are shown with length equal to 9999.00.

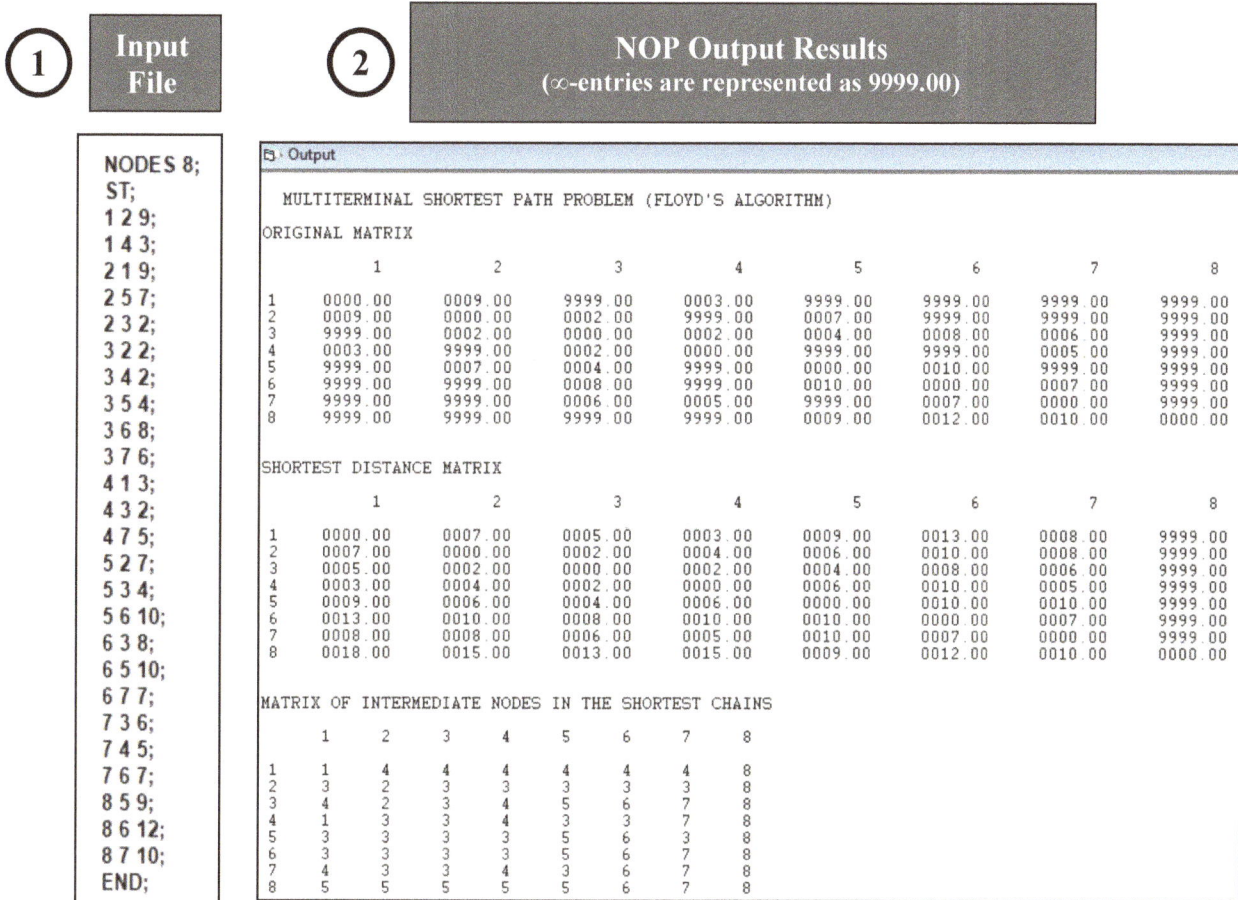

Figure 2.24. Sequence of Computer Screens for Shortest-Path with Turn Penalties Example.

2.8 SPECIAL CASES

A. Ford's algorithm to find the longest route from node s to node t
B. Most reliable route from node s to node t
C. Most-reliable route between each pair of nodes

2.8.1 Ford's Algorithm to Find the Longest Route

The procedure is a modified version of Dijkstra's method suitable for the consideration of negative arcs. *Positive* circuits are not allowed.

Step 0: For each arc (s,i) connecting the source node s to node i, set $\delta_i = c_{si}$; if there is no arc connecting node s to node i then $\delta_i = -\infty$.

Step 1: $[\delta] = \max_{i \in T} \{\delta_i\}$ where $\mathbf{T}$ is the set of nodes with non-boxed labels.

Step 2: new $\delta_i = \max \{$old $\delta_i; [\delta] + c_{ji}\}$ for all $i \in \mathbf{N}$. If a node with a boxed label can be reached from the last node to receive boxed label, its label is declared non-boxed.

Step 3: Terminate when all nodes have boxed labels and step 2 fails to alter any label. Otherwise, return to Step 1.

2.8.2 Most Reliable Route

In this case, we define the arc parameters as probabilities of arc success. An entire path is defined to be successful if each of its arcs is successful. A summary of the algorithm follows:

Step 0: $[\delta_s] = 1$; if an arc connects the source node s to node i, $\delta_i = c_{si}$. Otherwise, $\delta_i = 0$.

Step 1: $[\delta_j] = \max_{i \in T} \{\delta_i\}$ where j is the last node to get a permanent label, and $\mathbf{T}$ is the set of nodes

with temporary labels.

Step 2: If $[\delta]$ found, stop. Otherwise, go to Step 3.

Step 3: new $\delta_i = \max \{$old $\delta_i; [\delta] \times c_{ji}\}$ for $i \in \mathbf{L}$, where $\mathbf{L}$ is the set of temporarily labeled nodes reached from last permanently labeled node.

Step 4: Go to Step 1.

2.8.3 Most Reliable Route Between Each Pair of Nodes

The procedure to find the most reliable route between nodes in a directed network is a modified version of Floyd's algorithm.

Triple Operation

$$d_{ik}^{j} = max\{d_{ik}^{j-1}; d_{ij}^{j-1} \times d_{jk}^{j-1}\}, \quad i \neq j \neq k$$

Routing Methodology

$$r_{ik}^{j} = \begin{cases} r_{ik}^{j-1} & \text{if no improvement} \\ \\ r_{ik}^{j} = j & \text{otherwise} \end{cases}$$

2.9 THE *K*-SHORTEST PATH PROBLEM

In applications such as communication and transportation networks, it is sometimes desirable to have knowledge of several shortest paths, arranged in increasing order according to their lengths. As an example, the availability of good alternative routes can be used by transportation planners to model more realistically the flow of vehicular traffic on road networks when the shortest route is not operational. As a second example, the routing of messages through a communication network when some routes are temporarily obstructed can be based on the best alternatives available.

Thus, the identification of additional solutions provides an alternative approach for planning when the best solution is not available or is infeasible. Also, the knowledge of next-best solutions allows for a sensitivity analysis of the solutions with respect to external factors not included in the network model.

The general version of the *K*-shortest-path problem admits cycles in the paths and ties between the path lengths. Several relaxations of the problem are possible in order to provide alternative solution procedures to specific network problems of interest. Traditional approaches have been developed by several analysts and have been summarized and discussed in review papers, such as the one by Dreyfus [7]. Some new methods exploit a fairly strong analogy between the solution to the general *K*-shortest-path problem and the solution of a system of ordinary linear equations. One of these methods, known as the *double-sweep method* [29,30], simultaneously calculates the *K* shortest path lengths from a particular source node to all nodes in the network. This method will be the subject of our present discussion.

2.9.1 The Double-Sweep Method

Consider a directed network with nodes numbered 1 through *n* and suppose that for *each node* there is a vector with *estimates* of the *K* shortest *path lengths* from a given *source node*. Under the assumption that the initial estimates do not underestimate the path lengths, the double-sweep method successively reduces the estimates until the optimal vector of estimates is achieved in a finite number of iterations.

Each iteration consists of two passes. In the forward pass, the nodes are considered in increasing numerical order (i.e., $j = 1, 2, ..., n$). After identifying the list of nodes i incident to node j, such that $i < j$, the *K* shortest path lengths from the source to node j are successively examined to verify if shorter path lengths are possible through the incident nodes. If such path lengths exist, they will be used as new estimates in further iterations. A similar procedure is performed during the backward pass of the algorithm, but in this case the nodes are considered in decreasing numerical order (i.e., $j = n, n-1, ..., 1$) and only incident nodes $i > j$ are investigated.

The solution procedure of the double-sweep method involves the use of two special algebraic operations defined by Minieka and Shier [26]. These operations are performed on vectors rather than on single numbers, and for that reason are called *generalized operations*. The vectors to be considered must all have the same dimension, and can have finite as well as infinite elements, but it is required that the finite elements be in *strictly* increasing order and precede the infinity elements. A result of this requirement all the finite elements of the vector are numerically different. The following are examples of acceptable vectors:

$$\mathbf{A} = [-4, 0, 7, \infty]$$
$$\mathbf{B} = [3, 4, \infty, \infty]$$

On the other hand, the following vectors are not acceptable:

$$\mathbf{C} = [-4, 3, 3, 9]$$
$$\mathbf{D} = [-9, 0, \infty, 9]$$

The two generalized operations are now discussed. The first operation is referred to as a *generalized minimization* and the second one as a *generalized addition.* These generalized operations are performed on only two vectors at a time.

The generalized minimization defines a set formed with the elements of two given vectors of equal dimension, and then constructs a third vector, of same dimension, by including elements of the set in strictly increasing order of magnitude, starting with the lowest value of the set. If the number of finite elements in the third vector is less than its dimension, the vector is completed with ∞-entries. As an illustration, let us consider the vectors $\mathbf{A} = [-4, 0, 1, \infty]$ and $\mathbf{B} = [1, 7, 8, 9]$. The set formed with the elements of vectors $\mathbf{A}$ and $\mathbf{B}$ is given by $\{-4, 0, 1, \infty, 1, 7, 8, 9\}$. The elements of this set can be arranged in strictly increasing order as -4, 0, 1, 7, 8, 9, ∞. Since the vectors under consideration have dimension equal to 4, the generalized minimization of $\mathbf{A}$ and $\mathbf{B}$ is given by the vector $\mathbf{C} = [-4, 0, 1, 7]$, which contains the four minimal elements of the set under consideration.

The generalized addition defines a set formed with the cross sums of the elements of two given vectors of equal dimension, and then constructs a third vector, of same dimension, with its entries being cross sums arranged in strictly increasing order starting with the lowest cross sum of the set. As in the case of the generalized minimization, if the number of finite elements in the third vector is less than its dimension, the vector is completed with ∞-entries. As an illustration, let us consider the two vectors $\mathbf{A}$ and $\mathbf{B}$ used to explain the generalized addition. The cross sums of the elements of these two vectors are shown as follows:

		Elements of $\mathbf{A}$			
		-4	0	1	∞
	1	-3	1	2	∞
	7	3	7	8	∞
Elements of $\mathbf{B}$	8	4	8	9	∞
	9	5	9	10	∞

The set consisting of all values for the cross sums is $\{-3, 1, 2, 3, 7, 8, 4, 9, 5, 10, \infty\}$. Also, the elements of this set can be arranged in strictly increasing order as follows: -3, 1, 2, 3, 4, 5, 7, 8, 9, 10, ∞. Then, the generalized addition of $\mathbf{A}$ and $\mathbf{B}$ is equal to $\mathbf{C} = [-3, 1, 2, 3]$. A formal definition of these operations is now presented in order to support further discussion of the double-sweep algorithm. The following notation will be used

$\mathbf{R}$ the set of real numbers

$\mathbf{R}_\infty$ the set consisting of the elements of $\mathbf{R}$ plus the element ∞

K the desired number of path lengths

$\mathbf{S}(K)$ the set of vectors of dimension K with elements from $\mathbf{R}_\infty$, arranged in strictly increasing order

Min_j the operation of identifying the *jth* minimal element of a given set

Since by definition the elements of $\mathbf{S}(K)$ are vectors that have their entries in strictly increasing order, we must adopt the convention that $\infty < \infty$, when there are several ∞-entries in the vectors.

Generalized minimization. Let $\mathbf{A}$, $\mathbf{B}$, and $\mathbf{C}$ be three vectors from $\mathbf{S}(K)$. That is,

$$\mathbf{A} = [a_1, a_2, \cdots, a_K], \; a_1 < a_2, \cdots, < a_K; \; a_i \in \mathbf{R}_\infty$$
$$\mathbf{A} = [b_1, b_2, \cdots, b_K], \; b_1 < b_2, \cdots, < b_K; \; b_i \in \mathbf{R}_\infty$$
$$\mathbf{A} = [c_1, c_2, \cdots, c_K], \; c_1 < c_2, \cdots, < c_K; \; c_i \in \mathbf{R}_\infty$$

Let $\mathbf{T}_+$ be the set formed with the elements of the vectors $\mathbf{A}$ and $\mathbf{B}$. The generalized minimization $\oplus$ is defined by $\mathbf{A} \oplus \mathbf{B} = \mathbf{C}$, where $c_j = \text{Min}_j [\mathbf{T}_+]$, for $1, 2, \ldots, K$.

Generalized addition. Let $\mathbf{A}$, $\mathbf{B}$, and $\mathbf{C}$ be three vectors from $\mathbf{S}(K)$. Let $\mathbf{T}_x$ be the set formed with the cross sums of the elements of the vectors $\mathbf{A}$ and $\mathbf{B}$. The generalized addition $\otimes$ is defined by $\mathbf{A} \otimes \mathbf{B} = \mathbf{C}$, where $c_j = \text{Min}_j \{\mathbf{T}_x\}$, for $j = 1, 2, \ldots, K$.

Computational methodology. Let $\mathbf{A}$ be any vector in $\mathbf{S}(K)$, and define $\mathbf{F} = [0, \infty, \infty, \ldots, \infty]$ and $\mathbf{V} = [\infty, \infty, \ldots, \infty]$, both in $\mathbf{S}(K)$; then it is obvious that the following relations always hold:

$$\mathbf{A} \oplus \mathbf{V} = \mathbf{A}$$
$$\mathbf{A} \otimes \mathbf{V} = \mathbf{V}$$
$$\mathbf{A} \otimes \mathbf{F} = \mathbf{A}$$

The computational aspects of the double-sweep method developed by Shier [47] can be summarized as follows. Assume that all nodes in the network are numbered 1 through n, and that the length of each arc (i, j) is equal to d_{ij}. Now, define:

$$\mathbf{D}_{ij} = [d_{ij}, \infty, \infty, \ldots, \infty] \in \mathbf{S}(K)$$
$$\mathbf{D} = [\mathbf{D}_{ij}]$$
$$\mathbf{L} = [\mathbf{L}_{ij}], \text{ where } \mathbf{L}_{ij} = \mathbf{D}_{ij} \text{ for } i > j; \; \mathbf{L}_{ij} = \mathbf{V} \text{ for } i \leq j$$
$$\mathbf{U} = [\mathbf{U}_{ij}], \text{ where } \mathbf{U}_{ij} = \mathbf{D}_{ij} \text{ for } i < j; \; \mathbf{U}_{ij} = \mathbf{V} \text{ for } i \geq j$$

In these definitions $\mathbf{D}_{ij}$, $\mathbf{L}_{ij}$, and $\mathbf{U}_{ij}$ are vectors from $\mathbf{S}(K)$, and $\mathbf{D}$, $\mathbf{L}$, and $\mathbf{U}$ are matrices whose entries are *vectors* from $\mathbf{S}(K)$. If the network under consideration has more than one arc directly joining two nodes i and j, we can redefine $\mathbf{D}_{ij}$ as

$$\mathbf{D}_{ij} = [d_{ij}^1, d_{ij}^2, \ldots, d_{ij}^t, \infty, \infty, \ldots, \infty] \in \mathbf{S}(K)$$

where d_{ij}^1, d_{ij}^2, ..., d_{ij}^t are the lengths of the arcs joining i and j. Note that if $t > k$, there will be no ∞-entries in $\mathbf{D}_{ij}$.

83

Let $\mathbf{E_{0m}}$ be a vector in $\mathbf{S}(K)$ containing the initial estimates of the K shortest path lengths from the origin to node m. It is noted that the first element of $\mathbf{E_{0m}}$, is equal to zero when m is the source node. All the vectors $\mathbf{E_{0m}}$, $m = 1, 2, ..., n$, can be arranged in an array $\mathbf{E}(0)$ as

$$\mathbf{E}(0) = [\mathbf{E_{01}}, \mathbf{E_{02}}, ..., \mathbf{E_{0n}}]$$

Note that $\mathbf{E}(0)$ is actually a vector whose entries are elements of $\mathbf{S}(K)$. At the wth single sweep, the double-sweep algorithm constructs a vector $\mathbf{E}(w)$, defined as

$$\mathbf{E}(w) = [\mathbf{E_{w1}}, \mathbf{E_{w2}}, ..., \mathbf{E_{wn}}]$$

where $\mathbf{E_{wm}}$ is an element of $\mathbf{S}(K)$ containing the current estimates of the K shortest path lengths from the origin to node m. The generation of vectors of estimates is accomplished by the following pair of recursive relationships:

$$\mathbf{E}(2r+1) = \mathbf{E}(2r) \oplus \ \mathbf{E}(2r+1) \otimes \mathbf{L} \qquad (2\text{-}58)$$
$$\mathbf{E}(2r+2) = \mathbf{E}(2r+1) \oplus \mathbf{E}(2r+2) \otimes \mathbf{U} \qquad (2\text{-}59)$$

These sweeps are performed in alternating fashion for each value of $r = 0, 1, 2, ...$. The operation defined in Eq. (2-58) is called a backward *sweep* and the one defined in Eq. (2-59) is called a forward *sweep*. In both of these equations, the generalized addition is performed first. The solution is obtained when the results are unchanged after two successive single sweeps, once the first *complete* double-sweep has been performed. Note that $\mathbf{L}$ and $\mathbf{U}$ are the lower and upper triangular portions of $\mathbf{D}$ such that $\mathbf{D} = \mathbf{L} \oplus \mathbf{U}$.

Once the path lengths are obtained, a tracing procedure is used to identify the corresponding paths. In order to describe the tracing procedure, suppose that it is desired to find the path (or paths) from the origin to node i, corresponding to the m-shortest path length. Let H_{mi} be this path length, and let j be a node incident to node i. Then,

$$H_{mi} = H_{tj} + d_{ji} \qquad (2\text{-}60)$$

where, as defined before, d_{ji} is the length of the arc (j, i) and H_{tj} is the t-shortest path length, $t \leq m$, corresponding to node j. Thus, the tracing procedure performs a node search at each node i in order to identify the node j for which Eq. (2-60) is satisfied. Once j is found, the same procedure is repeated until the origin node is reached.

If only paths without cycles are desired, the previous tracing procedure should be modified. Essentially, the modification consists of a test to inspect if a node has already been considered by the backward relationship of Eq. (2-60) when the node becomes a candidate to belong in the path under consideration.

2.9.2 Sample Problem for the Double-Sweep Method

Consider the network represented in Figure 2.25. It is desired to investigate the three shortest path lengths from node 1 to nodes 1, 2, 3 and 4. For this example, $K = 3$ and $n = 4$. An acceptable arbitrary choice for the vectors of initial estimates is given by $\mathbf{E_{01}} = [0, 21, 22]$, $\mathbf{E_{02}} = \mathbf{E_{03}} = \mathbf{E_{04}} = [20, 21, 22]$. If choosing finite estimates is difficult or impossible, all the entries can be set equal to ∞, with the exception of the estimate of the shortest path length for the source node, which must be always equal to zero.

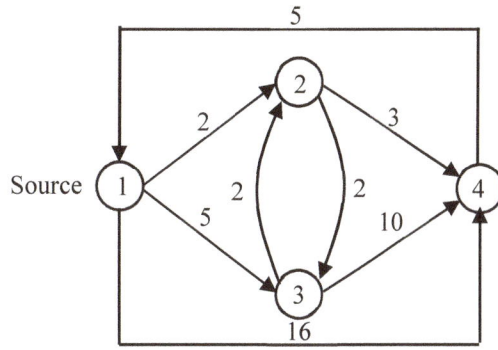

Figure 2.25. Network for sample problem.

The distance matrix **D**, and the matrices **L** and **U** consist of 16 elements, each element being a vector from **S**(3), arranged as follows:

$$\mathbf{D}=\begin{bmatrix}\mathbf{D}_{ij}\end{bmatrix}=\begin{pmatrix}\mathbf{D}_{11}&\mathbf{D}_{12}&\mathbf{D}_{13}&\mathbf{D}_{14}\\\mathbf{D}_{21}&\mathbf{D}_{22}&\mathbf{D}_{23}&\mathbf{D}_{24}\\\mathbf{D}_{31}&\mathbf{D}_{32}&\mathbf{D}_{33}&\mathbf{D}_{34}\\\mathbf{D}_{41}&\mathbf{D}_{42}&\mathbf{D}_{43}&\mathbf{D}_{44}\end{pmatrix}\quad\mathbf{L}=\begin{bmatrix}\mathbf{L}_{ij}\end{bmatrix}=\begin{pmatrix}\mathbf{V}&\mathbf{V}&\mathbf{V}&\mathbf{V}\\\mathbf{D}_{21}&\mathbf{V}&\mathbf{V}&\mathbf{V}\\\mathbf{D}_{31}&\mathbf{D}_{32}&\mathbf{V}&\mathbf{V}\\\mathbf{D}_{41}&\mathbf{D}_{42}&\mathbf{D}_{43}&\mathbf{V}\end{pmatrix}\quad\mathbf{U}=\begin{bmatrix}\mathbf{U}_{ij}\end{bmatrix}=\begin{pmatrix}\mathbf{V}&\mathbf{D}_{12}&\mathbf{D}_{13}&\mathbf{D}_{14}\\\mathbf{V}&\mathbf{V}&\mathbf{D}_{23}&\mathbf{D}_{24}\\\mathbf{V}&\mathbf{V}&\mathbf{V}&\mathbf{D}_{34}\\\mathbf{V}&\mathbf{V}&\mathbf{V}&\mathbf{V}\end{pmatrix}$$

The $\mathbf{D}_{ij}$ vectors can be obtained by enlarging the entries of $[d_{ij}]$ to accommodate $K\text{-}1 = 2$ infinite entries. The array $[d_{ij}]$ is obtained directly from the network in Figure 2.25:

$$\begin{bmatrix}d_{ij}\end{bmatrix}=\begin{pmatrix}0&2&5&16\\\infty&0&2&3\\\infty&2&0&10\\5&\infty&\infty&0\end{pmatrix}$$

The **D**, **L**, and **U** matrices are given as follows

D				L				U			
0	2	5	16	∞	∞	∞	∞	∞	2	5	16
∞	∞	∞	∞	∞	∞	∞	∞	∞	∞	∞	∞
∞	∞	∞	∞	∞	∞	∞	∞	∞	∞	∞	∞
∞	0	2	3	∞	∞	∞	∞	∞	∞	2	3
∞	∞	∞	∞	∞	∞	∞	∞	∞	∞	∞	∞
∞	∞	∞	∞	∞	∞	∞	∞	∞	∞	∞	∞
∞	2	0	10	∞	2	∞	∞	∞	∞	∞	∞
∞	∞	∞	∞	∞	∞	∞	∞	∞	∞	∞	∞
∞	∞	∞	∞	∞	∞	∞	∞	∞	∞	∞	∞
5	∞	∞	0	5	∞	∞	∞	∞	∞	∞	∞
∞	∞	∞	∞	∞	∞	∞	∞	∞	∞	∞	∞
∞	∞	∞	∞	∞	∞	∞	∞	∞	∞	∞	∞

A. Algorithmic Procedure

As an illustration, consider the case $r=0$. The backward sweep produces a vector of estimates, **E**(1), given by

$$\mathbf{E}(1) = \mathbf{E}(0) \oplus \mathbf{E}(1) \otimes \mathbf{L} \qquad (2\text{-}61)$$

The generalized addition **E**(1)⊗**L** in Eq. (2-61) is equal to

$$\begin{pmatrix} 25 & 22 & \infty & \infty \\ 26 & 23 & \infty & \infty \\ 27 & 24 & \infty & \infty \end{pmatrix}$$

After performing the backward sweep, the result obtained is $\mathbf{E}(1) = \mathbf{E}(0)$. The computational details of this evaluation will now be discussed. The steps involved in the operation $\mathbf{E}(1) \otimes \mathbf{L}$ resemble those of the multiplication of a vector by a matrix, with the elements of the product being identified in a *backward* fashion. In our case, however, the arrays consist of K-tuples from $\mathbf{S}(K)$ instead of single elements from $\mathbf{R}_\infty$, and the $\oplus$ and $\otimes$ operations are used instead of the standard arithmetic $+$ and $\times$ operations, respectively. Once a given element of $\mathbf{E}(1) \otimes \mathbf{L}$ is computed, it is compared against the corresponding element of $\mathbf{E}(0)$ by means of a generalized minimization, in order to produce the corresponding element of $\mathbf{E}(1)$. In Figure 2.26 we display the arrays $\mathbf{E}(0)$, $\mathbf{L}$, $\mathbf{E}(1) \otimes \mathbf{L}$, and indicate how to obtain the elements of $\mathbf{E}(1)$ in the backward sweep under consideration.

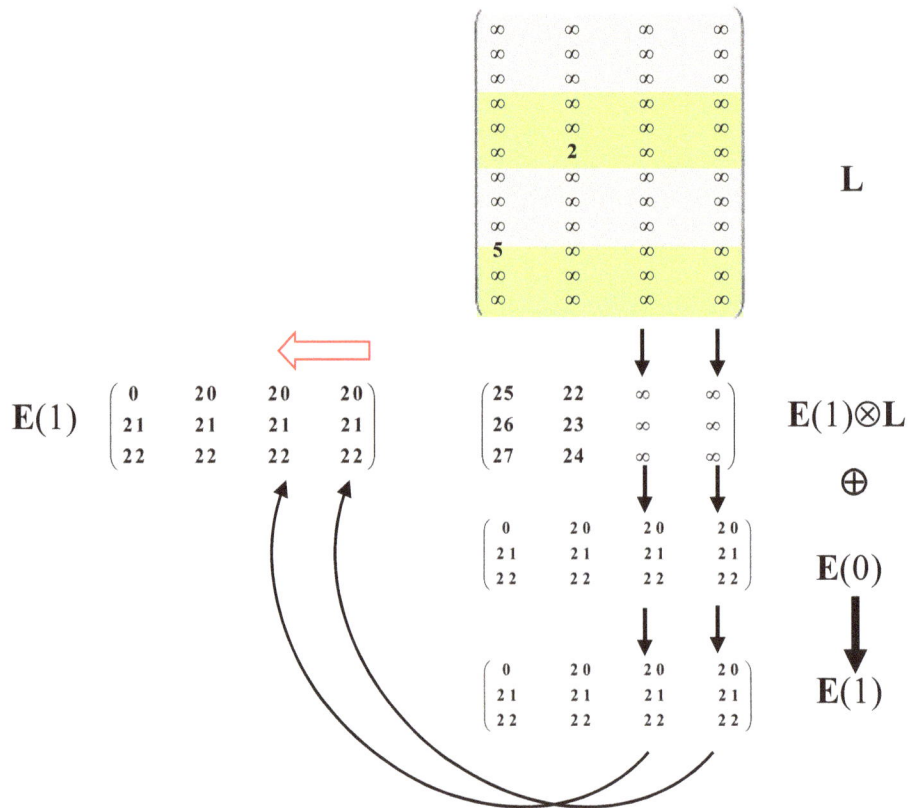

Figure 2.26. Backward sweep for sample problem.

The first forward sweep is performed next. The corresponding operation is indicated by

$$\mathbf{E}(2) = \mathbf{E}(1) \oplus \mathbf{E}(2) \otimes \mathbf{U} \tag{2-62}$$

Similarly, the generalized addition $\mathbf{E}(2) \otimes \mathbf{U}$ in Eq. (2-62) is given by the following vector of elements in $\mathbf{S}(3)$:

86

$$\begin{pmatrix} \infty & 2 & 4 & 5 \\ \infty & 23 & 5 & 14 \\ \infty & 24 & 22 & 15 \end{pmatrix}$$

Then, the new vector of estimates, $\mathbf{E}(2)$, is found to be

$$\begin{pmatrix} 0 & 2 & 4 & 5 \\ 21 & 20 & 5 & 14 \\ 22 & 21 & 20 & 15 \end{pmatrix}$$

In this case, we perform a sequence of steps similar to the ones described for the generalized addition $\mathbf{E}(1) \otimes \mathbf{L}$, but considering $\mathbf{E}(2) \otimes \mathbf{U}$ instead, and computing the elements in a *forward* fashion, as shown in Figure 2.27.

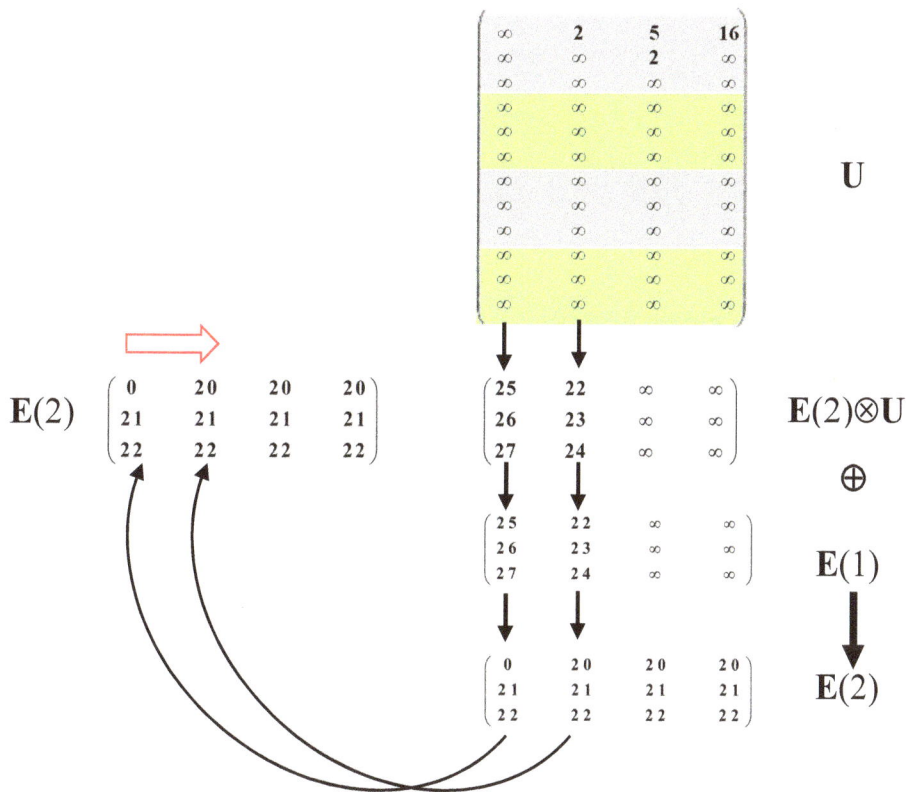

Figure 2.27. Forward sweep for sample problem.

Proceeding in this fashion, it is possible to obtain the results summarized in Table 2.15. For $r=2$, both the backward and forward sweeps yield the same solution. As previously explained, this indicates that the solution is optimal. As an illustration, Figure 2.28 shows the paths from node 1 to node 4 corresponding to the three shortest path-lengths for the network of Figure 2.25. It is noted that the second path has a circuit formed with arcs (2,3) and (3,2).

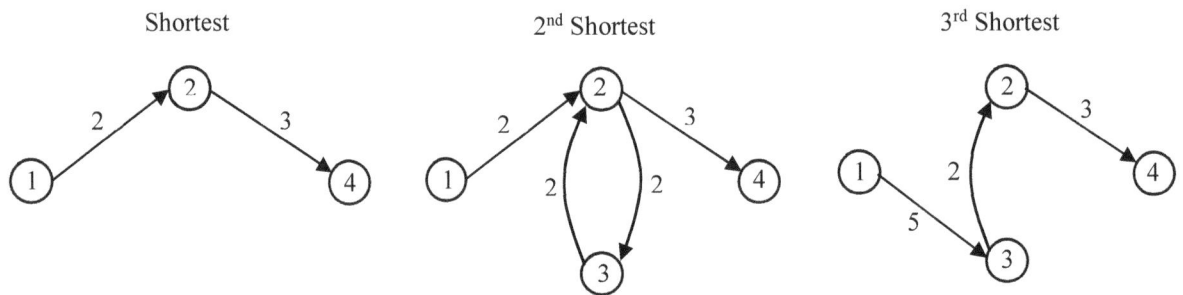

Figure 2.28. Paths for sample problem.

Table 2.15. RESULTS FOR SAMPLE PROBLEM – DOUBLE SWEEP METHOD

r	Type of Sweep	Vectors of Estimates
0	Backward	$\begin{pmatrix} 0 & 20 & 20 & 20 \\ 21 & 21 & 21 & 21 \\ 22 & 22 & 22 & 22 \end{pmatrix}$
0	Forward	$\begin{pmatrix} 0 & 2 & 4 & 5 \\ 21 & 20 & 5 & 14 \\ 22 & 21 & 20 & 15 \end{pmatrix}$
1	Backward	$\begin{pmatrix} 0 & 2 & 4 & 5 \\ 10 & 6 & 5 & 14 \\ 19 & 7 & 20 & 15 \end{pmatrix}$
1	Forward	$\begin{pmatrix} 0 & 2 & 4 & 5 \\ 10 & 6 & 5 & 9 \\ 19 & 7 & 8 & 10 \end{pmatrix}$
2	Backward	$\begin{pmatrix} 0 & 2 & 4 & 5 \\ 10 & 6 & 5 & 9 \\ 14 & 7 & 8 & 10 \end{pmatrix}$
2	Forward	$\begin{pmatrix} 0 & 2 & 4 & 5 \\ 10 & 6 & 5 & 9 \\ 14 & 7 & 8 & 10 \end{pmatrix}$

Tracing Procedure

The steps of the tracing procedure will be illustrated using the results given in Table 2.15. The 2nd path shown in Figure 2.28 corresponding to the second shortest path-length from node 1 to node 4 for the network given in Figure 2.25 will be traced. The optimal *labels* for the nodes of the network are the lengths of the shortest and second shortest paths:

Node 1:	0	10
Node 2:	2	6
Node 3:	4	5
Node 4:	5	9

Step 1: arcs (1,4), (2,4) and (3,4) are connected to node 4. The lengths of these arcs are 16, 3, and 10, respectively. For nodes 1, 2, and 3, we can compute the following tentative labels:

88

Node 1: 9 - 16 = -7
Node 2: 9 - 3 = 6
Node 3: 9 - 10 = -1

The optimal labels that need to be considered are:

Node 2:   2   6

Therefore, node 2 is chosen. This means that the last arc of the second shortest path is arc (2,4).

Step 2: arcs (1,2) and (3,2) are connected to node 2. The lengths of these arcs are equal to 2 and 2. Thus the tentative labels are:

Node 1: 6 - 2 = 4
Node 3: 6 - 2 = 4

The optimal labels are for nodes 1 and 3 are:

Node 1:   0   10
Node 3:   4   5

Therefore, node 3 is chosen. Then the last two arcs of the path are (3,2) and (2,4).

Step 3: arcs (1,3) and (2,3) are connected to node 3. The lengths of these arcs are 5 and 2, respectively. For these nodes we can compute the following tentative labels:

Node 1: 4 - 5 = -1
Node 2: 4 - 2 = 2

Thus, the optimal labels that need to be considered are:

Node 2:   2   6

Therefore, node 2 is chosen. Then the last three arcs of the path are (2,3), ((3,2) and (2,4).

Step 4: arcs (1,2) and (3,2) are connected to node 2. The lengths of these arcs are 2 and 2, respectively. Thus, the tentative labels are computed as follows:

Node 1: 2 - 2 = 0
Node 3: 2 - 2 = 0

The optimal labels to be considered are:

Node 1:   0   10
Node 3:   4   5

Therefore, node 1 is chosen. The complete path is traced as consisting of the arcs (1,2), (2,3), (3,2), (2,4). Note the length of this path is 2+2+2+3 = 9. This problem is now solved by NOP.

A. Solution by NOP

For a specified *source node* in a network with *n* nodes, the Double-Sweep method finds *K* pathlengths in increasing order of magnitude for *each node* of the network (including the source node). Therefore, the total number of pathlengths to be traced is equal to *Kn*. It is formally understood that the shortest pathlength for the source node itself is equal to zero and corresponds to a one-node path

Once the *Kn* pathlengths are found, NOP traces each pathlength to determine the corresponding paths using Eq. (2-60). Since it is possible to have *multiple paths* with the same length, NOP requests the user to provide a maximum number of paths (PMAX) to be traced for each pathlength. Figure 2.29 shows the computer screens for the NOP solution: (1) input file, (2) pairs of nodes for which paths are traced and upper bound on the number of paths, (3) NOP output results.

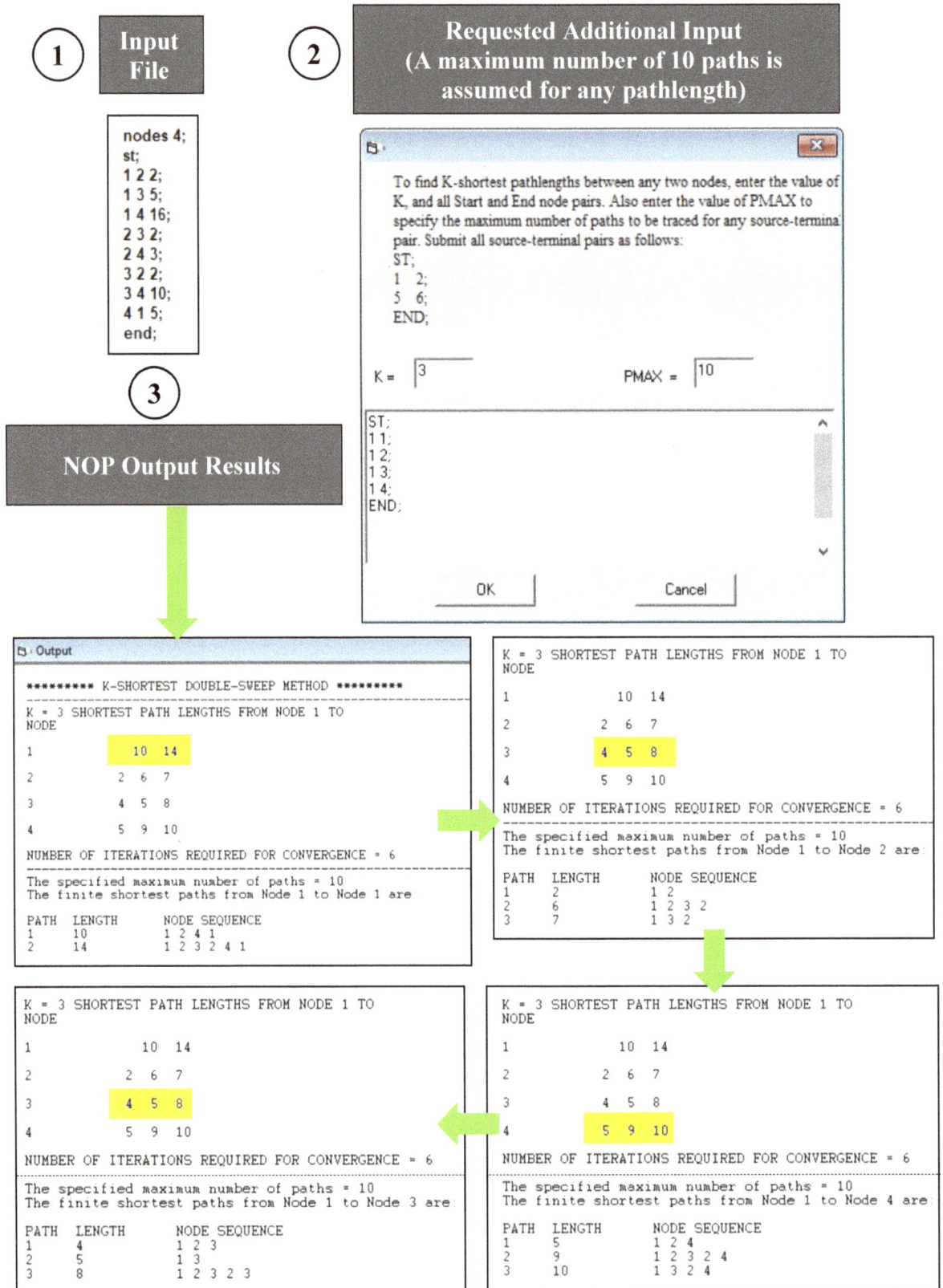

Figure 2.29. Sequence of Computer Screens for *K*-Shortest-Path Example.

90

2.10 COMPUTATIONAL COMPLEXITY OF SELECTED SHORTEST-PATH ALGORITHMS

The reasons for analyzing algorithms can be classified into two important categories: practical reasons and theoretical reasons. Practical reasons can be summarized as a need to obtain estimates or upper bounds on the computer memory and/or execution-time requirements involved in the implementation of algorithms. Perhaps the most important theoretical reason for the analysis of algorithms is the desirability of quantitative standards, such as running times, that would permit the comparison of two or more algorithms designed to solve the same problem.

The purpose of this section is to provide upper bounds on the amount of computational work involved in the application of Dijkstra's method, Floyd's algorithm, and the double-sweep method. In the algorithms by Dijkstra and by Floyd there are only two types of elementary operations: additions and comparisons. It is usually assumed that an addition and a comparison require approximately the same amount of time. By worst-case conditions for the execution of an algorithm, it is meant that the required number of elementary operations to terminate the algorithm is maximum. This upper bound on amount of computation is a function of both the size of the network and the desired number of solutions (paths) and is referred to as the *computational complexity* of the algorithm. In the case of the double-sweep method, the computational complexity will be given in terms of the number of generalized operations defined in Section 2.9.1. A summary of results developed in this section is shown below:

a. Dijkstra's algorithm: $3n(n-1)/2$ elementary operations, $O(n^2)$.
b. Floyd's algorithm: $2n^2(n-3)$ elementary operations, $O(n^3)$.
c. Double-Sweep algorithm: Kn^3 generalized operations.

2.10.1 Computational Complexity of Dijkstra's Method

Let $G = (N, A)$ be the network under consideration, with n nodes in set N. In the worst-case condition the terminal node is the nth node to receive a permanent label. Assume that at any given time of the execution of the algorithm there are m nodes with permanent labels, and $n-m$ nodes with temporary labels. To identify the $(m+1)$st node that must be given a permanent label, we update the $n - m$ temporary labels by performing an addition and then one comparison for each label. Once the temporary labels are updated, one more minimization is needed to identify the label that becomes permanent. This minimization operation implies $n-m-1$ comparisons. Therefore, the number of elementary operations required to assign the $(m+1)$st permanent label is equal to $3(n - m)-1 \approx 3(n - m)$.

The total number of elementary operations needed for the termination of the algorithm, under worst-case conditions, is given by

$$\sum_{m=1}^{n} 3(n-m) = 3\big((n-1)+(n-2)+\cdots 1\big) = \frac{3n(n-1)}{2}$$

Using a Fibonacci heap implementation of Dijkstra's algorithm due to Fredman and Tarjan [14] a shortest path problem can be solved in $O(m+n\log n)$ time, where n is the number of nodes and m is the number of arcs.

2.10.2 Computational Complexity of Floyd's Algorithm

Let $G = (\mathbf{N}, \mathbf{A})$ be the network under consideration, with $\mathbf{N}$ defined as the set $\mathbf{N} = \{1, 2, \ldots, n\}$. At any iteration of the algorithm, the total number of entries to be evaluated using the triple operation defined in Eq. (2-56) can be obtained by reasoning as follows:

1. The total number of entries in the matrix is equal to n^2.
2. The number of entries in either the pivot row and/or the pivot column is equal to $2n-1$.
3. The number of zero entries in the main diagonal is equal to n and need not be reevaluated.
4. One element of the pivot row and one element of the pivot column lie in the main diagonal.
5. The analysis of each entry implies one addition and one comparison according to the triple operation.
6. The maximum total number of operations at a given iteration is then *approximately* equal to $2n(n-3)$.
7. Since the number of iterations is equal to the number of nodes, n, the total number of elementary operations required to terminate the algorithm is at most equal to $2n^2(n-3)$.

2.10.3 Computational Complexity of the Double-Sweep Method

Let $G = (\mathbf{N}, \mathbf{A})$ be the network under consideration, with $\mathbf{N}$ defined as the set $\mathbf{N} = \{1, 2, \ldots, n\}$. As an illustration of the procedure that can be followed to find the number of *generalized* operations, let us consider a backward sweep. As indicated in Section 2.6.1, the nodes in a backward sweep are considered according to the order in the sequence $n, n-1, \ldots, 1$. The number of generalized additions $\otimes$ and generalized minimizations $\oplus$ required to perform a backward sweep is shown in Table 2.16.

Table 2.16. NUMBER OF GENERALIZED OPERATIONS: BACKWARD SWEEP

Node	Number of Generalized Additions $\otimes$	Number of Generalized Minimizations $\oplus$
n	0	0
$n-1$	1	1
$n-2$	2	2
$n-3$	3	3
$\vdots$	$\vdots$	$\vdots$
1	$n-1$	$n-1$
Total	$n(n-1)/2$	$n(n-1)/2$

Therefore, according to the results in Table 2.16, the number of generalized operations for a backward sweep is equal to $n(n-1)$. Similarly, it is shown that the total number of generalized operations in a forward sweep is also equal to $n(n-1)$. Then the total number of generalized operations in a double sweep is given by $2n(n-1)$. If K is the number of path lengths, the total number of estimates produced at each single sweep is equal to nK. Since at least one estimate is improved at each single sweep, under worst-case conditions the number of double sweeps is equal to $\frac{1}{2}nK$. Therefore, the total number of generalized operations needed for the termination of the double-sweep method is given by $(\frac{1}{2}K)(2n)(n-1) = Kn^2(n-1) \approx Kn^3$.

2.11 MINIMAL SPANNING TREE PROBLEM

An undirected network is considered in this section. Recall that a *tree* is a connected graph with no cycles and having at least two nodes. A *spanning tree* is a tree that contains all the nodes of a network. If each arc has a parameter defined as its *length* or *cost*, then we can develop the notion of a *minimal* (or maximal) spanning tree. A minimal spanning tree has the property that the sum of all its arc parameters is a minimum over the set of all possible spanning trees.

The minimal spanning tree problem is found to be useful in many real-life applications. A typical application is that of interconnecting several points (towns, buildings, etc.) in such a way that the total length of all arcs (roads, telephone lines, etc.) used is minimized. Minimal spanning trees are also often used as decomposition strategies in the optimization or sub-optimization of larger more complex networks.

2.11.1 Algorithm

All algorithms previously discussed in this textbook have utilized recursive computational schemes that iterate to an optimal solution. As it happens, the minimal spanning tree problem is one of the few problems in operations research that can be solved via a *greedy* approach that requires a minimal amount of effort. In order to describe the algorithm, the following notation will be used. Let $G = (N, A)$ be an *undirected* network. Let S and $\bar{S}$ be two mutually exclusive sets such that $N = S \cup \bar{S}$, where S is defined as the set of *connected* nodes, and $\bar{S}$ as the set of *unconnected* nodes. The algorithm can be divided into the following steps:

Step 0: $\bar{S} = N, S = \varnothing$.

Step 1: Choose any node in $\bar{S}$ and connect it to its nearest neighbor. Put connected nodes in S.

Step 2: Identify the node i in $\bar{S}$ which is most closely connected to any node in S. Transfer node i from set $\bar{S}$ to set S.

Step 3: If $\bar{S} = \varnothing$, stop. Otherwise, go to Step 2.

Note that arc (i,j) cannot be in a minimal spanning tree if its length is larger than or equal to *each arc-length* of the path linking nodes i and j in the tree.

2.11.2 Example

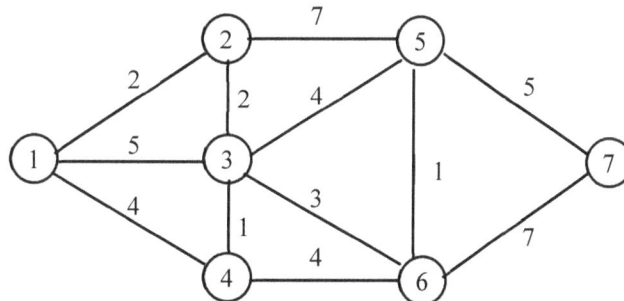

Figure 2.30. Sample network for minimal spanning tree problem.

It is desired to find a minimal spanning tree for the undirected network shown in Figure 2.30. The numbers adjacent to the arcs are their corresponding lengths.

A. Algorithmic Procedure

Step 0: $\overline{S} = \{1,2,3,4,5,6,7\}$, $S = \varnothing$.

Step 1: Choose node 6. Its nearest neighbor is node 5. Therefore, both nodes 5 and 6 are removed from $\overline{S}$ and inserted in set S. That is, $\overline{S} = \{1,2,3,4,7\}$, $S = \{6,5\}$.

Step 2: This step is performed five times, as indicated below:

    a. Choose node 3. $\overline{S} = \{1,2,4,7\}$, $S = \{6,5,3\}$

    b. Choose node 4. $\overline{S} = \{1,2,7\}$, $S = \{6,5,3,4\}$

    c. Choose node 2. $\overline{S} = \{1,7\}$, $S = \{6,5,3,4,2\}$

    d. Choose node 1. $\overline{S} = \{7\}$, $S = \{6,5,3,4,1,2\}$

    e. Choose node 7. $\overline{S} = \varnothing$, $S = \{6,5,3,4,1,2\}$.

Step 3: The execution of the algorithm is terminated, since $\overline{S} = \varnothing$. The minimal spanning tree is shown in Figure 2.31. The total *weight* of this tree is equal to 14.

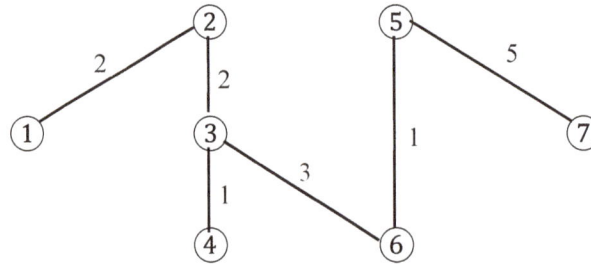

Figure 2.31. Minimal spanning tree.

B. Solution by NOP

Figure 2.32 shows the computer screens for the NOP solution: (1) input file, (2) output.

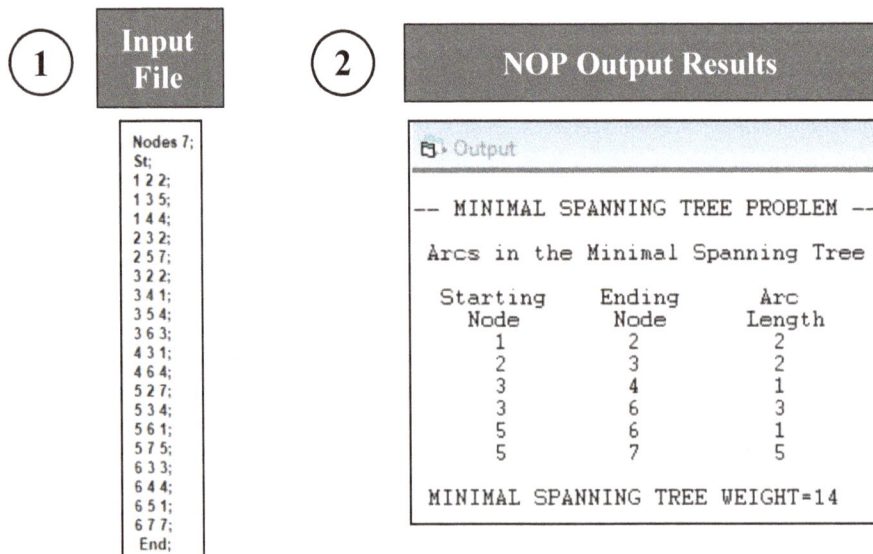

Figure 2.32. Sequence of Computer Screens for Minimal Spanning Tree Example.

2.12 THE MAXIMUM-FLOW PROBLEM

Let G = (**N, A**) be a directed and capacitated network (each arc has a finite capacity) with node s being the source and node t being the terminal. Under the assumption that there is an infinite availability of flow at node s, and that the flow conservation condition holds for all the intermediate nodes of the network, it is desired to find the maximum flow into node t.

The maximum-flow problem was formulated as a linear programming model in Section 2.1.1, and hence can be solved by the standard simplex method. It is the purpose of the present section to discuss an alternative and more efficient solution procedure. The algorithm starts with a feasible solution and then uses a labeling procedure developed by Ford and Fulkerson [11,12] in order to produce another feasible solution with a better flow value. In the algorithm, nodes are viewed as flow transshipment points and arcs as distribution channels. Two fundamental concepts are necessary for a formal discussion of the algorithm: the notion of *labeling*, and the definition of a *flow-augmenting path*.

The maximum-flow problem was formulated as a linear programming model in Section 2.1.1, and hence can be solved by the standard simplex method. It is the purpose of the present section to discuss an alternative and more efficient solution procedure. The algorithm starts with a feasible solution and then uses a labeling procedure developed by Ford and Fulkerson [11,12] in order to produce another feasible solution with a better flow value. In the algorithm, nodes are viewed as flow transshipment points and arcs as distribution channels. Two fundamental concepts are necessary for a formal discussion of the algorithm: the notion of *labeling*, and the definition of a *flow-augmenting path*.

Nodes are labeled from other nodes along directed arcs. As a result of labeling, arcs become either forward or reverse. More specifically, the purpose of labeling a node of a directed arc from its other node is to indicate both the amount and the origin of a flow shipment that causes a change in the current arc flow. In symbols, if a shipment of q_j units of flow is sent from node i to node j and it causes an increase in the flow on arc (i,j), we say that node j is labeled from node i by $+q_j$ along the forward arc (i,j). In this case, the label $[+q_j, i]$ is assigned to node j. This situation is illustrated in Figure 2.32.

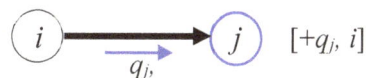

Figure 2.32. Labeling along a forward arc.

Similarly, if a flow shipment of q_j from node i to node j causes a reduction in the flow on arc (j,i), we say that node j is labeled from node i by $-q_j$ along the reverse arc (j,i). In this case, the label $[-q_j, i]$ is assigned to node j. This situation is illustrated in Figure 2.33.

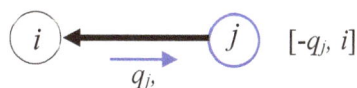

Figure 2.33. Labeling along a reverse arc.

95

A *flow-augmenting path* is a path from s to t with all nodes labeled. The flow on each arc in this path is adjusted by adding q_t to the flow on each forward arc and subtracting q_t from the flow on each reverse arc. The process of determining a flow-augmenting path is called a *breakthrough*.

2.12.1 Labeling Procedure

The maximum-flow problem is frequently encountered in practice and it is not uncommon to formulate models with thousands of nodes and arcs in a real-world setting. Hence, it is necessary that a computationally efficient procedure be used to solve such problems. Because of the simple structure of a maximum-flow problem, an efficient solution algorithm can be developed which recursively seeks an optimal (maximum flow) solution through labeling procedures.

The methodology will now be presented. Let (i,j) be a directed arc from node i to node j with flow f_{ij} and capacity equal to U_{ij}. Furthermore, let the label of node i include a flow amount of q_i. The flow on forward arc (i,j) can be increased by at most the minimum between the *residual* capacity of the arc and the amount of flow available at node i. Therefore, the label for node j is $[+q_j, i]$, where

$$q_j = min \; \{q_i, \; U_{ij} - f_{ij}\} \qquad\qquad (2\text{-}63)$$

The same logic enables us to label node j using reverse arc $(j\; i)$. Here we must determine if a decrease in flow across are (j,i) *is* possible. Evidently, it will be possible only if $f_{ji} > 0$. The flow on reverse arc (j,i) can be reduced by at most the minimum between the value of the current flow and the amount that we could move away from node i, that is, q_i. Hence the label for node j is $[-q_j, i]$, where

$$q_j = min \; \{q_i, \; f_{ij}\} \qquad\qquad (2\text{-}64)$$

In essence, the labeling procedure consists of the following four steps:

1. Determining an initial solution that satisfies flow conservation at each node.
2. Labeling nodes, starting with the source node and attempting to identify a flow-augmenting path connecting the source node to the terminal node.
3. Adjusting flows if there is a flow-augmenting path.
4. Deciding if the current set of arc flows results in maximal flow.

The source node is labeled with $[\infty,-]$ to indicates the there is an infinite supply available. All other nodes are initially unlabeled. We then seek a flow-augmenting path from the source node to the sink node proceeding across forward and reverse arcs, labeling nodes as we move along, seeking to reach the sink node. One of two events can occur:

1. The sink node t is labeled $[+q_t, k]$. Therefore, a flow-augmenting path has been found and each arc flow along this path can be increased, or decreased, by an amount q_t. After the flows are changed, the current labels are erased, and the entire procedure is repeated.
2. The sink node t cannot be labeled. This implies that no flow-augmenting path can be found. Hence, the present flow values represent an optimal (maximum-flow) solution.

2.12.2 Example of the Labeling Procedure

As an illustration of the labeling algorithm, the maximum flow that can be shipped from node s to node t will be found for the network shown in Figure 2.34. Each arc $(i\ j)$ is assigned a label $[f_{ij}, U_{ij}]$, where f_{ij} is the current value of the arc flow and U_{ij} is the capacity of the arc. Any set of flows satisfying the flow conservation condition at each intermediate node is an acceptable initial solution. In this example, all flows are initially set equal to zero.

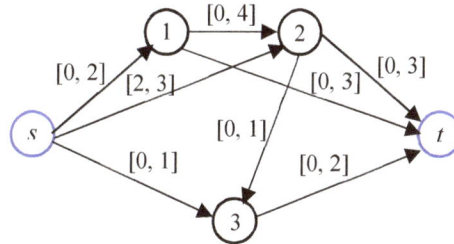

Figure 2.34. Capacitated Network.

A. Algorithmic Procedure

The numerical example under consideration is solved in six iterations. In the last iteration no labeling procedure is possible; this iteration is just a verification of the existence of a non-breakthrough. The labeling procedures are summarized in Tables 2.17 through 2.22. The arc flows after each iteration are shown in Figures 2.35(a) through 2.35(e). The steps at each iteration correspond to labeling nodes and changing flows when there is a breakthrough. In the last iteration the non-existence of the breakthrough indicates that the current set of flows is optimal, that is, it results in maximal flow into the terminal node.

Table 2.17. FIRST ITERATION

Steps for Labeling	Description See Figure 2.35 (a)
1	Label s with $[\infty,-]$, and choose any acceptable arc for labeling from s
2	Label 2 with $[+3,s]$
3	Label t with $[+2,2]$
4	Adjust flows: $f_{s2}=2$, $f_{2t}=2$

Table 2.18. SECOND ITERATION

Steps for Labeling	Description See Figure 2.35 (b)
5,6	Label s with $[\infty,-]$; label 1 with $[+2,s]$
7,8	Label 2 with $[+2,1]$; label 3 with $[+1,2]$
9	Label t with $[+1,3]$
10	Adjust flows: $f_{s1}=1$, $f_{12}=1$, $f_{2t}=2$, $f_{23}=1$, $f_{3t}=1$

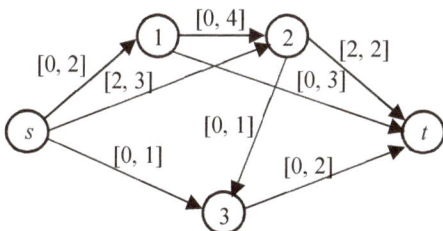

Figure 2.35 (a). Flows after 1st iteration.

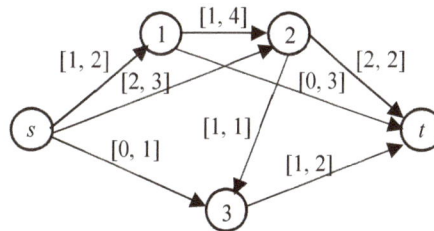

Figure 2.35 (b). Flows after 2nd iteration.

Table 2.19. THIRD ITERATION

Steps for Labeling	Description See Fig. 2.35 (c)
11,12	Label s with [∞,-]; label 1 with [+1,s]
13	Label 2 with [+2,1]; label 3 with [+1,2]
14	Adjust flows: $f_{s1} = 2$, $f_{1t} = 1$

Table 2.20. FOURTH ITERATION

Steps for Labeling	Description See Fig. 2.35 (d)
15,16,17	Label s with [∞,-]; label 2 with [+1,s], label 1 with [-1,2]
18	Label t with [+1,1]
19	Adjust flows: $f_{s2} = 3$, $f_{12} = 0$,

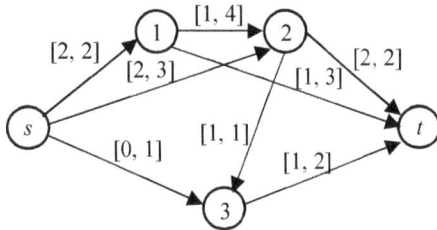

Figure 2.35 (c). Flows after 3rd iteration.

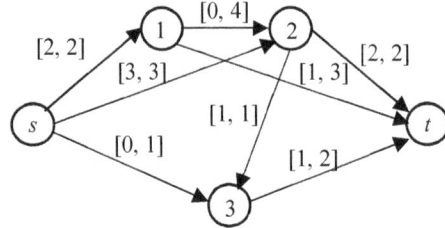

Figure 2.35 (d). Flows after 4th iteration.

Table 2.21. FIFTH ITERATION

Steps for Labeling	Description See Fig. 2.35 (e)
20,21,22	Label s with [∞,-]; label 3 with [+1,s], label t with [+1,3]
23	Adjust flows: $f_{s3} = 1$, $f_{3t} = 2$ Note that all arcs from s in Figure 2.35 (e) are *saturated*

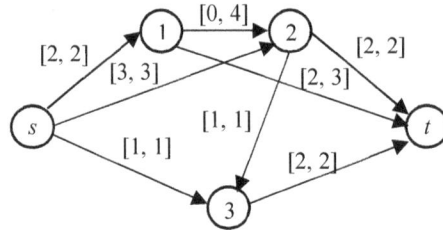

Figure 2.35 (e). Flows after 5th iteration.

Table 2.22 shows that a non-breakthrough occurs at the sixth iteration since all arcs emanating from the source node are now saturated and no further labeling is possible.

Table 2.22. SIXTH ITERATION

Steps for 6th iteration	Description See Fig. 2.35 (e)
24	Label s with [∞,-]
25	Labeling cannot be continued since all arcs starting at the source node are saturated. The maximum flow is equal to 6. Arc flows are shown in Fig. 3.35 (e).

It is noted that if arcs $(s,1)$, $(s,2)$, and $(s,3)$ in Figure 2.35 (e) are removed the source node becomes disconnected from the terminal node. This signifies that these arcs form a cut. The value of this cut is equal to $2+3+1 = 6$, which is the value of the maximum flow. Therefore, this cut is a *minimum cut*.

B. Solution by NOP

First all nodes are renumbered to comply with the assumption that the source node is node 1. The new numbers (labels) for the nodes are 1, 2, 3, 4, and 5. Node 5 is the terminal node and nodes 2, 3 and 4 are the new labels for the original nodes 1, 2, and 3, respectively.   Figure 2.36 shows the computer screens for the NOP solution: (1) input file, (2) NOP output results.

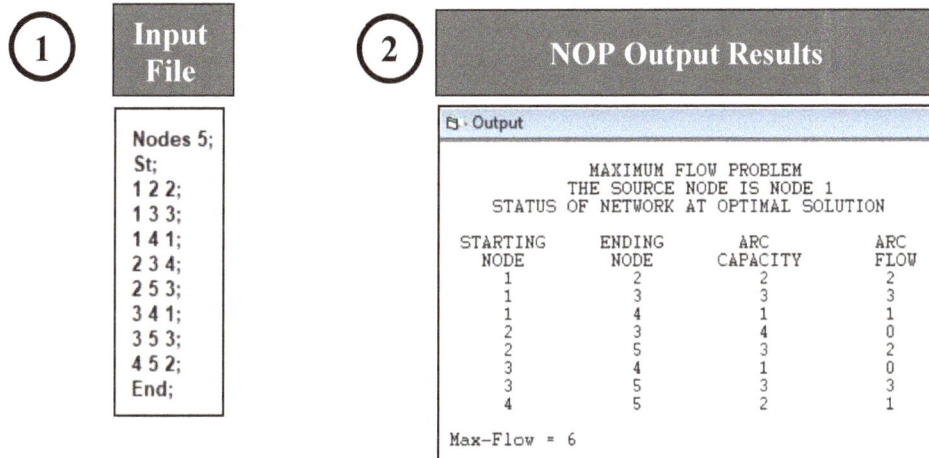

Figure 2.36. Sequence of Computer Screens for Maximal Flow Example.

2.12.3 The Maximum Flow / Minimum Cut Theorem

We consider a capacitated network G = (**N, A**). Each arc (i,j) has a lower bound L_{ij} and an upper bound U_{ij} on its flows f_{ij}.   Let $C = (\mathbf{S, N\text{-}S})$ be a cut separating a source node s from a terminal node t. Furthermore, let us use the following notation:

$$C^+ = \{(i,j)\in \mathbf{A} \mid i\in \mathbf{S}, j\in \mathbf{N\text{-}S}\} \tag{2-65}$$
$$C^- = \{(i,j)\in \mathbf{A} \mid i\in \mathbf{N\text{-}S}, j\in \mathbf{S}\} \tag{2-66}$$

The total net flow coming out of **S** is equal to

$$
\begin{aligned}
F(C) &= \sum_{(i,j)\in C^+} f_{ij} - \sum_{(i,j)\in C^-} f_{ij} \\
&= \sum_{i\in S} (\sum_{\{j|(i,j)\in A\}} f_{ij} - \sum_{\{j|(j,i)\in A\}} f_{ji}) \\
&\leq \sum_{(i,j)\in C^+} U_{ij} - \sum_{(i,j)\in C^-} L_{ij}
\end{aligned}
\tag{2-67}
$$

Let F^* be the value of the feasible maximal flow out of a source node s and into a terminal node t. Let $\mathbf{F}^*$ be the corresponding vector of arc flows. There cannot be a *non-saturated* path from s to t because by increasing the flow through this path up to its capacity, the value of the flow would increase. This implies that there is a cut C^* *saturated* with respect to $\mathbf{F}^*$ separating s from t. Thus,

99

$$F(C^*) = \sum_{(i,j) \in C^+} U_{ij} - \sum_{(i,j) \in C^-} L_{ij} \qquad \text{(2-68)}$$

2.13 MULTI-TERMINAL MAXIMUM FLOW PROBLEM

There are numerous problems in engineering and business where the systems under consideration can be approximated by deterministic *multi-terminal* flow models. A few examples are (1) transportation systems, where the highways are represented by the arcs of the network model and the capacities represent upper bounds on vehicular traffic intensities; (2) telephone systems, where the lines are represented by the arcs of the network and the capacities indicate the maximal number of calls allowed at any time; and (3) electric power distribution systems, where the transmission elements are represented by arcs and the capacities specify the maximal amounts of energy that can be absorbed by those elements of the system. In all these systems it is assumed that there are several sources at which a given commodity is available, and that the amount that can be transported to several terminals is limited only by the capacities of the distribution links.

Let us consider an *undirected* capacitated network, that is, one in which the total flow on a given arc must be less than or equal to its capacity. In Section 2.11.1, we discussed a labeling procedure to solve the one-source one-terminal case, where it is assumed that there is an infinite availability of the commodity at the source. The objective of the problem was to find the maximal amount of commodity that could be shipped from the source to the terminal through the arcs of the network, without violating the capacity conditions. Several analysts have studied the problem of finding the maximal flow for all pairs of nodes in a given undirected capacitated network [12,18]. This problem can be viewed as a generalization of the one-source one-terminal problem and, accordingly, can be solved by applying the procedure of Section 2.12.1 to each pair of nodes. A more elegant and efficient approach is due to Gomory and Hu [18]. Our presentation uses the fundamental results developed in their paper and will include a justification of the algorithm.

Assuming that the capacity of any arc is the same regardless of the direction in which the arc is traversed, and that each pair of nodes can be defined as a source-destination pair, the total number of maximal-flow problems that would have to be solved is equal to $\frac{1}{2} n(n-1)$, where n is the number of nodes in the network. The Gomory-Hu algorithm solves the same problem in only $n-1$ maximal-flow determinations.

2.13.1 The Gomory-Hu Algorithm

Let $G = (\mathbf{N}, \mathbf{A})$ be an *undirected* network with $\mathbf{N} = \{1, 2, \ldots, n\}$ and let c_{ij} be the capacity of each $(i, j) \in \mathbf{A}$. If this arc is replaced by two directed arcs with opposite orientations, then $c_{ij} = c_{ji}$. It is desired to find the maximum flow between each pair of nodes. The Gomory-Hu algorithm [18] constructs a spanning tree with arcs representing minimum-cuts and arc weights representing the corresponding maximal flow values. The algorithm uses a *condensation* property and a *maximal spanning tree* property to iteratively include branches in the *cut tree*.

A flow-chart of the algorithm is given in Figure 2.37. Once the cut-tree is constructed, the *maximum* flow between i and j is equal to the *minimum* value along the sequence of branches connecting i and j. This way we can construct a max-flow matrix for all pairs of source/terminal nodes. It should be noted that these maximal flows are not to be considered simultaneously, but only separately for specific pairs.

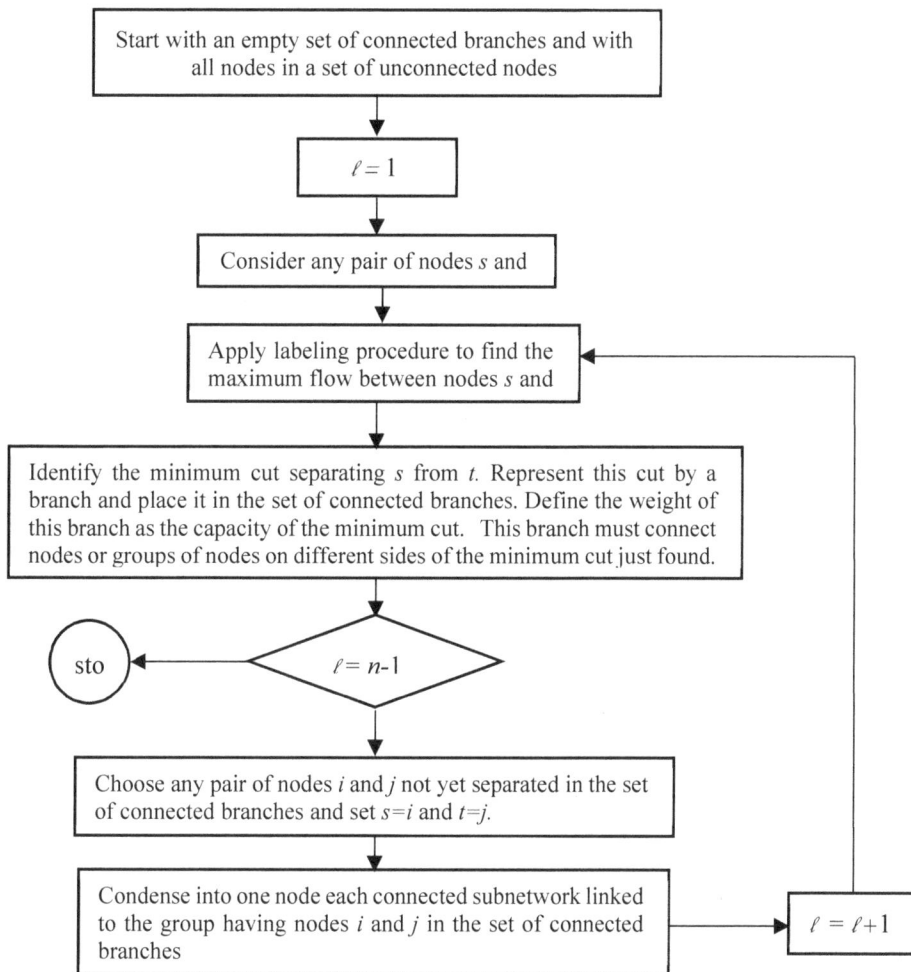

Figure 2.37. Flowchart for Gomory-Hu Algorithm.

2.13.1.1 *Condensation Property*

After finding the maximum flow between any two nodes of G either of the two sides of the corresponding minimum cut can be condensed into one equivalent node, and the maximum flow between any two nodes on the non-condensed side of the cut can be computed more efficiently considering the condensed network, instead of the original network. The condensation process can be described as follows:

Step 1. Choose a pair of nodes s and t and find v_{st} using the labeling procedure of Section 2.11.1. For this purpose, each undirected arc must be replaced by two directed arcs with the same capacity but opposite orientations. Identify the minimum cut from the solution to the maximum flow problem.

101

Step 2. Condense one side of the minimum cut found in Step 1. Let this cut be represented by $(\mathbf{X}, \overline{\mathbf{X}})_{st}$ where $s \in \mathbf{x}$ and $t \in \overline{\mathbf{X}}$. The condensed network can be symbolically represented by the notation $\overline{\mathbf{G}}_{st} = (\mathbf{N}_{st}, \mathbf{A}_{st})$. This network can be constructed as follows:

Case 1: If $\mathbf{X}$ is condensed, all arcs connecting nodes $i \in \mathbf{X}$ to the same node $j \in \overline{\mathbf{X}}$ are represented by a single arc $(\mathbf{X}, j)$ with capacity $\sum_{i \in \mathbf{X}} c_{ij}$.

Case 2: If $\overline{\mathbf{X}}$ is condensed, all arcs connecting the same node $i \in \mathbf{X}$ to nodes $j \in \overline{\mathbf{X}}$ are represented by a single arc $(i, \overline{\mathbf{X}})$ with capacity $\sum_{j \in \mathbf{X}} c_{ij}$.

2.13.1.2 *Maximal Spanning Tree Property*

For any arc (i, j) not to belong to a maximal spanning tree of a network, the following condition must be satisfied:

$$w_{ij} \leq min\left\{ w_{ii_1}, w_{ii_2}, \cdots, w_{i_r j} \right\} \tag{2-69}$$

where i, i_1, i_2, ..., i_r, j is a unique sequence of nodes connecting branches of the tree, and where the w_{ij} is the *weight* of arc (i, j).

2.13.1.3 *Justification of the Algorithm*

Let $G = (\mathbf{N}, \mathbf{A})$ be an *undirected* network with arc capacities such that for any arc joining nodes i and j the capacity of the arc when traversed in either direction is the same. That is, $c_{ij} = c_{ji}$ for all the arcs in $\mathbf{A}$. Now, let us consider three nodes i, j and k in $\mathbf{N}$. The maximal flow v_{ij} from node i to node j is equal to the capacity of a minimal cut separating these nodes. If the minimal cut is represented by $(\mathbf{X}, \overline{\mathbf{X}})_{ij}$ and its cut value by $C(\mathbf{X}, \overline{\mathbf{X}})_{ij}$ then $v_{ij} = C(\mathbf{X}, \overline{\mathbf{X}})_{ij}$. Furthermore, if node k is in $\overline{\mathbf{X}}$, then $v_{ik} \leq C(\mathbf{X}, \overline{\mathbf{X}})_{ij}$. Alternatively, if node k is in $\mathbf{X}$ then $v_{kj} \leq C(\mathbf{X}, \overline{\mathbf{X}})_{ij}$. Therefore, $v_{ij} \geq v_{ik}$ or $v_{ij} \geq v_{kj}$, which implies that

$$v_{ij} \geq min\{v_{ik}, v_{kj}\}$$

If a similar argument is repeated for v_{ik} and v_{kj}, we can conclude that $v_{ik} \geq min\{v_{ip}, v_{pk}\}$ and $v_{kj} \geq min\{v_{kq}, v_{qj}\}$, where $\{i, p, k, q, j\}$ is a connected sequence of nodes in $\mathbf{N}$. Therefore,

$$v_{ij} \geq min \{v_{ip}, v_{pk}, v_{kq}, v_{qj}\}$$

In general,

$$v_{ij} \geq min\left\{ v_{ii_1}, v_{i_1 i_2}, \cdots, v_{i_r j} \right\} \tag{2-70}$$

where $\{i, i_1, i_2, ..., i_r, j\}$ is a connected sequence of nodes in $\mathbf{N}$.

The property formulated in Eq. (2-69) will be rewritten for a maximal spanning tree with weights $w_{ij} = v_{ij}$. Specifically, the condition for arc (i, j) not being in the tree is:

$$v_{ij} \leq min\left\{ v_{ii_1}, v_{i_1 i_2}, \cdots, v_{i_r j} \right\} \tag{2-71}$$

where $\{i, i_1, i_2, ..., i_r, j\}$ is a connected sequence of nodes in the tree along a path from i to j. From Eqs. (2-70) and (2-71) it is concluded that for any arc (i, j) not in the cut-tree the maximal flow v_{ij} from node i to node j is given by

$$v_{ij} = min\left\{v_{ii_1}, v_{i_1i_2}, \cdots, v_{i_rj}\right\} \qquad (2\text{-}72)$$

The maximal spanning tree satisfying the property defined in Eq. (2-72) is called a *cut-tree* because each branch corresponds to a cut and the weight of the branch is equal to the capacity of the cut. If we want to find the maximal value of the flow between any two nodes, we just trace the path connecting the two nodes in the tree and then find the minimal weight in the path. This minimal weight is equal to the maximal value of flow between n the nodes being considered.

2.13.2 Example

Consider the network shown in Figure 2.38. The numbers assigned to the arcs represent flow capacities, and it is desired to find the maximal value of the flow between any two nodes of the network. This problem is solved in $n - 1 = 7 - 1 = 6$ iterations of the Gomory-Hu algorithm. If the labeling procedure were applied to each pair of nodes, we would need to solve 21 maximum-flow problems. The minimum cut identified at each iteration corresponds to arcs with residual capacities equal to zero and will be found by inspection. Furthermore, for presentation purposes, the results from the individual maximum-flow labeling procedures will be omitted.

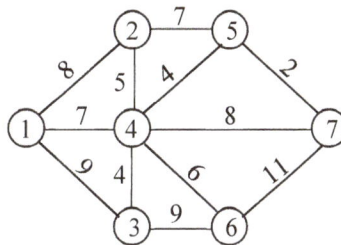

Figure 2.38. Network for Illustration of the Gomory-Hu Algorithm.

A. Algorithmic Procedure

Iteration 1: Consider $s = 2$ and $t = 5$. The maximal-flow value is equal to 13. Therefore, $v_{25} = v_{52} = 13$. The cut with minimal capacity indicates that we can start building the cut tree with a branch joining node 5 with a condensed node consisting of nodes 1, 2, 3, 4, 6, 7 [Figure 2.39(a)]. The weight of this branch is equal to 13.

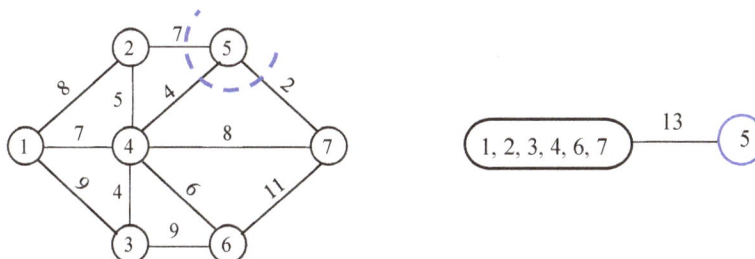

Figure 2.39(a). Gomory-Hu Algorithm: First iteration.

Iteration 2: Consider $s = 1$ and $t = 2$. The maximal-flow value is equal to 19. Therefore, $v_{12} = v_{21} = 19$. The minimal cut indicates that nodes 2 and 5 are on one side and the other nodes on the other side [Figure 2.39(b)]. The weight of the branch joining node 2 with the condensed node consisting of 1, 3, 4, 6, 7 is equal to 19.

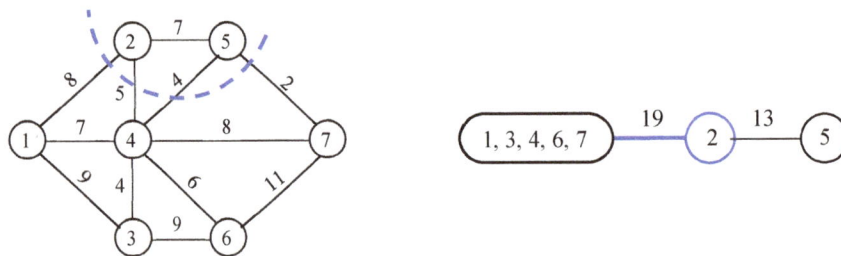

Figure 2.39(b). Gomory-Hu Algorithm: Second iteration.

Iteration 3: Consider $s = 6$ and $t = 7$. The value of the maximal flow is equal to 21. Therefore, $v_{67} = v_{76} = 21$. The minimal cut indicates that in the cut-tree node 7 is connected to the condensed node consisting of 1, 3, 4, 6 by an arc of weight equal to 21. The minimal cut also indicates that nodes 2 and 5 are on one side of the condensed node (1, 3, 4, 6) and node 7 is on the other side [Figure 2.39(c)].

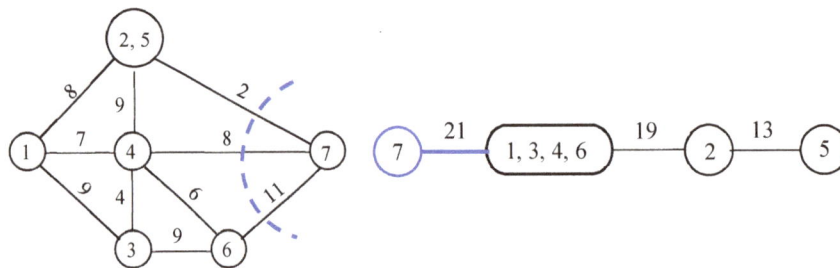

Figure 2.39(c). Third iteration.

Iteration 4: Consider $s = 4$ and $t = 6$. The maximal-flow value is equal to 25. Therefore, $v_{46} = v_{64} = 25$. The minimal cut shows that nodes 6 and 7 are on the same side of node (1, 3, 4) in the cut tree [Figure 2.39(d)].

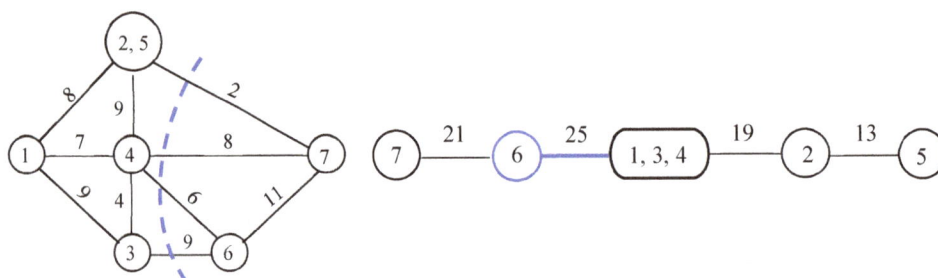

Figure 2.39(d). Fourth iteration.

Iteration 5: Consider $s = 1$ and $t = 4$. The maximal-flow value is equal to 24. Therefore, $v_{14} = v_{41} = 24$. The minimal cut removes node 1 from node (1, 3, 4) and places it on the side of (3, 4) opposite to all other nodes [Figure 2.39(e)]. The corresponding arc in the cut tree has a weight equal to 24.

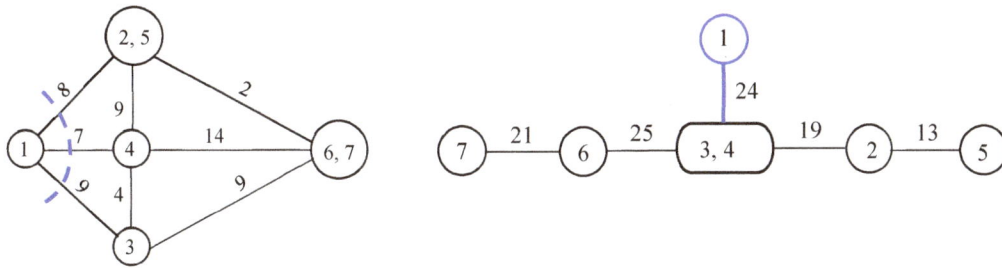

Figure 2.39(e). Fifth iteration.

Iteration 6: Consider $s = 3$ and $t = 4$. The maximal flow value is 22. Therefore, $v_{34} = v_{43} = 22$. The minimal cut removes node 3 and links it to 4 in the cut tree with an arc with weight equal to 22. Now the cut tree is complete, that is, consists of six arcs, and then the procedure is terminated [Figure 2.39(f)].

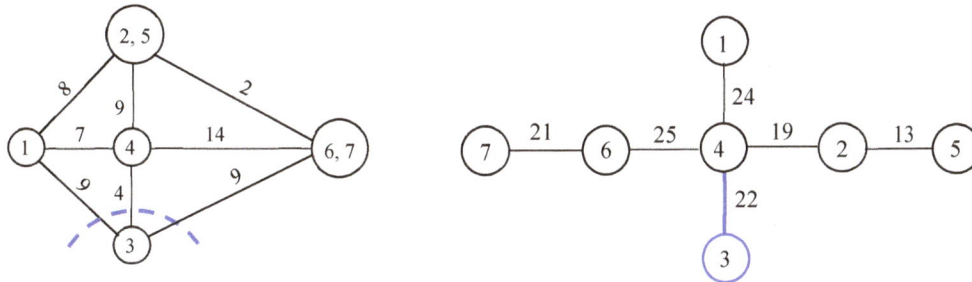

Figure 2.39(f). Sixth iteration.

The maximal-flow values can be arranged into matrix $\mathbf{V} = [v_{ij}]$ as follows:

$$\mathbf{V} = \begin{pmatrix} - & 19 & 22 & 24 & 13 & 24 & 21 \\ 19 & - & 19 & 19 & 13 & 19 & 19 \\ 22 & 19 & - & 22 & 13 & 22 & 21 \\ 24 & 19 & 22 & - & 13 & 25 & 21 \\ 13 & 13 & 13 & 13 & - & 13 & 13 \\ 24 & 19 & 22 & 25 & 13 & - & 21 \\ 21 & 19 & 21 & 21 & 13 & 21 & - \end{pmatrix}$$

B. Solution by NOP

Figure 2.40 shows the computer screens for the NOP solution: (1) input file, (2) NOP output results. The computer output consists of the maximal flow and the minimal cut for each iteration of the algorithm plus the maximal flow matrix. It is noted that the choices for the source-terminal pairs in part A are different from those selected by NOP.

105

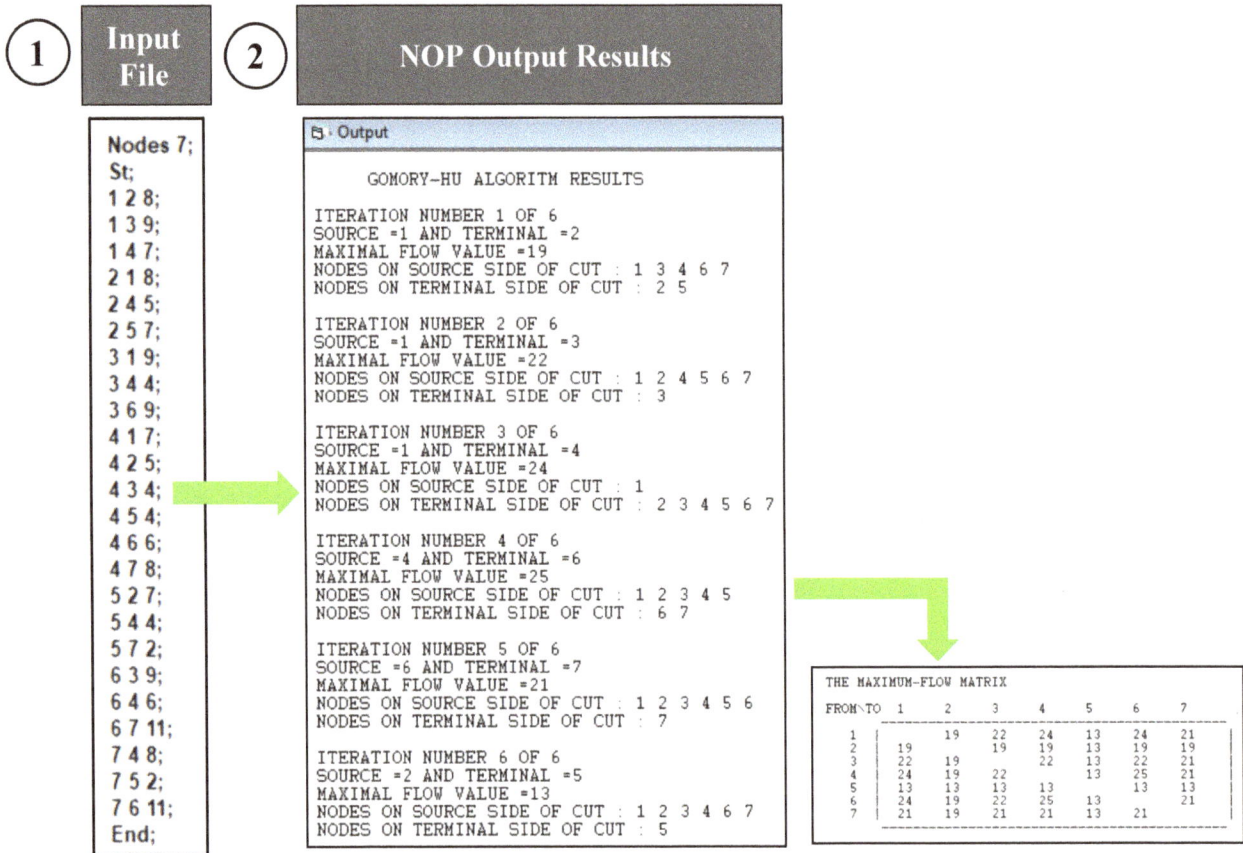

① Input File	② NOP Output Results

Input File:

```
Nodes 7;
St;
1 2 8;
1 3 9;
1 4 7;
2 1 8;
2 4 5;
2 5 7;
3 1 9;
3 4 4;
3 6 9;
4 1 7;
4 2 5;
4 3 4;
4 5 4;
4 6 6;
4 7 8;
5 2 7;
5 4 4;
5 7 2;
6 3 9;
6 4 6;
6 7 11;
7 4 8;
7 5 2;
7 6 11;
End;
```

Output:

```
             GOMORY-HU ALGORITM RESULTS

ITERATION NUMBER 1 OF 6
SOURCE =1 AND TERMINAL =2
MAXIMAL FLOW VALUE =19
NODES ON SOURCE SIDE OF CUT :  1 3 4 6 7
NODES ON TERMINAL SIDE OF CUT :  2 5

ITERATION NUMBER 2 OF 6
SOURCE =1 AND TERMINAL =3
MAXIMAL FLOW VALUE =22
NODES ON SOURCE SIDE OF CUT :  1 2 4 5 6 7
NODES ON TERMINAL SIDE OF CUT :  3

ITERATION NUMBER 3 OF 6
SOURCE =1 AND TERMINAL =4
MAXIMAL FLOW VALUE =24
NODES ON SOURCE SIDE OF CUT :  1
NODES ON TERMINAL SIDE OF CUT :  2 3 4 5 6 7

ITERATION NUMBER 4 OF 6
SOURCE =4 AND TERMINAL =6
MAXIMAL FLOW VALUE =25
NODES ON SOURCE SIDE OF CUT :  1 2 3 4 5
NODES ON TERMINAL SIDE OF CUT :  6 7

ITERATION NUMBER 5 OF 6
SOURCE =6 AND TERMINAL =7
MAXIMAL FLOW VALUE =21
NODES ON SOURCE SIDE OF CUT :  1 2 3 4 5 6
NODES ON TERMINAL SIDE OF CUT :  7

ITERATION NUMBER 6 OF 6
SOURCE =2 AND TERMINAL =5
MAXIMAL FLOW VALUE =13
NODES ON SOURCE SIDE OF CUT :  1 2 3 4 6 7
NODES ON TERMINAL SIDE OF CUT :  5
```

THE MAXIMUM-FLOW MATRIX

FROM\TO	1	2	3	4	5	6	7
1		19	22	24	13	24	21
2	19		19	19	13	19	19
3	22	19		22	13	22	21
4	24	19	22		13	25	21
5	13	13	13	13		13	13
6	24	19	22	25	13		21
7	21	19	21	21	13	21	

Figure 2.40. Sequence of Computer Screens for Multiterminal Maximal Flow Example.

2.14 MAXIMAL-CAPACITY PATH PROBLEM

The algorithm of Section 2.13.1 determines the maximum possible flow between all pairs of nodes. It is clear that the maximum flow between any two nodes might involve flow that forms *multiple* paths or chains from source to sink. Consider the simplified network shown in Figure 2.41 where the number on each arc represents the capacity or upper bound on the arc flow.

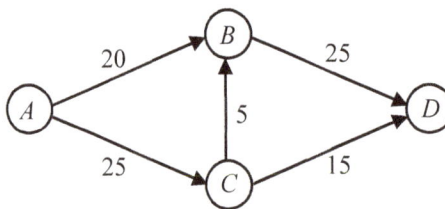

Figure 2.41. Maximum capacity network.

The maximum possible flow between nodes A and D is equal to 40 and corresponds to the following pattern of flows: $f_{AB} = 20$, $f_{AC} = 20$, $f_{CB} = 5$, $f_{BD} = 25$, and $f_{CD} = 15$. From the figure we can see that the following three chains are used in the distribution of the maximum flow:

$$(A,B), (B,D)$$
$$(A,C), (C,B), (B,D)$$
$$(A,C), (C,D)$$

It can be verified that the above chains are moving 20, 5, and 15 units of flow, respectively. In the maximum-capacity path problem we are only interested in a *single* path that moves as much flow as possible from the source node to the terminal node of a given network. In this example the maximum amount of flow that can move along a single path is 20, and the path itself corresponds to the connected sequence of arcs (A,B), (B,D).

2.14.1 Maximal-Capacity Chain for Networks with Single Source and Terminal Nodes

This is a straightforward modification of Dijkstra's method. In this case we define the labels as capacities. The capacity of a chain is defined as the minimum of its arc capacities. If an arc does not exist, its capacity can be defined to be equal to 0 (or $-\infty$). It is assumed that node s is the source and node t is the terminal. A summary of the algorithm follows:

Step 0: $[\delta_s] = \infty$; $\delta_i = c_{si}$

Step 1: $[\delta_j] = max \{\delta_i\}$, where j is the node that received the last permanent label, and **T** is the set $i \in \mathbf{T}$

of nodes with temporary labels.

Step 2: If $[\delta_t]$ found, stop. Otherwise, go to Step 3.

Step 3: new $\delta_i = max \{$old δ_i ; $min ([\delta_j],c_{ji})\}$ for $i \in \mathbf{L}$, where **L** is the set of temporarily labeled nodes reached from the last permanently labeled node.

Step 4: Go to Step 1.

As an exercise, the student is asked to find the maximal-capacity chain from node A to node D in the network shown in Figure 2.41.

2.14.2 Maximal-Capacity Chains for Networks with Multiple Sources and Terminals

Hu [24] has developed an efficient computational procedure based on a modification of the triple operation used in the multi-terminal multi-source shortest-path problem. The algorithm can be described as follows. Given a directed capacitated network G = (**N**, **A**), **N** = {1,2, ..., n}, it is desired to find the chain with maximum flow capacity for each pair of nodes in the network. The following notation will be used in the formulation of an algorithm to determine the chain:

c_{ik} capacity of directed arc (i, k)

d_{ik}^{j} maximum capacity of a chain from node i to node k containing intermediate nodes from the set $\{1, 2, ..., j\}$

r_{ik}^{j} intermediate node of a maximum-capacity chain from i to k containing nodes from the set $\{1, 2, ..., j\}$

Starting with $d_{ik}^{0} = c_{ik}$ and $r_{ik}^{0} = k$, the triple operation used in Floyd's algorithm is modified as indicated below and it is performed sequentially for $j = 1, 2, ..., n$ noting that $i \neq j \neq k$:

$$d_{ik}^{j} = max\left\{d_{ik}^{j-1}, min\left[d_{ik}^{j-1}, d_{ik}^{j-1}\right]\right\}\qquad(2\text{-}73)$$

The routing matrix is modified according to the following relationship:

$$r_{ik}^{j} = \begin{cases} r_{ik}^{j-1} & if\ \ d_{ik}^{j} \le d_{ik}^{j-1} \\ j & otherwise \end{cases}\qquad(2\text{-}74)$$

Computational savings similar to those previously discussed for Floyd's algorithm are also possible for this algorithm. As an exercise the student is asked to summarize them.

2.14.3 Example

East Texas Freight Company (ETFC) operates a trucking firm that specializes in moving large, bulky equipment. ETFC is planning the moving of bulk freight between seven NASA installations in the Houston area. This particular operation calls for the movement of several classes of NASA equipment resulting in extremely *high* loads when placed on a truck. A major problem is the maximum allowable vertical clearance allowed by the bridges and underpasses on a route.

In this application, the nodes of the network represent the seven NASA installations and the arcs correspond to the available routes; the maximum height of the load allowed on an arc is viewed as the *capacity* of the arc. Those arcs not allowed on the route can be assigned a capacity equal to zero, or, more generally, and arbitrarily large negative value (i.e. $-\infty$).

Vertical bridge clearances, in excess of a specified minimum (30 feet), are shown in Table 2.23 for approved routes connecting the seven installations.

Table 2.23. EXCESS CLEARANCE ALLOWED IN MOVING BULK FREIGHT

		To						
		1	2	3	4	5	6	7
	1	0	11	30	-	-	-	-
	2	11	0	-	12	2	-	-
	3	30	-	0	19	-	4	-
From	4	-	12	19	0	11	9	-
	5	-	2	-	11	0	-	-
	6	-	-	4	9	-	0	-
	7	-	-	-	20	1	1	0

ETFC wishes to know the maximum height allowed (including the height of the vehicle used) of a load to be moved between seven specified shipping and receiving points. The maximum height of the load allowed to move from a specified source node to a specified terminal node can now be determined as the capacity of the maximum-capacity route between the two nodes.

A. Algorithmic Procedure

The maximal-capacity matrix $\mathbf{D^0}$ and the route matrix $\mathbf{R^0}$ can be initialized as indicated below. If the capacities are assumed to be non-negative, all the $-\infty$ entries can be replaced by zeros. The NOP software assumes non-negative arc capacities.

$$
\mathbf{D^0} = \begin{array}{c} \\ 1 \\ 2 \\ 3 \\ 4 \\ 5 \\ 6 \\ 7 \end{array}
\begin{array}{c} 1 \quad 2 \quad 3 \quad 4 \quad 5 \quad 6 \quad 7 \\
\left(\begin{array}{ccccccc}
- & 11 & 30 & -\infty & -\infty & -\infty & -\infty \\
11 & - & -\infty & 12 & 2 & -\infty & -\infty \\
30 & -\infty & - & 19 & -\infty & 4 & -\infty \\
-\infty & 12 & 19 & - & 11 & 9 & -\infty \\
-\infty & 2 & -\infty & 11 & - & -\infty & -\infty \\
-\infty & -\infty & 4 & 9 & -\infty & - & -\infty \\
-\infty & -\infty & -\infty & 20 & 1 & 1 & -
\end{array} \right)
\end{array}
\qquad
\mathbf{R^0} = \begin{array}{c} \\ 1 \\ 2 \\ 3 \\ 4 \\ 5 \\ 6 \\ 7 \end{array}
\begin{array}{c} 1 \quad 2 \quad 3 \quad 4 \quad 5 \quad 6 \quad 7 \\
\left(\begin{array}{ccccccc}
1 & 2 & 3 & 4 & 5 & 6 & 7 \\
1 & 2 & 3 & 4 & 5 & 6 & 7 \\
1 & 2 & 3 & 4 & 5 & 6 & 7 \\
1 & 2 & 3 & 4 & 5 & 6 & 7 \\
1 & 2 & 3 & 4 & 5 & 6 & 7 \\
1 & 2 & 3 & 4 & 5 & 6 & 7 \\
1 & 2 & 3 & 4 & 5 & 6 & 7
\end{array} \right)
\end{array}
$$

Iteration 1: Here $j = 1$ is defined as the pivot node. Hence, we delete the first row and the first column of matrix $\mathbf{D^\circ}$. Additionally, columns 4, 5, 6, and 7 have $-\infty$ entries in the pivot row; and rows 4, 5, 6, and 7 have $-\infty$ entries in the pivot column. Therefore, only the entries d^0_{23} and d^0_{32} of $\mathbf{D^\circ}$ which are shown in Figure 2.42 need to be analyzed with the triple operation. The triple operation for each of these entries finds the minimum of the *projections* of the entry on the pivot row and the pivot column and selects the maximum between the minimal projection and the current value of the entry.

	1	2	3	4	5	6	7
1	-	11	30				
2	11	-	$-\infty$				
3	30	$-\infty$	-				
4							
5							
6							
7							

Figure 2.42. Entries to be checked in Iteration 1.

The results for the triple operations for the two entries are:

$$ d^1_{32} = max\left[d^0_{32}, min(11,30) \right] = max\left[-\infty, 11 \right] = 11 $$
$$ d^1_{23} = max\left[d^0_{23}, min(30,11) \right] = max\left[-\infty, 11 \right] = 11 $$

109

After the first iteration the maximal-capacity and route matrix are as follows:

$$\mathbf{D^1} = \begin{array}{c} \\ 1 \\ 2 \\ 3 \\ 4 \\ 5 \\ 6 \\ 7 \end{array}
\begin{array}{ccccccc}
1 & 2 & 3 & 4 & 5 & 6 & 7 \\
\left(\begin{array}{ccccccc}
- & 11 & 30 & -\infty & -\infty & -\infty & -\infty \\
11 & - & 11 & 12 & 2 & -\infty & -\infty \\
30 & 11 & - & 19 & -\infty & 4 & -\infty \\
-\infty & 12 & 19 & - & 11 & 9 & -\infty \\
-\infty & 2 & -\infty & 11 & - & -\infty & -\infty \\
-\infty & -\infty & 4 & 9 & -\infty & - & -\infty \\
-\infty & -\infty & -\infty & 20 & 1 & 1 & -
\end{array} \right)
\end{array}$$

$$\mathbf{R^1} = \begin{array}{c} \\ 1 \\ 2 \\ 3 \\ 4 \\ 5 \\ 6 \\ 7 \end{array}
\begin{array}{ccccccc}
1 & 2 & 3 & 4 & 5 & 6 & 7 \\
\left(\begin{array}{ccccccc}
1 & 2 & 3 & 4 & 5 & 6 & 7 \\
1 & 2 & 1 & 4 & 5 & 6 & 7 \\
1 & 1 & 3 & 4 & 5 & 6 & 7 \\
1 & 2 & 3 & 4 & 5 & 6 & 7 \\
1 & 2 & 3 & 4 & 5 & 6 & 7 \\
1 & 2 & 3 & 4 & 5 & 6 & 7 \\
1 & 2 & 3 & 4 & 5 & 6 & 7
\end{array} \right)
\end{array}$$

Iteration 2: Here $j = 2$ is defined as the pivot node. Hence, we delete the second row and the second column of matrix $\mathbf{D^1}$. Additionally, columns 4, 5, 6, and 7 have $-\infty$ entries in the pivot row; and rows 4, 5, 6, and 7 have $-\infty$ entries in the pivot column. Therefore, only the entries d_{13}^1, d_{14}^1, d_{15}^1, d_{31}^1, d_{34}^1, d_{35}^1, d_{41}^1, d_{43}^1, d_{45}^1, d_{51}^1, d_{53}^1, and d_{54}^1 of $\mathbf{D^1}$ which are shown in Figure 2.43 need to be analyzed with the triple operation.

	1	2	3	4	5	6	7
1	-	11	30	-∞	-∞		
2	11	-	11	12	2		
3	30	11	-	19	-∞		
4	-∞	12	19	-	11		
5	-∞	2	-∞	11	-		
6							
7							

Figure 2.43. Entries to be checked in Iteration 2.

After applying the triple operation to each of the entries shown in Figure 2.43 the corresponding results after the second iteration for the maximal-capacity and route matrix are as follows:

110

$$
D^2 = \begin{array}{c} \\ 1 \\ 2 \\ 3 \\ 4 \\ 5 \\ 6 \\ 7 \end{array}
\begin{array}{c}
\begin{array}{ccccccc} 1 & 2 & 3 & 4 & 5 & 6 & 7 \end{array} \\
\begin{pmatrix}
- & 11 & 30 & 11 & 2 & -\infty & -\infty \\
11 & - & 11 & 12 & 2 & -\infty & -\infty \\
30 & 11 & - & 19 & 2 & 4 & -\infty \\
11 & 12 & 19 & - & 11 & 9 & -\infty \\
2 & 2 & 2 & 11 & - & -\infty & -\infty \\
-\infty & -\infty & 4 & 9 & -\infty & - & -\infty \\
-\infty & -\infty & -\infty & 20 & 1 & 1 & -
\end{pmatrix}
\end{array}
\qquad
R^2 = \begin{array}{c} \\ 1 \\ 2 \\ 3 \\ 4 \\ 5 \\ 6 \\ 7 \end{array}
\begin{array}{c}
\begin{array}{ccccccc} 1 & 2 & 3 & 4 & 5 & 6 & 7 \end{array} \\
\begin{pmatrix}
1 & 2 & 3 & 2 & 2 & 6 & 7 \\
1 & 2 & 1 & 4 & 5 & 6 & 7 \\
1 & 1 & 3 & 4 & 2 & 6 & 7 \\
2 & 2 & 3 & 4 & 5 & 6 & 7 \\
2 & 2 & 2 & 4 & 5 & 6 & 7 \\
1 & 2 & 3 & 4 & 5 & 6 & 7 \\
1 & 2 & 3 & 4 & 5 & 6 & 7
\end{pmatrix}
\end{array}
$$

It can be verified that only the third and fourth iterations will result in maximal-capacity improvements. Figure 2.44 and shows the entries to be checked in these two iterations.

	1	2	3	4	5	6	7
1	-	11	30	11	2	-∞	
2	11	-	11	12	2	-∞	
3	30	11	-	19	2	4	-∞
4	-∞	12	19	-	11	9	
5			-∞				
6	-∞	-∞	4	9	-∞	-	
7			-∞				

Iteration 3

	1	2	3	4	5	6	7
1	-	12	30	19	11	9	
2	12	-	12	12	11	9	
3	30	12	-	19	11	9	
4	19	12	19	-	11	9	-∞
5	11	11	11	11	-	9	
6	9	9	9	9	9	-	
7	19	12	19	20	11	9	

Iteration 4

Figure 2.44. Entries to be checked in Iterations 3 and 4.

Remaining Iterations: Proceeding in this fashion, the corresponding capacity and route matrices for the remaining iterations of the algorithm are shown in Table 2.24. The maximum capacity allowed on a chain connecting two specified source and terminal nodes is equal to the entry associated with the selected pair of nodes in the *capacity matrix* found for iteration 7 and shown in Table 2.24. The corresponding sequence of arcs in the route can be recovered by tracing arcs through the final *route matrix* also shown in Table 2.24.

As an illustration, the maximum-capacity chain from node 1 to node 4 is equal to $d^*_{14} = 19$. The route is traced as follows:

$r^*_{14} = 3$ (move from node 1 to node 4 through node 3)
$r^*_{34} = 4$ (move from node 3 to node 4 directly)
$r^*_{13} = 3$ (move from node 1 to node 3 directly)

This means that the chain consists of the following arcs: (1,3), (3,4). The maximal total clearance allowed on this chain is equal to 19 + 30 = 49 feet.

Table 2.24. RESULTS FOR REMAINING ITERATIONS

Maximum Capacity Matrix	Routing Matrix
Iteration 3	
$\mathbf{D}^3$	$\mathbf{R}^3$

$$
\mathbf{D}^3 =
\begin{array}{c}
\\1\\2\\3\\4\\5\\6\\7
\end{array}
\begin{array}{ccccccc}
1 & 2 & 3 & 4 & 5 & 6 & 7\\
- & 11 & 30 & 19 & 2 & 4 & -\infty\\
11 & - & 11 & 12 & 2 & 4 & -\infty\\
30 & 11 & - & 19 & 2 & 4 & -\infty\\
19 & 12 & 19 & - & 11 & 9 & -\infty\\
2 & 2 & 2 & 11 & - & 2 & -\infty\\
4 & 4 & 4 & 9 & 2 & - & -\infty\\
-\infty & -\infty & -\infty & 20 & 1 & 1 & -
\end{array}
$$

$$
\mathbf{R}^3 =
\begin{array}{c}
\\1\\2\\3\\4\\5\\6\\7
\end{array}
\begin{array}{ccccccc}
1 & 2 & 3 & 4 & 5 & 6 & 7\\
1 & 2 & 3 & 3 & 2 & 3 & 7\\
1 & 2 & 1 & 4 & 5 & 3 & 7\\
1 & 1 & 3 & 4 & 2 & 6 & 7\\
3 & 2 & 3 & 4 & 5 & 6 & 7\\
2 & 2 & 2 & 4 & 5 & 3 & 7\\
3 & 3 & 3 & 4 & 3 & 6 & 7\\
1 & 2 & 3 & 4 & 5 & 6 & 7
\end{array}
$$

Iterations 4, 5, 6 and 7	
$\mathbf{D}^4 = \mathbf{D}^5 = \mathbf{D}^6 = \mathbf{D}^7$	$\mathbf{R}^4 = \mathbf{R}^5 = \mathbf{R}^6 = \mathbf{D}^7$

$$
\begin{array}{c}
\\1\\2\\3\\4\\5\\6\\7
\end{array}
\begin{array}{ccccccc}
1 & 2 & 3 & 4 & 5 & 6 & 7\\
- & 12 & 30 & 19 & 11 & 9 & -\infty\\
12 & - & 12 & 12 & 11 & 9 & -\infty\\
30 & 12 & - & 19 & 11 & 9 & -\infty\\
19 & 12 & 19 & - & 11 & 9 & -\infty\\
11 & 11 & 11 & 11 & - & 9 & -\infty\\
9 & 9 & 9 & 9 & 9 & - & -\infty\\
19 & 12 & 19 & 20 & 11 & 9 & -
\end{array}
$$

$$
\begin{array}{c}
\\1\\2\\3\\4\\5\\6\\7
\end{array}
\begin{array}{ccccccc}
1 & 2 & 3 & 4 & 5 & 6 & 7\\
1 & 4 & 3 & 3 & 4 & 4 & 7\\
4 & 2 & 4 & 4 & 4 & 4 & 7\\
1 & 4 & 3 & 4 & 4 & 4 & 7\\
3 & 2 & 3 & 4 & 5 & 6 & 7\\
4 & 4 & 4 & 5 & 5 & 4 & 7\\
4 & 4 & 4 & 4 & 4 & 6 & 7\\
4 & 4 & 4 & 4 & 4 & 4 & 7
\end{array}
$$

The problem is now solved by NOP as shown below.

B. Solution by NOP

Figure 2.45 shows the computer screens for the NOP solution: (1) Input file, (2) NOP output results. The NOP output results include the following:

1. The arc capacity matrix for the given directed network
2. The max-cap matrix
3. The routing matrix which shows intermediate nodes to sequentially identify a required path from a given origin to a specified destination.

112

Input File

```
Nodes 7;
St;
1 2 11;
1 3 30;
2 1 11;
2 4 12;
2 5 2;
3 1 30;
3 4 19;
3 6 4;
4 2 12;
4 3 19;
4 5 11;
4 6 9;
5 2 2;
5 4 11;
6 3 4;
6 4 9;
7 4 20;
7 5 1;
7 6 1;
End;
```

Output

**** MULTITERMINAL MAXIMAL CAPACITY PROBLEM ****

ARC CAPACITY MATRIX

	1	2	3	4	5	6	7
1	00000.00	00011.00	00030.00	00000.00	00000.00	00000.00	00000.00
2	00011.00	00000.00	00000.00	00012.00	00002.00	00000.00	00000.00
3	00030.00	00000.00	00000.00	00019.00	00000.00	00004.00	00000.00
4	00000.00	00012.00	00019.00	00000.00	00011.00	00009.00	00000.00
5	00000.00	00002.00	00000.00	00011.00	00000.00	00000.00	00000.00
6	00000.00	00000.00	00004.00	00009.00	00000.00	00000.00	00000.00
7	00000.00	00000.00	00000.00	00020.00	00001.00	00001.00	00000.00

MULTITERMINAL MAXIMAL-CAPACITY MATRIX

	1	2	3	4	5	6	7
1	00000.00	00012.00	00030.00	00019.00	00011.00	00009.00	00000.00
2	00012.00	00000.00	00012.00	00012.00	00011.00	00009.00	00000.00
3	00030.00	00012.00	00000.00	00019.00	00011.00	00009.00	00000.00
4	00019.00	00012.00	00019.00	00000.00	00011.00	00009.00	00000.00
5	00011.00	00011.00	00011.00	00011.00	00000.00	00009.00	00000.00
6	00009.00	00009.00	00009.00	00009.00	00009.00	00000.00	00000.00
7	00019.00	00012.00	00019.00	00020.00	00011.00	00009.00	00000.00

MATRIX OF INTERMEDIATE NODES ON THE MAX-CAP PATH

	1	2	3	4	5	6	7
1	1	4	3	3	4	4	7
2	4	2	4	4	4	4	7
3	1	4	3	4	4	4	7
4	3	2	3	4	5	6	7
5	4	4	4	4	5	4	7
6	4	4	4	4	4	6	7
7	4	4	4	4	4	4	7

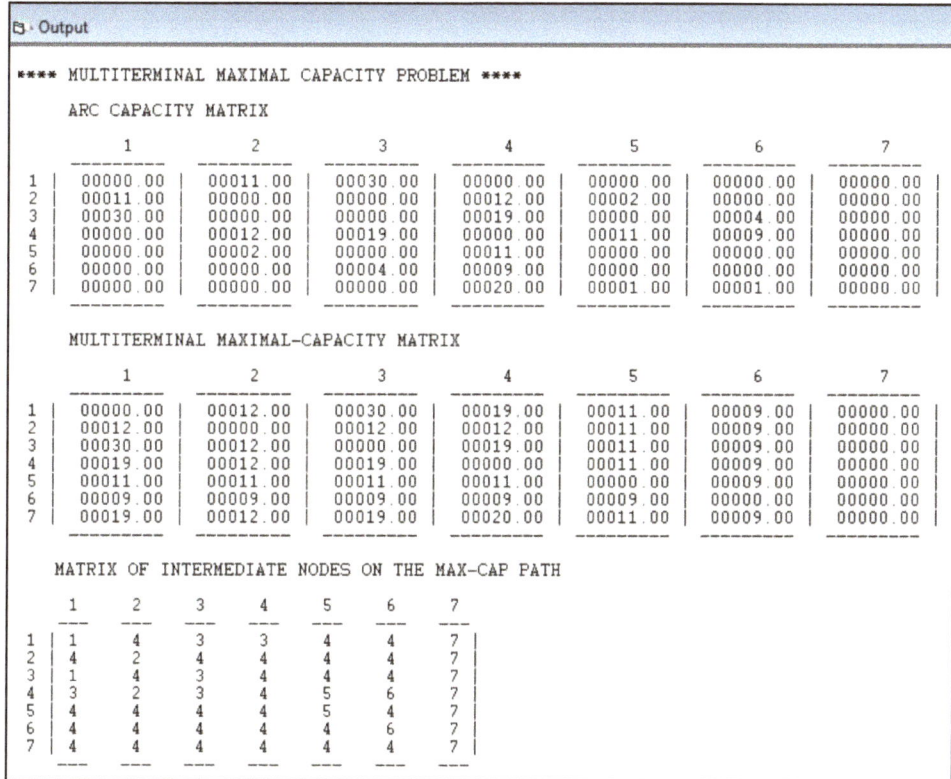

Figure 2.45. Sequence of Computer Screens for Maximal Capacity Example.

EXERCISES

1. Indicate how we can use Dijkstra's method to obtain an arborescence with a given root.

2. Discuss under which conditions a longest-path model can be converted into an equivalent shortest-path model. Explain how to do this.

3. Find the length of the *longest* path from node 1 to node 5 in the given directed network. Find the corresponding sequence of nodes.

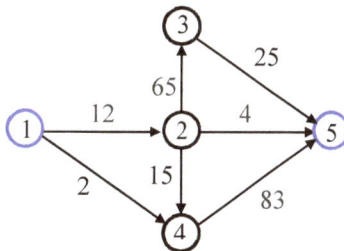

4. Find the *shortest* path from node 1 to node 4 in the network shown (a) by inspection; (b) using Dijkstra's method. (c) Using Ford's method. Explain your results.

113

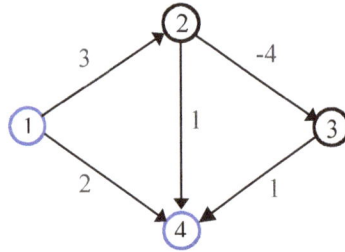

5. Dijkstra's method assumes that all arc parameters are nonnegative. Ford's algorithm is a modification of Dijkstra's algorithm that allows the consideration of arcs with negative *length*. What condition is of critical importance in this case?

6. Explain why the number of iterations of Floyd's algorithm is exactly the same as the number of nodes in the network.

7. Can we have negative arc parameters in Floyd's algorithm and the double-sweep method? What assumption is important?

8. Is it possible for cycles to appear in the paths corresponding to those path-lengths found by the double-sweep method? Discuss what you would do if only cycleless paths are wanted.

9. Explain why the order in which the nodes are numbered may affect the computational efficiency of Floyd's algorithm and the double-sweep method. Discuss each case separately.

10. Demonstrate that in a directed circuitless network with arcs (i, j) such that $i < j$ the double-sweep method reduces itself to one sweep. Which sweep is eliminated? Consider the same exercise for the case where the arcs (i, j) are such that $j < i$.

11. Under which conditions an undirected arc connecting nodes i and j can be and cannot be replaced by two directed arcs (i, j) and (j, i) with same capacity as that of the original arc. Illustrate your explanations with examples.

12. Show how the shortest-route problem can be solved by the transshipment model. Illustrate your result using a small numerical example.

13. In the turn-penalty algorithm, why it is not possible to add the penalty to each arc parameter and then solve the problem by a regular shortest-path algorithm?

14. Consider the network consisting of the following arcs with the given arc lengths:

(i, j)	(1,2)	(1,3)	(2,3)	(2,4)	(3,2)	(3,4)
d_{ij}	3	6	2	-8	-1	2

Determine the shortest path from node 1 to node 4 using the following turn penalties:

i-j-k	1-2-3	1-3-4	1-3-2	2-3-4	3-2-4
p_{ijk}	1	2	2	1	2

15. Consider the generalized network with the following arc multipliers:

(i, j)	(1,2)	(1,3)	(2,3)	(2,4)	(3,2)	(3,4)
A_{ij}	3	6	2	2/3	1/2	1/3

Write the constraint matrix for the net flow conditions for the nodes of the network.

16. Consider the network data given below:

Arc	Flow Bounds
(1,2)	[2,0]
(1,4)	[4,0]
(2,3)	[4,0]
(2,4)	[1,0]
(3,4)	[2,1]
(4,2)	[1.0]

The first number for each arc is its upper bound, and the second number is its lower bound on arc flow. (a) Use the labeling procedure to determine the maximal flow from node 1 to node 3, and all the corresponding arc flows; (b) by inspection identify a minimal cut and its value.

17. For the network shown, in which the arc parameters represent arc costs, find:

(a) The cheapest route from node 1 to node 13 using Dijkstra's algorithm.
(b) The cheapest route from each node to every other node using Floyd's algorithm.
(c) The $k = 3$ cheapest path-lengths and corresponding paths from node 2 to node 13.

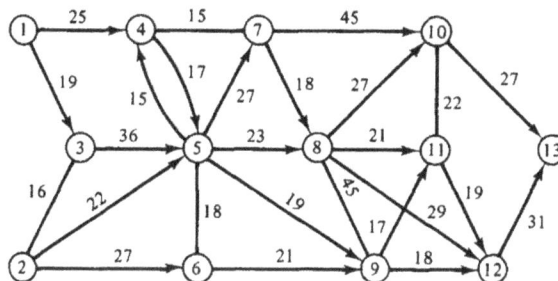

18. In Exercise 17 find the cheapest route from node 1 to node 13 assuming that arc (3,5) must be included in the path.

19. Consider the network shown, in which each arc parameter represents the capacity of the corresponding arc. Find each of the following using the appropriate algorithm.

(a) The maximum flow from node 2 to node 10.
(b) The maximum-capacity chain from node 2 to node 10.
(c) The maximum flow between each pair of nodes, assuming that the network is now *undirected.*

115

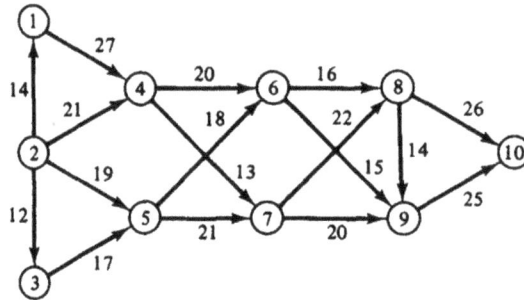

20. Find (by inspection) the minimum cut corresponding to the maximum flow found in Exercise 19, part (a).

21. In the following network representing a particular scenario of the employment-scheduling problem discussed in this chapter, the arc values are flows and the node values indicate supply/demand requirements. (A) Write a linear programming model, assuming that the cost per employee and per period is constant and equal to $25. (B) Write the equivalent minimum-cost flow model. (C) State specific actions taken at each period for the given scenario.

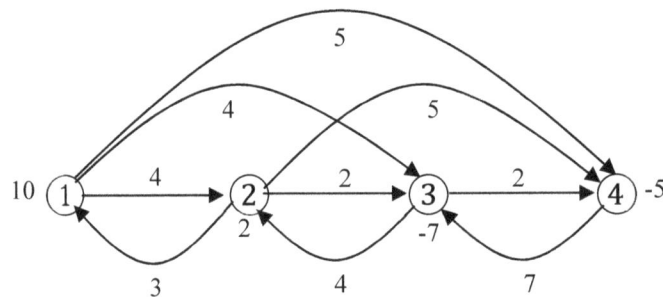

22. A businessman has the option of investing in two plans, A and B. Plan A guarantees that each dollar invested will earn $0.70 a year hence, while Plan B guarantees that each dollar invested will earn $2.00 two years hence. In Plan B only every two years investments are allowed. Suppose that the businessman has a total of $100,000. How should he allocate his sum in order to maximize his earnings at the end of three years? Formulate a network representation for this problem. In particular, indicate: (a) meaning of nodes, (b) meaning of arcs, (c) arc parameters, (d) network diagram, and (e) how to recognize the optimal solution (no need to find it).

23. A certain device consists of eight components as shown in the figure. The probability of failure for each component is as given in the graph. The device will function properly if all the elements in a path from s to t function satisfactorily. Furthermore, assume that once block 2 is used, after using block 1, it is not possible to use block 1 again. Formulate this problem as a shortest-path model to identify the most reliable path for the given device.

116

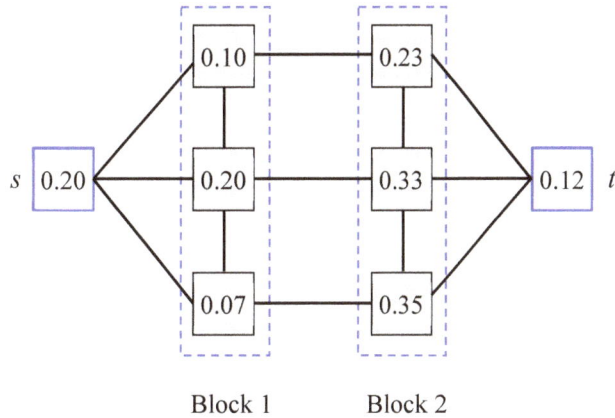

Block 1   Block 2

24. (a) Verify the max-flow min-cut theorem on the network shown. The two values given for each arc are the lower and upper bounds on arc flow. (b) Explain how the min-cut can be feasible.

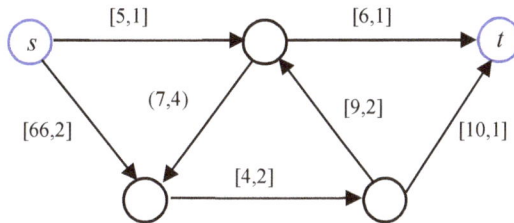

25. Show that the following matrices are or are not totally unimodular:

$$\begin{pmatrix} -1 & +1 & 0 & 0 \\ 0 & +1 & -1 & 0 \\ 0 & 0 & +1 & -1 \\ +1 & 0 & 0 & -1 \end{pmatrix} \quad \begin{pmatrix} -1 & +1 & 0 & +1 \\ -1 & 0 & 0 & +1 \\ +1 & -1 & +1 & +1 \end{pmatrix}$$

26. Is the following matrix totally unimodular or not? Prove your result.

$$\begin{pmatrix} 1 & 1 & 0 & 0 & 0 \\ -1 & 0 & -1 & 0 & 0 \\ 0 & 1 & -1 & 0 & 0 \\ 0 & 0 & 1 & 1 & 1 \end{pmatrix}$$

27. Determine if the following matrices are or are not totally unimodular:

(a)

$$A = \begin{bmatrix} 1 & 1 & 0 & 0 & 0 & 0 & 0 \\ 1 & 0 & 1 & 0 & 1 & 0 & 0 \\ 0 & 1 & 0 & 1 & 0 & 0 & 0 \\ 0 & 0 & 1 & 1 & 0 & 1 & 0 \\ 0 & 0 & 0 & 0 & 1 & 0 & 1 \\ 0 & 0 & 0 & 0 & 0 & 1 & 1 \end{bmatrix}$$

(b)

$$A = \begin{bmatrix} 1 & 1 & 0 & 0 & 0 & 0 & 0 \\ 1 & 0 & 1 & 0 & 1 & 0 & 0 \\ 0 & 1 & 0 & 1 & 0 & 0 & 0 \\ 0 & 0 & 1 & 1 & 0 & 1 & 0 \\ 0 & 0 & 0 & 0 & 1 & 0 & 1 \\ 0 & 0 & 0 & 0 & -1 & 1 & 1 \end{bmatrix}$$

117

28. Prove the following theorem. Let **A** be an integer matrix with linearly independent rows. Then the following three conditions are equivalent: (a) The determinant of every basis matrix **B** of **A** is equal to +1 or -1. (b) Every basic feasible solution of $\mathbf{AX} = \mathbf{b}, \mathbf{X} \geq \mathbf{0}$, is integer for any integer vector **b**. (c) Every basis matrix **B** of **A** has an integer inverse $\mathbf{B}^{-1}$. Hint: Show that (a) implies (b), (b) implies (c), and (c) implies (a). Note: we may say that matrix **A** as defined above is *unimodular* if it satisfies condition (a).

29. What can we conclude from the following system of equations where both b_1 and b_2 are integer?

$$5x_1 + 2x_2 = b_1$$

$$2x_1 + x_2 = b_2$$

30. Consider the 4-node network corresponding to the following matrix, where the entry on the *i*th row and the *j*th column is the probability that arc (i,j) fails:

$$\begin{pmatrix} - & 0.3 & 0.7 & 0.9 \\ - & - & 0.1 & 0.3 \\ - & 0.3 & - & 0.5 \\ - & - & - & - \end{pmatrix}$$

Find the most reliable chain from node 1 to node 4. Compute the probability of success. Use Dijkstra's algorithm.

31. Consider the network defined by the following connectivity and arc length lists:

Node	1	2	3	4	
Number of exiting arcs	2	2	1	0	
End-nodes of arcs	2	4	3	4	4
Arc lengths	3	2	-4	1	1

   (a) Use *Ford's algorithm* to find the length of the shortest path from node 1 to node 4. Find the arcs in the path.
   (b) Use *Floyd's algorithm* to find the length of the shortest path between any two nodes. Find the arcs in the path from node 1 to node 4.
   (c) Use the *Double Sweep Method* to find the two shortest path lengths between node 1 and every node in the network. Based on the results, find the shortest path from 1 to 4.

32. Find a minimal spanning tree for the networks shown. What condition is true for any arc not to be in the minimal spanning tree?

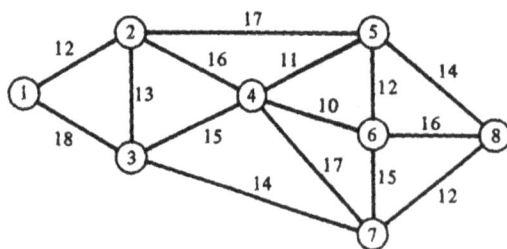

(a)

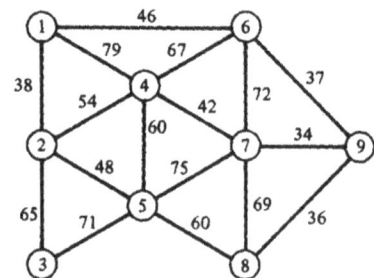

(b)

118

33. A manufacturing company is planning to produce air conditioners involving three main components: cabinets, fans, and motors. It can tool up to fabricate the cabinets for $20,000, fans for $50,000, and motors for $80,000. However, once the fans are in production, the cost of tooling up for the cabinets and motors will be reduced by 5 %. If the motors are put into production first, the costs of tooling up for the other units will be reduced by 10%. If the cabinets are in production first, the other costs are reduced by 5 %. After 2 units are in production there is an extra reduction of 5 % for the third unit. Draw the network, formulate, and solve as a shortest-path problem with one source and one terminal. Define the nodes, the arcs, and the arc parameters.

34. A conveyor manufacturer is designing a conveyor system to link point B to point C on a factory floor. The recommended system configuration is as shown. The numbers on each arc represent construction costs from point to point. Every junction point requires an additional amount of money to install if a turn is to be made. Points 2, 6, 10, and 9 cost 1 unit extra. Points 3, 5, 8, 7, and 4 cost 2 units extra. Find the least-cost route from point B to point C.

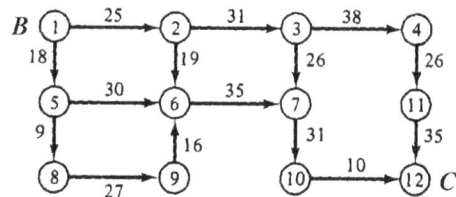

35. Find the shortest path between each pair of nodes in the network shown.

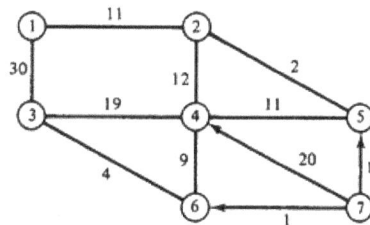

36. In the network having the net-flow coefficient matrix shown below, node 1 has a supply of four units of flow and node 2 has a demand of one unit. (a) Draw the network showing all arc data. (b) What kind of network is this? (c) If all supply available is used, how many units of flow will be available at node 4 after following the path 1-3-2-4? (d) Write the flow-conservation condition for node 2.

$$\begin{pmatrix} 1 & 1 & 0 & 0 & 0 & 0 \\ -3 & 0 & 1 & 1 & -1/2 & 0 \\ 0 & -6 & -2 & 0 & 1 & 1 \\ 0 & 0 & 0 & -2/3 & 0 & -1/3 \end{pmatrix}$$

37. Consider the network corresponding to the following matrix, where the entry on the ith row and the jth column is the probability that arc (i,j) fails:

$$\begin{pmatrix} - & 0.3 & 0.7 & 0.9 \\ 0.6 & - & 0.1 & 0.3 \\ 0.3 & 0.3 & - & 0.5 \\ 0.6 & 0.1 & 0.7 & - \end{pmatrix}$$

    (a) *Formulate* the triple operation used by Floyd's algorithm to find the maximum probability of success between any two nodes.

    (b) *Reformulate* a triple operation that uses the probability information *directly* to solve this problem.

    (c) Execute the steps of Floyd's algorithm for each case.

38. A ship fleet operator wishes to plan the fleet schedule so as to maximize ship utilization. The information about number of ships, port of origin, port of delivery, and delivery date is summarized in the following table.

Ships	From	To	Date
2	A	C	3
1	A	C	9
1	B	D	3
1	C	B	6

The travel times in any direction between the ports of interest for both loaded and unloaded ships are given below.

	Loaded			Unloaded	
	C	D		C	D
A	3	2	A	2	1
B	2	3	B	1	1

Draw a complete *circulation* network to model this problem as a min-cost network flow problem.

39. Use Dijkstra's algorithm to find the shortest path from node 1 to node 5 in the directed network shown, where the numbers associated with the arcs are the corresponding arc lengths.

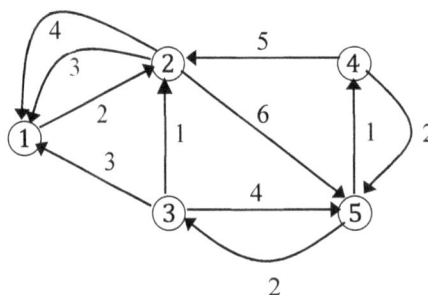

40. Find the three shortest path-lengths and corresponding paths from node 2 to every node in the network shown.

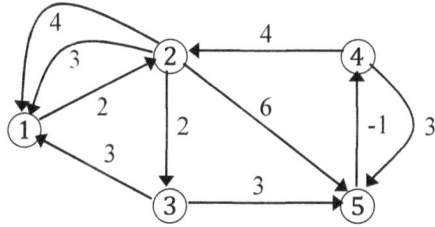

41. The Krumley family has painted almost every house in Buda, Texas (population 498). The family is large enough that there are three crews of workers available throughout the spring and summer months. Albert E. Krumley, the family patriarch and head of the company, has contracted for five jobs during this season. Each job may be started on or after specified dates and must be completed on or before agreed-upon dates.

Job	Start Date	Completion Date	Time (weeks)
A	April 1	April 28	2
B	April 14	May 12	4
C	April 1	May 26	7
D	April 21	May 26	3
E	April 1	May 26	6

Jobs need not be worked on continuously from start to finish. However, Mr. Krumley makes it a practice to assign his crews to in such a way that in any given week they will be working on only one job. Further, no two crews ever work on the same job at the same time. It is desired to know if all these jobs can be finished by their completion dates.

42. Á company supplies clerical employees upon demand to several business concerns in a metropolitan area. The best available estimates of demand for the next four months, as well as the costs to hire an employee at the start of a month and to terminate the employment at the end of each month, are shown below:

Month	1	2	3	4
Demand	22	15	32	8
$/hire	170	290	150	80
$/end	220	230	110	120

All employees are paid at the rate of $1000 per month. The best estimates of possible new hires in the four months are 25, 10, 24, and 10 for months 1, 2, 3, 4, respectively. It is assumed that no employee will be kept after the end of month 4. Formulate the problem of finding the most cost-effective policy as a network model (show a picture with all relevant information).

44. Consider the following *undirected* capacitated network with arc capacities as shown next to each arc.

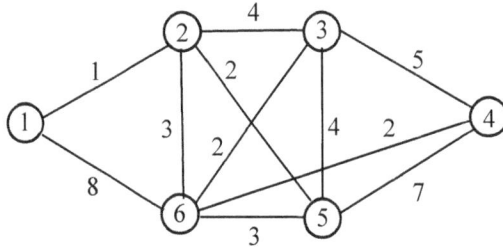

(a) Use the Gomory-Hu algorithm. At each iteration show the condensed network, the maximal flow, the minimal cut, and the current status of the cut graph. Consider the following source/terminal (s,t) pairs: (2,6), (1,2), (3,6), (4,5), and (3,5), respectively. Find max-flows and min-cuts by inspection. Write the max-flow matrix.

(b) Find maximal-capacity path from node 1 to node 4.

(c) Find the value of the maximal capacity between each pair of nodes.

45. Consider the transportation network shown in Figure 2.5. Assuming that $s_1 = 12$, $s_2 = 18$, $d_1 = 15$, $d_2 = 5$, $d_3 = 10$, $c_{11} = 1$, $c_{12} = 2$, $c_{13} = 3$, $c_{21} = 41$, $c_{22} = 3$, $c_{23} = 5$, design a *circulation network* model that can be used to minimize total transportation costs. Each arc of the network must be assigned three parameters: lower bound, upper bound, and per-unit flow cost.

46. Consider the following *undirected* capacitated network (arc capacities shown next to each arc).

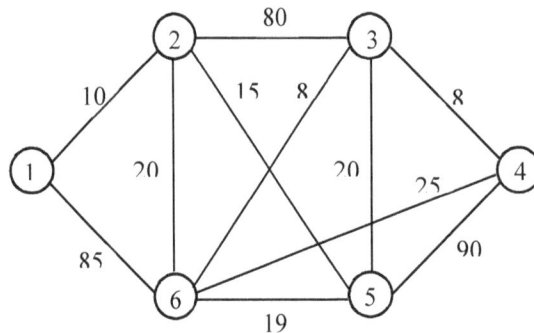

(a) Use the Gomory-Hu algorithm, showing the *condensed network* at each iteration, along with the maximal flow (*by inspection*), the minimal cut, and the current status of the *min-cut tree*. Consider the following source/terminal (s,t) pairs for iterations 1, 2, ..., 5: (2,6), (1,4), (5,4), (1,6), and (2,3), respectively. Find the max-flow matrix.

(b) Find the maximal flow in iteration 2 using the *labeling procedure* (for directed networks).

47. Terminate the application of the Gomory-Hu algorithm to find the complete min-cut tree. Use the *labeling procedure* to determine max-flows and find min-cuts by inspection.

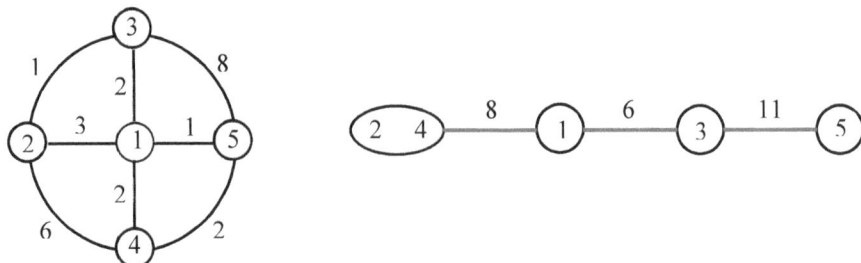

REFERENCES

[1]   BELLMORE, M., G. BENNINGTON, AND S. LUHORE, "A Multivehicle Tanker Scheduling Problem," *Transportation Science*, 5, 36-47 (1971).

[2]   BENNINGTON, G. E., "Applying Network Analysis," *Journal of Industrial Engineering,* 6 (1), 17-25 (1974).

[3]   BRADLEY, G. H., "Survey of Deterministic Networks," *AIIE Transactions,* 7, 222-234 (1975).

[4]   DANTZIG, G. B., W. BLATTNER, AND M. R. RAO, "Finding a Cycle in a Graph with a Minimum Cost to Time Ratio with Applications to a Ship Routing Problem," *Theory of Graphs International Symposium, pp.* 77-83. Paris: Dunod, and New York: Gordon and Breach, 1966.

[5]   DANTZIG, G.B. AND FULKERSON, "Minimizing the Number of Tankers to Meet a Fixed Schedule," *Naval Research Logistics Quarterly*, 1, 217-222 (1954).

[6]   DIJKSTRA, E. W., "A Note on Two Problems in Connection with Graphs," *Numerishe Mathematik,* 1, 269-271 (1959).

[7]   DREYFUS, S. E., "An Appraisal of Some Shortest Path Algorithms," *Operations Research,* 17, 395-412 (1969).

[8]   DREYFUS, S. E., "A Generalized Equipment Replacement Study," *Journal of the Society for Industrial and Applied Mathematics,* 8, 425-435 (1960).

[9]   EVANS, J. R., "Network Modeling in Production Planning," *Proceedings of the 1978 AIIE National Systems Conference,* Montreal, Canada, 1978.

[10] FLOYD, R. W., "ALGORITHM 97: Shortest Path," *Communications of the ACM*, 5, 345 (1962).

[11] FORD, L. R., AND D. R. FULKERSON, "Maximal Flow through a Network," *Canadian Journal of Mathematics*, 18, 399-404 (1956).

[12] FORD, L. R., AND D. R. FULKERSON, *Flows in Networks.* Princeton, N.J.: Princeton University Press, 1962.

[13] FRANK, H., AND I. T. FRISCH, *Communication, Transmission and Transportation Networks.* Reading, Mass.: Addison-Wesley, Publishing Co., Inc., 1971.

[14] FREDMAN AND TARJAN, "Fibonacci Heaps and their Use in Improved Network Optimization Algorithms," *Journal of ACM,* 34, 596-615 (1987).

[15] GARCIA-DIAZ, A., "A Network Flow Approach to Airline Fuel Allocation Problems," *The Annals of the Society of Logistics Engineers*, Vol. 2, No. 1, 39-53, 1990.

[16] GARFINKEL, R. S., AND G. L. NEMHAUSER, *Integer Programming.* New York: John Wiley & Sons, Inc., 1972.

[17] GLOVER, F., AND D. KLINGMAN, "Network Applications in Industry and Government," *AIIE Transactions,* 9 (4) (1977).

[18] GOMORY, R. E., AND T. C. HU, "Multi-terminal Network Flows," SIAM *J. Soc. Indust. Appl. Math,* 9, 551-571 (1971).

[19] HELLER, L, AND C. B. TOMPKINS, "An Extension of a Theorem of Dantzig's," in *Linear Inequalities and Related Systems,* ed. H. Kuhn and A. W. Tucker. Princeton, N.J.: Princeton University Press, 1956.

[20] HITCHCOCK, F. L., "The Distribution of a Product from Several Sources to Numerous Localities," *Journal of Mathematics and Physics,* 20, 224-230 (1941).

[21] HOFFMAN, A. J., AND S. WINOGRAD, "Finding All Shortest Distances in a Directed Network," *IBM Journal of Research and Development,* 16, 412-414 (1972).

[22] HU, T. C., *Integer Programming and Network Flows.* Reading, Mass.: Addison Wesley Publishing Co., Inc., 1969.

[23] HU T. C., "The Maximum Capacity Route Problem," *Operations Research,* 9, 898-900 (1961).

[24] JACOBS, W. W., "The Caterer Problem," *Naval Research Logistics Quarterly,* 1, 154-165 (1954).

[25] LAWLER, E. L., "Optimal Cycles in Doubly Weighted Linear Graphs," In *Theory of Graphs: International Symposium,* Paris: Dunod, New York: Gordon and Breach, 209-213 (1966).

[26] MINIEKA, E. T., AND D. R. SHIER, "A Note on an Algebra for the *K* Best Routes in a Network," *Journal of the IMA,* 11, 145-149 (1973).

[27] NEMHAUSER, G.L., AND L.A. WOLSEY, *Integer and Combinatorial Optimization,* Wiley-Interscience, John Wiley & Sons, Inc., New York, 1988.

[28] POLLACK, M., "The Maximum Capacity Route through a Network," *Operations Research,* 8, 733-736 (1960).

[29] SHIER, D. R., "Iterative Methods for Determining the *K* Shortest Paths in a Network," *Networks,* 6, 205-230 (1976).

[30] SHIER, D. R., "Computational Experience with an Algorithm for Finding the K Shortest Paths in a Network," *Journal of Research, National Bureau of Standards,* 78B, 139-165 (July-September 1974).

[31] SMYTHE, W. R., AND L. JOHNSON, *Introduction to Linear Programming with Applications.* Englewood Cliffs, N.J

Chapter 3

THE OUT-OF-KILTER ALGORITHM

*"We shouldn't surrender so readily to only half understanding all kinds of things.
Try harder to understand, and then when you have understood
take a little time out maybe to explain to others."*

Irving Kaplansky
In *More Mathematical People*
Harcourt Brace Jovanovich, Publishers,1990

The out-of-kilter algorithm (OKA) represents a significant contribution to the theory of network analysis, and for a number of years it was considered to be the most general and widely used algorithm for minimizing flow costs in capacitated, deterministic networks. Although the network specialization of the primal simplex procedure (to be studied in Chapter 4) has been found to be a more computationally efficient method, the broad applicability and unifying framework provided by the out-of-kilter algorithm warrants the special attention to which we will now address our efforts.

The out-of-kilter algorithm will be discussed in three parts. Part I contains the underlying theory of the algorithm and demonstrates its use by means of a small numerical example. Part II addresses the modeling versatility of the circulation-network concept and the applicability of the out-of-kilter algorithm. Part III shows a few selected formulations of real-world applications.

The primary objective of Part I is the development of the out-of-kilter algorithm using the concepts of linear programming duality theory and complementary slackness condition. Essentially, the algorithm consists of two procedures to alter flows on arcs and change the dual variables of the node flow conservation constraints. These procedures are iteratively executed until all arcs of the network satisfy linear programming optimality conditions. These two general types of strategies are integrated into an elegant algorithmic framework that allows a rare visualization of the optimization method itself. The actual mechanics of the solution procedure will be explained using a rather simplistic although effective graphical approach that streamlines the application of labeling procedures and dual-variable adjusting methods. A small numerical example will be used to illustrate all steps of the out-of-kilter algorithm

Part I

Network-Flow Optimization with the OKA - Fundamental Concepts

3.1 RELEVANT TERMINOLOGY

To provide a common ground for further discussion, the following terminology relevant to network flows will be considered in this section. Some definitions have been previously introduced, but will be repeated here for clarity and exposition purposes.

1. A node is the basic entity of a network diagram and usually represents a physical origin or termination point, such as a factory, retailer, contract or source of manpower, and so on. A

node that generates flow (such as a warehouse) is called a source node. A node that consumes flow (such as a customer) is called a sink or terminal node.

2. An arc (sometimes called branch) is a *generic channel* used for the transmission of flow. A directed arc is one in which flow is only permitted in a designated direction. Depending on the direction along which a directed arc is traversed, it can be made either forward or reverse. Flow is increased along forward arcs, and decreased along reverse arcs.

3. A network is a connected set of arcs and nodes, generally representing a physical process or system in which units of a commodity (flow) move from source(s) to sink(s). A typical network is given in Figure 3.1, in which node 1 is the source node and node 4 is the sink or terminal node. A network may have multiple sources and sinks such as that given in Figure 3.2, where nodes 2 and 3 are source nodes, and nodes 5 and 6 are sink nodes. When there are multiple sources and multiple terminals it may be necessary in some applications to create a *super source* node and connect it to the original source nodes, and a *super sink* node (or super terminal node) and connect the original terminal nodes to this node. In Figure 3.2, nodes 1 and 7, connected with dotted arcs, are the *super source* and *super sink* nodes, respectively.

4. If an arc must have its flow being less than or equal to a specified quantity known as arc *capacity,* it is said to be a *capacitated* arc. Usually, a capacitated arc has a lower bound specified on its flow, as well. A network with capacitated arcs is called a *capacitated network*.

5. The out-of-kilter algorithm assumes that the network is a *circulation network*. A circulation network is one in which there is flow conservation at each node. In other words, all nodes of the network are *intermediate* nodes. In a number of applications, it is necessary to transform the original network into an equivalent circulation network. As an illustration, for the networks of Figures 3.1 and 3.2, an additional arc is required to connect the sink with the source. This arc is referred to as a *return arc* and is illustrated in Figure 3.3. The nodes and parameters of the return arc depend upon each specific situation. Several selected applications will be discussed in some detail in Part II of this chapter.

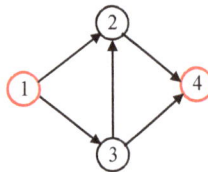

Figure 3.1. Directed network.

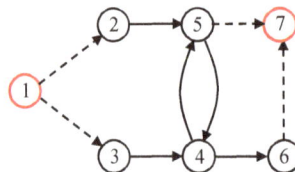

Figure 3.2. Super source and super sink.

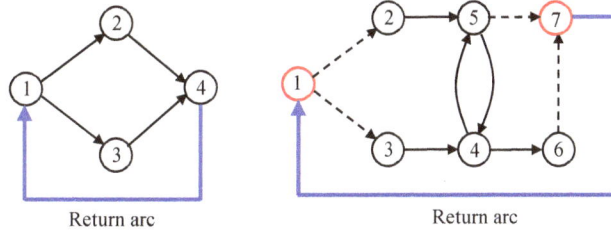

Figure 3.3. Circulation networks.

3.2 PRIMAL AND DUAL MODEL FORMULATIONS

Given a directed *circulation* network $G = (\mathbf{N}, \mathbf{S})$ with each arc having parameters (U_{ij}, L_{ij}, c_{ij}) representing an upper bound, a lower bound, and a cost per flow unit, respectively, it is desired to identify all the arc flow assignments resulting in minimum total arc flow costs. There are three types of constraints to be satisfied in the minimization of these costs. The first type of constraints is the flow-conservation condition for each node. The second type is a condition establishing that each arc flow must be between two specified values. The third type is the non-negativity of arc flows. The linear programming model is formulated in Eqs. (3-1) through (3-4).

$$\text{Minimize} \quad \sum_{(i,j) \in \mathbf{S}} c_{ij} f_{ij} \tag{3-1}$$

subject to

$$\sum_{j \in N} f_{ij} - \sum_{j \in N} f_{ji} = 0, \text{ for all } i \in \mathbf{N} \tag{3-2}$$

$$L_{ij} \leq f_{ij} \leq U_{ij}, \text{ for all } (i,j) \in \mathbf{S} \tag{3-3}$$

$$f_{ij} \geq 0, \text{ for all } (i,j) \in \mathbf{S} \tag{3-4}$$

This model can be equivalently reformulated as in Eqs. (3-5) through (3-8).

$$\text{Maximize} \quad \sum_{(i,j) \in \mathbf{S}} -c_{ij} f_{ij} \tag{3-5}$$

subject to

$$\sum_{j \in N} f_{ij} - \sum_{j \in N} f_{ji} = 0, i \in \mathbf{N} \tag{3-6}$$

$$f_{ij} \leq U_{ij}, (i,j) \in \mathbf{S} \tag{3-7}$$
$$-f_{ij} \leq -L_{ij}, (i,j) \in \mathbf{S}$$

$$f_{ij} \geq 0, (i,j) \in \mathbf{S} \tag{3-8}$$

The dual model corresponding to the primal model under consideration is formulated as in Eqs. (3-9) through (3-13).

$$\text{Minimize} \quad \sum_{(i,j) \in \mathbf{S}} (U_{ij} \alpha_{ij} - L_{ij} \delta_{ij}) \tag{3-9}$$

subject to

$$\pi_i - \pi_j + \alpha_{ij} - \delta_{ij} \geq -c_{ij}, \quad (i,j) \in \mathbf{S} \tag{3-10}$$

$$\alpha_{ij} \geq 0, (i,j) \in \mathbf{S} \tag{3-11}$$

$$\delta_{ij} \geq 0, (i,j) \in \mathbf{S} \tag{3-12}$$

$$\pi_i \text{ unrestricted}, i \in \mathbf{N} \tag{3-13}$$

127

As an example, suppose that it is desired to ship 3 units of product from the source node 1 to the sink node 4 through the arcs and nodes of the network shown in Figure 3.4, where the triplet (U_{ij}, L_{ij}, c_{ij}) on each arc (i,j) represents the upper bound, lower bound, and per-unit cost.

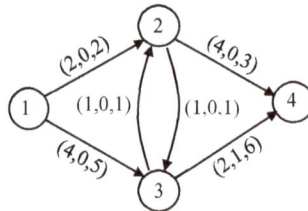

Figure 3.4. Open capacitated network.

To solve this problem as a *circulation network*, the original network must be *closed* by adding the *return arc* $(4,1)$. The amount of flow on this arc will be equal to the amount shipped from node 1 to node 4 and the cost associated with the return arc must be equal to zero. Thus, we set $L_{41} = U_{41} = 3$ and $c_{41} = 0$. The complete circulation network is shown in Figure 3.5.

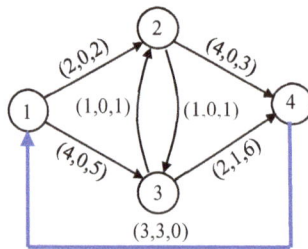

Figure 3.5. Circulation Network.

The primal and dual model formulations for this problem are shown in Tables 3.1 and 3.2, respectively.

Table 3.1. PRIMAL MODEL FORMULATION

Maximize $-2f_{12} - 5f_{13} - f_{23} - 3f_{24} - f_{32} - 6f_{34}$								
$f_{12} + f_{13}$					$- f_{41}$	$=$	0	
$- f_{12}$	$+ f_{23} + f_{24} - f_{32}$					$=$	0	Nodes
	$- f_{13} - f_{23}$		$+ f_{32} + f_{34}$			$=$	0	
		$- f_{24}$		$- f_{34} + f_{41}$		$=$	0	
f_{12}						$\leq$	2	
f_{13}						$\leq$	4	
	f_{23}					$\leq$	1	
		f_{24}				$\leq$	4	Upper bounds
			f_{32}			$\leq$	1	
				f_{34}		$\leq$	2	
					f_{41}	$\leq$	3	
				$- f_{34}$		$\leq$	-1	Lower bounds
					$- f_{41}$	$\leq$	-3	
					f_{ij}	$\geq$	0	All arcs

128

Table 3.2. DUAL MODEL FORMULATION

Minimize $2\alpha_{12} + 4\alpha_{13} + 1\alpha_{23} + 4\alpha_{24} + 1\alpha_{32} + 2\alpha_{34} + 3\alpha_{41} - \delta_{34} - 3\delta_{41}$										
π_1 - π_2			$+\ \alpha_{12}$					$\geq$	-2	
π_1	- π_3		$+\ \alpha_{13}$					$\geq$	-5	
	π_2 - π_3		$+\ \alpha_{23}$					$\geq$	-1	
	π_2	- π_4	$+\ \alpha_{24}$					$\geq$	-3	
- π_2 + π_3			$+\ \alpha_{32}$					$\geq$	-1	
	π_3	- π_4		$+\ \alpha_{34}$	- δ_{34}			$\geq$	-6	
-π_1		+ π_4		$+\ \alpha_{41}$		- δ_{41}		$\geq$	0	
						α_{ij}		$\geq$	0	$(i,j) \in$ **S**
						δ_{ij}		$\geq$	0	$(i,j) \in$ **S**
		π_1	π_2	π_3	π_4	unrestricted				

We now focus our attention on the primal and dual formulations to develop a technique that will always yield optimal solutions (when they exist) and yield optimal solutions more efficiently than the general-purpose linear programming algorithms. Conditions for optimality will be developed in Section 3.3 using well-known results from primal-dual linear programming theory.

3.3 OPTIMALITY CONDITIONS

From the complementary slackness conditions of linear programming, the following results can be established for optimal flows:

(a) $\pi_i - \pi_j + \alpha_{ij} - \delta_{ij} > -c_{ij} \Rightarrow f_{ij} = 0$

(b) $\alpha_{ij} > 0 \Rightarrow f_{ij} = U_{ij}$

(c) $\delta_{ij} > 0 \Rightarrow f_{ij} = L_{ij}$

Note that, in general, α_{ij} and δ_{ij} cannot be simultaneously positive, since $L_{ij} \neq Y_{uji}$.

A well-known result of linear programming analysis is that the feasibility of the dual problem is equivalent to the optimality of the primal model when pairs of complementary solutions (having the same value of the objective function) are considered. Defining an *adjusted cost* for arc (i,j) as

$$\bar{c}_{ij} = c_{ij} + \pi_i - \pi_j$$

the feasibility of the dual problem can be described in terms of the following relationships

$$\bar{c}_{ij} \geq \delta_{ij} - \alpha_{ij}$$

$$\alpha_{ij} \geq 0$$

$$\delta_{ij} \geq 0$$

Combining the complementary slackness conditions (b) and (c) with the above dual feasibility conditions, the following three *optimality relationships* are established:

    I. $\bar{c}_{ij} > 0$ and $f_{ij} = L_{ij}$

    II. $\bar{c}_{ij} = 0$ and $L_{ij} \leq f_{ij} \leq U_{ij}$

    III. $\bar{c}_{ij} < 0$ and $f_{ij} = U_{ij}$

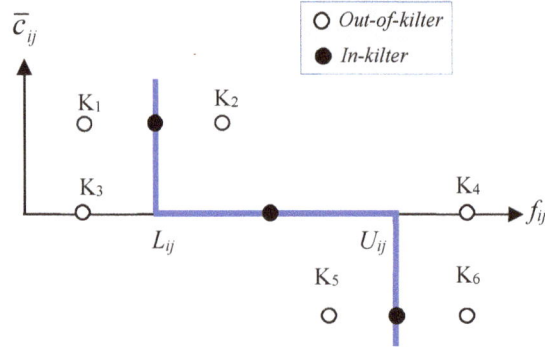

Figure 3.6. In-kilter line for arc (i,j).

Figure 3.6 illustrates the six possible states (K_1, K_2, K_3, K_4, K_5, and K_6) for arc (i,j) to be out-of-kilter. States K_1, K_3, and K_5 correspond to arcs represented by points to the left of the in-kilter line. Alternatively, states K_2, K_4, and K_6 correspond to arcs represented by points to the right of the in-kilter line. Note that appropriate *increases* of flow (*horizontal* displacements to the *right*) would cause arcs in states K_1, K_3, or K_5 to become in kilter, since in this case the points representing the arcs can be moved to coincide with the in-kilter line. Similarly, appropriate *reductions* in flow (*horizontal* displacements to the *left*) would cause arcs in states K_2, K_4, or K_6 to become in kilter, since in this case the points representing the arcs can be moved to coincide with the in-kilter line.

An optimal solution to the network circulation problem exists when: (a) the flow conservation condition is satisfied by each node; and (b) each arc satisfies one of the optimality conditions I, II, or III. Any arc satisfying one of these conditions is said to be *in-kilter*; otherwise, it is said to be *out-of-kilter*. It can be shown that the optimal values for the arc dual variables are

IV. $\alpha_{ij} = max\ \{0; -\bar{c}_{ij}\}$

V. $\delta_{ij} = max\ \{0; \bar{c}_{ij}\}$

Moreover, an arc (i,j) in state K_2 can become in-kilter if its $\bar{c}_{ij}$ value is *decreased* to *zero*, since in this case the point representing the arc would be *downward vertically* moved to coincide with the in-kilter line. Also, an arc (i,j) in state K_5 can become in-kilter if its $\bar{c}_{ij}$ value is *increased* to *zero*, since in this case the point representing the arc would be *upward vertically* moved to coincide with the in-kilter line.

Figure 3.7 summarizes the appropriate directions for horizontal displacements in (a) and (b), and vertical displacements in (c) and (d). The out-of-kilter algorithm sequentially changes arc flows and node dual-variable values until either each arc is represented by a point *on* its in-kilter line, or it is evident that a feasible solution does not exist.

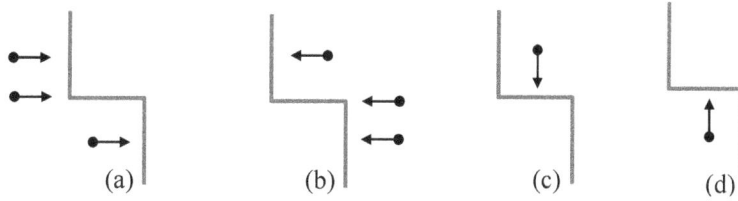

Figure 3.7. Valid directions for horizontal and vertical displacements.

3.4 CORRECTIVE ACTIONS FOR OUT-OF-KILTER ARCS

As already indicated, for specified values of arc flow and node dual-variable values, arc (i,j) can be represented as a point in the coordinate system shown in Figure 3.6. If the arc is in-kilter, the point $(f_{ij}, \bar{c}_{ij})$ falls *on* the three-segment line. If the arc is out-of-kilter, the corresponding point falls somewhere else, so there is a need to take one of two possible corrective courses of action:

1. Modification of the flow (f_{ij}).
2. Modification of the adjusted cost $(\bar{c}_{ij})$.

Flow modifications correspond to *horizontal displacements* in the graphs of Figure 3.7(a),(b). Furthermore, modifications of adjusted costs correspond to *vertical displacements* in the graphs of Figure 3.7(c),(d). Section 3.5 addresses the rules to control the size of the horizontal and vertical displacements. The rules for the horizontal displacements are equivalent to those for the labeling procedure. The rules for the vertical displacements are equivalent to those for changing the dual values.

3.5 LABELING PROCEDURE FOR FLOW CHANGES

Let us consider any arc (i,j) whose condition is out of kilter. Let this arc be represented by the point $(f_{ij}, \bar{c}_{ij})$. This point can lie to the left or right of either of the three segments of the in-kilter line. If the point lies to the left, it needs to be moved to the right, which means that the arc flow will be increased. On the other hand, if the point lies to the right, it needs to be moved to the left, which means that the arc flow will be reduced. These flow alterations can be monitored using labeling procedures, as indicated below:

1. Determine the flow change h_{ij} needed for the point to be moved to coincide with the in-kilter line. In this case, arc (i,j) would become in kilter if the flow alteration is made. Let h_{ij} be positive in the case of flow increases, and negative in case of flow reductions. If the flow must be increased, node j is labeled $[h_{ij}, i^+]$; if the flow must be reduced, node i is labeled $[-h_{ij}, j^-]$. The flow alteration in either case is only possible when an alternative path, known as a *flow-augmenting path*, can be identified to route the *flow used in labeling* along a loop containing arc (i,j). That is, we need to find a path from j to i if flow f_{ij} is going to be increased, or from i to j if flow f_{ij} is going to be reduced. Figure 3.8 shows illustrating flow-augmenting paths for increasing and reducing flow. These paths must have all their nodes labeled to allow flow to move along their arcs.

131

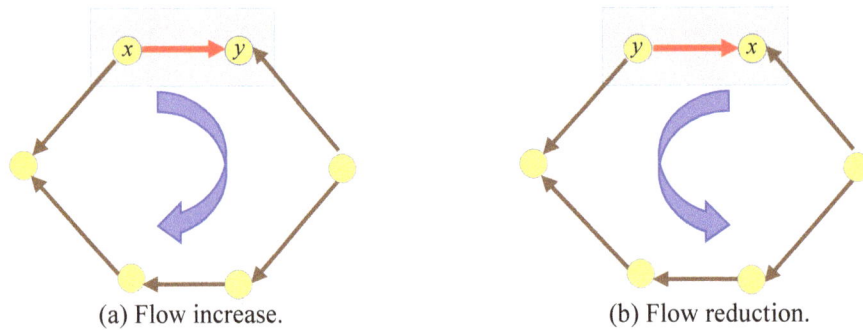

(a) Flow increase.   (b) Flow reduction.

Figure 3.8. Flow-augmenting paths.

2. To find a flow-augmenting path, we label nodes sequentially, proceeding as follows. Let x be a labeled node and y an unlabeled node. Two possibilities exist: either label node y from node x along a forward arc (x,y) or label node y from node x along reverse (y,x). From the in-kilter line of the arc chosen for labeling, we can determine the flow change needed for the arc to be in-kilter. This change will be represented by either h_{xy} or h_{yx}, for arc (x,y) or (y,x). Note that in the case of a flow increase the arc is a forward arc, and in the case of a reduction it is a reverse arc.

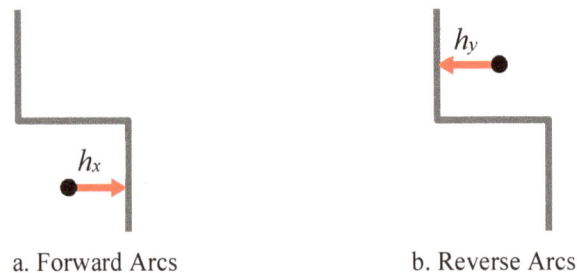

a. Forward Arcs   b. Reverse Arcs

Figure 3.9. Flow alterations on forward and reverse arcs.

3. The label for node y is given by

    i. $[q_y, x^+]$, where $q_y = min\ [q_x, h_{xy}]$, if (x,y) is forward

    ii. $[q_y, x^-]$, where $q_y = min\ [q_x, -h_{yx}]$, if (y,x) is reverse

4. Starting with either node i as node x if h_{ij} is negative, or node j as node x if h_{ij} is positive, the above labeling procedure is repeated until all arcs in the alternative path being sought are found, or until it is established that this path does not exist. If the path exists, increase (decrease) the flow on each forward (reverse) arc by q_i if $h_{ij} > 0$, or by q_j if $h_{ij} < 0$.

3.6 CHANGES IN NODE DUAL-VARIABLE VALUES

This procedure is followed when the labeling procedure fails to end in a breakthrough. In this case the *nodes* can be classified into two categories, either labeled or unlabeled nodes. Let **A**

132

be the set of labeled nodes, and $\overline{\mathbf{A}}$ be the set of unlabeled nodes. After identifying all arcs (x,y) or (y,x) such that $x \in \mathbf{A}$ and $y \in \overline{\mathbf{A}}$, it is possible to construct the following two sets of *arcs*:

$$\mathbf{B} = \{(x, y) \mid x \in \mathbf{A}, y \in \overline{\mathbf{A}}, \overline{c}_{xy} > 0, f_{xy} \leq U_{xy}\}$$
$$\overline{\mathbf{B}} = \{(y, x) \mid x \in \mathbf{A}, y \in \overline{\mathbf{A}}, \overline{c}_{yx} < 0, f_{yx} \geq L_{yx}\}$$

Let

$$\zeta_1 = \min_{(x,y) \in \mathbf{B}} \{\overline{c}_{xy}\}, \quad \zeta_2 = \min_{(y,x) \in \overline{\mathbf{B}}} \{-\overline{c}_{yx}\}$$

and $\zeta = \min \{\zeta_1, \zeta_2\}$. If $\mathbf{B} = \varnothing$, $\zeta_1 = \infty$; and if $\overline{\mathbf{B}} = \varnothing$, $\zeta_2 = \infty$. If $\zeta = \infty$, the problem has no feasible solution; otherwise, the node dual values of the *unlabeled* nodes can be changed by adding ζ to the current values.

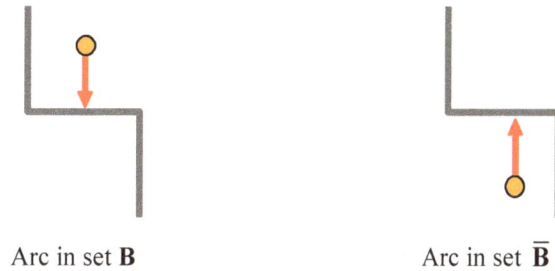

Arc in set **B**   Arc in set $\overline{\mathbf{B}}$

Figure 3.10. Typical arcs in sets **B** and $\overline{\mathbf{B}}$.

3.7 DESCRIPTION OF THE OUT OF KILTER ALGORITHM

The steps of the OKA can be briefly described as follows. The algorithm starts with any set of arc flows that satisfies flow-conservation for each node. Then it proceeds to modify arc flows and node-numbers (dual values) until all arcs are in-kilter or no solution exists.

1. Starting solution (flow feasibility).
2. Determine the state (in-kilter or out-of-kilter) of each arc. If all arcs are in-kilter, stop with an optimal solution.
3. If all arcs are not in-kilter, choose any arc out-of-kilter and determine the possible flow correction needed. In order to add or subtract the required flow quantity and preserve the flow conservation conditions, the labeling procedure is performed.
4. If the labeling procedure yields a breakthrough, adjust flows. Otherwise perform the node-number change procedure.
5. Find new node-numbers and go to Step 2. If all–finite node numbers are not possible, stop with an infeasible solution.

3.8 NUMERICAL EXAMPLE

The out-of-kilter algorithm will now be used to completely solve the minimum-cost circulation problem given in Figure 3.5. Only two steps need to be taken in order to apply the algorithm; (1) choose initial values for the dual variables, π_i, $i = 1, 2, 3, 4$; and (2) choose an initial set of arc flows satisfying flow conservation at each node. For convenience, the dual-variable node

133

values are chosen as $\pi_1 = \pi_2 = \pi_3 = \pi_4 = 0$. Furthermore, the initial arc flows are set equal to $f_{12} = 0$, $f_{13} = 2$, $f_{23} = 0$, $f_{24} = 2$, $f_{32} = 2$, $f_{34} = 0$, and $f_{41} = 2$.

A. Algorithmic Procedure

It is intuitively obvious that if near-optimal flows are chosen initially, the algorithm will terminate more rapidly. Note that the initial solution chosen is not *feasible*. The initial state for each arc is shown in Figure 3.11. As can be seen in this figure, arcs (1,2) and (2,3) are in-kilter (represented by green dots) and the remaining arcs are out-of-kilter (represented by red dots).

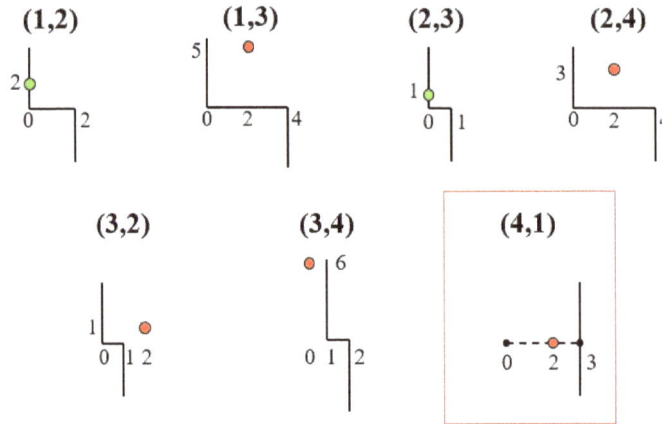

Figure 3.11. Initial arc flow states.

Iteration 1 (First Non-breakthrough)

1. Pick out-of-kilter arc (4,1), as indicated in Figure 3.11 by framing its in-kilter line.
2. The state of this arc (as indicated in Figure 3.6) is K_3. Therefore, its flow can be increased. As shown in Figure 3.11, the size of the increase is equal to 1.
3. Find path from node 1 to node 4 by using the labeling procedure.

Labeling procedure

As illustrated below, there is no breakthrough. This means that horizontal displacements are not the appropriate corrective action for the arcs. Instead, vertical displacements will be tried.

Node	Label
1	$[1,4^+]$
2	Cannot be labeled (arc is in kilter)
3	Cannot be labeled (a flow increase will drive arc further out of kilter)

The corresponding sets of nodes and arcs associated with the non-breakthrough are shown below.

$\mathbf{A} = \{1\}$ $\mathbf{B} = \{(1,2),(1,3)\}$ $\zeta_1 = \min_{\mathbf{B}} \{\bar{c}_{ij}\} = \min [2,5] = 2$

$\overline{\mathbf{A}} = \{2,3,4\}$ $\overline{\mathbf{B}} = \varnothing$ $\zeta_2 = \min_{\mathbf{B}} \{-\bar{c}_{ji}\} = \min \varnothing = \infty$ $\Big\} \quad \therefore \; \zeta = 2$

134

The new values of the dual variables (node numbers) are $\pi_1 = 0$, $\pi_2 = 2$, $\pi_3 = 2$, $\pi_4 = 2$. The new states for the arcs of the network, after adjusting the node dual-variable values, are shown in Figure 3.12.

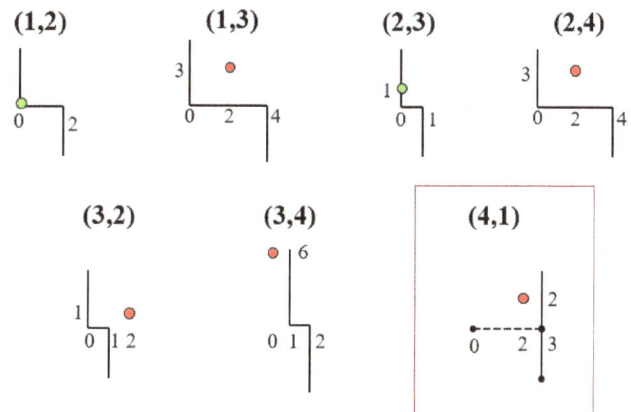

Figure 3.12. Arc flow states after iteration 1.

Iteration 2 (Breakthrough)

1. Arc (4,1) is still out-of-kilter, and will be selected again, as indicated in Figure 3.12.
2. The state of this arc has become K_1. Therefore, an increase of one unit in its flow is allowed.
3. Find path from node 1 to node 4 by using the labeling procedure.

Labeling procedure

Node	Label
1	$[1,4^+]$
2	$[1,1^+]$
3	$[1,2^-]$ Flow can be decreased to upper bound without driving arc out of kilter.
4	$[1,3^+]$

A breakthrough has occurred.

As a result of the breakthrough, it is possible to circulate one unit of flow from node 4 to itself, forwarding arcs (4,1) and (1,2), reversing (3,2), and forwarding (3,4). The new arc flows are as follows: $f_{12} = 1$, $f_{13} = 2$, $f_{23} = 0$, $f_{24} = 2$, $f_{32} = 1$, $f_{34} = 1$, and $f_{41} = 3$. The new arc states are shown in Figure 3.13. Note that arc (4,1) is now in-kilter.

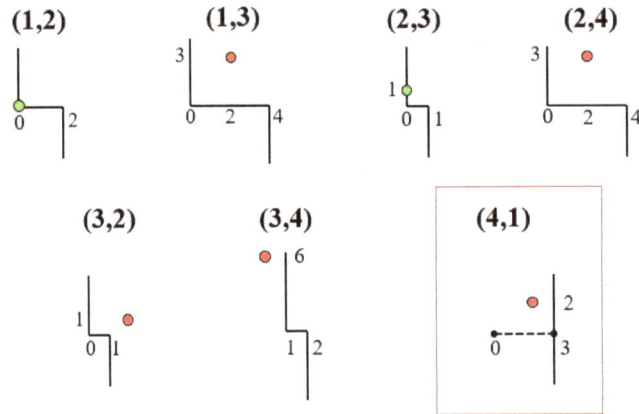

Figure 3.13. Arc flow states after iteration 2.

Iteration 3 (Breakthrough)

1. Pick out-of-kilter arc (1,3), as shown in Figure 3.13.
2. The state of this arc is K_2. Therefore, its flow can be reduced, in this case, by *up to* two units.
3. Find path from node 1 to node 3 by using the labeling procedure.

Labeling procedure

Node	Label
1	$[2,3^-]$
2	$[1,1^+]$
3	$[1, 2^-]$ Flow can be increased in arc (1,2) by one without driving arc out of kilter.

A breakthrough has occurred.

As a result of the breakthrough, one unit of flow will circulate from node 3 to itself, reversing arc (1,3), forwarding arc (1,2), and finally reversing arc (3,2).

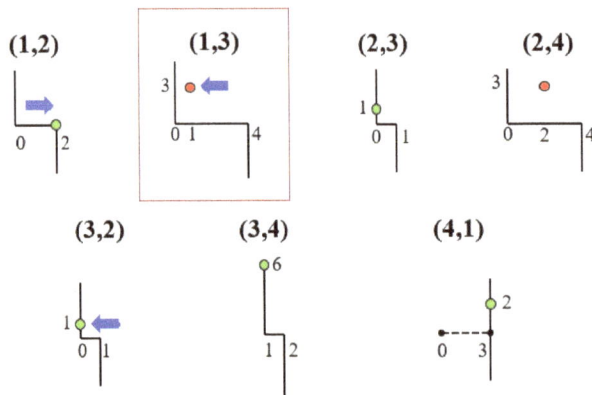

Figure 3.14. Arc states after iteration 3.

136

The new arc flows are as follows: $f_{12} = 2$, $f_{13} = 1$, $f_{23} = 0$, $f_{24} = 2$, $f_{32} = 0$, $f_{34} = 1$, and $f_{41} = 3$. The new arc states are shown in Figure 3.14. Only two arcs are still out-of-kilter.

Iteration 4 (Second Non-breakthrough)

1. Pick out-of-kilter arc (1,3) as indicated in Figure 3.14.
2. The state of this arc is K_2. Therefore, it is possible to reduce its flow. The size of the reduction is equal to one unit, as shown in Figure 3.14.
3. Find path from node 1 to node 3 by using the labeling procedure.

Labeling procedure

As illustrated below, there is no breakthrough. This means that horizontal displacements are not the appropriate corrective action for the arcs. Instead, vertical displacements will be tried.

Node	Label
1	[1,3⁻]
2	Cannot be labeled
4	Cannot be labeled

The corresponding sets of nodes and arcs associated with the non-breakthrough are shown below.

$$A = \{1\} \qquad B = \{(1,3)\} \qquad \zeta_1 = \min_B \{\overline{c}_{ij}\} = min\ \{3\} = 3$$
$$\overline{A} = \{2,3,4\} \qquad \overline{B} = \varnothing \qquad \zeta_2 = \min_{\overline{B}} \{-\overline{c}_{ji}\} = min\ \varnothing = \infty$$
$$\left. \right\} \therefore \zeta = 3$$

The new values of the dual variables (node numbers) are $\pi_1 = 0$, $\pi_2 = 5$, $\pi_3 = 5$, $\pi_4 = 5$. The new states for the arcs, after updating the node dual-variable values, are shown in Figure 3. 15.

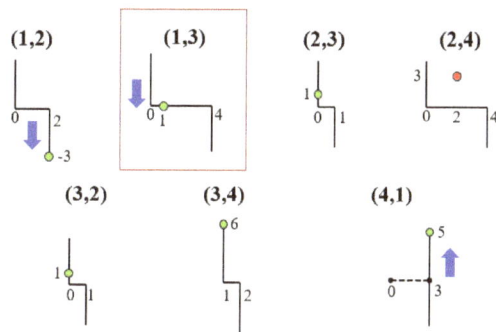

Figure 3.15. Arc flow states after iteration 4.

Iteration 5 (Third Non-breakthrough)

1. The current node-numbers (dual values for node flow conservation constraints) are $\pi_1 = 0$, $\pi_2 = \pi_3 = \pi_4 = 5$. The new states, as previously mentioned, are graphically represented in Figure 3.15.

137

2. Arc (2,4) is the only arc still out-of-kilter. After choosing this arc, it can be verified that no flow-augmenting path is possible since none of the nodes can be labeled.

3. Hence, a non-breakthrough is obtained. The corresponding sets of nodes and arcs associated with the non-breakthrough are shown below.

$$A = \{2\} \qquad B = \{(2,3),(2,4)\} \qquad \zeta_1 = \min_{B} \{\overline{c}_{ij}\} = 1$$
$$\overline{A} = \{1,3,4\} \qquad \overline{B} = \{(1,2)\} \qquad \zeta_2 = \min_{\overline{B}} \{-\overline{c}_{ji}\} = 3 \qquad \therefore \zeta = 1$$

The new values of the node dual variables along with the corresponding adjusted costs are:

$$\pi_2 = 5; \ \pi_1 = 1; \ \pi_3 = 6; \ \pi_4 = 6$$
$$\overline{c}_{12} = -2; \ \overline{c}_{13} = 0; \ \overline{c}_{23} = 0; \ \overline{c}_{32} = 2; \overline{c}_{24} = 2 \ ; \overline{c}_{34} = 6; \overline{c}_{41} = 5.$$

The new states are graphically represented in Figure 3.16.

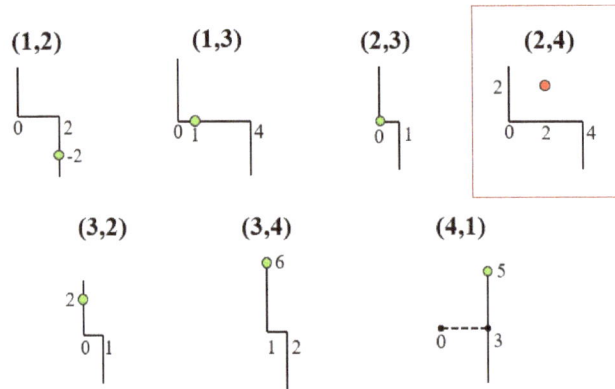

Figure 3.16. Arc states after iteration 5.

Iteration 6 (Fourth Non-breakthrough)

1. As can be seen in Figure 3.16, arc (2,4) is still out-of-kilter.

2. Again, a non-breakthrough occurs. The sets of nodes and arcs associated with the non-breakthrough are shown below.

$$A = \{1,2,3\} \qquad B = \{(2,4),(3,4)\} \qquad \zeta_1 = \min_{B} \{\overline{c}_{ij}\} = 2$$
$$\overline{A} = \{4\} \qquad \overline{B} = \varnothing \qquad \zeta_2 = \min_{\overline{B}} \{-\overline{c}_{ji}\} = \infty \qquad \therefore \zeta = 2$$

The new values of the node dual variables along with the corresponding adjusted costs are:

$$\pi_4 = 8; \ \pi_1 = 1; \ \pi_2 = 5; \ \pi_3 = 6$$
$$\overline{c}_{12} = -2; \ \overline{c}_{13} = 0; \ \overline{c}_{23} = 0; \ \overline{c}_{32} = 2; \ \overline{c}_{24} = 0; \ \overline{c}_{34} = 4; \ \overline{c}_{41} = 7.$$

The optimal values of the arc dual variables are determined using the results IV and V stated in Section3.3: $\alpha_{ij} = \max \{0; -\overline{c}_{ij}\}, \ \delta_{ij} = \max \{0; \overline{c}_{ij}\}$.

$$\alpha_{12} = 2, \ \alpha_{13} = 0, \ \alpha_{23} = 0, \ \alpha_{32} = 0, \ \alpha_{24} = 0, \ \alpha_{34} = 0, \ \alpha_{41} = 0$$
$$\delta_{12} = 0, \ \delta_{13} = 0, \ \delta_{23} = 0, \ \delta_{32} = 2, \ \delta_{24} = 0, \ \delta_{34} = 4, \ \delta_{41} = 7$$

As shown in Figure 3.17, all arcs are now in kilter. Figure 3.18 shows the optimal arc flows. The minimal total cost for this set of arc flows is equal to 21.

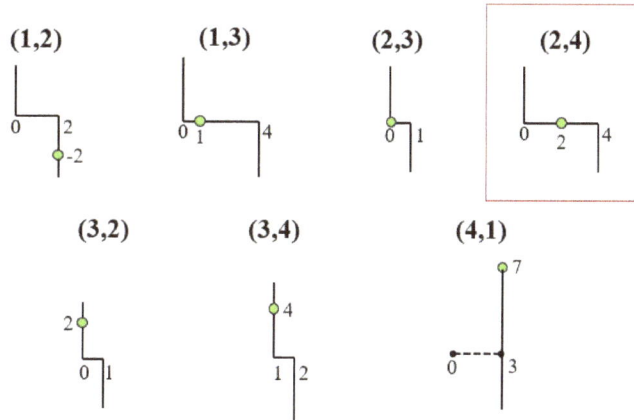

Figure 3.17. Final arc flow states.

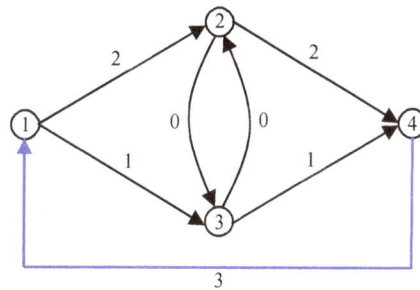

Figure 3.18. Optimal flows.

B. Solution by NOP

It is first noted that NOP starts the Out-of-Kilter algorithm with all flows and all node dual variables set equal to zero. On the other hand, the numerical example started the algorithm with the dual-variable node values $\pi_1 = \pi_2 = \pi_3 = \pi_4 = 0$ and the initial arc flows $f_{12} = 0$, $f_{13} = 2$, $f_{23} = 0$, $f_{24} = 2$, $f_{32} = 2$, $f_{34} = 0$, and $f_{41} = 2$.

It is also noted that the dual model has multiple optimal solutions, although the optimal solution to the primal model is unique. The numerical procedure and the NOP program show two of the multiple optimal solutions for the node dual variables. These solutions are listed below:

(a) From the numerical procedure, $\pi_1 = 1$, $\pi_2 = 5$, $\pi_3 = 6$ and $\pi_4 = 8$.
(b) NOP finds the values $\pi_1 = 0$; $\pi_2 = 4$; $\pi_3 = 7$ and $\pi_4 = 7$.

Figure 3.19 shows the NOP screens for this example: (1) Input data file, (2) NOP Output results. The output results include the optimal arc flows, the optimal dual arc values, and the optimal dual node values.

139

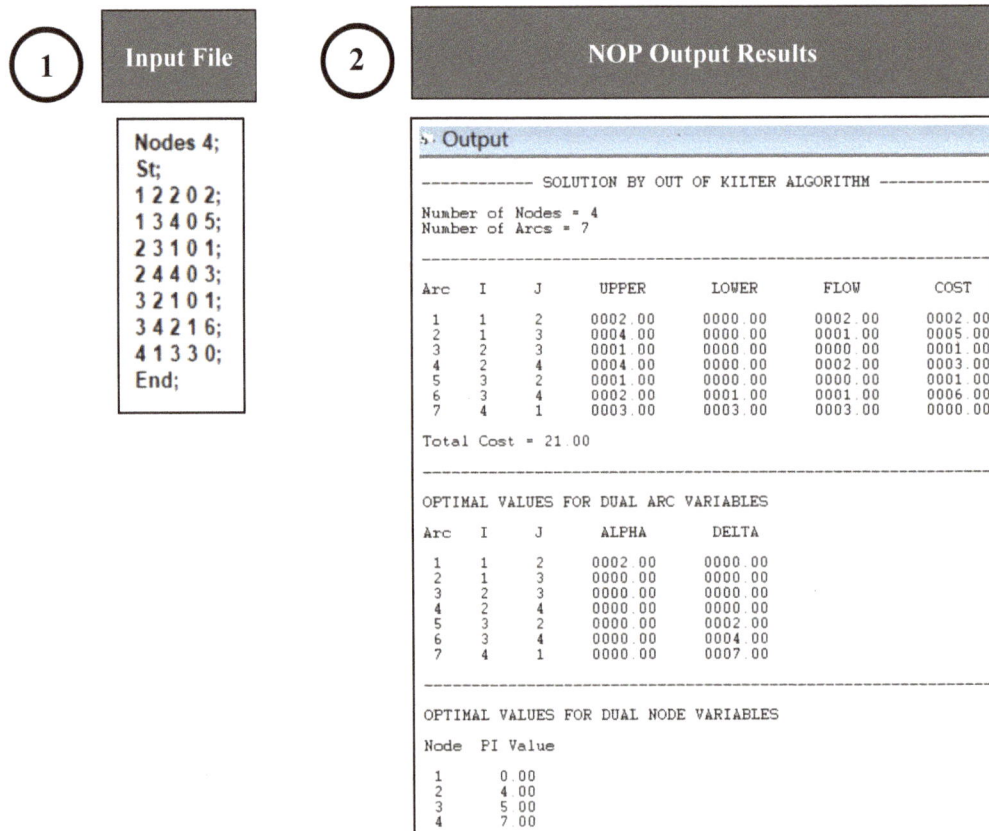

| ① | **Input File** | ② | **NOP Output Results** |

Input File:

```
Nodes 4;
St;
1 2 2 0 2;
1 3 4 0 5;
2 3 1 0 1;
2 4 4 0 3;
3 2 1 0 1;
3 4 2 1 6;
4 1 3 3 0;
End;
```

NOP Output Results:

```
⑀ Output

------------ SOLUTION BY OUT OF KILTER ALGORITHM ------------

Number of Nodes = 4
Number of Arcs = 7


Arc    I    J      UPPER        LOWER        FLOW        COST
 1     1    2     0002.00      0000.00      0002.00     0002.00
 2     1    3     0004.00      0000.00      0001.00     0005.00
 3     2    3     0001.00      0000.00      0000.00     0001.00
 4     2    4     0004.00      0000.00      0002.00     0003.00
 5     3    2     0001.00      0000.00      0000.00     0001.00
 6     3    4     0002.00      0001.00      0001.00     0006.00
 7     4    1     0003.00      0003.00      0003.00     0000.00

Total Cost = 21.00


------------

OPTIMAL VALUES FOR DUAL ARC VARIABLES

Arc    I    J      ALPHA        DELTA
 1     1    2     0002.00      0000.00
 2     1    3     0000.00      0000.00
 3     2    3     0000.00      0000.00
 4     2    4     0000.00      0000.00
 5     3    2     0000.00      0002.00
 6     3    4     0000.00      0004.00
 7     4    1     0000.00      0007.00


------------

OPTIMAL VALUES FOR DUAL NODE VARIABLES

Node   PI Value

 1        0.00
 2        4.00
 3        5.00
 4        7.00
```

Figure 3.19. Sequence of Computer Screens for Out-of-Kilter Example.

Part II
Network-Flow Optimization with the OKA - Modeling Concepts

The primary objective of Part II is to demonstrate the applicability of the out-of-kilter algorithm to a variety of well-known capacitated network-flow problems. The network structures for solving the following capacitated network-flow problems are discussed:

1. The transportation problem.
2. The assignment problem.
3. Maximum-flow problem.
4. The shortest-path tree Problem.
5. The transshipment problem.
6. Minimum-cost flow problem.

In addition to the above formulations, two cases will be discussed:

(a) The approximation of nonlinear costs in a network with flow conservation at each node and arc between specified bounds.

(b) The solution to a production planning problem using the OKA. The economic interpretation of the dual variables will be illustrated in this case.

3.9 SPECIAL CASES

To apply the out-of-kilter algorithm, two steps are essential in the problem formulation:

1. Proper representation of the problem as a circulation capacitated network-flow model. This is illustrated in Figure 3.20.
2. Initial values for the dual node variables, and initial values for all arc flows in the network, satisfying the conservation-of-flow conditions for each node.

In the circulation network of step 1 the source node and a sink or terminal node shown may in fact be super source and super sink nodes. The *return arc* connects the terminal node t to the source node s, as illustrated in Figure 3.20. The values of the parameters of the return arc, represented by the capacity-cost triplet (U_{ts}, L_{ts}, c_{ts}), depend on each specific application, as illustrated in the following formulations.

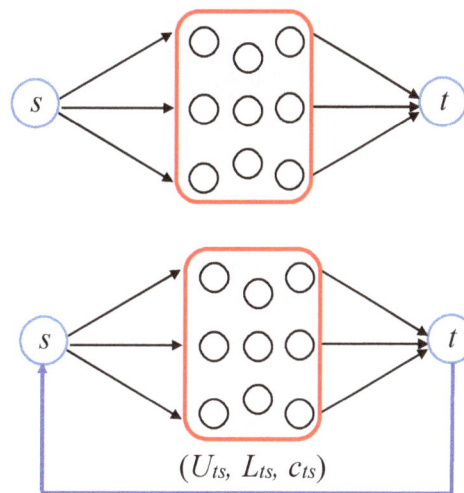

Figure 3.20. Non-Circulation and Circulation networks.

3.9.1 Transportation Problem

Suppose that there are m warehouses and n sales outlets. The available quantity of a particular product is known at each warehouse, and the corresponding demand is known for each outlet. The cost of shipping one unit of this product from each warehouse to each outlet is also known. The object is to find the shipping policy from warehouses to outlets that minimizes the total shipping cost, while at the same time meeting all demands without exceeding the existing supplies. In this formulation the following notation is used:

f_{ij} = amount shipped from warehouse i to outlet j
a_i = supply at warehouse i
b_j = demand at outlet j
c_{ij} = cost of shipping one item from warehouse i to outlet j

141

To solve a transportation problem using the out-of-kilter algorithm, the procedural guidelines outlined below can be followed:

(a) Create a super-source s, and super-sink t. Connect node s to each of m original sources, and each of n original terminals to node t. Also add the return arc (t,s).
(b) Set $(U_{ij}, L_{ij}, c_{ij}) = (\infty, 0, c_{ij})$ for each original arc (i,j).
(c) Set $(U_{si}, L_{si}, c_{si}) = (a_i, 0, 0)$ for each arc (s,i), and $(U_{jt}, L_{jt}, c_{jt}) = (\infty, b_j, 0)$ for each arc (j,t).
(d) For the return arc (t,s), set $(U_{ts}, L_{ts}, c_{ts}) = (\sum_j b_j, \sum_j b_j, 0)$. Here it is assumed that $\sum_i a_i \geq \sum_j b_j$.
(e) Set the starting conditions as $f_{ij} = 0$ for each arc $(i,,j)$, and $\pi_k = 0$ for each node k.

3.9.2 Assignment Problem

This is a special case of the transportation problem in which $m = n$, and $a_i = 1$ for each source $i = 1, 2, ..., m$; and $b_j = 1$ for each destination $j = 1, 2, ..., n$. The procedural guidelines outlined below can be followed:

(a) Create a super-source s, and super-sink t. Connect node s to each of the m original sources, and each of m original terminals to node t. Also add the return arc (t,s).
(b) Set $(U_{ij}, L_{ij}, c_{ij}) = (1, 0, c_{ij})$ for each original arc (i,j).
(c) Set $(U_{si}, L_{si}, c_{si}) = (1, 0, 0)$ for each arc (s,i), and $(U_{jt}, L_{jt}, c_{jt}) = (1, 0, 0)$ for each arc (j,t).
(d) For the return arc (t,s), set $(U_{ts}, L_{ts}, c_{ts}) = (m, m, 0)$.
(e) Set the starting conditions as $f_{ij} = 0$ for each arc $(i,,j)$, and $\pi_k = 0$ for each node k.

3.9.3 Maximum Flow Problem

It is desired to find the value of the maximum flow of a specified commodity from a source node s to a terminal node t through the capacitated arcs of a directed network. The capacity of arc (i,j) is represented by k_{ij}. The following steps can be taken to solve this problem using the out-of-kilter algorithm:

(a) Create arc (t,s) joining the terminal node t to the source node s. Set $(U_{ts}, L_{ts}, c_{ts}) = (\infty, 0, -M)$, where M is a large positive number. The maximum flow is equal to the flow on the return arc.
(b) Set $(U_{ij}, L_{ij}, c_{ij}) = (k_{ij}, 0, 0)$ for each original arc (i,j).
(c) Starting conditions: $f_{ij} = 0$ for each (i,j), and $\pi_k = 0$ for each node k. Note that every arc is in-kilter except arc (t,s).

3.9.4 Shortest Path Problem

It is desired to find the shortest path from a source node s to a terminal node t in a directed network with each arc (i,j) having length equal to d_{ij}. The *length* of each arc can be interpreted as the *cost* or *time* (or any other path-wise additive measure of effectiveness) needed to traverse the arc. We assume that d_{ij} is a real number and that negative circuits are not allowed in the shortest path. To find the shortest path, using the out-of-kilter algorithm, we proceed as follows:

(a) Create arc (t,s) and set $(U_{ts}, L_{ts}, c_{ts}) = (1, 1, 0)$.

(b) $(U_{ij}, L_{ij}, c_{ij}) = (1,0,d_{ij})$ for each original arc (i,j).
(c) Starting conditions: $f_{ij} = 0$ for all (i,j), and $\pi_k = 0$ for all k.

At the termination of the out-of-kilter algorithm, the shortest path is found by tracing it from node s to node t over all arcs with flow equal to 1.

An alternative formulation is to view the shortest-path problem as a minimum-cost flow problem. In this case, we follow the guidelines given below:

(a) Create arc (t,s) and set $(U_{ts}, L_{ts}, c_{ts}) = (1,0, -\infty)$.
(b) $(U_{ij}, L_{ij}, c_{ij}) = (1,0,d_{ij})$ for each original arc (i,j).
(c) Starting conditions: $f_{ij} = 0$ for all (i,j), and $\pi_k = 0$ for all k.

The algorithm will send exactly 1 unit of flow through the network, and the shortest path includes all arcs with flow equal to 1.

3.9.5   The Shortest-Path Tree Problem

In a shortest-path tree the branches of the tree are connected to represent the shortest path from an arbitrary node k to every other node in a selected group of r nodes in a network containing n nodes (note that $r \leq n-1$). This can be accomplished by the following steps.

(a) Create an arc directed from each node i in the selected group to node k.
(b) Set (U_{ik}, L_{ik}, c_{ik}) on each of these r arcs equal to $(1,1,0)$.
(c) For each *original* arc (i, j) in the network, set $(U_{ij}, L_{ij}, c_{ij}) = (\infty, 0, d_{ij})$. Here d_{ij} is the length of arc (i,j).
(d) Starting conditions: $f_{ij} = 0$ for all (i,j), and $\pi_k = 0$ for all k.
(e) The resulting network has r return arcs ending at k.

This method yields an *arborescent* tree in which the path from node k to each node in the tree is a shortest path. The arcs that are in the tree are those with positive flow at termination. This is not the same as the minimal-spanning-tree problem, in which the sum of the lengths of the arcs in the tree is minimized.

3.9.6 The Transshipment Problem

Consider a distribution system in which there are m warehouses and n retail outlets. Warehouse i has a_i units of a product available and retail outlet j requires b_j units of the product. As in the *transportation* problem, units of product can be shipped from any warehouse to any retail outlet.   However, it assumed that units of product can also be shipped through alternative routes from each warehouse to any other warehouse, and from each retail outlet to any other retail outlet.   The shipping costs involved must necessarily be determined from existing economic conditions and problem configuration.   This *transshipment* problem can be modeled as a transportation problem after a few modifications.

In a transportation problem, the amount *available* at each retail outlet and the amount *required* at each outlet is equal to zero; hence, no transshipment would be possible [6]. It is also conceivable that the cost of shipping 1 unit of product from source i to outlet j may not be equal to the cost of shipping 1 unit from outlet j to source i. To overcome the first problem, choose an amount θ large enough to cover any possible transshipment and add this to each supply quantity and each required amount. A logical choice for the value of θ would be $min\{\sum_{i=1}^{m} a_i, \sum_{j=1}^{n} b_j\}$.

As an illustration, consider a transshipment problem with two sources s_1 and s_2 supplying two retail outlets o_1 and o_2 with the following cost matrix.

	s_1	s_2	o_1	o_1	Available
s_1	0	4	6	2	$5+\theta$
s_2	2	0	8	3	$4+\theta$
o_1	6	7	0	3	θ
o_2	5	2	9	0	θ
Required	θ	θ	$3+\theta$	$6+\theta$	

In the above matrix, $\theta = min\{\sum_{i=1}^{m} a_i, \sum_{j=1}^{n} b_j\} = min\ \{9,9\} = 9$. Therefore, the *adjusted* supplies for the *rows* are 14, 13, 9, and 9. Similarly, the *adjusted* demands for the *columns* are 9, 9, 12, and 15. Note that the *flow* in the network is equal to 45.

The *transshipment* problem has been transformed into a classical *transportation* problem, and the out-of-kilter steps to problem solution are exactly the same as those for the transportation problem. In this particular example (a) $m = n$ and (b) $\sum_{i=1}^{m} a_i = \sum_{j=1}^{n} b_j$, but in general this will not be true. When not true, the return arc should be forced to carry all *flow* allowed, which is the minimum between the sum of *adjusted* supplies and the sum of *adjusted* demands.

3.10 NONLINEAR ARC FLOW COSTS

Certain problems with nonlinear arc costs can be solved with the out-of-kilter algorithm by using a piecewise linear approximation. As an illustration, Figure 3.21 shows a convex nonlinear cost function for arc (x,y) and the corresponding piecewise linear approximation.

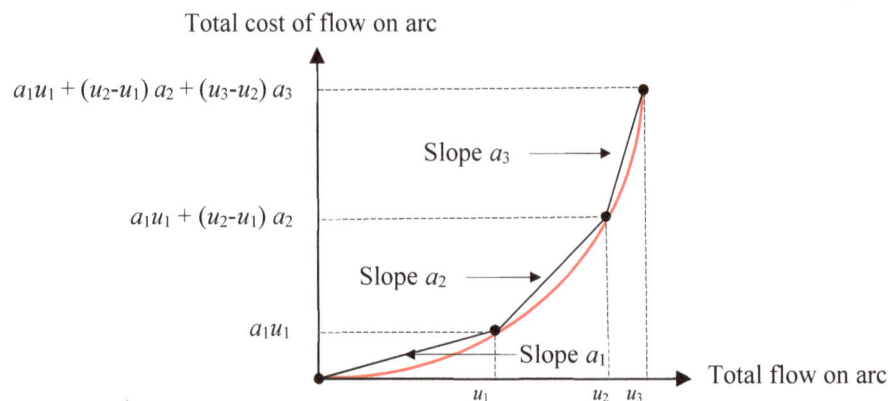

Figure 3.21. Nonlinear costs.

The *approximated* cost function consisting of three *linear segments* is represented in the network by three arcs from x to y labeled (1), (2), (3) in Figure 3.22. Because the original cost function is convex, the slopes of the linear segments satisfy the relationship $a_1 < a_2 < a_3$. Thus, as flow is increased from x to y by the algorithm, it is first assigned to arc (1). When arc (1) is at full flow, flow is assigned to arc (2). When arc (2) is full, flow is assigned to arc (3).

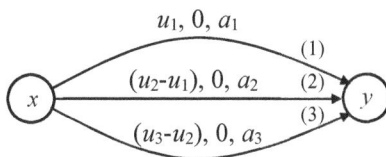

Figure 3.22. Network representation of the piecewise approximation.

If the cost vs. flow relationship is a concave function as in Figure 3.23, it cannot be approximated by a linear network. The out-of-kilter algorithm cannot therefore be used if cost relationships such as this are present. The reader is invited to see Jensen and Barnes for treatment of concave cost flows [5].

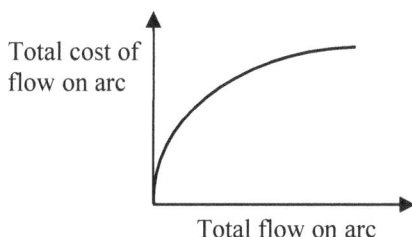

Figure 3.23. Concave nonlinear cost function.

3.11 A PRODUCTION-DISTRIBUTION PROBLEM

This section illustrates an application of the out-of-kilter algorithm in the solution of a manufacturing problem developed by Jensen and Phillips [8]. A chair manufacturing company has four plants located around the country. The manufacturing costs per chair, *excluding raw materials*, and the minimum and maximum monthly production levels for each plant are shown in Table 3.3.

Table 3.3. MANUFACTURING COSTS (EXCLUDING RAW MATERIALS) AND PRODUCTION LEVELS

Plant	Cost per Chair	Maximum Production	Minimum Production
1	$150	500	0
2	170	750	400
3	130	1000	500
4	140	250	250
		2500	1150

Twenty pounds of wood is required to make each chair. The company obtains wood from two sources. The sources can supply any amount to the company, but contracts specify that the company must buy at least 16,000 pounds (two tons) of wood from each supplier. This requirement corresponds to at least $16,000/20 = 800$ chairs. Additionally, the cost per pound of

145

obtaining wood at Source 1 is $2.00 and at Source 2 is $1.50, and shipping costs in dollars per pound of wood from the sources to the plants are given below.

		Plant			
		1	2	3	4
Wood	1	0.1	0.2	0.4	0.4
Source	2	0.4	0.3	0.2	0.2

Since each chair requires twenty pounds, the costs of making wood available at the sources correspond to $40 per chair for Source 1 and $30 for Source 2. Additionally, shipping costs in dollars per chair are given as follows.

		Plant			
		1	2	3	4
Wood	1	2	4	8	8
Source	2	8	6	4	4

If we add the cost of making wood available at each source to the shipping costs shown above, we obtain the following total costs.

		Plant			
		1	2	3	4
Wood	1	42	44	48	48
Source	2	38	36	34	34

The chairs are sold in four major cities: New York, Austin, San Francisco and Chicago. Transportation costs in dollars per chair from the plants to these cities are shown below.

		City			
		NY	A	SF	C
	1	10	10	20	0
Plant	2	30	60	70	30
	3	30	10	50	30
	4	80	20	10	40

The maximum and minimum demand and the selling price for chairs in each city are shown in Table 3.4.

Table 3.4. PRICE AND DEMAND DATA

City	Selling Price	Maximum Demand	Minimum Demand
NY	$320	2000	500
A	315	400	100
SF	320	1500	500
C	318	1500	500

Using the out-of-kilter algorithm, find the strategy resulting in maximum profit. In particular, the management of the company wants to find the most effective measures for the following components of the overall business strategy:

1. Where each plant should buy wood.
2. How many chairs should be manufactured in each plant.
3. How many chairs should be sold in each city.
4. Where each plant should ship its product.

The circulation network formulation for this problem is given in Figure 3.24. The arcs of the network are classified according to five categories:

o Purchase requirements from two wood sources
o Raw material costs from wood sources to production plants
o Production cost for four plants
o Shipping cost from plants to four cities
o Demand conditions for each city

A summary of the arc parameters is given in Table 3.5. The first column shows the arc counter for each of the 35 arcs of the network; the second column shows a description for each arc; the third and fourth columns show the initial and terminal node for each directed arc; the fifth and sixth columns show the lower and upper bounds for each arc; finally, the seventh column shows the per-unit flow cost of each arc. Note that all the costs are in dollars per chair, and all the upper and lower bounds are given as number of chairs. Additionally, revenues (selling prices) are treated as negative costs.

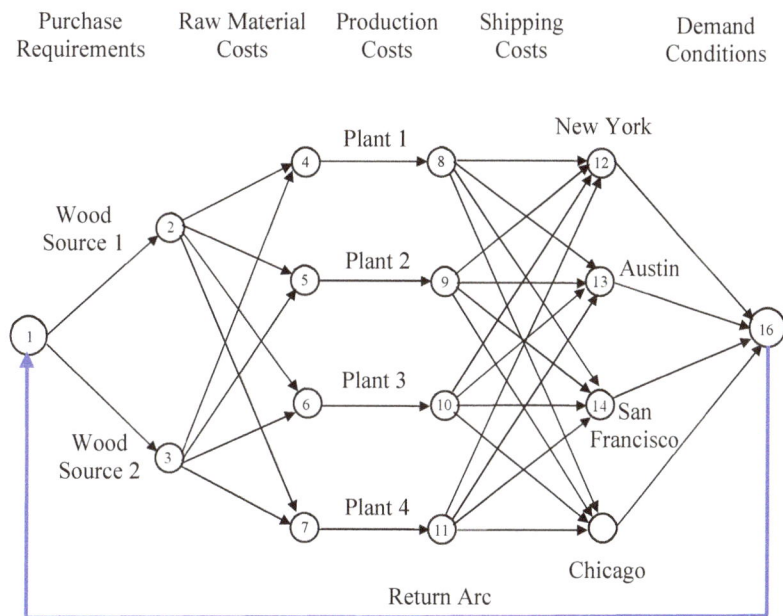

Figure 3.24. Production/distribution network.

Table 3.5. ARC PARAMETERS[1]

Arc	Arc Description	From Node	To Node	Upper Bound	Lower Bound	Cost ($)
1	Purchase	1	2	2500	800	0
2	Requirements	1	3	2500	800	0
3		2	4	2500	0	42
4		2	5	2500	0	44
5		2	6	2500	0	48
6	Raw Material	2	7	2500	0	48
7	Costs	3	4	2500	0	38
8		3	5	2500	0	36
9		3	6	2500	0	34
10		3	7	2500	0	34
11		4	8	500	0	150
12	Manufacturing	5	9	750	400	170
13	Costs	6	10	1000	500	130
14		7	11	250	250	140
15		8	12	500	0	10
16		8	13	500	0	10
17		8	14	500	0	20
18		8	15	500	0	0
19		9	12	750	0	30
20		9	13	750	0	60
21		9	14	750	0	70
22	Shipping Costs	9	15	750	0	30
23		10	12	1000	0	30
24		10	13	1000	0	10
25		10	14	1000	0	50
26		10	15	1000	0	30
27		11	12	250	0	80
28		11	13	250	0	20
29		11	14	250	0	10
30		11	15	250	0	40
31		12	16	2000	500	-320
32	Demand	13	16	400	100	-315
33	Conditions	14	16	1500	500	-320
34		15	16	1500	500	-318
35	Return Arc	16	1	5400	1600	0

The NOP code was used to obtain the optimal solution from the out-of-kilter algorithm. This solution is summarized in Table 3.6. The total *cost* is -$ 284,600, indicating a corresponding *profit* of $ 284,600.

---

[1] In this example there are no physical upper bounds on the flow along arcs 1 through 10 and arc 35. The upper bounds on theses arcs were set equal to an arbitrarily large number.

Table 3. 6. NOP SOLUTION BY THE OUT-OF-KILTER ALGORITHM

Arc (i,j)	Arc Description	i	j	Flow (Chairs)	Unit Cost ($)
1	Purchase	1	2	800	0
2	Requirements	1	3	1700	0
3		2	4	500	42
4		2	5	300	44
5		2	6	0	48
6	Raw Material	2	7	0	48
7	Acquisition	3	4	0	38
8		3	5	450	36
9		3	6	1000	34
10		3	7	250	34
11		4	8	500	150
12	Manufacturing	5	9	750	170
13	Activities	6	10	1000	130
14		7	11	250	140
15		8	12	0	10
16		8	13	0	10
17		8	14	0	20
18		8	15	500	0
19		9	12	750	30
20		9	13	0	60
21		9	14	0	70
22	Shipping	9	15	0	30
23	Strategies	10	12	350	30
24		10	13	400	10
25		10	14	250	50
26		10	15	0	30
27		11	12	0	80
28		11	13	0	20
29		11	14	250	10
30		11	15	0	40
31		12	16	1100	-320
32	Demand	13	16	400	-315
33	Conditions	14	16	500	-320
34		15	16	500	-318
35	Return Arc	16	1	2500	0

The original questions can now be answered in terms of the optimal flows obtained from the out-of-kilter algorithm.

1. Where should each plant buy its raw materials?

   a. Plant 1 should buy 10,000 lb (500 chairs) from wood source 1 and 0 lb from source 2.
   b. Plant 2 should buy 6,000 lb (300 chairs) from wood source 1 and 9,000 lb (450 chairs) from wood source 2.
   c. Plant 3 should buy 0 lb from wood source 1 and 20,000 lb (1,000 chairs) from source 2.
   d. Plant 4 should buy 0 lb from wood source 1 and 5000 lb (250 chairs) from wood source 2.

149

2. How many chairs should be made at each plant?
   a. Plant 1 should make 500 chairs.
   b. Plant 2 should make 750 chairs.
   c. Plant 3 should make 1000 chairs.
   d. Plant 4 should make 250 chairs.

3. How many chairs should be sold at each city?
   a. New York should sell 1100 chairs.
   b. Austin should sell 400 chairs.
   c. San Francisco should sell 500 chairs.
   d. Chicago should sell 500 chairs.

4. Where should each plant ship its product?
   a. Plant 1 ships 500 chairs to Chicago.
   b. Plant 2 ships 750 chairs to New York.
   c. Plant 3 ships 350 chairs to New York, 400 to Austin, and 250 to San Francisco.
   d. Plant 4 ships 250 chairs to San Francisco.

Recall that the out-of-kilter algorithm utilizes the complementary slackness conditions of linear programming, and that the dual variables α_{ij} and δ_{ij} are chosen such that

$$\alpha_{ij} = max\,\{0; -c_{ij} - \pi_i + \pi_j\}$$
$$\delta_{ij} = max\,\{0; c_{ij} + \pi_i - \pi_j\}$$

Based on the NOP output, Table 3.7 shows the list of arcs not having both dual variables equal to zero, and Table 3.8 shows the values of the node dual variables.

Table 3.7. ARC DUAL VARIABLES

Arc	i	j	α_{ij}	δ_{ij}
5	2	6	0	6
6	2	7	0	6
7	3	4	0	4
11	4	8	136	0
12	5	9	84	0
13	6	10	126	0
14	7	11	156	0
15	8	12	0	10
16	8	13	0	30
20	9	13	0	50
21	9	14	0	20
27	11	12	0	90
28	11	13	0	50
30	11	15	0	50
32	13	16	15	0
33	14	16	0	20
34	15	16	0	2

Table 3.8. NODE DUAL VARIABLES

Node i	π_i
1	258
2	250
3	258
4	292
5	294
6	292
7	292
8	578
9	548
10	548
11	588
12	578
13	558
14	598
15	578
16	258

As an illustration, let us consider arc 11 or (4,8). This arc, according to Figure 3.24, represents the manufacturing operation in plant 1. From Table 3.5, the unit cost for this arc is 150. From Table 3.7, the node dual values are $\pi_4 = 292$ and $\pi_8 = 578$. Therefore,

$$\alpha_{48} = max\ \{0; -150 - 292 + 578\} = 136$$
$$\delta_{48} = max\ \{0; 150 + 292 - 578\} = 0$$

Note that these are the values listed in Table 3.7. Furthermore, if the upper bound is increased from 500 to 501 chairs the out-of-kilter algorithm (using NOP) will find and optimal profit equal to \$284,736. The increase in profit is equal to \$284,736 - \$ 284,600 = \$136, which is exactly equal to α_{48}. Because of the economic interpretation of the dual variables they are often referred to as *pricing values*.

3.12 SUMMARY AND CONCLUSIONS

The out-of-kilter algorithm is a linear network optimization procedure that uses the dual formulation of a linear programming model that minimizes costs subject to flow conservation conditions for nodes and arc capacity (lower and upper bounds) for the arcs. The algorithm consists of a labeling procedure to modify arc flows and a second procedure to modify the node dual values of the flow conservation conditions. These procedures iteratively improve the status of at least one arc until all arcs satisfy the optimality conditions of linear programming or until it is concluded that no optimal solution exists.

The out-of-kilter algorithm is applicable to any problem that can be described as a minimal-cost circulation problem. This does not require that the physical process itself involve a circulation. Certainly, the transportation problem involves only goods shipped one way, from sources to destinations. The circulation is provided by adding a *return arc*. The flow in this arc does not necessarily have physical significance.

It is also unnecessary that flow be explicitly present in the problem. The assignment and the shortest-path problems, for instance, are *combinatorial problems*, not *flow problems*. Flow introduced by the circulation model is used as a device to find the optimal solution.

The key clue to a possible application of the out-of-kilter algorithm is the network. So many situations seem to be easily described by networks that the user with his new tool in hand, rather than having difficulty finding applications, is often driven to misapplication.

The user should recognize that the only modeling tool required to apply the out-of-kilter algorithm is the formulation of a directed network model. Each arc is described by three numbers: cost per unit flow, lower bound on flow, and upper bound on flow. If there are any side conditions relating the flows on different arcs (except conservation of flow), the out-of-kilter algorithm is ruled out as a solution tool. Often such problems can be modeled as general linear programs and solved using readily available linear programming codes or, in some cases, the problems can be transformed into equivalent problems amiable to the out-of-kilter structure.

Part III

Network-Flow Optimization with the OKA – Selected Applications[2]

3.13 BOTTLENECK ASSIGNMENT PROBLEM

As a first application, let us consider the *bottleneck assignment problem*; operational researchers usually have neat-sounding names for their problems. Let us initially describe a commonly encountered assignment problem. Given a group of workers, a group of machines to be operated by these workers, and a known efficiency for each worker operating each machine, the *standard* assignment problem is to assign exactly one worker to each machine (or workers to jobs, teachers to classes, etc.) so that the *sum* of the efficiencies resulting from the assignment is as large as possible. On the other hand, the objective of the *bottleneck* assignment problem is to maximize the *minimum* efficiency resulting from an assignment of operators to machines. This objective is often realistic as, for example, in series-type assembly lines with different workstations. In such lines, the worker with the minimum efficiency (the bottleneck) will affect the output rate of the entire assembly line.

Note that merely assigning each worker to that machine on which he is most efficient is not necessarily a solution to the bottleneck assignment problem, as several workers may be most efficient on the same machine, and the solution calls for exactly one worker being assigned to each machine. It is likely, in the standard assignment model, to assign workers to machines on which they are not most efficient, in order to maximize the *sum* of efficiencies.

Network Formulation

As shown in Part II for the standard assignment model, the *circulation* network consists of one super source, one super terminal, two columns of nodes, arcs joining nodes in the first column to nodes in the second one, and a return arc. Nodes in the first column represent workers, and those in the second column represent machines. Arcs joining nodes from the first column to nodes in the second column represent all possible assignments. Furthermore, the return arc is used to carry the optimal flow, whose value is equal to the number of assignments made. Figure 3.25 shows a network to model an assignment problem with three operators (represented by nodes 1, 2, and 3) and three machines (represented by nodes 4, 5, and 6).

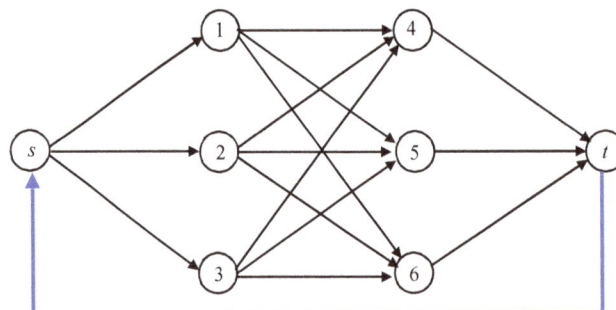

Figure 3.25. Assignment of workers to machines.

---

[2] Portions of Part III are taken from material published in Industrial Engineering with a permission of the American Institute of Industrial Engineers. The examples are due to Swanson, Woolsey, and Hillis [ll].

Solution Procedure

The bottleneck assignment problem has a network solution. In fact, this problem can be solved by sequentially solving related standard assignment problems. The procedure for doing this is described as follows:

(1) Solve the bottleneck assignment problem as if it were an ordinary assignment problem, using the out- of-kilter algorithm. We minimize the sum of inefficiencies (negatives of efficiencies) to maximize the sum of efficiencies.

(2) From the optimal solution obtained in Step (1), find the assignment with the greatest inefficiency (lowest efficiency). If there are ties, pick one of the assignments arbitrarily.

(3) Remove from the network used in Step (1) those arcs that have inefficiencies worse (larger) than or equal to the inefficiency determined in Step (2).

(4) Go to Step (1) with the network obtained in Step (3).

(5) Repeat until the OKA indicates that there is no feasible solution, as more and more arcs are removed from the original network. This means that the *last feasible* assignment is the optimal solution to the bottleneck assignment problem. The justification for this claim is that the OKA is forced to solve a sequence of assignment problems in which the *minimum* value of the objective function is continually *increased*, until an infeasible solution occurs.

An Illustration

Five workers must be assigned to five jobs in such a way that the lowest-efficiency assignment is as high as possible. The entry a_{ij} in the matrix given in Table 3.9 represents the efficiency (number of units processed per hour) if worker i is assigned to job j. The zero entries indicate that workers 1 and 3 are both unqualified for job 5.

Table 3.9. EFFICIENCY MATRIX FOR BOTTLENECK ASSIGNMENT EXAMPLE

		Job				
		1	2	3	4	5
	1	1	3	2	6	0
	2	4	2	3	8	3
Worker	3	8	4	1	5	0
	4	3	5	4	8	8
	5	2	6	9	5	2

Solving the bottleneck problem as if it were an ordinary assignment problem (12 nodes, 36 arcs), the OKA produces the following assignment:

Assignment 1

Worker 1	⇒	Job 2	with efficiency of	3
Worker 2	⇒	Job 4	with efficiency of	8
Worker 3	⇒	Job 1	with efficiency of	8
Worker 4	⇒	Job 5	with efficiency of	8
Worker 5	⇒	Job 3	with efficiency of	9

153

The sum of the efficiencies is 36; the lowest efficiency is 3. Thus, delete those entries in the matrix (and corresponding arcs in the network) with efficiencies less than or equal to 3. Now there are 12 nodes and 24 arcs comprising the network. Solving this assignment problem with the OKA results in the following solution:

Assignment 2

Worker 1	⟹	Job 4	with efficiency of	6
Worker 2	⟹	Job 1	with efficiency of	4
Worker 3	⟹	Job 2	with efficiency of	4
Worker 4	⟹	Job 5	with efficiency of	8
Worker 5	⟹	Job 3	with efficiency of	9

The sum of the efficiencies has now decreased to 31, but the smallest efficiency has increased to 4. To continue, delete those arcs of the assignment network with efficiencies less than or equal to 4. There are now 12 nodes and 21 arcs. The OKA now indicates that there is no feasible solution to this new network. Therefore, assignment 2 is the optimal solution for the bottleneck assignment problem. There is a 33⅓ % increase in output with this solution over the original assignment, even though the *sum* of the efficiencies is smaller (the bottleneck is larger).

This increase in output is related to the number of times one is forced to delete arcs from the network and resolve the problem. Thus, not only does one obtain an optimal solution to the bottleneck assignment problem, but one also reinforces the proverbial work ethic: the more you work, the greater the reward.

3.14 SCHEDULING WORKERS TO TIME-DEPENDENT TASKS

The objective of this model is to determine the minimum number of workers necessary to accomplish a fixed schedule of partially overlapping tasks. Assume that the following information is known: number of tasks, their start and end times, and an array with the (i,j)-entry indicating the time taken to go from task i to task j.

A different setting for the same problem is to determine the minimum number of machines needed to perform a schedule of tasks with known setup times, or to determine the minimum number of airplanes needed to service a flight schedule [1]. Another scenario for this problem is to determine the minimum number of buses required to service a proposed route and schedule plan [4].

Network Formulation

Initially we introduce the definition that job i is related to job j if job i can be completed and job j can be set up before it is scheduled to be started. To generate the graph of the network we start with the source and sink nodes and two columns of nodes. The first column contains one node for each task that has at least one relation. The second column contains as many nodes as there are tasks. The arcs are then determined by whether or not two given nodes (tasks) are related. Specifically, a relation is indicated by one arc directed from a node in the first column to a node in the second column. Thus, arc (i,j) exists if the task represented by node i in column 1 is related to the task represented by node j in column 2. Before describing the solution procedure, the following considerations are offered to support the formulation of this problem as a maximal-flow model:

(1) One unit of flow for a feasible relation means that the two jobs can be given to the same worker.
(2) All connected relations, each traversed by a separate unit of flow, correspond to jobs that can be done by the same worker.
(3) The maximal-flow solution forces each sequence to have as much flow as possible. Therefore, the number of workers is minimized.

Solution Procedure

(1) Create a return arc (t,s) with $(U,L,c) = (N\text{-}n,0,\text{-}c)$ where N is the total number of tasks; n is the number of unrelated tasks; and c is any positive number.
(2) For all other arcs, $(U,L,c) = (1,0,0)$.
(3) Solve the network problem for maximum flow. The solution to the maximum-flow problem will determine arc flows that result in a minimum number of workers necessary to accomplish the tasks. The arc flows resulting in maximum flow from the source node to the terminal node are found using the OKA.

An Illustration

This example was originally designed by P. Jensen (University of Texas, Department of Mechanical Engineering). Ten tasks are to be performed. Their start and end times are shown in Table 3.10. This array shows the time it takes to go from one task to any other (setup time). We are asked to find the minimum number of workers to perform the ten tasks.

Table 3.10. TASK-TIME MATRIX

Task	Start	End	1	2	3	4	5	6	7	8	9	10
1	1:00 p.m.	1:30 p.m.	-	60	10	230	180	20	15	40	120	30
2	6:00 p.m.	8:00 p.m.	10	-	40	75	40	5	30	60	5	15
3	10:30 p.m.	11:00 p.m.	70	30	-	0	70	30	20	5	120	70
4	4:00 p.m.	5:00 p.m.	0	50	75	-	20	15	10	20	60	10
5	4:00 p.m.	7:00 p.m.	200	240	150	70	-	15	5	240	90	65
6	12:00 p.m.	1:00 p.m.	20	15	20	75	120	-	30	30	15	45
7	2:00 p.m.	5:00 p.m.	15	30	60	45	30	15	-	10	5	0
8	11:00 p.m.	12:00 a.m.	20	35	15	120	75	30	45	-	20	15
9	8:10 p.m.	9:00 p.m.	25	60	15	10	100	70	80	60	-	120
10	1:45 p.m.	3:00 p.m.	60	60	30	30	120	40	50	60	70	-

From Table 3.10 we can identify the following *relationships:*

Job 1 is related to jobs 2,3,7,8,9
Job 2 is related to jobs 3,8,9
Job 3 is not related to any job
Job 4 is related to jobs 2,3,8,9

155

Job 5 is related to jobs 3,8
Job 6 is related to jobs 2,3,4,5,7,8,9,10
Job 7 is related to jobs 2,3,8,9
Job 8 is not related to any job
Job 9 is related to jobs 3,8
Job 10 is related to jobs 2,3,4,8,9

The network in Figure 3.28 shows the arcs indicating the *relationships for job* 1 and the *optimal unit arc flows* (on red arcs) resulting in a *maximal flow* equal to 7. These flows can be obtained using the OKA.

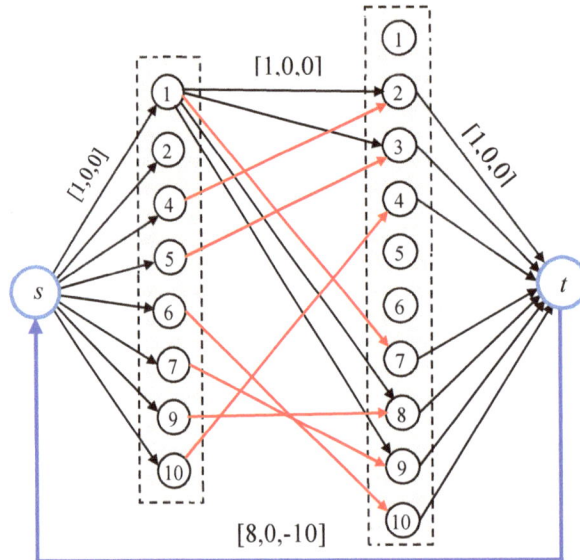

Figure 3.28. Optimal flows and relationships for job 1.

The *assignment arcs* corresponding to the following task connections have flows equal to one unit at optimality:

Task 1	⟹	Task 7
Task 4	⟹	Task 2
Task 5	⟹	Task 3
Task 6	⟹	Task 10
Task 7	⟹	Task 9
Task 9	⟹	Task 8
Task 10	⟹	Task 4

From the above results, we can identify the following three sequences of connected related tasks:

(1) Task 1- Task 7 – Task 9 – Task 8
(2) Task 6 – Task 10 – Task 4 – Task 2
(3) Task 5 – Task 3

The maximal flow solution forces each sequence of connected job relations to have as many jobs as possible. Each sequence of jobs can be done by the same worker. Therefore, the minimum number of required workers is equal to 3.

3.15 A WHOLESALE STORAGE AND MARKETING PROBLEM

Consider the following inventory problem. The owner of a wholesale outlet bought 50 table radios for $20 each. Because of changing demand, the selling price of these radios fluctuates from month to month. Because of the size of the radios and storage conditions, storage costs also fluctuate from month to month. The relevant data for the problem are given in Table 3.11.

Table 3.11. STORAGE COSTS AND SELLING PRICES

Number of radios	0-10	11-25	26-45	46 or More
Storage Costs per Radio ($)				
Month 1	1.80	1.60	1.40	1.20
Month 2	2.60	2.40	2.20	2.00
Month 3	2.40	2.20	2.00	1.80
Selling Price per Radio ($)				
Month 1	45.00	40.00	35.00	30.00
Month 2	40.00	35.00	30.00	25.00
Month 3	55.00	50.00	45.00	40.00
Month 4	50.00	45.00	40.00	35.00

The decision that needs to be made is how many radios to sell in each of the months 1, 2, 3 and 4 to maximize profit. Although the data show this particular problem to be nonlinear, this problem and those of identical structure have a fortunate characteristic in that the optimal solution is obtained when all radios are sold in the same month. The problem then reduces to finding the proper month in which to sell.

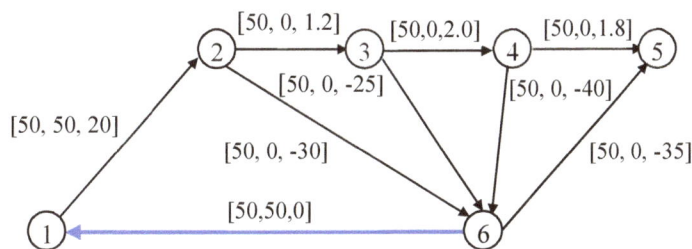

Figure 3.29. Circulation network for marketing example.

The network representation of this decision-making process is given in Figure 3.29. Arc (1,2) represents the purchase of the 50 radios. Arcs (2,3), (3,4), and (4,5) model the activities of holding radios in months 1, 2, and 3, respectively. Arcs (2,6), (3,6), (4,6), and (5,6) correspond to selling the radios in month 1, 2, 3, and 4, respectively. Note the selling price is shown as a negative cost, so that in a minimization problem the solution corresponds to maximum revenue from selling the radios. Arc (6,1) is the return arc needed for the out of kilter algorithm. The OKA can now be used to find the *longest path* from node 1 to need 6. The optimal policy in this case is to hold all radios in inventory until month 3 and then liquidate the entire stock. The profit is $(40-20-1.2-2.0)(50) = $840.

157

EXERCISES

1. Define and formulate the following problems as flow-network circulation models: (a) The transportation problem (five sources and four sinks). (b) The assignment problem (three sources and three sinks). (c) The maximum-flow problem (general framework). (d) The shortest-path problem (general framework). (e) The shortest-path tree problem (four nodes in the network with each connected to one another). (f) The transshipment problem (four sources and four sinks).

2. Consider a manufacturer faced with the problem of meeting the demand for a product over the next three periods. Two production alternatives, regular time and overtime, are available in each period. The following data are provided. The cost of carrying 1 unit in inventory from one period to the next is $1.00. The inventory level at the start of period 1 is 15 units. Formulate and solve this problem as a circulation network using the out-of-kilter algorithm.

Period	Capacity (units)		Unit Production cost		Anticipated Demand (units)
	Regular Time	Overtime	Regular Time	Overtime	
1	100	20	$14	$18	60
2	100	10	17	22	80
3	60	20	17	22	140

3. Find the maximum flow from node 1 to node 5 in the following circulation network starting with a zero flow for all arcs and zero dual values for all nodes. Each arc is labeled with a triplet, (L_{ij}, U_{ij}, c_{ij}). Repeat assuming that all costs are equal to zero. Compare the two solutions.

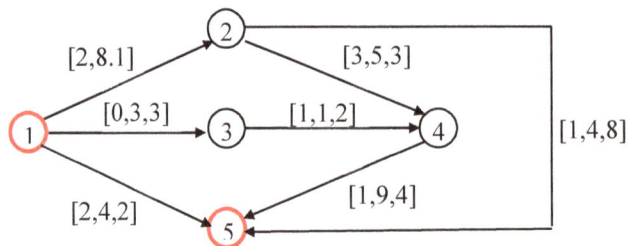

4. For the network given in Problem 3, find the arc flows resulting in total minimum cost.

5. Two manufacturing plants supply units of a product to three distribution centers. Formulate and solve the transportation problem shown using the out-of-kilter algorithm.

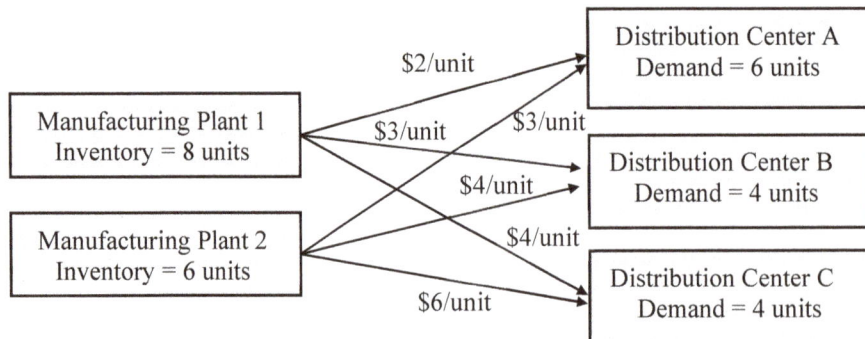

158

6. Formulate the primal and dual models for the following network. Develop the optimality conditions (in-kilter conditions). For every arc, assume a triplet of values (U_{ij}, L_{ij}, c_{ij}).

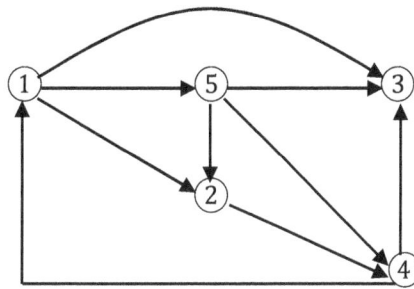

7. Consider the capacitated network shown. (a) Find the minimum-cost path from node 1 to node 11. (b) Find the arc flows necessary to supply 50 units of flow to node 11 assuming that nodes 1 and 2 are sources.

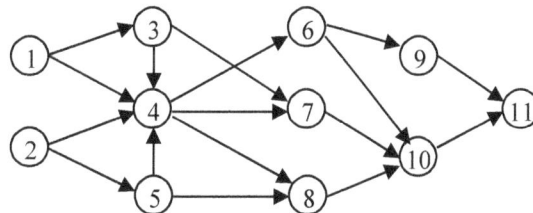

Arc	Upper Bound	Lower Bound	Cost per Unit Flow ($)
(1,3)	20	0	12
(1,4)	15	0	8
(2,4)	17	0	13
(2,5)	32	0	9
(3,4)	41	0	10
(3,7)	27	0	15
(4,6)	∞	0	0
(4,7)	16	0	7
(4,8)	19	0	9
(5,4)	23	0	11
(5,8)	29	0	5
(6,9)	31	0	7
(6,10)	14	0	10
(7,10)	19	0	8
(8,10)	28	0	14
(9,11)	34	0	12
(10,11)	29	0	11

8. Solve Problem 21, Chapter 2, using the out-of-kilter algorithm.

9. Solve Problem 42, Chapter 2, using the out-of-kilter algorithm.

10. Solve the caterer problem given in Figure 2.7(a) of Chapter 1, using the following data: $d_1 = 200$, $d_2 = 500$, $d_3 = 150$, $a_1 = a_2 = a_3 = 0.50$, $b_1 = b_2 = b_3 = 0.10$, $c_1 = c_2 = c_3 = 0.15$.

11. Consider the marketing problem described in Section 3.15, assuming 60 radios instead of 50. (a) Find the optimal solution showing intermediate results, as well as the final results on an in-

159

kilter diagram for each arc at each iteration. (b) Solve using the network optimization program (NOP).

12. Use the out-of-kilter algorithm to find the *longest* path from node 1 to node 5 in the following network:

Arc (i,j)	(1,3)	(1,4)	(2,5)	(3,2)	(3,4)	(3,5)	(4,5)
Length	12	9	12	6	15	4	23

13. For the given arc lengths, find the *longest* path from node 1 to node 4 using the out-of-kilter algorithm. As starting solution, consider the path 1-3-2-4 and node dual values as shown in the graph.

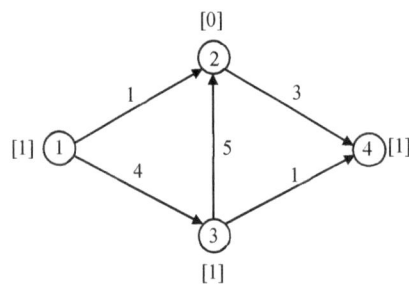

14. The following data are for a capacitated network.

(i,j)	U_{ij}	L_{ij}	c_{ij}
(1,2)	3	1	3
(1,3)	4	2	1
(2,3)	2	0	1
(2,4)	5	2	1
(3,4)	6	1	3

(a) Draw a circulation network for finding the maximal flow from node *1* to node *4* through the arcs of the network.
(b) Set all initial values equal to zero and draw the in-kilter lines for all arcs.
(c) Choose arc (2,4) and solve using the out-of-kilter algorithm.

15. Use the out-of-kilter algorithm to determine how to optimize profit in the following crude oil production and distribution problem. Crude oil is obtained from two sources. The first source has a cost of $ 70.15 and the second source $76.45. Both sources can provide three refineries with a minimum of 300,000 gallons per day. The refineries ship gasoline to three cities. Find the optimal solution using the network optimization program (NOP). Refining costs and production capacity, shipping costs, and daily demand and price data (excluding taxes) are given below.

Refinery	Refining Cost ($/gallon)	Maximum Production (gallons)	Minimum Production (gallons)
1	1.50	1,000,000	250,000
2	2.34	1,500,000	300,000
3	1.89	1,250,000	350,000

Refinery	City A ($/gallon)	City B ($/gallon)	City C ($/gallon)
1	0.34	0.25	9.50
2	0.23	0.19	0.33
3	0.40	0.29	0.20

City	Maximum Demand (gallon)	Minimum Demand (gallon)	Selling Price ($/gallon)
A	480,000	350,000	2.05
B	450,000	400,000	2.50
C	400,00	370,000	2.98

16. It is desired to find the minimum number of workers necessary to accomplish the following tasks. The array shows the time in minutes it takes to go from one task to any other (setup time).

Task	Start	End	1	2	3	4
1	1:00 pm	1:30 pm	-	60	230	10
2	6:00 pm	8:00 pm	10	-	75	40
3	4:00 pm	5:00 pm	0	50	-	75
4	10:30 pm	11:00 pm	70	30	0	-

17. There are two supply points and three demand points. The maximum availability figures for the supply points are 100 and 150. Demand point 1 requires any quantity between 20 and 50; demand point 2 requires at least 45; demand point 3 requires at most 60. Each unit from supply point costs $40. Each unit from supply point 2 costs $38. Each unit arriving at demand point 1, 2, and 3 results in a handling cost of $1. The given table provides transportation costs per unit.

2	3	1
1	2	3

(a) *Formulate* this model as a *circulation network* to minimize total costs. Assign to each arc an upper bound, a lower bound, and a per-unit cost.

(b) Solve using the out-of-kilter algorithm. Start with all-zero arc flows and all-zero node dual values. Show all relevant results on separate sets of in-kilter lines at each iteration.

18. Consider the following network:

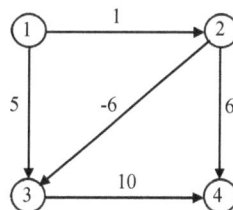

161

(a) Draw the circulation network to find the shortest path from node 1 to node 4. Include all arc parameters in your network.

(b) Assuming zero flows and zero node-numbers as the initial solution, determine which arcs are in-kilter or out-of-kilter. Show the status of each arc on a separate in-kilter line. Considering the return arc as the initial out-of-kilter arc, solve this problem using the OKA.

19. In the following network node 1 is a source with a supply equal to 1 and node 5 a terminal with a demand equal to 1.

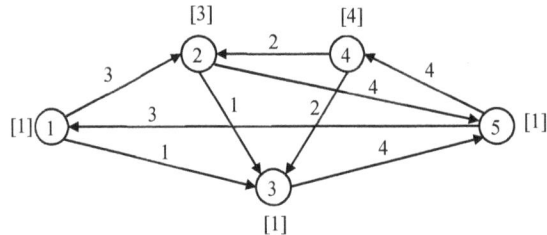

Perform one iteration of the Out-of-Kilter algorithm, using the given starting solution. Current node dual values are shown in brackets next to each node, and flow values are shown next to each arc.

Arc (i,j)	c_{ij}	U_{ij}	L_{ij}
(1,2)	2	5	2
(1,3)	1	4	1
(2,3)	2	6	1
(2,5)	1	6	2
(3,5)	3	7	1
(4,2)	1	7	2
(4,3)	1	4	2
(5,1)	-3	15	3
(5,4)	-2	5	0

20. Formulate the max-flow model for the network shown in Figure 1.16 (Chapter 1) as a circulation network and solve using the Out-of-Kilter algorithm. Verify your solution using NOP. Compare the optimal arc flows to those shown in Figure 1.17 (a).

21. In the network shown in Exercise 19 ignore lower bounds and costs. (a) Do one iteration of the Out-of-Kilter algorithm to determine the maximum flow from node 1 to node 5 starting with zero flows and zero node numbers. (b) Solve using NOP. (c) From the NOP solution determine the minimum cut.

22. Consider the marketing problem discussed in Section 3.15. It is desired to sell all the radios in one month, except the third moth. Use the OKA to find the optimal solution.

REFERENCES

[1] DURBIN, E. P., "The Out-of-Kilter Algorithm: A Primer," Rand Corporation, Santa Monica, California, December 1967.

[2] FORD, L. R., AND D. R. FULKERSON, "Maximal Flow through a Network," Canadian Journal of Mathematics (August 1956).

[3] FORD, L. R., AND D. R. FULKERSON, Flows in Networks. Princeton, N.J.: Princeton University Press, 1962.

[4] FULKERSON, D. R., "The Out-of-Kilter Method for Minimal Cost Flow Problems," Journal of Applied Mathematics, 9 (1) (March 1961).

[5] JENSEN, P. A., AND W. BARNES, Network Flow Programming. New York: John Wiley & Sons, Inc., 1979.

[6] PHILLIPS, D. T., AND P. A. JENSEN, "Network Flow Optimization with the Out-of-Kilter Algorithm, Part I-Theory," Research Memorandum 71-2, Purdue University, February 1971.

[7] PHILLIPS, D. T., AND P. A. JENSEN, "Network Flow Optimization with the Out-of-Kilter Algorithm, Part II-Applications," Research Memorandum 71-3, Purdue University, February 1971.

[8] PHILLIPS, D. T., AND P. A. JENSEN, "Network Flow Analysis: The Out-of-Kilter Algorithm," Industrial Engineering (February 1974). Portions reproduced by permission of the authors and the American Institute of Industrial Engineers.

[9] PHILLIPS, D. T., A. RAVINDRAN, AND J. J. SOLBERG, Operations Research: Principles and Practice. New York: John Wiley & Sons, Inc., 1977.

[10] SARA, J. L., "An Algorithm for Bus Scheduling Problems," Operational Research Quarterly, 21 (4) (December 1970).

[11] SWANSON, H. S., R. E. D. WOOLSEY, AND H. HILLIS, "Using the Out-of-Kilter Algorithm," Industrial Engineering (March 1974). Portions reproduced by permission of the authors and the American Institute of Industrial Engineers.

[12] SWANSON, H. S., AND R. E. D. WOOLSEY, "An Out-of-Kilter Network Tutorial," ACM SIGMAP Bulletin, January 1973.

[13] VAIDA, Mathematical Programming. Reading, Mass.: Addison-Wesley Publishing Co., Inc., 1961.

Chapter 4

PRIMAL SIMPLEX PROCEDURES FOR NETWORK OPTIMIZATION

"Perhaps it hasn't one," Alice ventured to remark.
"Tut, tut, child" said the Duchess.
"Everything's got a moral, if only you can find it."

From
Alice's Adventures in Wonderland
Lewis Carroll

This chapter is an introduction to linear network optimization procedures using the network specialization of the simplex method [1, 4]. The method is an adaptation of the bounded variable primal simplex algorithm to fit the structure of a network flow model. The chapter is divided into four in sections. Section 4.1 provides an illustration of both *fixed* and *variable* external flows for the nodes of the network. The network specialization procedure developed in this chapter assumes that only *fixed* external flows are defined for the nodes, and that the flows on all arcs have lower bounds equal to *zero. Variable* external flows and *positive* lower bounds on arc flows are eliminated by means of a transformation procedure explained in Section 4.2. The purpose of Section 4.3 is to indicate how the primal simplex algorithm is adapted to the special structure of a pure network flow cost minimization problem. An extension of this methodology is provided in Section 4.4 to include generalized network problems. Detailed numerical examples are presented in Sections 4.3 and 4.4 to illustrate all relevant steps of the network specialization of the primal simplex method for bounded variables. The presentation of the network specialization procedure shown in this chapter for the bounded-variable simplex method closely follows that given by Jensen [2, 3].

4.1 EXTERNAL FLOWS

For most problems there are nodes at which flows enter or leave the network. These are called *external* flows. A node may have three parameters: a fixed external flow, a variable or slack external flow, and a per-unit cost for the variable external flow. The demand and supply information for the nodes of a network can be represented by means of the external flow parameters. The signs + and - are used to represent whether flows are in or out, with + describing flow a node and - indicating flow out of the node.

A fixed external flow represents a fixed requirement that is forced into or out of a node. . Costs are not defined for fixed flows since these quantities are not variable and costs or revenues associated with them are irrelevant to the optimization. In addition to a fixed flow, there may be a variable external or slack flow. This is an additional quantity that represents a requirement that is subject to variation depending on the flow requirements and arc parameters of a network. The *absolute magnitude* of the variable flow indicates the maximum amount that can be inserted or withdrawn at the node in addition to the fixed flow. The sign indicates the direction of flow.

As an illustration of the meaning of both fixed and variable flows, an adaptation of a similar example developed by Jensen [1] will be considered. A particular item is manufactured in three plants in cities B, E, and H and shipped to cities A, C, D, F, and G. These cities are represented by nodes and the available shipping routes are represented by arcs in the network shown in Figure 4.1.

164

As can be seen in the figure, Cities A and D are both transshipment points and destinations for product shipments.

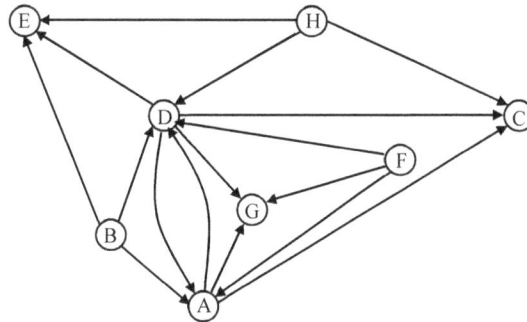

Figure 4.1. Network with routes between cities.

The information concerning the demands for the product and production costs at the plant are given below for the network of Figure 4.1.

City A: This plant is to be discontinued. The entire inventory of 410 units must be shipped or sold for scrap. The scrap value is $7 per unit.
City B: There are 120 units left over from previous shipments. These can only be sold in City B. No firm demand, but up to 220 units can be sold at $23 each.
City C: A firm demand of 200 units which must be received.
City D: All regular-time production of 300 units must be shipped. An additional 140 units can be produced using overtime at a cost of $14 per unit.
City E: Contracted demand for 300 units. No fewer than 180 units must be received. A penalty of $4 must be paid for each unit less than 300 received.
City F: Minimum demand of 210 units. Additional 90 units could be sold if available, with revenue of $19 per unit.
City G: Maximum production of 250 units with manufacturing cost of $12 per unit. There is no minimum shipment required from City G.
City H: Demand for 130 units that must be met.

Based on the definitions of both external fixed and variable (slack) flows, the parameters defining these flows for the example problem are listed in Table 4.1. Note that the fixed and variable external flows may have different signs.

Table 4.1. PARAMETERS FOR EXTERNAL FLOWS

Node	Fixed	Variable	Cost
A	410	-410	-7
B	120	-220	-23
C	-200	0	0
D	300	140	14
E	-300	120	4
F	-210	-90	-19
G	0	250	12
H	-130	0	0

165

4.2 NETWORK TRANSFORMATION PROCEDURE

Consider the minimum cost flow model formulated in Eqs. (4-1) to (4-3) using matrix notation.

$$\textit{Minimize } \mathbf{CF^*} \qquad\qquad (4\text{-}1)$$
$$\mathbf{DF^*} = \mathbf{b^*} \qquad\qquad (4\text{-}2)$$
$$\mathbf{L^*} \leq \mathbf{F^*} \leq \mathbf{U^*} \qquad\qquad (4\text{-}3)$$

The *original* network problem may have: (a) *fixed* external flow requirements, and (b) *slack* or *variable* external flow requirements for each node of the network. However, the formulation provided in Eqs. (4-1) through (4-3) has only fixed external flow requirements. To achieve this, all slack external flows are eliminated by transforming the original network to allow the inclusion of an additional node and several additional arcs. The additional node, represented by s, is referred to as the *slack node*. The additional arcs are directed from node s to each node with a positive slack flow, or from each node with a negative slack flow to node s. More specifically, a positive slack flow θ_i is represented by an arc from the slack node s to node i with an upper bound set equal to θ_i and cost equal to a specified quantity h_{si}. On the other hand, a negative slack flow θ_i is represented in the network by an arc from node i to the slack node s, with upper bound equal to $-\theta_i$ and cost h_{is}. A fixed-flow constraint is not needed for the slack node because conservation of flow is not required there, although in a pure network flows will be automatically conserved at the slack node without a constraint.

The transformation $\mathbf{F} = \mathbf{F^*} - \mathbf{L^*}$ can be used to reformulate the model given in Eqs. (4-1) – (4-3) as an equivalent min-cost model with *zero* lower bounds on arc flows:

$$\textit{Minimize } \mathbf{CF} + \mathbf{CL^*} \qquad\qquad (4\text{-}4)$$
$$\mathbf{DF} = \mathbf{b} \qquad\qquad (4\text{-}5)$$
$$\mathbf{0} \leq \mathbf{F} \leq \mathbf{U} \qquad\qquad (4\text{-}6)$$

where $\mathbf{b} = \mathbf{b^*} - \mathbf{DL^*}$ and $\mathbf{U} = \mathbf{U^*} - \mathbf{L^*}$.

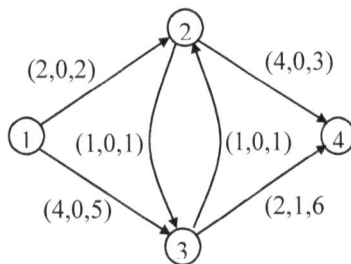

Figure 4.2 (a). Original network.   **Figure 4.2 (b).** Transformed network.

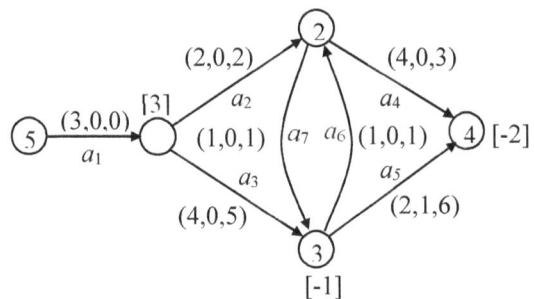

As an illustration of the type of problems that can be solved, the network shown in Figure 3.4 in Chapter 3 will be considered again. This network is shown in this chapter in Figure 4.2 (a). The purpose of this example is to identify the most cost-effective way to deliver three units of flow from node 1 to node 4. For this network the following scenario will be investigated: node 1 has a fixed flow requirement equal to 3, nodes 2 and 3 have fixed external flows equal to zero;

166

and node 4 has a fixed external flow requirement equal to -3. Also, note that all arcs, except arc (3,4), have zero lower bounds. The appropriate transformation for arc (3,4) is $f_{34} = f_{34}^* - 1$, where f_{34}^* is the flow on arc (3,4) in the original network. For this transformed flow, $f_{34} \leq 1$. This also implies that the net flow requirements for nodes 3 and 4 must be adjusted as $b_3 = 0 - 1 = -1$ and $b_4 = -3 + 1 = -2$. Note that $m = 4$; thus node $m+1=5$ will represent the slack node); and $n = 7$, after adding arc (5,1), which is not actually necessary as a device to eliminate slack flows in this example. The parameters for arc (5,1) are assumed to be (3,0,0). The transformed network is shown in Figure 4.2 (b). The formulation of the model follows:

$$\text{Minimize } 2f_{12} + 5f_{13} + f_{23} + 3f_{24} + f_{32} + 6f_{34} + 0f_{51}$$

subject to

$$
\begin{aligned}
f_{12} + f_{13} && - f_{51} &= 3 \\
-f_{12} && + f_{23} + f_{24} - f_{32} &= 0 \\
&-f_{13} - f_{23} && + f_{32} + f_{34} &= -1 \\
&&& -f_{24} && - f_{34} &= -2 \\
f_{12} &&&&&&& \leq 2 \\
f_{13} &&&&&&& \leq 4 \\
f_{23} &&&&&&& \leq 1 \\
f_{24} &&&&&&& \leq 4 \\
f_{32} &&&&&&& \leq 1 \\
f_{34} &&&&&&& \leq 1 \\
f_{51} &&&&&&& \leq 3 \\
\end{aligned}
$$
$$f_{ij} \geq 0, \text{ for all arcs } (i,j)$$

For this model we have:

$$\mathbf{F}^T = \begin{pmatrix} f_{12} & f_{13} & f_{23} & f_{24} & f_{32} & f_{34} & f_{51} \end{pmatrix}$$

$$\mathbf{D} = \begin{pmatrix} 1 & 1 & 0 & 0 & 0 & 0 & -1 \\ -1 & 0 & 1 & 1 & -1 & 0 & 0 \\ 0 & -1 & -1 & 0 & 1 & 1 & 0 \\ 0 & 0 & 0 & -1 & 0 & -1 & 0 \end{pmatrix}$$

$$\boldsymbol{\beta}^T = \begin{pmatrix} 3 & 0 & -1 & -2 \end{pmatrix}$$

$$\mathbf{C} = \begin{pmatrix} 2 & 5 & 1 & 3 & 1 & 6 & 0 \end{pmatrix}$$

$$\mathbf{U}^T = \begin{pmatrix} 2 & 4 & 1 & 4 & 1 & 1 & 3 \end{pmatrix}$$

4.3 PURE NETWORK FLOW COST MINIMIZATION PROBLEM

The formulation of the pure network cost minimization model with all lower bounds equal to zero is given in Eqs. (4-7) to (4-9) using matrix notation.

$$\text{Minimize } \mathbf{CF} \tag{4-7}$$
$$\mathbf{DF} = \mathbf{b} \tag{4-8}$$
$$0 \leq \mathbf{F} \leq \mathbf{U} \tag{4-9}$$

It is assumed that the rank of matrix $\mathbf{D}$ in Eq. (4-8) is equal to m. Matrix $\mathbf{D}$ has m rows (nodes) and n columns (arcs). It is noted that node $m+1$ represents the slack node, even in cases where no slack node is actually present in the original network model. In this case, we will connect the slack node to any node with a fixed requirement. The above model can now be solved using the bounded-variable methodology of the primal simplex approach. The simplifications resulting from the network specialization being considered will improve significantly the computational efficiency of the optimization procedure.

As an illustration of the type of problem that can be formulated, the network shown in Figure 4.3 will be considered. Each arc in this network has a length equal to the number shown next to the arc. It is desired to find the *longest* path from node 1 to node 6.

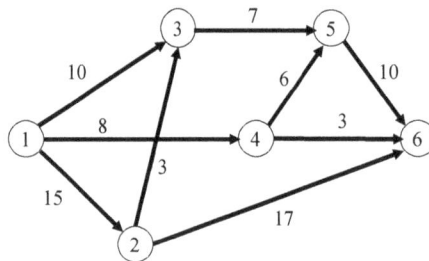

Figure 4.3. Network for longest-path application.

The primal linear programming model to find the desired longest path is formulated as follows:

maximize $15x_{12} + 10x_{13} + 8x_{14} + 3x_{23} + 17x_{26} + 7x_{35} + 6x_{45} + 3x_{46} + 10x_{56}$

subject to

$$
\begin{array}{rcl}
x_{12} + x_{13} + x_{14} & = & 1 \\
-x_{12} + x_{23} + x_{26} & = & 0 \\
-x_{13} - x_{23} + x_{35} & = & 0 \\
-x_{14} + x_{45} + x_{46} & = & 0 \\
-x_{35} - x_{45} + x_{56} & = & 0 \\
-x_{26} - x_{46} - x_{56} & = & -1
\end{array}
$$

The optimal basis for this model is shown in Figure 4.4. As can be seen, the basis can be represented as an arborescent spanning tree rooted at the source node. This result is of fundamental importance in the developments presented in this chapter.

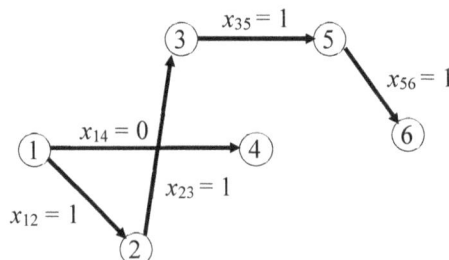

Figure 4.4. Optimal basis.

168

4.3.1 OVERVIEW OF THE NETWORK SPECIALIZATION OF THE SIMPLEX METHOD

1. Find an initial basic feasible solution. Choose a basis (set of basic arc flows) and specify the non-basic arc flows set at their lower bound and those set at their upper bounds. Construct the basis tree. Evaluate the basic flows.

2. Evaluate the dual variables.

3. Compute the reduced cost for the non-basic arcs and test the solution for optimality. If the solution is optimal, stop.

4. If the solution is not optimal select an arc to enter the basis using the "most negative value" rule or some other method. Trace the cycle that is formed with the basic tree and the entering arc.

5. Determine the arc to leave the basis and compute the new arc flows. If the entering arc and leaving arc are not the same, change the basis by recomposing the sets of basic and non-basic arc flows (identifying those set at their lower bound and those at their upper bound). Redraw the basis tree and go to step 2.

4.3.2 Basic Feasible Solution

Every valid *basis subnetwork* corresponds to a spanning tree. For computational purposes it is convenient, although not necessary, to consider a *spanning arborescent* tree rooted at the slack node as a basis tree. In order for this to be possible, sometimes it is necessary to replace some arcs by other arcs joining the same nodes but having opposite orientation. This means that the *original* arcs are *reversed.* Usually the *new* arcs are referred to as *mirror arcs.* When arc a_k is replaced by an arc joining the same two nodes but having opposite orientation, the corresponding mirror arc is represented by the notation a_{-k}. This is illustrated below in Figure 4.5. Figure 4.5 (a) shows a network consisting of 5 nodes (including the slack node). Figure 4.5 (b) shows a basis tree. Figure 4.5 (c) shows the corresponding spanning arborescent tree containing one mirror arc.

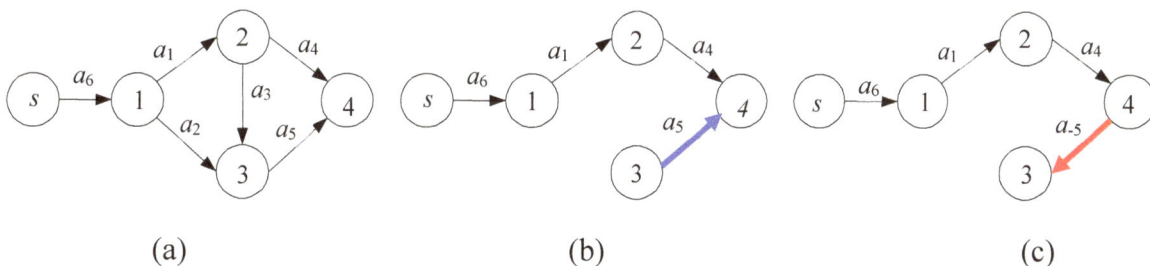

(a)   (b)   (c)

Figure 4.5. Basis subnetwork with and without mirror arcs.

If **B** is a basis matrix, that is, a matrix with columns corresponding to the constraint coefficients in matrix **D** for all basic arcs, $\mathbf{B}^{-1}$ can be found using the following rules:

1. Associate the rows and columns of the inverse matrix with basic arcs and the nodes of the network, respectively.

2. Identify the arcs in the path (reverse chain) from node i to the slack node in the basis tree. Column i of the inverse matrix has a value equal to -1 in the row corresponding to an original arc a_k or +1 in the row corresponding to a mirror arc a_{-k}; otherwise, the value is equal to zero.

It should be clear that matrix $\mathbf{B}$ has no row, and matrix $\mathbf{B^{-1}}$ no column, associated with the slack node. With the above rules, the determination of the inverse matrix becomes the simple task of identifying paths in an arborescent tree. This is now illustrated for the basis tree shown in Figure 4.5 (c). The basic arcs are considered in the order a_1, a_4, a_{-5}, a_6. Nodes are considered in the order 1, 2, 3, 4. Both the basis matrix and its inverse matrix are shown below:

$$
\begin{array}{c}
\begin{array}{cccc} a_1 & a_4 & a_{-5} & a_6 \end{array} \\
\mathbf{B} = \begin{pmatrix} +1 & 0 & 0 & -1 \\ -1 & +1 & 0 & 0 \\ 0 & 0 & +1 & 0 \\ 0 & -1 & -1 & 0 \end{pmatrix} \begin{array}{c} 1 \\ 2 \\ 3 \\ 4 \end{array}
\end{array}
\qquad
\begin{array}{c}
\begin{array}{cccc} 1 & 2 & 3 \end{array} \\
\mathbf{B^{-1}} = \begin{pmatrix} 0 & -1 & -1 & -1 \\ 0 & 0 & -1 & -1 \\ 0 & 0 & +1 & 0 \\ -1 & -1 & -1 & -1 \end{pmatrix} \begin{array}{c} a_1 \\ a_4 \\ a_{-5} \\ a_6 \end{array}
\end{array}
$$

A basic feasible solution consists of basic variables (arc flows) $\mathbf{F_B}$ and non-basic variables $\mathbf{F_N}$. For the bounded-variable simplex procedure, the non-basic flows are classified into two sets. The first set $\mathbf{F_0}$ consists of non-basic flows at their lower bounds (all equal to zero), and the second set $\mathbf{F_1}$ of those non-basic flows at their upper bounds. Let us formulate again the bounded-variable linear programming problem under consideration:

> *Minimize* $\mathbf{CF}$
> $\mathbf{DF} = \mathbf{b}$
> $0 \le \mathbf{F} \le \mathbf{U}$

Let $\mathbf{F} = [\mathbf{F_B}\ \mathbf{F_0}\ \mathbf{F_1}]^T$, where $\mathbf{F_B}$, $\mathbf{F_0}$, and $\mathbf{F_1}$ are vectors consisting of the basic flows, zero-nonbasic flows, and nonbasic flows at their upper bound, respectively, for a given basis matrix $\mathbf{B}$. The matrix of constraint coefficients $\mathbf{D}$ can be partitioned into three submatrices $\mathbf{B}$, $\mathbf{D_0}$ and $\mathbf{D_1}$ corresponding to $\mathbf{F_B}$, $\mathbf{F_0}$, and $\mathbf{F_1}$, respectively. Solving for the basic flows in terms of the non-basic flows, we obtain the following result:

$$\mathbf{F_B} = \mathbf{B^{-1}}(\mathbf{b} - \mathbf{D_0 F_0} - \mathbf{D_1 F_1}) \tag{4-10}$$

Setting $\mathbf{F_0} = \mathbf{0}$, and $\mathbf{F_1} = \mathbf{U_1}$, where $\mathbf{U_1}$ is the vector of upper bounds on arc flows in $\mathbf{F_1}$, it can be easily verified that

$$\mathbf{b'} = \mathbf{b} - \mathbf{D_1 U_1} = \mathbf{b} - \sum_{j \in \mathbf{N_1}} u_j \mathbf{d}_j \tag{4-11}$$

where $\mathbf{N_1}$ is the set of index j values such that $f_j \in \mathbf{F_1}$ and $\mathbf{d}_j$ is the column vector of $\mathbf{D}$ corresponding to the non-basic flow f_j. As we know, this column vector has an entry equal to +1 in the row corresponding to the starting node and an entry of -1 in the row corresponding to the ending node of arc a_j. Therefore, we conclude from Eq. (4-11):

170

$$b_i' = b_i - \sum_{j \in \alpha_i \cap N_1} u_j + \sum_{j \in \beta_i \cap N_1} u_j \qquad (4\text{-}12)$$

where α_i and β_i are the sets of arcs directed out of, and into node i, respectively.

The basic flows can now be computed from $\mathbf{F_B} = \mathbf{B^{-1}b'}$. It can be shown that the results from this computation are as follows, where j is the terminal node of arc a_k or a_{-k}, and $\mathbf{S}(j)$ is the set of nodes in the subtree (arborescent subtree) rooted at node j:

$$f_k = -\sum_{i \in \mathbf{S}(j)} b_i' \text{ for original (forward) arcs } a_k \qquad (4\text{-}13)$$

$$f_{-k} = \sum_{i \in \mathbf{S}(j)} b_i' \text{ for mirror arcs } a_{-k} \qquad (4\text{-}14)$$

Any row of $\mathbf{B^{-1}}$ has only 1's and 0's for a_{-k}'s and -1's and 0's for a_k's. When Eq. (4-14) is used, the flow on the original arc is obtained as $f_k = -f_{-k}$.

4.3.3   Dual Variables Computation

The model formulated in Eqs. (4-7) through (4-9) can be rewritten as a maximization problem:

$$\begin{aligned} &\textit{Maximize } \mathbf{-CF} &(4\text{-}15)\\ &\mathbf{DF = b} &(4\text{-}16)\\ &0 \le \mathbf{F} \le \mathbf{U} &(4\text{-}17) \end{aligned}$$

The dual model of the linear programming model given in Eqs. (4-15) – (4-17) is formulated below:

$$\begin{aligned} &\textit{Minimize } \prod \mathbf{b} + \gamma\, \mathbf{U} &(4\text{-}18)\\ &\prod \mathbf{D} + \gamma \ge \mathbf{-C} &(4\text{-}19)\\ &\gamma \ge 0 &(4\text{-}20) \end{aligned}$$

The optimal vector of dual variables for the net-flow constraints is given by $\prod = \mathbf{-C_B B^{-1}}$. The tree representation of the basis again simplifies this computation. It can be verified that

$$\pi_i = \sum_{k \in \mathbf{K}(i)} c_k + \sum_{k \in \mathbf{K}(i)} c_{-k} \qquad (4\text{-}21)$$

where $\mathbf{K}(i)$ is the set of arcs on the path from the slack node to node i, and it is understood that $c_{-k} = -c_k$. In other words, π_i is the cumulative cost from node s to node i along the path.

4.3.4   Optimality Condition and Selection of Entering Variable

It can be easily proved, after partitioning $\mathbf{D}$ and $\mathbf{F}$, that the total cost of any primal solution is equal to:

$$\mathbf{CF} = \mathbf{-\prod b} + (\mathbf{C_0} + \prod \mathbf{D_0})\, \mathbf{F_0} + (\mathbf{C_1} + \prod \mathbf{D_1})\, \mathbf{F_1} \qquad (4\text{-}22)$$

To determine if this is an optimal solution, we need to investigate the effect of increasing a single non-basic variable in the set $\mathbf{F_0}$ from 0 or decreasing a non-basic variable from the set $\mathbf{F_1}$ from its upper bound. The term associated with a non-basic variable f_k in the expression for Eq. (4-22) is

$$W_k = c_k + \prod \mathbf{d}_k \tag{4-23}$$

In the network model, let arc a_k = arc (i,j) be the nonbasic arc to be considered. Therefore, $\prod \mathbf{d}_k = \pi_i - \pi_j$. Thus, $w_k = c_k + \pi_i - \pi_j$ (or, equivalently, $w_{ij} = c_{ij} + \pi_i - \pi_j$) and the current basic solution will be optimal if the following condition is satisfied by each nonbasic flow:

$$\text{If } f_k = 0, \text{ then } w_k \geq 0 \text{ for each } f_k \in \mathbf{F_0} \tag{4-24}$$
$$\text{If } f_k = u_k, \text{ then } w_k \leq 0 \text{ for each } f_k \in \mathbf{F_1} \tag{4-25}$$

When there are nonbasic arcs that do not satisfy the above condition, the arc that violates the optimality condition the most can be chosen as the entering variable. The next section describes the procedure for selecting the corresponding leaving variable.

4.3.5 Selection of the Leaving Variable

The *reduction* in *basic* flows caused by the *increase* of the *nonbasic* flow f_k by *one* unit of flow, is given by

$$\mathbf{Y}_k = \mathbf{B}^{-1}\mathbf{d}_k \tag{4-26}$$

Since in Eq. (4-26) the vector $\mathbf{d_k}$ representing arc (i,j) has an entry equal to $+1$ in row i and an entry equal to -1 in row j, $\mathbf{Y}_k$ is equal to the vector resulting from subtracting the *j-th* column of $\mathbf{B}^{-1}$ from its *i-th* column. The element of $\mathbf{Y}_k$ corresponding to the ith basic arc flow, $i = 1,2,...,m$, is equal to $+1$, -1, or 0. If it is equal to $+1$, it means that this arc should be reversed to reduce its flow; it is equal to -1, it means that this arc should become a forward arc to increase its flow. It is noted at this point that if a mirror arc becomes a reverse (forward) arc, the original arc associated with it becomes a forward (reverse) arc.

All flow alterations correspond to a displacement of flow along a cycle. The flow along the cycle should be selected in such a way that flow-conservation condition remains satisfied for each node in the cycle, and the flow around the cycle should be as much as possible without causing any arc to violate its own lower (zero) and upper bound constraints on arc flow. Once the cycle of flow alterations is recognized, the variable associated with the flow moved along the cycle leaves the basis and becomes non-basic with its flow becoming an element of vector $\mathbf{F_0}$ or vector $\mathbf{F_1}$. When arc a_k is selected to enter the basis graph by increasing its flow from zero, the maximum increase in f_k is equal to

172

$$\delta = min \begin{cases} min\left\{ \dfrac{f'_{Bi}}{y_{ki}} \mid y_{ki} > 0 \right\} \\[4mm] min\left\{ -\dfrac{U_{Bi} - f'_{Bi}}{y_{ki}} \mid y_{ki} < 0 \right\} \\[4mm] U_k \end{cases} \qquad (4\text{-}27)$$

Alternatively, when the arc is selected to enter the basis graph by decreasing its flow from its upper bound, the amount of flow change is equal to

$$\delta = min \begin{cases} min\left\{ -\dfrac{f'_{Bi}}{y_{ki}} \mid y_{ki} < 0 \right\} \\[4mm] min\left\{ \dfrac{U_{Bi} - f'_{Bi}}{y_{ki}} \mid y_{ki} > 0 \right\} \\[4mm] U_k \end{cases} \qquad (4\text{-}28)$$

The arc flow which leaves the basis is the basic flow corresponding to the value of δ in Eq. (4-27) or Eq. (4-28).

4.3.6 Changing the Basis

After the variable substitution process is completed a new arborescent spanning tree rooted at the slack node is obtained. The new tree is the result of adding the entering arc and removing the leaving arc. The entire procedure followed in the network specialization of the primal simplex method for pure networks will be illustrated in Section 4.3.7 using a numerical example.

4.3.7 Example

We will consider again the numerical example corresponding to Figure 3.4. The purpose of this example is to determine the least-cost distribution of flows resulting in a supply of three units at node 1 being sent to the terminal node 4. For convenience, the network is shown again in Figure 4.6 (a) and Figure 4.6 (b) shows the corresponding transformed network, including the slack node as node 5 and arc (5,1).

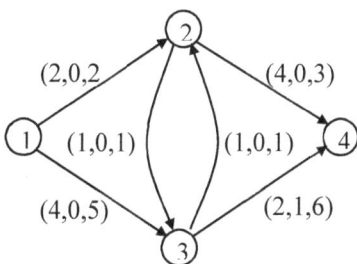

Figure 4.6 (a). Original network. **Figure 4.6 (b).** Transformed network.

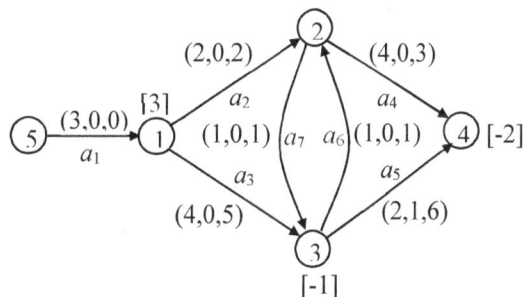

A. Algorithmic Procedure

The flow vector, constraint matrix, and net-flow vector for the network shown in Figure 4.5 (b) are given below:

$$\mathbf{F}^T = \begin{pmatrix} f_{12} & f_{13} & f_{23} & f_{24} & f_{32} & f_{34} & f_{51} \end{pmatrix}$$

$$\mathbf{D} = \begin{pmatrix} 1 & 1 & 0 & 0 & 0 & 0 & -1 \\ -1 & 0 & 1 & 1 & -1 & 0 & 0 \\ 0 & -1 & -1 & 0 & 1 & 1 & 0 \\ 0 & 0 & 0 & -1 & 0 & -1 & 0 \end{pmatrix}$$

Iteration **1**

The spanning arborescent tree rooted at the slack node and shown in Figure 4.7 will be considered as the initial basis tree.

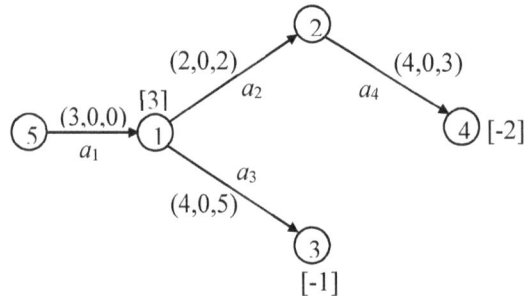

Figure 4.7. Initial basis tree.

The following matrices and vectors are defined for the initial basis shown in Figure 4.7:

$$\mathbf{B} = \begin{pmatrix} \mathbf{d}_1 & \mathbf{d}_2 & \mathbf{d}_3 & \mathbf{d}_4 \end{pmatrix} = \begin{pmatrix} -1 & 1 & 1 & 0 \\ 0 & -1 & 0 & 1 \\ 0 & 0 & -1 & 0 \\ 0 & 0 & 0 & -1 \end{pmatrix}$$

$$\mathbf{F}_\mathbf{B}^T = \begin{pmatrix} f_1 & f_2 & f_3 & f_4 \end{pmatrix}$$

$$\mathbf{D}_0 = \begin{pmatrix} \mathbf{d}_5 & \mathbf{d}_6 & \mathbf{d}_7 \end{pmatrix}$$

$$\mathbf{F}_0^T = \begin{pmatrix} f_5 & f_6 & f_7 \end{pmatrix}$$

$$\mathbf{N}_1 = \varnothing$$

Since $\mathbf{N}_1 = \varnothing$, then $\mathbf{b}' = \mathbf{b} = (3\ 0\ -1\ -2)^T$. According to the rules developed for constructing the inverse matrix, the reverse chain from node *i* to the slack node *5* is identified in the basis tree, and column *i* of the matrix is assigned a value equal to -1 in each row associated with original arcs; a value equal to +1 in each row associated with mirror arcs; and a value equal to 0, otherwise. These rules will produce the following inverse matrix:

$$\mathbf{B}^{-1} = \begin{pmatrix} -1 & -1 & -1 & -1 \\ 0 & -1 & 0 & -1 \\ 0 & 0 & -1 & 0 \\ 0 & 0 & 0 & -1 \end{pmatrix}$$

174

From the current basis tree shown in Figure 4.7 it is possible to identify the sets of nodes in each subtree rooted at each node:

$$\mathbf{S}(1) = \{1 \;\; 2 \;\; 3 \;\; 4\}$$
$$\mathbf{S}(2) = \{2 \;\; 4\}$$
$$\mathbf{S}(3) = \{3\}$$
$$\mathbf{S}(4) = \{4\}$$

According to the rule for calculating *basic* arc flows, the above information on each set $\mathbf{S}(j)$ and $\mathbf{b'}$ will produce the following results:

$$f_1 = -(3+0-1-2) = 0$$
$$f_2 = -(0-2) = 2$$
$$f_3 = -(-1) = 1$$
$$f_4 = -(-2) = 2$$

Again, from the basis tree shown in Figure 4.6, the following sets can be identified as sets of arcs in each path from the slack node to each other node:

$$\mathbf{K}(1) = \{a_1\}$$
$$\mathbf{K}(2) = \{a_1 \;\; a_2\}$$
$$\mathbf{K}(3) = \{a_1 \;\; a_3\}$$
$$\mathbf{K}(4) = \{a_1 \;\; a_2 \;\; a_4\}$$

Using the rules for calculating dual values (cumulative costs from node s to node i) we obtain the following results:

$$\pi_1 = 0$$
$$\pi_2 = 0+2 = 2$$
$$\pi_3 = 0+5 = 5$$
$$\pi_4 = 0+2+3 = 5$$

Now the optimality conditions for the three non-basic arc flows can be tested:

For arc $a_5 = (3,4)$, $w_5 = 6+5-5 = 6$ (optimal)
For arc $a_6 = (2,3)$, $w_6 = 1+2-5 = -2$ (not optimal)
For arc $a_7 = (3,2)$, $w_7 = 1+5-2 = 4$ (optimal)

Arcs a_6 is the only arc that is not optimal. The flow f_6 on arc $a_6 = (2,3)$ will be selected as the entering variable. For this arc flow we can now compute $\mathbf{Y}_6$ as shown below:

$$\mathbf{Y}_6 = \begin{pmatrix} -1 \\ -1 \\ 0 \\ 0 \end{pmatrix} - \begin{pmatrix} -1 \\ 0 \\ -1 \\ 0 \end{pmatrix} = \begin{pmatrix} 0 \\ -1 \\ +1 \\ 0 \end{pmatrix}$$

Recalling that $\mathbf{B} = (\mathbf{d}_1 \; \mathbf{d}_2 \; \mathbf{d}_3 \; \mathbf{d}_4)$, the above result indicates that an increase of *one* unit of flow on arc $a_6 = (2,3)$ will cause the following *independent* flow alterations: (a) no change in the flow on arc a_1; (b) an increase of one unit of flow on arc a_2; (c) a reduction of one unit of flow on arc a_3; and (d) no change in the flow on arc a_4. Since $\mathbf{F_B}^T = (f_1 \; f_2 \; f_3 \; f_4) = (0 \; 2 \; 1 \; 2)$ and $\mathbf{U}^T = (3 \; 2 \; 4 \; 4)$, the collective pattern of arc flow alterations along the *cycle* shown in Figure 4.8 is:

(a) Forward at most one unit of flow from 2 to 3 along arc $a_6 = (2,3)$ since the arc upper bound is equal to 1.

(b) Reverse at most one unit of flow from 3 to 1 along arc $a_3 = (1,3)$ since the current arc flow is equal to 1.

(c) Forward 0 units of flow from 1 to 2 along arc $a_2 = (1,2)$ since the current flow is at its upper bound of 2.

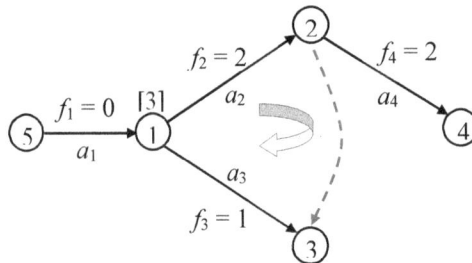

Figure 4.8. Flow alterations resulting from iteration 1.

The maximum flow around the cycle of flow alterations associated with this iteration of the simplex method is given by $\delta = min \{1\text{-}0, 1/1, 2\text{-}2\} = 0$. The values used for the evaluation of δ correspond to flow alterations on arc a_6. The first value is the amount of flow needed to reach its upper bound. The second value is the amount of flow required to cause the flow on arc a_3 to reach the value of zero (lower bound). The third value is the amount of flow needed for arc a_2 to become saturated (equal to upper bound). As a result of this analysis, arc a_2 may be selected as the leaving variable. It is now noted that this arc becomes a nonbasic arc with its flow set at its upper bound. The new arc flows can be computed as follows:

$$f_1 = 0$$
$$f_2 = 2 + (1)(0) = 2 \text{ (leaving variable)}$$
$$f_3 = 1 - 1(0) = 1$$
$$f_4 = 2$$
$$f_6 = 0 + 0 = 0 \text{ (entering variable)}$$

Iteration 2

The basis tree for this iteration is shown in Figure 4.9. Figure 4.9 (a) does not have mirror arcs. Figure 4.9 (b) has a mirror arc for the basis subnetwork to become an arborescent tree rooted at the slack node (node 5).

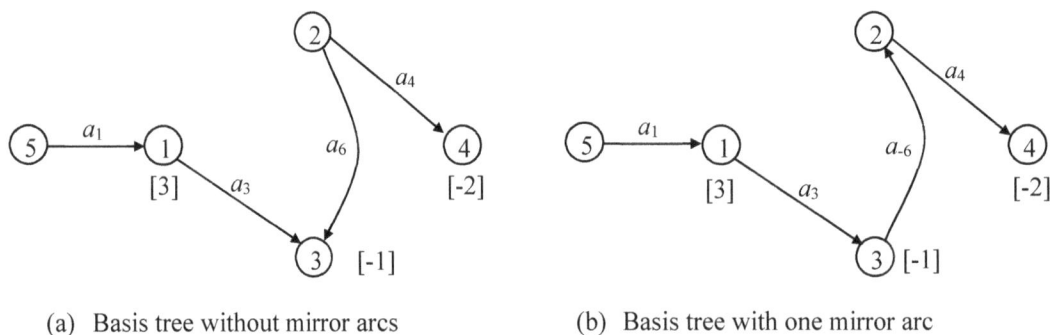

(a) Basis tree without mirror arcs   (b) Basis tree with one mirror arc

Figure 4.9. Basis tree for iteration 2.

Proceeding in the same manner as shown for iteration 1, the following results are obtained.

176

Basis and Inverse Matrices

$$\mathbf{B} = (\mathbf{d}_1 \quad \mathbf{d}_6 \quad \mathbf{d}_3 \quad \mathbf{d}_4) = \begin{pmatrix} -1 & 0 & -1 & 0 \\ 0 & 1 & 0 & 1 \\ 0 & -1 & -1 & 0 \\ 0 & 0 & 0 & -1 \end{pmatrix} \qquad \mathbf{B}^{-1} = \begin{pmatrix} -1 & -1 & -1 & -1 \\ 0 & +1 & 0 & +1 \\ 0 & -1 & -1 & 0 \\ 0 & 0 & 0 & -1 \end{pmatrix}$$

Non-basic Arcs

$$\mathbf{D}_0 = (\mathbf{d}_5 \ \mathbf{d}_7)$$
$$\mathbf{D}_1 = (\mathbf{d}_2)$$
$$\mathbf{N}_1 = \{2\}$$

Flows on Basic Arcs Using Figure 4.8 (b)

$$\mathbf{b}' = = (3 \ 0 \ -1 \ -2)^{\mathrm{T}} - 2(1 \ -1 \ 0 \ 0)^{\mathrm{T}} = (1 \ 2 \ -1 \ -2)^{\mathrm{T}}$$
$$\mathbf{S}(1) = \{1 \ 2 \ 3 \ 4\}$$
$$\mathbf{S}(2) = \{2 \ 4\}$$
$$\mathbf{S}(3) = \{2 \ 3 \ 4\}$$
$$\mathbf{S}(4) = \{4\}$$
$$f_1 = -(1+2-1-2) = 0$$
$$f_{-6} = +(2-2) = 0, \quad f_6 = -f_{-6} = 0$$
$$f_3 = -(2-1-2) = 1$$
$$f_4 = -(-2) = 2$$

Note that these values are exactly the same ones computed at the end of iteration 1. These computations, therefore, were not necessary but were done anyway to provide more insight into the ongoing illustration.

Values of Dual Variables

$$\mathbf{K}(1) = \{a_1\}$$
$$\mathbf{K}(2) = \{a_1 \ a_3 \ a_{-6}\}$$
$$\mathbf{K}(3) = \{a_1 \ a_3\}$$
$$\mathbf{K}(4) = \{a_1 \ a_3 \ a_{-6} \ a_4$$
$$\pi_1 = 0$$
$$\pi_2 = 0+5-1 = 4$$
$$\pi_3 = 0+5 = 5$$
$$\pi_4 = 0+5-1+3 = 7$$

Optimality Condition

$$w_2 = 2+0-4 = -2 \text{ (optimal)}$$
$$w_5 = 6+5-7 = +4 \text{ (optimal)}$$
$$w_7 = 1+5-4 = +2 \text{ (optimal)}$$

Optimal Flows in Original Network

Basic flows

$$f_6 = 0 \implies f^*_{23} = 0$$
$$f_3 = 1 \implies f^*_{13} = 1$$
$$f_4 = 2 \implies f^*_{24} = 2$$
$$f_1 = 0 \implies f^*_{51} = 0$$

Non-basic flows

$$f_5 = 0 \implies f^*_{34} = 0 + 1 = 1$$
$$f_7 = 0 \implies f^*_{32} = 0$$
$$f_2 = 2 \implies f^*_{12} = 2$$

Total Cost = 21

B. Solution by NOP

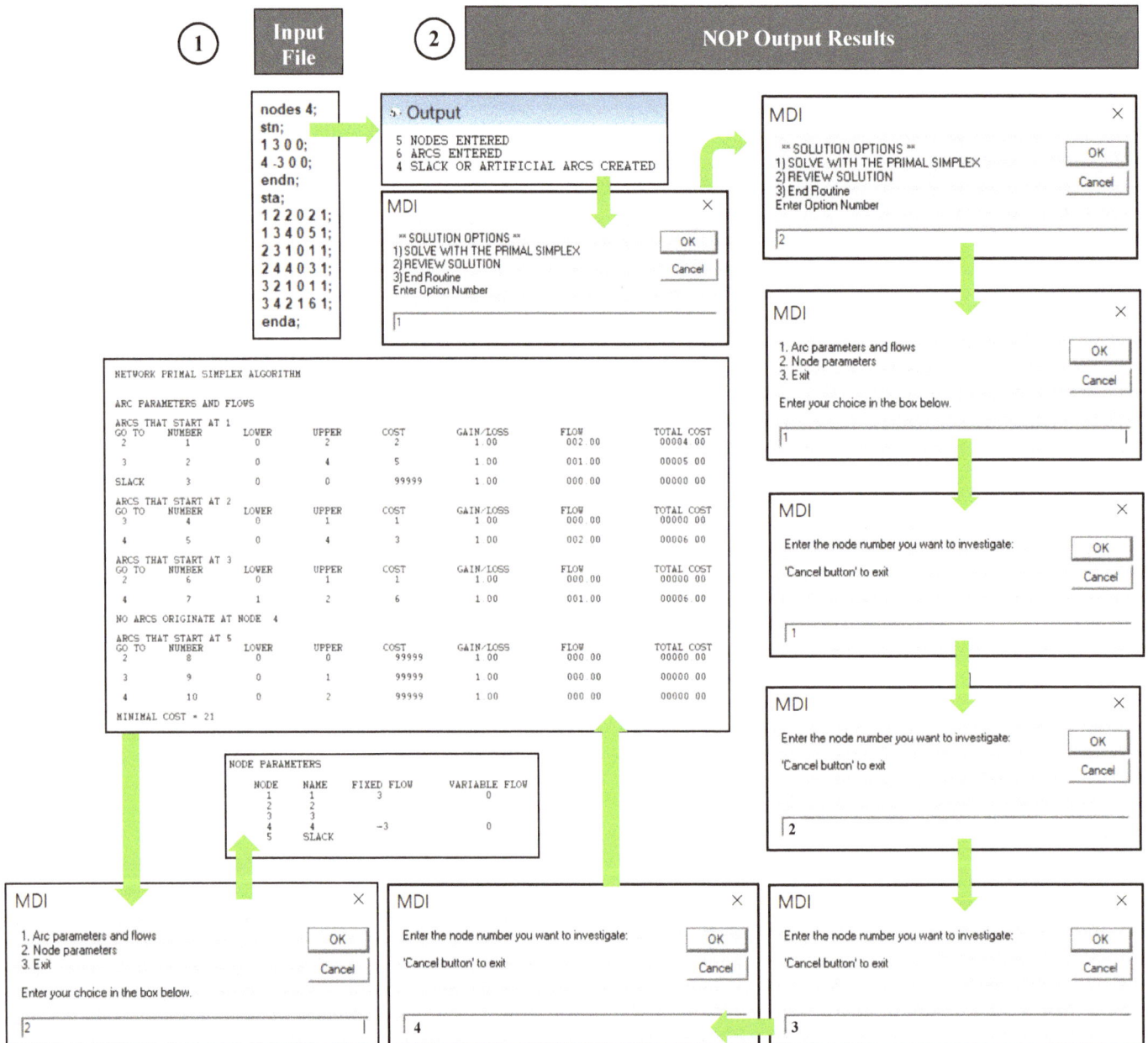

Figure 4.10. Computer Screens for Pure Network Problem.

Figure 4.10 shows the NOP screens for this example: (1) Input data file, (2) NOP Output results. Figure 4.10 also shows MDI screens requesting from the user additional input.   It is also noted that NOP generates an initial basis following a systematic procedure outlined in Section 4.5.

4.3.8   Final Remarks

Evidently, the *order* in which the basic arc flows are considered must be consistent in both the basis and inverse matrices. Furthermore, when mirror arcs are created, it is possible to consider a mirror arc flow instead of the original arc flow. As an illustration, consider

$$\mathbf{B} = \begin{bmatrix} \mathbf{d}_{-6} & \mathbf{d}_3 & \mathbf{d}_4 & \mathbf{d}_1 \end{bmatrix} = \begin{pmatrix} 0 & 1 & 0 & -1 \\ -1 & 0 & 1 & 0 \\ 1 & -1 & 0 & 0 \\ 0 & 0 & -1 & 0 \end{pmatrix}$$

The above basis matrix corresponds to a spanning arborescent tree rooted at the slack node. Its inverse matrix, shown below, can be used to determine f_{-6}, f_3, f_4, and f_1.

$$\mathbf{B}^{-1} = \begin{pmatrix} 0 & -1 & 0 & -1 \\ 0 & -1 & -1 & -1 \\ 0 & 0 & 0 & -1 \\ -1 & -1 & -1 & -1 \end{pmatrix}$$

4.4 GENERALIZED NETWORK FLOW COST MINIMIZATION PROBLEM

This section summarizes the specialization of the bounded-variable primal simplex approach to the generalized network flow cost minimization model. It can be shown that the subnetwork defining a basis consists of a tree rooted at the slack node and any number of disconnected components, with each component consisting of a tree and one additional arc. The total number of arcs in the subnetwork is always equal to m. The tree rooted at the slack node can have from zero (a degenerate tree) to m arcs (a spanning tree). In the case of a spanning tree the subnetwork defining a basis has no additional disconnected components.

For many purposes it is useful to consider a mirror arc a_{-k} as an arc joining the same nodes but having opposite orientation to that of arc a_k. The mirror arc has a gain/loss factor defined as $1/A_k$ and cost defined as $-c_k/A_k$. This is equivalent to making the transformation $f_{-k} = -f_k A_k$. In order to present the general results, two cases are considered.

Case 1: A basis with a spanning tree rooted at the slack node

Case 2: A basis with a non-spanning tree rooted at the slack node and any number of components, each containing a single cycle

Case 1: A basis with a spanning tree rooted at the slack node is illustrated in Figure 4.11.

179

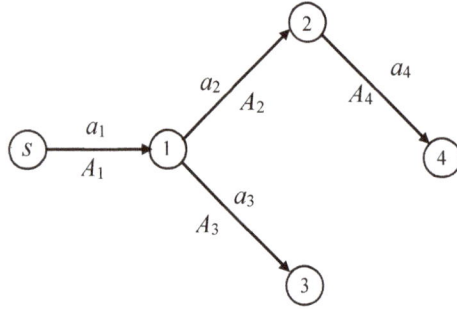

Figure 4.11. An arborescent spanning tree basis.

In this case the basis matrix is a diagonal matrix with the gain/loss factors appearing on the diagonal and all entries below the diagonal being equal to zero. In general, for a basis component describing a tree rooted at the slack node, the determinant of the basis matrix is

$$|\mathbf{B}| = \pm \prod_{k \in \mathbf{S_B}} A_k = \pm \left(A_1 A_2 \cdots A_m \right) \qquad (4\text{-}29)$$

where $\mathbf{S_B}$ is the set of basic arc flows.

Case 2: A basis with a non-spanning tree rooted at the slack node and any number of components, each containing a single cycle, is illustrated in Figure 4.12.

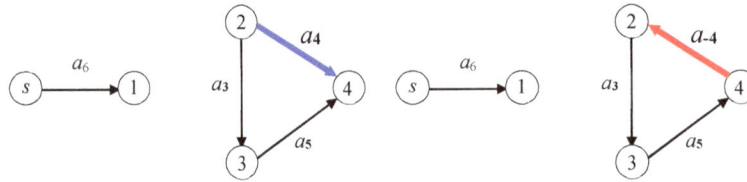

Figure 4.12. Non-spanning tree basis with one component.

The components containing cycles are sometimes referred to as 1-trees because they contain one more arc than a tree. The tree rooted at the slack node can be degenerate in the sense that it contains no branches. The total number of arcs in the collection of 1-trees and the single non-spanning tree rooted at the slack node is equal to m. In general, for any 1-tree, the determinant of the basis will be proportional to the product of the gain/loss factors for those arcs not on the cycle multiplied by the cycle gain/loss factor β minus 1. If $\mathbf{T}$ is the set of basic arcs not on the cycle, the determinant is equal to

$$|\mathbf{B}| = \pm (\beta\text{-}1) \prod_{k \in \mathbf{T}} A_k \qquad (4\text{-}30)$$

where

$$\beta = \prod_{k \in \mathbf{S_B} - \mathbf{T}} A_k \qquad (4\text{-}31)$$

4.4.1   The Inverse of the Basis Matrix

As in the pure network flow cost minimization model, the inverse of the basis matrix can be found after identifying the path to each node in the basis tree or in the component containing one tree. The entries of the inverse matrix can be determined using the following two rules:

Rule 1: This rule applies to the case where the basis consists of one tree rooted at the slack node. There is one directed path from the root to each node i. For each arc a_k in this path proceed as follows. The entry in column i associated with arc a_k is equal to the negative of the inverse of the product of the gain/loss factors on the path from the starting node of arc a_k to node i. For arcs not in the directed path to node i the entries in column i are equal to 0. Figure 4.13 illustrates how to find an entry of the inverse matrix. For those arcs not in the directed path to node i the entries in column i are equal to 0.

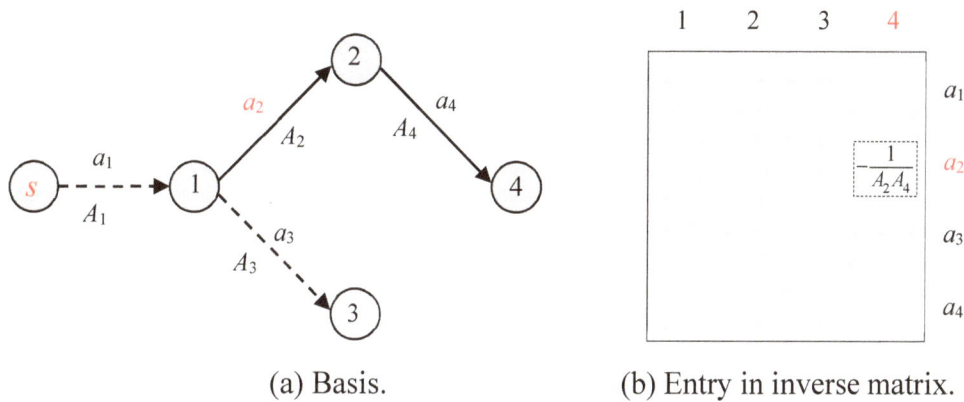

(a) Basis.   (b) Entry in inverse matrix.

Figure 4.13. Illustration for a spanning tree basis.

Rule 2: This rule applies to the case where the basis contains a component rooted at a cycle. In this case the unique directed path terminating in any specified node may contain cycles. The determination of the inverse matrix proceeds in a similar manner to that of the case considered in Rule 1. The only difference is that the elements representing arcs on the cycle are multiplied by the factor $\eta = \beta/(\beta-1)$.

4.4.2   Basic Feasible Solution

The external flow for node i adjusted for the bounded flows is equal to

$$\mathbf{b}_i' = \mathbf{b}_i - \sum_{j \in P_i} u_j + \sum_{j \in Q_i} A_j \, u_j \qquad (4\text{-}32)$$

where $P_i = \alpha_i \cap N_1$ and $Q_i = \beta_i \cap N_1$. Here α_i and β_i are the sets of arcs out of and into node i, respectively. An alternative and more efficient way to evaluate the basic flows $\mathbf{F_B} = \mathbf{B}^{-1}\mathbf{b}'$ is to first determine the columns of the inverse matrix associated with nodes having non-zero external flows, and then use Eq. (4-33) where $\mathbf{a}_i'$ is the column associated with node i.

$$\mathbf{F_B} = \sum_{i=1}^{m} b_i' \mathbf{a}_i' \qquad (4\text{-}33)$$

181

4.4.3 Dual Variables Computation

For the part of the basis consisting of a tree rooted at the slack node, the dual variables are determined by first defining $\pi_{m+1} = 0$ and then for each basic arc $a_k = (i,j)$ in the tree assign the dual values so that

$$\pi_j = (\pi_i + c_k)/A_k \qquad\qquad (4\text{-}34)$$

For components rooted at a cycle it can be shown that for node i the value of π_i is given by the cost of routing one unit around the cycle starting at node i divided by β-1. Once one of the dual variables on the cycle is determined, all others can be recursively computed using the general condition $\pi_j = (\pi_i + c_k)/A_k$.

4.4.4 Optimality Condition

For each non-basic arc $a_k = (i,j)$, let us define $w_k = c_k + \pi_i - A_k\pi_j$. If the solution is optimal, one of the following conditions must hold for each non-basic flow:

$$\text{If } f_k = 0 \text{ then } w_k \geq 0 \qquad\qquad (4\text{-}35)$$

$$\text{If } f_k = u_k \text{ then } w_k \leq 0 \qquad\qquad (4\text{-}36)$$

If at least one non-basic flow violates the optimality condition, that one corresponding to the *largest* violation can be considered as the entering variable.

4.4.5 Selection of the Leaving Variable

The reduction in the basic flows associated with the increase of the nonbasic flow f_k by one unit of flow is given by $\mathbf{Y}_k = \mathbf{B}^{-1} \mathbf{d}_k$. Since the vector $\mathbf{d}_k$ containing the coefficients of arc (i,j) has an entry equal to +1 in row i and an entry equal to $-A_k$ in row j, the above matrix multiplication is the equal to the sum of column i of the inverse with the negative of column j multiplied by the arc gain/loss factor. Once $\mathbf{Y}_k$ is determined, the procedures of the bounded-variable simplex method are used to select the arc which leaves the basis. In the case of a generalized network, the leaving variable does not necessarily correspond to an arc on the cycle formed by the arc associated with the entering variable and the arcs associated with basic flows.

4.4.6 Changing the Basis

After the entering and leaving arc flows are selected, the new basis is constructed by changing the contents of $\mathbf{F_B}$, $\mathbf{N_0}$, and $\mathbf{N_1}$. The arc associated with the leaving arc flow is deleted, the one associated with the entering arc flow is included, and, if needed, the remaining arcs are redirected to form a directed tree rooted at the slack node, and one or more components rooted at cycles. All arcs in each cycle must be oriented in the same direction.

4.4.7 Example

Consider the network given in Figure 4.14. All lower bounds on arc flows are equal to zero and all upper bounds are equal to 3. Arc a_k, $k = 1, 2, 3, 4, 5$ has per-unit costs $c_k = 2, 1, -1,$

3, ½, respectively. Arc gain/loss multipliers A_k's and node fixed external flows are as shown in the network. Assume the initial basis given in Figure 4.15. Arc (1,2) is a *non-basic* arc with its flow set at its *upper bound* of 3. Use the network specialization of the Simplex method to find optimal flows.

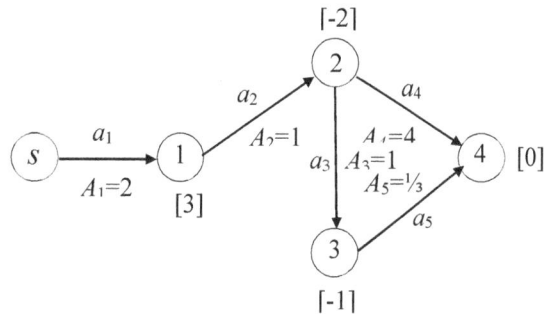

Figure 4.14. Network for example problem.

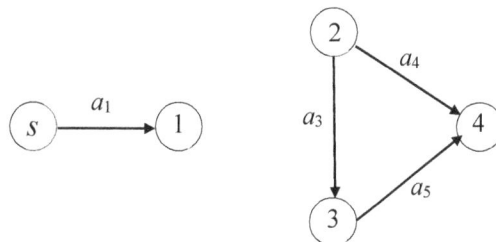

Figure 4.15. Initial basis for network of example problem.

A. Algorithmic Procedure

The following vectors and matrix are defined based on Figures 4.14 and 4.15. It is noted that A is here the vector of arc multipliers for the original network and A_{-k} is the vector of multipliers for the network with mirror arcs.

$$\mathbf{D} = \begin{pmatrix} -2 & 1 & 0 & 0 & 0 \\ 0 & -1 & 1 & 1 & 0 \\ 0 & 0 & -1 & 0 & 1 \\ 0 & 0 & 0 & -4 & -\frac{1}{3} \end{pmatrix}$$

$\mathbf{b}^T = \begin{pmatrix} 3 & -2 & -1 & 0 \end{pmatrix}$

$\mathbf{C} = \begin{pmatrix} 2 & 1 & -1 & 3 & \frac{1}{2} \end{pmatrix}$ $\mathbf{C}_{-k} = \begin{pmatrix} -1 & -1 & 1 & -\frac{3}{4} & -\frac{3}{2} \end{pmatrix}$

$A^T = \begin{pmatrix} 2 & 1 & 1 & 4 & \frac{1}{3} \end{pmatrix}$ $A_{-k}^T = \begin{pmatrix} \frac{1}{2} & 1 & 1 & \frac{1}{4} & 3 \end{pmatrix}$

The vectors of upper bounds and lower bound for the original network and for the network with mirror arcs are those shown below:

$\mathbf{U} = \begin{pmatrix} 3 & 3 & 3 & 3 & 3 \end{pmatrix}$ $\mathbf{U}_{-k} = \begin{pmatrix} 0 & 0 & 0 & 0 & 0 \end{pmatrix}$

$\mathbf{L} = \begin{pmatrix} 0 & 0 & 0 & 0 & 0 \end{pmatrix}$ $\mathbf{L}_{-k} = \begin{pmatrix} -6 & -3 & -3 & -12 & -1 \end{pmatrix}$

Iteration **1**

We proceed to calculate the inverse of the basis matrix. To do this task, mirror arc a_{-4} is created. The corresponding basis is shown in Figure 4.16.

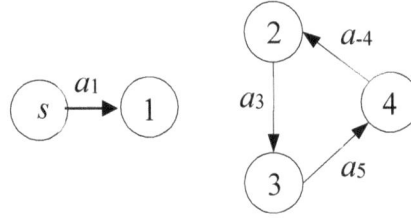

Figure 4.16. Initial basis with mirror arc.

The gain factor for the cycle of the basis is equal to $\beta = A_3 A_5 A_{-4} = \frac{1}{12}$.

Based on this result, we now calculate $\eta = \dfrac{\beta}{\beta - 1} = \dfrac{\frac{1}{12}}{\frac{1}{12} - 1} = -\dfrac{1}{11}$.

It is noted that the arcs are considered in the order a_1, a_3, a_{-4}, a_5. For this ordering of arcs, the basis matrix is shown on the right.

$$\mathbf{B} = \begin{pmatrix} -2 & 0 & 0 & 0 \\ 0 & 1 & -\frac{1}{4} & 0 \\ 0 & -1 & 0 & 1 \\ 0 & 0 & 1 & -\frac{1}{3} \end{pmatrix}$$

The inverse of the basis matrix is given as follows based on developments in Section 4.4.1:

$$B^{-1} = \begin{pmatrix} -\frac{1}{A_1} & 0 & 0 & 0 \\ 0 & -\frac{\eta}{A_{-4} A_5 A_3} & -\frac{\eta}{A_3} & -\frac{\eta}{A_5 A_3} \\ 0 & -\frac{\eta}{A_{-4}} & -\frac{\eta}{A_{-4} A_3} & -\frac{n}{A_5 A_3 A_{-4}} \\ 0 & -\frac{\eta}{A_{-4} A_5} & -\frac{\eta}{A_3 A_{-4} A_5} & -\frac{\eta}{A_5} \end{pmatrix} = \begin{pmatrix} -\frac{1}{2} & 0 & 0 & 0 \\ 0 & \frac{12}{11} & \frac{1}{11} & \frac{3}{11} \\ 0 & \frac{4}{11} & \frac{4}{11} & \frac{12}{11} \\ 0 & \frac{12}{11} & \frac{12}{11} & \frac{3}{11} \end{pmatrix}$$

The remaining relevant matrices and sets are:

$$\mathbf{D}_0 = \varnothing$$
$$\mathbf{D}_1 = (d_2)$$
$$\mathbf{N}_0 = \varnothing$$
$$\mathbf{N}_1 = \{2\}$$

The vector of flows is calculated as follows:

$$\mathbf{b}' = \mathbf{b} - \mathbf{D}_1 \mathbf{U}_1 = \begin{pmatrix} 3 \\ -2 \\ -1 \\ 0 \end{pmatrix} - \begin{pmatrix} 1 \\ -1 \\ 0 \\ 0 \end{pmatrix}(3) = \begin{pmatrix} 0 \\ 1 \\ -1 \\ 0 \end{pmatrix} \qquad \mathbf{F_B} = \mathbf{B}^{-1}\mathbf{b}' = \begin{pmatrix} 0 \\ \frac{12}{11} \\ \frac{4}{11} \\ \frac{12}{11} \end{pmatrix} - \begin{pmatrix} 0 \\ \frac{1}{11} \\ \frac{4}{11} \\ \frac{12}{11} \end{pmatrix} = \begin{pmatrix} 0 \\ 1 \\ 0 \\ 0 \end{pmatrix}$$

Dual Variable Computation

$R_2 = c_3 + A_3 c_5 + A_3 A_5 c_{-4} = -1 + 1(1/2) + (1)(1/3)(-3/4) = -3/4$. Also, $\beta = 1/12$. Thus, $\pi_2 = 9/11$.

$$\pi_5 = 0 \quad \pi_1 = \frac{0+2}{2} = 1 \quad \pi_2 = \frac{R_2}{\beta - 1} \quad \pi_3 = \frac{\pi_2 + c_3}{A_3} \quad \pi_4 = \frac{\pi_3 + c_5}{A_5}$$

$$\boldsymbol{\pi} = \begin{pmatrix} \pi_1 & \pi_2 & \pi_3 & \pi_4 \end{pmatrix} = \begin{pmatrix} 1 & \frac{9}{11} & \frac{-2}{11} & \frac{21}{22} \end{pmatrix}$$

Optimality Conditions

For non-basic arc a_2 the calculated value $w_2 = c_2 + \pi_1 - A_2 \pi_2 = \frac{13}{11} \geq 0$. This means that the arc flow does not satisfy the optimality condition, since arc a_2 is at its upper bound. Therefore, arc a_2 enters the basis and its flow is set to be equal at its upper bound. The **reductions** per unit of flow on arc a_2 are:

$$\mathbf{Y}_2 = \mathbf{B}^{-1}\mathbf{d}_2 = \begin{pmatrix} -\frac{1}{2} \\ 0 \\ 0 \\ 0 \end{pmatrix} - \begin{pmatrix} 0 \\ \frac{12}{11} \\ \frac{4}{11} \\ \frac{12}{11} \end{pmatrix} = \begin{pmatrix} -\frac{1}{2} \\ -\frac{12}{11} \\ -\frac{4}{11} \\ -\frac{12}{11} \end{pmatrix}$$

Iteration 2

The new basis is formed by a_1, a_2, a_3 and a_4 is shown in Figure 4.17.

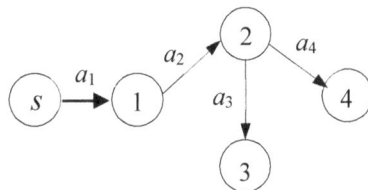

Figure 4.17. Second basis for example problem.

A cycle in the tree of Figure 4.17 is formed with arcs a_3, a_{-4}, a_5. Now, let us *reverse* the cycle going from 2 to 4 to 3 to 2. The only positive arc flow is $f_3 = 1$. The entering arc has its flow equal to the upper bound of 3. Therefore, considering the arcs in the order a_2, a_1, a_3, a_{-4}, a_5, the value of δ is equal to

$$\delta = min \ \{3, 0/(1/2), 1/(12/11), 0/(4/11), 0/(12/11)\} = 0$$

This means that either arc a_1, arc a_{-4}, or arc a_5 can leave the basis. If we choose a_5 then f_5 becomes non-basic with flow set equal to its lower bound.

$$\mathbf{B} = \begin{pmatrix} -2 & 1 & 0 & 0 \\ 0 & -1 & 1 & 1 \\ 0 & 0 & -1 & 0 \\ 0 & 0 & 0 & -4 \end{pmatrix} \quad \mathbf{B}^{-1} = \begin{pmatrix} -\frac{1}{2} & -\frac{1}{2} & -\frac{1}{2} & -\frac{1}{8} \\ 0 & -1 & -1 & -\frac{1}{4} \\ 0 & 0 & -1 & 0 \\ 0 & 0 & 0 & -\frac{1}{4} \end{pmatrix}$$

185

$$\mathbf{D}_0 = (d_5)$$
$$\mathbf{D}_1 = \varnothing$$
$$\mathbf{N}_1 = \varnothing$$
$$\mathbf{N}_0 = \{5\}$$

$$\mathbf{b}' = \mathbf{b} \qquad \mathbf{F}_\mathbf{B} = \mathbf{B}^{-1}\mathbf{b}' = \begin{pmatrix} -\frac{1}{2} & -\frac{1}{2} & -\frac{1}{2} & -\frac{1}{8} \\ 0 & -1 & -1 & -\frac{1}{4} \\ 0 & 0 & -1 & 0 \\ 0 & 0 & 0 & -\frac{1}{4} \end{pmatrix} \begin{pmatrix} 3 \\ -2 \\ -1 \\ 0 \end{pmatrix} = 3\begin{pmatrix} -\frac{1}{2} \\ 0 \\ 0 \\ 0 \end{pmatrix} - 2\begin{pmatrix} -\frac{1}{2} \\ -1 \\ 0 \\ 0 \end{pmatrix} - \begin{pmatrix} -\frac{1}{2} \\ -1 \\ -1 \\ 0 \end{pmatrix} = \begin{pmatrix} 0 \\ 3 \\ 1 \\ 0 \end{pmatrix}$$

Dual Variable Computation

$$\boldsymbol{\pi} = \begin{pmatrix} \pi_1 & \pi_2 & \pi_3 & \pi_4 \end{pmatrix} = \begin{pmatrix} 1 & 2 & 1 & \frac{5}{4} \end{pmatrix}$$

Optimality Condition

$$w_5 = c_5 + \pi_3 - A_5\pi_4 = \tfrac{1}{2} \geq 0 \quad \text{(Optimal)}$$

Optimal flows

$f_1 = 0,\ f_2 = 3,\ f_3 = 1,\ f_4 = 0,\ f_5 = 0.$

$\text{Cost} = \mathbf{C_B F_B} + \mathbf{C_O F_O} + \mathbf{C_1 F_1} = 2.$

B. NOP Solution

The multiplier for arc a_5 joining node 3 to node 4, as can be seen in Figure 4.14 is equal to 1/3. For the computer solution the corresponding multiplier is approximated by 0.3333. It is also noted that the software generates an initial basis following a systematic procedure outlined in Section 4.5. Figure 4.18 shows the computer screen for this example: (1) Input data file, (2) Output results. The output results include the following:

1. Number of nodes, arcs and artificial arcs created.
2. Solution Option: Solve by the Primal Simplex method.
3. Solution Option: Review the Solution.
4. Arcs parameters and flows.
5. Table of arc parameters, flows, and total cost for each arc.

```
nodes 5;
stn;
1 3 0 0;
2 -2 0 0;
3 -1 0 0;
4 0 0 0;
endn;
sta;
5 1 3 0 2 2;
1 2 3 0 1 1;
2 3 3 0 -1 1;
2 4 3 0 3 4;
3 4 3 0 0.5 0.3333;
enda;
```

```
6  NODES ENTERED
5  ARCS ENTERED
5  SLACK OR ARTIFICIAL ARCS CREATED
```

MDI ✕

```
** SOLUTION OPTIONS **
1) SOLVE WITH THE PRIMAL SIMPLEX
2) REVIEW SOLUTION
3) End Routine
Enter Option Number
```
OK Cancel

```
1
```

MDI ✕

```
** SOLUTION OPTIONS **
1) SOLVE WITH THE PRIMAL SIMPLEX
2) REVIEW SOLUTION
3) End Routine
Enter Option Number
```
OK Cancel

```
2
```

```
NETWORK PRIMAL SIMPLEX ALGORITHM
```

MDI ✕

```
1. Arc parameters and flows
2. Node parameters
3. Exit

Enter your choice in the box below.
```
OK Cancel

```
1
```

```
ARC PARAMETERS AND FLOWS

ARCS THAT START AT 1
GO TO   NUMBER    LOWER   UPPER   COST    GAIN/LOSS   FLOW     TOTAL COST
2       1         0       3       1       1.00        003.00   00003.00

SLACK   2         0       0       99999   1.00        000.00   00000.00

ARCS THAT START AT 2
GO TO   NUMBER    LOWER   UPPER   COST    GAIN/LOSS   FLOW     TOTAL COST
3       3         0       3       -1      1.00        001.00   -00001.00

4       4         0       3       3       4.00        000.00   00000.00

ARCS THAT START AT 3
GO TO   NUMBER    LOWER   UPPER   COST    GAIN/LOSS   FLOW     TOTAL COST
4       5         0       3       0.5     0.33        000.00   00000.00

NO ARCS ORIGINATE AT NODE  4

ARCS THAT START AT 5
GO TO   NUMBER    LOWER   UPPER   COST    GAIN/LOSS   FLOW     TOTAL COST
1       6         0       3       2       2.00        000.00   00000.00

ARCS THAT START AT 6
GO TO   NUMBER    LOWER   UPPER   COST    GAIN/LOSS   FLOW     TOTAL COST
2       7         0       2       99999   1.00        000.00   00000.00

3       8         0       1       99999   1.00        000.00   00000.00

4       9         0       0       99999   1.00        000.00   00000.00

5       10        0       0       99999   1.00        000.00   00000.00

MINIMAL COST = 2
```

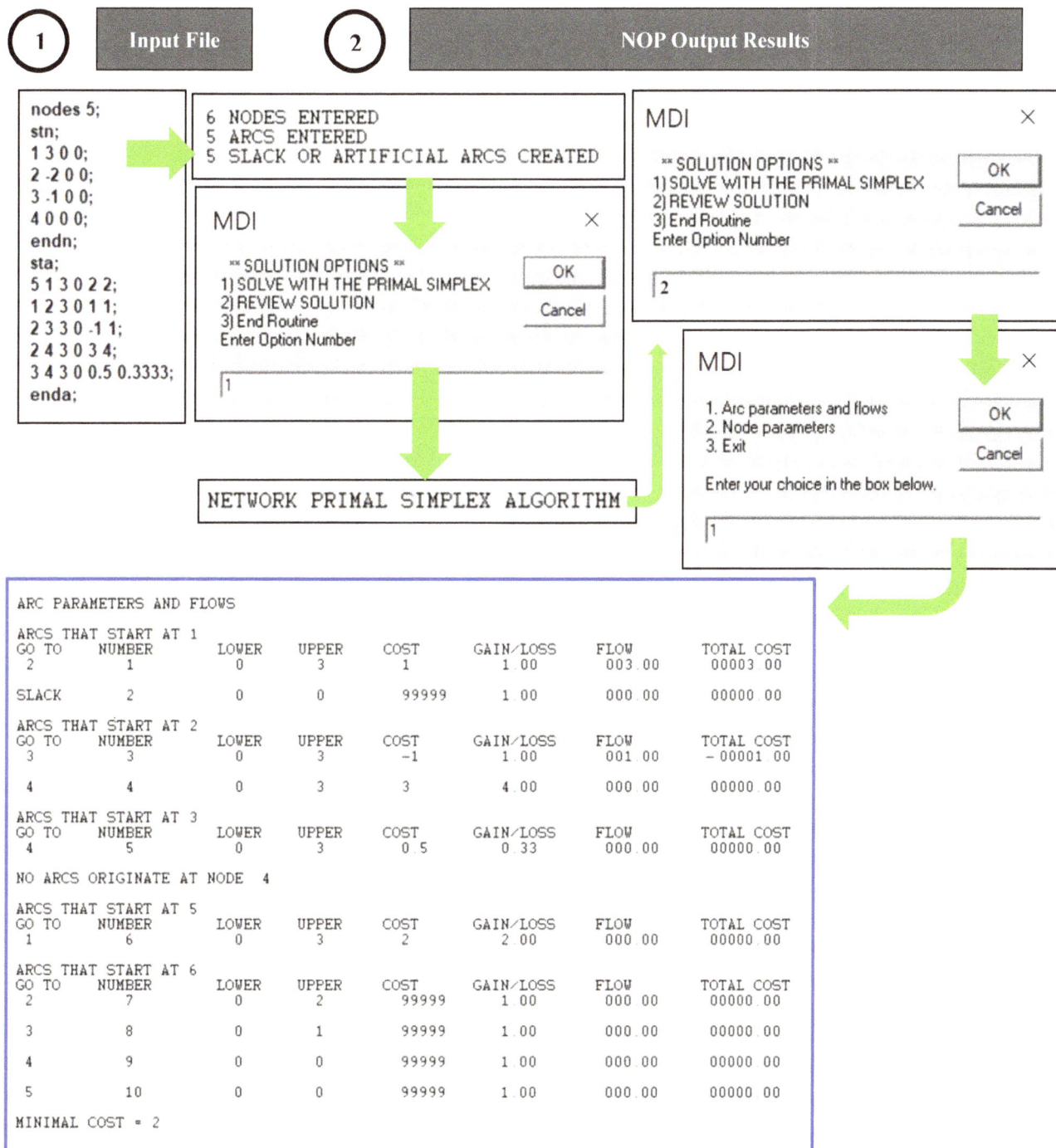

Figure 4.18. Sequence of NOP Screens for Generalized Network Example.

4.5 FINDING AN INITIAL FEASIBLE BASIS

4.5.1 Creating Basic Arcs Connected to the Slack Node

The process of finding an initial feasible basis starts by identifying all the arcs that originate or terminate at the slack node and are connected to nodes having external flow requirements (supply or demand). For each arc it is verified if it can be used to satisfy *some* or *all* of the fixed external flow requirement. If only *some* of the flow can be accommodated, the flow on the arc is set equal to its capacity, and the fixed external flow of the node is accordingly reduced. As an illustration, the network shown in Figure 4.19 will be considered. Node 5 is the slack node.

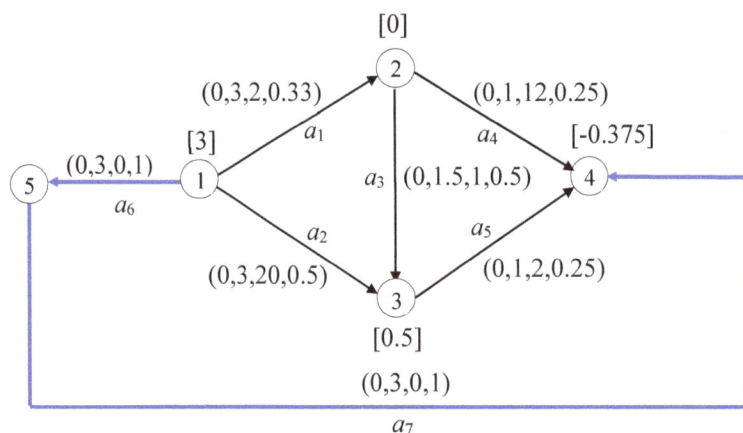

Figure 4.19. Generalized network with node *5* as slack node.

Arcs that originate or terminate at the slack node are node 1 with a positive external flow equal to 3 and node 4 with a negative external flow equal to –0.375. Node 1 can provide all its external flow to the slack node along arc a_6. Therefore, the flow on arc a_6 is set equal to the magnitude of the external flow on node 1. That is, $f_6 = 3$. Furthermore, arc a_7 can provide all of the fixed external flow required at node 4. Therefore, the flow on arc a_7 is set to the magnitude of the external flow on node 4, that is $f_7 = 0.375$. The fixed external flows on node 1 and node 4 are reset to be equal to zero. Both arcs a_6 and a_7 become basic arcs connected to node 1 and node 4, respectively.

4.5.2 Creating Artificial Basic Arcs Connected to the Slack Node

Artificial arcs are created for those nodes that have not been assigned basic arcs yet. An artificial basic arc is created in the direction required to provide the external flow remaining at the node. If the fixed external flow of a node is positive, an artificial arc directed from the node to the slack node is created and used as a basic arc. Alternatively, in the case of a negative external flow, the artificial arc is directed from the slack node to the node. The arc capacity and flow are set to the absolute value of the fixed external flow and the gain is 1. A large positive cost is placed on the arc. Any primal procedure will eventually cause the flow on the arc to become zero if a feasible solution exists. If the external flow is equal to zero either direction of the arc connecting the slack node with node *i* is correct.

188

Node 2 and node 3 have not been assigned basic arcs yet. Then the artificial basic arcs are created for both nodes in the direction to provide external flow remaining at the node. Node 3 has a positive fixed external flow; therefore, its artificial arc is created from node 3 to the slack node. Since the fixed external flow of node 2 is zero, its artificial arc is created from the slack node to node 2. For each artificial arc the capacity and flow are set equal to the absolute value of the fixed external flow and the multiplier is set equal to 1. A large positive cost (equal to 10,000 in Figure 4.20) is placed on each arc. The primal simplex procedure will eventually cause the flow on the artificial arcs to become zero if a feasible solution exists.

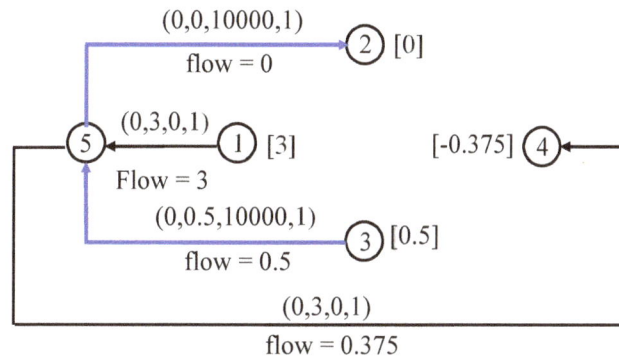

Figure 4.20. Initial Basis for network of Figure 4.15.

EXERCISES

1. Find the transformed network to eliminate the positive lower bound on the flow along arc (1,3). The parameters shown for this arc are the lower bound, the upper bound, the per-unit cost and the multiplier.

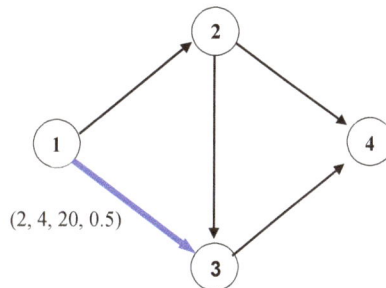

2. Transform the given network to represent variable external flows by means of arcs. Each arc has two values corresponding to the upper bound and the cost. Each node has three values representing its fixed flow, variable flow, and cost per unit of variable flow. The single number given for node 4 corresponds to its fixed flow.

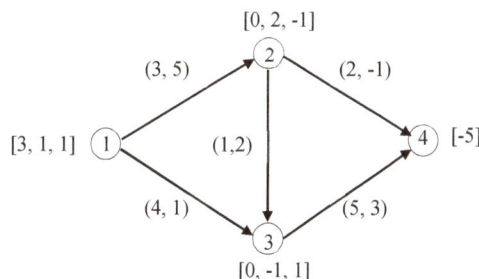

189

3. Solve the minimum-cost problem for the network given in Problem 2, using the network specialization of the primal simplex method.

4. Discuss how to identify feasible initial solutions to both pure and generalized network problems. Illustrate your procedures using small networks.

5. Find the parameters for each city given the following information:
   City A: This plant is to be discontinued. The entire inventory of 400 units must be shipped or sold for scrap. The scrap value is $8 per unit.
   City B: There are 100 units left over from previous shipments. These can only be sold in City B. No firm demand but up to 200 units can be sold at $25 each.
   City C: A firm demand of 200 units which must be received.
   City D: All regular-time production of 300 units be shipped. An additional 150 units can be produced using overtime at a cost of $13 per unit.
   City E: Contracted demand for 300 units. No fewer than 175 units must be received. A penalty of $5 must be paid for each unit less than 300 received.
   City F: Minimum demand of 200 units. An additional 100 units could be sold if available with a revenue of $21 per unit.
   City G: Maximum production of 250 units with manufacturing cost of $12 per unit. There is no minimum shipment required from New York.
   City H: Demand for 120 units that must be met.

6. Arc numbers are upper bound, lower bound, and per-unit cost. Arcs (2,4) and (3,4) have multipliers (gain/loss factors) equal to 3 and 1/2, respectively. The remaining arcs have multipliers equal to 1. Node 1 supplies 3 units at $10 each, plus up to 2 units each costing an *additional* $0.50. Node 4 needs 1 unit but can receive as many additional units as possible. It is desired to *formulate* the min-cost network with zero lower bounds and no slack external flows. Show all parameters in the transformed network.

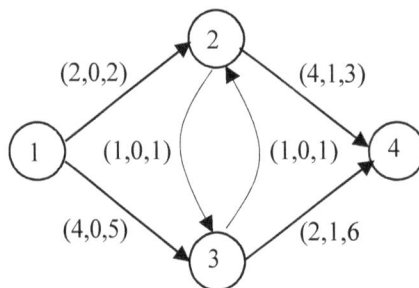

7. For the network shown, find the arc flows using the following basis matrix.

$$\mathbf{B} = \begin{pmatrix} \mathbf{d}_{-6} & \mathbf{d}_3 & \mathbf{d}_4 & \mathbf{d}_1 \end{pmatrix} = \begin{pmatrix} 0 & 1 & 0 & -1 \\ -1 & 0 & 1 & 0 \\ 1 & -1 & 0 & 0 \\ 0 & 0 & -1 & 0 \end{pmatrix}$$

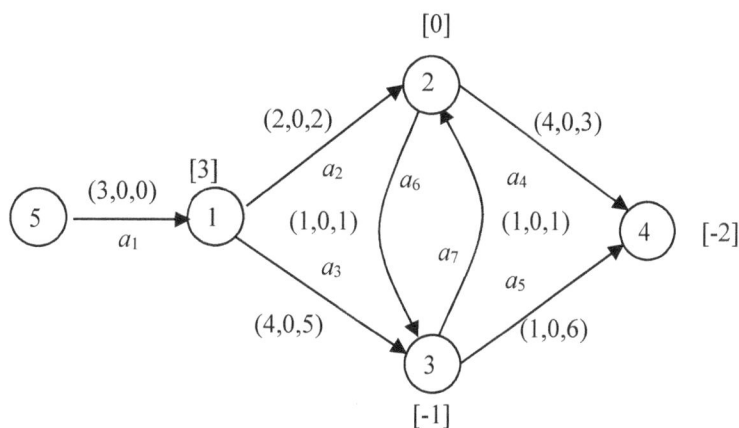

8. The following information is given for the network shown. External fixed flows for nodes 1, 2, 3, 4 are equal to 3, 0, 0, -3, respectively. Arcs (2,4) and (3,4) have multipliers (gain/loss factors) equal to 2 and 1/2, respectively. Transform the network into an equivalent one with all lower bounds set equal to zero. Show all parameters in the transformed network.

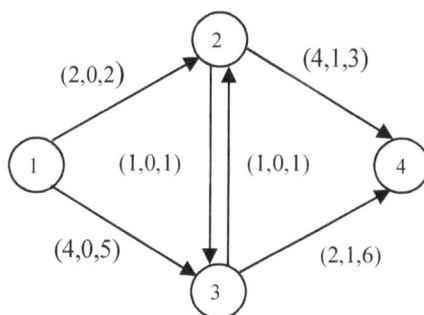

9. Consider the following network

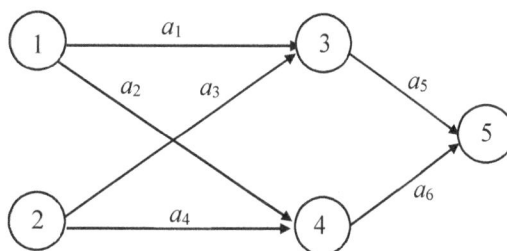

(a) External fixed flows for nodes 1, 2, 3, 4 are equal to 3, 0, 0, -3, respectively. Only arcs (1,3) and (2,3) are generalized with multipliers (gain/loss factors) equal to 2 and 1/2, respectively. Transform the network to an equivalent one with all lower bounds set equal to zero. Show all parameters in the transformed network.

(b) Draw an initial basis consisting of a *spanning arborescent tree* rooted at the slack node. Include arcs (1,3) and (2,4) and those arcs created to remove the variable flows.

(c) Write the corresponding basis matrix and its inverse.

(d) Repeat part (c) for the case where no *mirror* arcs are used to represent the same initial basis chosen in part (a).

191

10. Apply the network specialization of the simplex method. Non-zero fixed flows are as shown.

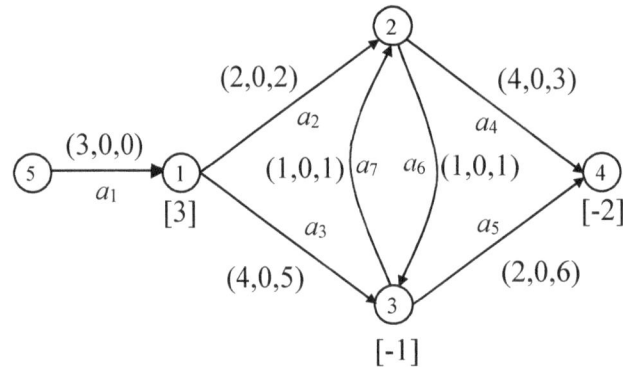

11. Non-zero external fixed flows are given within square brackets. For each arc three numbers are given within parentheses to indicate capacity, cost and gain/loss factor, respectively. Using the initial basis given, apply the network-specialization of the primal simplex.

12. Two types of products 1 and 2 can be manufactured from one raw material by a specialized process that can be carried out by either of two available operators. Given the following information for each unit (here 1 unit is equal to 1 pound) of both products, it is desired to formulate and solve using the network specialization of the primal simplex method the problem of deciding how many units of each product should be manufactured in order to maximize profits:

	Profit	Material	Operator 1	Operator 2
Product 1	$45/lb	3 pounds	2 hours	6 hours
Product 2	$50/lb	2 pounds	4 hours	7 hours

In the above table the number of hours required by each unit is given for both of the operators available, although each unit will be processed only by one operator. Note that these operators have different production rates. Assume that the maximum amount available of the raw material is 300 pounds, and that the maximum time Operator 1 and Operator 2 can be used are 100 and 200 hours, respectively. (a) Formulate this problem as a generalized network model. (b) Use the network specialization of the primal simplex to find the optimal solution.

13. Use the network specialization of the primal simplex to find the optimal flows in the given network. Label the arcs as follows: $a_1 = (1, 2)$, $a_2 = (1, 3)$, $a_3 = (2, 3)$, $a_4 = (2, 4)$, $a_5 = (3, 4)$, $a_6 = (3, 5)$, $a_7 = (5, 2)$, $a_8 = (5, 1)$.

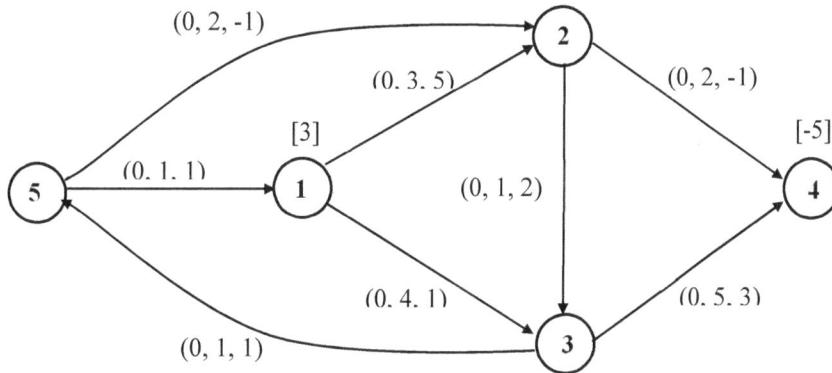

REFERENCES

[1] DANTZIG, G.B. (1951) "Application of the Simplex Method to a Transportation Problem," *Activity Analysis of Production and Allocation.* Koopmans, T.C., Ed., John Wiley and Sons, New York, 359-373, 1951.

[2] JENSEN, P., "Network Flow Programming," Department of Mechanical Engineering, The University of Texas at Austin, 1983.

[3] JENSEN P., "Microsolve Network Flow Programming," Department of Mechanical Engineering, The University of Texas at Austin, 1983.

[4] JOHNSON, E., "Networks and basic solutions", *Operations Research*, 14, 619–623, 1966.

Chapter 5

THE TRAVELING SALESMAN PROBLEM

"Come on, come on," Prew said.
"Whats holding things up? Let's get this show on the road."

James Jones
From Here to Eternity, 1951.

The *Traveling Salesman Problem* (TSP) and the *Multi-Traveling Salesman Problem* (MTSP) are combinatorial in nature. Their extremely simple description has attracted a significant number of analysts encouraged by the wide range of potential applications. However, through the years these problems have remained virtually unsolved in the case of large-scale applications, leaving most researchers in the field with the belief that it is indeed a very difficult problem to solve, as perhaps best summarized by the quotation *"Why, it's as hard as the traveling salesman problem!"* [6,8].

The traveling salesman problem can be stated as follows. There are *n* cities with distances known for trips between these cities. A salesman, starting at a given city, referred to as the *base* city, intends to visit each of the other *n*-1 cities exactly once and return to the starting city. It is desired to determine the *order* in which the cities must be visited to minimize the total distance traveled. Instead of distance, cost or time, any other measure of effectiveness that is additive along circuits of the city network can be used. The structure of the problem shows that there are (*n*-1)! *tours* to be considered in order to select one or more that should be optimal.

The multi-traveling salesman model allows any city to be visited by exactly one salesman, given that there are multiple salesmen visiting the cities. In this case, in addition to the travel cost associated with the trips between cities, there is a fixed cost for each salesman. The solution corresponds to a set of subtours, each starting and ending at the *base* city, resulting in minimal total distance, time or cost. Each subtour is assigned to exactly one salesman.

If some of the trips are not allowed, the corresponding arc *lengths* are set equal to infinity. In many cases one can assume that the distance between two cities is the same, regardless of the direction of the trip. However, in the algorithms that will be presented and discussed in this chapter this feature is not a requirement.

5.1 THE TRAVELING SALESMAN PROBLEM

Let G = (**N**, **A**) be a directed network with nodes 1, 2, ..., *n* representing cities and arcs representing trips between cities. City 1 is the *home city*. Furthermore, let d_{ij} be the non-negative length of arc (i,j) and let **D** = $[d_{ij}]$ be the associated *distance* matrix. It is desired to find the circuit $(1,j_1), (j_1,j_2), ..., (j_r,1)$ containing each node exactly once and having minimum total length. A *tour* is defined as a circuit including all nodes. A feasible tour includes all cities in a sequence which does not violate the specified order in which cities can be visited. A *subtour* is a circuit including city 1 and other cities but not all. Clearly, an *upper bound* on the length of the optimal tour is given by the length of *any* feasible tour.

5.1.1 Initial Lower Bound

The first step to determine the initial lower bound on the length of the optimal tour is known as *matrix reduction*. Consider the distance matrix $\mathbf{D} = [d_{ij}]$. Let C_i be the smallest element in row i; row reduction is defined as the subtraction of C_i from all entries in row i for $i = 1, 2, \ldots, n$. Now let Q_j be the smallest element in column j after all rows have been reduced; column reduction is defined as the subtraction of Q_j from all entries in column j, for $j = 1, 2, \ldots, n$. Note that if we could select a *zero-length* tour, that is, one having exactly one zero-entry in each row and each column of the reduced matrix, that would be the optimal (shortest) tour, with length equal to $H = \Sigma_i C_i + \Sigma_j Q_j$. Let $Z(\mathbf{T})$ be the length of a tour $\mathbf{T}$ *before* the reduction of the distance matrix and let $Z_1(\mathbf{T})$ be its length *after* the reduction. Therefore, $Z(\mathbf{T}) = H + Z_1(\mathbf{T})$, where $H = \Sigma_i C_i + \Sigma_j Q_j$. Since $Z_1(\mathbf{T})$ must be non-negative, then H represents a *lower bound* on the length of the optimal tour.

5.1.2 Branching

The partitioning of the set of all tours into disjoint subsets, as a result of branching, will be represented by a tree structure such as that shown in Figure 5.1. In this figure, the *root* node represents the set of all tours. Additionally, the node labeled (i,j) represents the subset of all tours that include the trip from city i to city j. The node labeled $\overline{(i,j)}$ represents the subset of all tours that do not include this trip. Furthermore, the node labeled (k,l) represents all tours that include both the trip from city i to city j and the trip from city k to city l. As before, the node labeled $\overline{(k,l)}$ represents all tours that include the trip from city i to city j but do not include the trip from city k to city l.

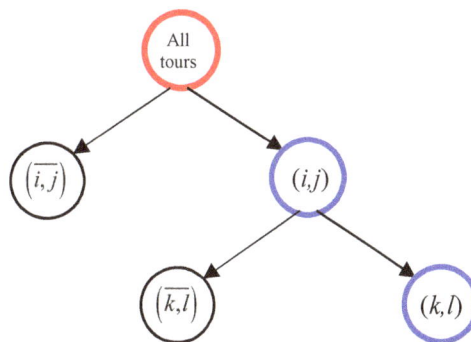

Figure 5.1. Partitioned subsets resulting from branching.

The reason for branching can easily be explained; if we partition the set of all tours into smaller and smaller subsets, we should eventually have a subset containing a single tour. The links or the city pairs that make up a single tour can then be recovered by tracing backward to the all-tours root node. The branching process is controlled by the lower-bound criterion. If at any iteration of the algorithm the lower bound for a subset of tours is larger than the lower bound for another subset, the first subset can be deleted from further consideration. That is, no additional branching is performed from the corresponding node.

In general, when branching is done on the set of tours X with respect to a specified trip Y, node X branches into two subsequent nodes. The node that includes the specified $\overline{(k,l)}$ trip will be represented by Y and the node that does not include the trip will be represented by $(\overline{Y})$.

5.1.3 Penalties

At the time just before branching, the most advantageous tours to be considered for Y are those containing a link or trip (i,j) for which $d_{ij} = 0$ in the reduced distance matrix. However, since there is at least one zero entry in each row and each column of the matrix, we must determine which one of all zero-links should be chosen for inclusion in Y. To do this we evaluate each zero-link, as follows. Let us first consider the possible tours for $\overline{Y}$, that is, the set of tours not including link (i,j). Since city i must connect to some city other than city j, these tours must include the link corresponding to the lowest entry in row i, excluding $d_{ij} = 0$. Let us denote the value associated with this entry by A_i. Similarly, since city j must now be reached from some city other than city i, the tours must also include the link corresponding to the lowest entry in column j other than $d_{ij} = 0$. Let us denote the value of this entry by B_j. As a result of the preceding analysis, the penalty (increase in length) for not including trip or link (i,j) is equal to at least $\Phi_{ij} = A_i + B_j$. Once this penalty is computed for each link such that $d_{ij} = 0$, the link corresponding to the *largest* penalty is chosen to perform the branching operation.

5.1.4 Distance Matrix Modifications

Note that if we include any link (k,l) in the tours we do not need to consider row k and column l any further. Moreover, if link (k,l) is in any tour that belongs to Y then link (l,k) cannot be in the same tour generated by Y. This can be enforced by setting $d_{lk} = \infty$ in all further iterations. Finally, there may be other links which if included for further consideration might create subtours not including all the n cities. These links are referred to as *forbidden links* and can be eliminated from consideration by setting their lengths equal to ∞ in the reduced matrix available at the time the forbidden link is detected. Now we will consider the set of tours $\overline{Y}$. Since link (k,l) is not included in the tours in Y, then the tours in $\overline{Y}$ must have $d_{kl} = \infty$.

After these modifications, the resulting distance matrix is a candidate for further reductions in row l and column k, since any column that had a single zero in row k and any row that had a single zero in column l can be further reduced. In addition, any other rows and columns that contain forbidden links might also be candidates for reduction. The lower bound on Y can now be computed as the sum of the new reducing constants and the previous lower bound.

5.2 LITTLE'S ALGORITHM

The algorithm presented in this chapter is generally referred to as a *branch-and-bound* algorithm, and was originally developed by Little, Murty, Sweene and Karel [10]. Following Little et al., the fundamental approach can be described in this fashion. The set of all tours (feasible solutions) is decomposed into increasingly small subsets by means of the strategy known as *branching*. At each step of this decomposition approach, a *lower bound* on the length of the optimal tour is obtained. Once it is determined that a subset cannot contain the optimal tour, it is not considered for further exploration. This determination, using the branch-and-bound terminology, is known as *fathoming*. The process of *bounding* subsets to eliminate sub-optimal alternatives and branching to more attractive subsets is the basis of the algorithm. Eventually the optimal tour is determined as a subset containing exactly one tour with its length being less than or equal to some lower bound for every tour.

5.2.1 Steps of the Algorithm

Little's procedure for solving the TSP consists of three main steps: (1) Identify link on which to base branching; this is based on a reduced matrix and penalty calculation for the 0-entries of the reduced matrix. (2) Compute new lower bounds; and (3) Examine the potential of each node in the tree to continue branching. Each step is summarized below:

Step 1: Calculation of Penalties

In this step all entries (i,j) in the *current* reduced distance matrix such that $d_{ij} = 0$ are first identified. Let X represent the set of all admissible tours in the branch-and-bound tree. For each 0-entry the minimum penalty for not choosing it is defined as

$$\Phi_{ij} = A_i + B_j$$

where A_i is the smallest element in row i and B_j is the smallest element in column j, excluding $d_{ij} = 0$ in both cases. Once the penalties are calculated for all 0-entries, we find the *largest* penalty:

$$\Phi^* = max \{\Phi_{ij}\}$$

Step 2: Branching and Computation of Lower Bounds

Branching is done with respect to the trip $Y = (i,j)$ having the largest penalty, Φ^*. Two new branches are included in the tree. One branch connects node X to node Y. The other branch connects node X to node $\overline{Y}$. The lower bounds for these nodes are computed as follows:

(a) $w(\overline{Y}) = w(X) + \Phi^*$

(b) $w(Y) = w(X) + H$

where H is the sum of reducing constants after modifying the distance matrix as follows: (a) eliminate row i and column j; (b) set $d_{ji} = \infty$; (c) set $d_{ab} = \infty$ for all links or branches (a,b) which may close subtours (circuits not including all cities).

Step 3: Branching Rule

Find $X = (i,j)$ or $X = (\overline{i,j})$ corresponding to the minimum lower bound. If $X = (i,j)$ is chosen, go to Step 1. If $X = (\overline{i,j})$ is chosen, set $d_{ij} = \infty$ and go to Step 1. If a *complete* tour is now available, stop the execution of the algorithm.

5.2.2 Example

Solve the traveling salesman problem with distance matrix given in Table 5.1.

Table 5.1. INITIAL DISTANCE MATRIX

	1	2	3	4
1	∞	9	8	5
2	2	∞	7	4
3	3	10	∞	4
4	7	5	4	∞

A. Algorithmic Procedure

Although it is not essential to choose first a feasible tour, we will consider the one consisting of the links (1,4), (4,2), (2,3), and (3,1). The length of this feasible tour is $Z_U = 5+5+7+3 = 20$. The value of Z_U is actually an *upper bound* on the length of the optimal tour **T***; that is, $Z(\mathbf{T}^*) \leq Z_U$. The evaluation of an upper bound often helps in the reduction of the computational effort involved, since any branch with a *lower bound* that exceeds the value of Z_U should be fathomed. Next, we proceed with the reduction of the distance matrix, first by rows and then by columns. The row reduction is shown in Table 5.2, where the C_i column contains the reducing constants.

Table 5.2. ROW REDUCTION

	1	2	3	4	C_i
1	∞	4	3	0	5
2	0	∞	5	2	2
3	0	7	∞	1	3
4	3	1	0	∞	4

Column reduction is now attempted. Upon inspection of Table 5.2, we see that each column except column 2 contains at least one 0-entry. Hence, only column 2 can be effectively reduced. We reduce column 2 by 1 unit to obtain the results shown in Table 5.3, where the Q_j row contains the reducing constants for each column. The intersection of the Q_j row and the C_i column shows the sum of the reducing constants, that is, $H = \Sigma_i \, C_i + \Sigma_j \, Q_j$. This is a lower bound $Z(\mathbf{T})$ on all tours for the problem.

Table 5.3. COLUMN REDUCTION

	1	2	3	4	C_i
1	∞	3	3	0	5
2	0	∞	5	2	2
3	0	6	∞	1	3
4	3	0	0	∞	4
Q_j	0	1	0	0	15

Clearly, an entry equal to zero in Table 5.3 represents a city pair that has a potential to be included in the optimal tour. In order to identify the trip on which to base the first branching operation, the penalty for not including the trip corresponding to each 0-entry must be computed. The A_i and B_j values are written in a separate column and a separate row, respectively, in addition to the reduced matrix, as shown in Table 5.4. Table 5.5 shows the penalty value (additional distance) for each zero entry of Table 5.4. In Table 5.5 we observe that the largest penalty is equal to 4 and corresponds to link (1,4). Thus, $\Phi^* = \Phi_{14} = 4$. Hence, we will base the next branching operation on link (1,4), and set $Y = (1,4)$ and $\overline{Y} = \left(\overline{1,4}\right)$. The lower bound on the length of tours in $\overline{Y}$ is equal to $15 + 4 = 19$. Before we can determine the new lower bound on the length of tours in Y, which includes link $(1,4)$, row 1 and column 4 in Table 5.4 will be deleted from further consideration. Also, d_{41} is set equal to ∞. No forbidden arcs are detected.

Table 5.4. FIRST DECISION MATRIX

	1	2	3	4	C_i	A_i
1	∞	3	3	0	5	3
2	0	∞	5	2	2	2
3	0	6	∞	1	3	1
4	3	0	0	∞	4	0
Q_j	0	1	0	0	15	
B_j	0	3	3	1		

Table 5.5. FIRST CALCULATION OF PENALTIES

Link (i,j)	$\Phi_{ij} = A_i + B_j$
(1,4)	4
(2,1)	2
(3,1)	1
(4,2)	3
(4,3)	3

Table 5.6 shows the first modified distance matrix. Table 5.7 summarizes the matrix reduction and penalty calculation results.

Table 5.6. FIRST MODIFIED DISTANCE MATRIX

	1	2	3
2	0	∞	5
3	0	6	∞
4	∞	0	0

Table 5.7. SECOND DECISION MATRIX

	1	2	3	C_i	A_i
2	0	∞	5	0	5
3	0	6	∞	0	6
4	∞	0	0	0	0
Q_j	0	0	0	0	
B_j	0	6	5		

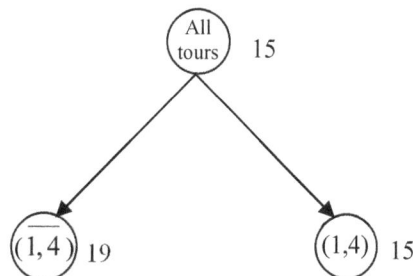

Figure 5.2. Initial branching.

199

The sum of the reducing constants in Table 5.7 is equal to zero. Therefore, the new lower bound on the length of tours in Y is equal to $15+0 = 15$. Figure 5.2 shows the initial branching with lower bounds. As can be seen in Figure 5.2, the value 15 is currently the smallest available lower bound among the nodes from which branching can be done in the branch-and-bound tree. Therefore, the next branching will be rooted at node $Y = (1,4)$. The 0-entries in Table 5.7 will be evaluated to identify the link on which to base this branching operation. In Table 5.8 we observe that the largest penalty is equal to 6 and corresponds to links $(3,1)$ and $(4,2)$. In order to proceed let us choose $\Phi^* = \Phi_{31} = 6$. Hence, we will base the next branching operation on link $(3,1)$, and set $Y = (3,1)$ and $\overline{Y} = \overline{(3,1)}$.

Table 5.8. SECOND CALCULATION OF PENALTIES

Link (i,j)	$\Phi_{ij} = A_i + B_j$
(2,1)	5
(3,1)	6
(4,2)	6
(4,3)	5

The lower bound on the length of tours in $\overline{Y}$ is equal to $15 + 6 = 21$. This lower bound exceeds the value of the **upper bound** initially set equal to 20. Therefore, this branch of the tree is fathomed. As before, to determine the new lower bound on Y, we delete row 3 and column 1 in Table 5.7. Note that it is not necessary to set $d_{13} = \infty$ from which branching can be done since row 1 was already removed. Up to this point of our analysis, two links have been chosen: $(1,4)$ and $(3,1)$. It is noted that link $(4,3)$ is a forbidden link since, if chosen, it would close the subtour $(1,4)$-$(4,3)$-$(3,1)$. Therefore, we set $d_{43} = \infty$. Table 5.9 shows the new modified distance matrix.

Table 5.9. SECOND MODIFIED DISTANCE MATRIX

	2	3
2	∞	5
4	0	∞

Table 5.10 shows the results from the reduction of the second modified distance matrix. As it can be seen in this table, the sum of the reducing constants is equal to 5. Therefore, the lower bound for the node $Y = (3,1)$ is equal to $15 + 5 = 20$. Figure 5.3 shows the branches and lower bounds of the current tree.

Table 5.10. THIRD DECISION MATRIX

	2	3	C_i	A_i
2	∞	0	5	∞
4	0	∞	0	∞
Q_j	0	0	5	
B_j	∞	∞		

200

In Figure 5.3 the subset of tours represented by $\bar{Y} = \overline{(1,4)}$ has a lower bound equal to 19. Since this is currently the smallest lower bound, we *backtrack* to this node in the tree and proceed reducing the corresponding distance matrix, which is actually the original distance matrix after setting $d_{14} = \infty$.

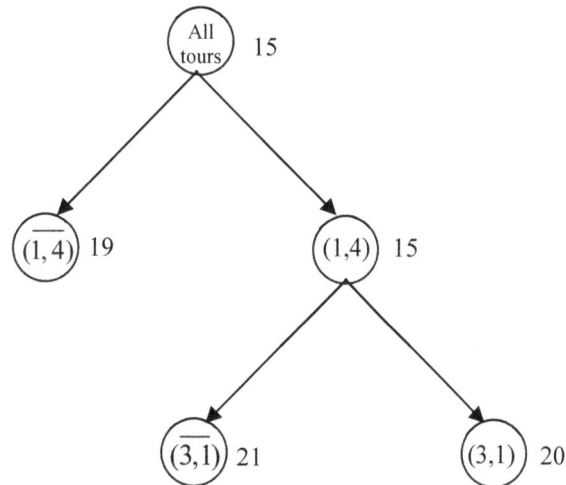

Figure 5.3. Second branching.

The reduction of the matrix and calculations needed for penalty evaluations are summarized in Table 5.11. The penalty values are shown in Table 5.12. From this table, we can choose between links (2,1) and (3,4). Choosing link (2,1), we conclude that the lower bound for node $\bar{Y} = \overline{(2,1)}$ in the branch-and-bound tree of Figure 5.3 is equal to 19+1=20.

Table 5.11. FOURTH DECISION MATRIX

	1	2	3	4	C_i	A_i
1	∞	0	0	∞	8	0
2	0	∞	5	1	2	1
3	0	6	∞	0	3	0
4	3	0	0	∞	4	0
Q_j	0	1	0	1	19	
B_j	0	0	0	1		

Table 5.12. THIRD CALCULATION OF PENALTIES

Link (i,j)	$\Phi_{ij} = A_i + B_j$
(1,2)	0
(1,3)	0
(2,1)	1
(3,1)	0
(3,4)	1
(4,2)	0
(4,3)	0

201

The selection of link (2,1) makes link (1,2) forbidden. After setting $d_{12} = \infty$, and deleting row 2 and column 1, the distance matrix shown in Table 5.11 is modified as shown in Table 5.13. Since the sum of the reducing constants is equal to zero, the lower bound for node $Y = (2,1)$ is equal to 19+0=19, as indicated in Figure 5.4.

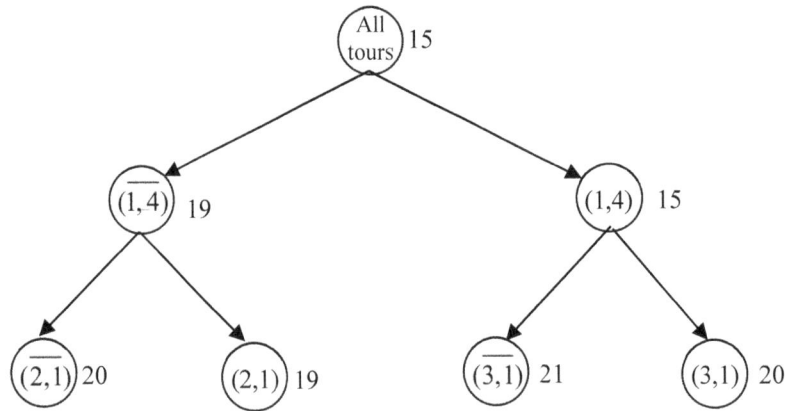

Figure 5.4. Third branching.

To continue with branching we need to inspect the current tree, shown in Figure 5.4, and identify the node having the smallest lower bound on tour length. It is seen that this node corresponds to the set of tours $Y = (2,1)$. The penalty values needed to select the link on which to base the next branching operation are given in Table 5.14. From the results shown in the table, we can choose either of the links (1,3) and (3,4). Choosing link (1,3), the lower bound for node $\overline{Y} = \overline{\left(1,3\right)}$ is equal to ∞. Evidently, this branch must now be fathomed.

Table 5.13. FIFTH DECISION MATRIX

	2	3	4	C_i	A_i
1	∞	0	∞	0	∞
3	6	∞	0	0	6
4	0	0	∞	0	0
Q_j	0	0	0	0	
B_j	6	0	∞		

Table 5.14. FOURTH CALCULATION OF PENALTIES

Link (i,j)	$\Phi_{ij} = A_i + B_j$
(1,3)	∞
(3,4)	∞
(4,2)	6
(4,3)	0

202

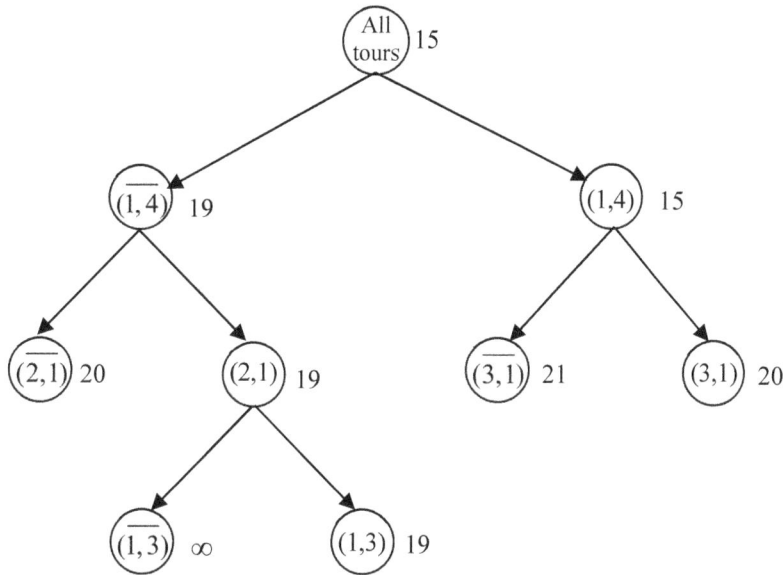

Figure 5.5. Fourth branching.

After choosing link (1,3) the distance matrix given in Table 5.13 is modified by deleting row 1 and column 3, and resetting $d_{32} = \infty$, since link (3,2) is a forbidden link. If not removed, this link would close the subtour (2,1)-(1,3)-(3,2). The results for the analysis of this link are shown in Table 5.15.

Table 5.15. LAST DECISION MATRIX

	2	4	C_i	A_i
3	∞	0	0	∞
4	0	∞	0	∞
Q_j	0	0	0	
B_j	∞	∞		

Since the sum of the reducing constants is equal to 0, the node $Y = (1,3)$ has a lower bound equal to 19+0=19, as shown in Figure 5.5. The results for the penalty evaluation shown in Table 5.15 indicate that both links (3,4) and (4,2) must be in the optimal tour. Once this is done, a complete-tour lower bound equal to 19 is obtained. Based on this result, the branches in Figure 5.6 having a lower bound larger than 19 are fathomed.

The complete tree is shown in Figure 5.6. As shown in this figure, the optimal tour consists of the links (2,1), (1,3), (3,4) and (4,2). This implies that the optimal route, starting and ending at the base city, is the sequence 1-3-4-2-1. The total length of this tour is equal to 19.

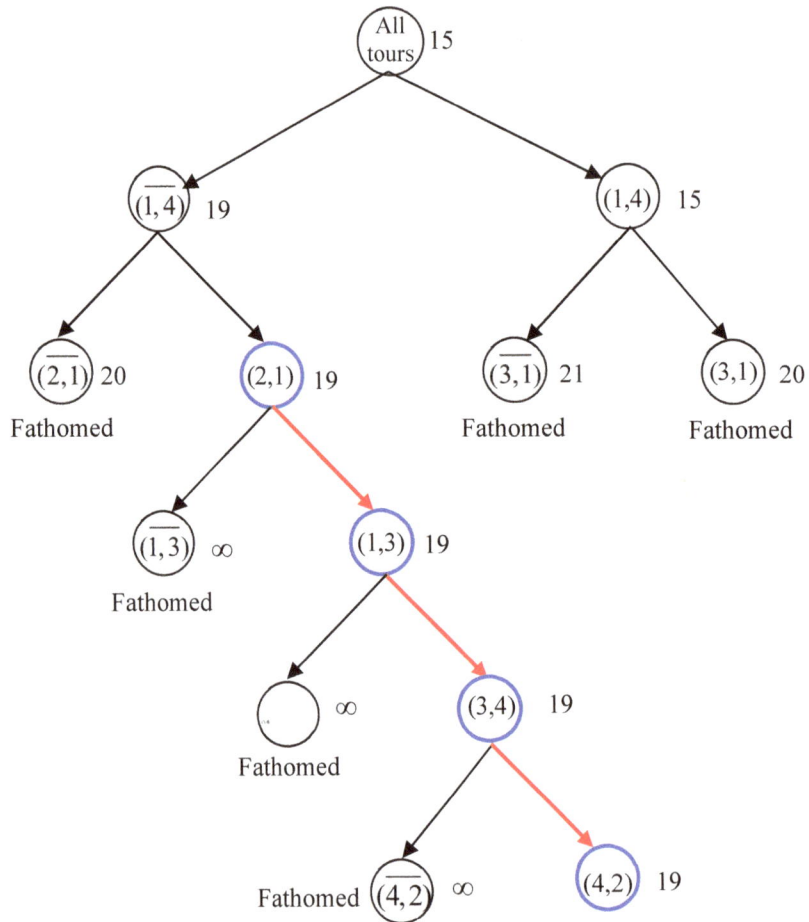

Figure 5.6. Complete branch-and-bound tree.

B. NOP Solution

Figure 5.7 shows the NOP screens for this example: (1) Input data file, (2) NOP Output results. The optimal tour is (1,3), (3,4), (4,2) and (2,1), with total length equal to 19.

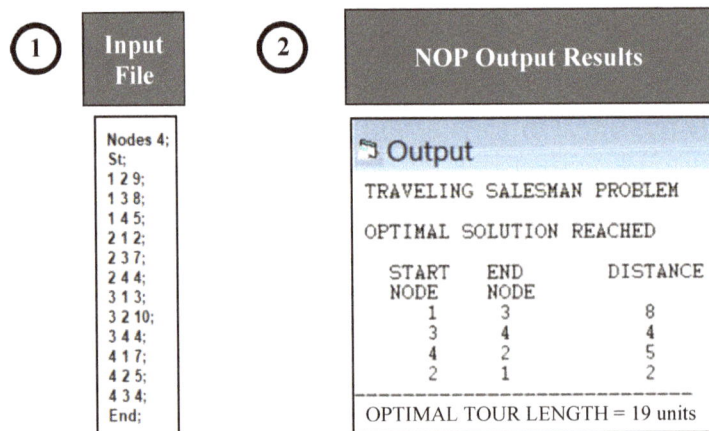

Figure 5.7. Sequence of Computer Screens for TSP Example.

204

5.3 THE MULTI-TRAVELING SALESMAN PROBLEM[1]

The Multi-Traveling Salesman Problem can be defined as follows. Given N cities and M available salesmen based at one of these cities, it is desired to visit the N cities in such a way that (a) each city different from the base city is visited exactly once, (b) all tours (closed chains containing the base city) start and finish at the base city, and (c) each tour is assigned to exactly one salesman. There is a fixed cost associated with each salesman and a travel cost associated with a trip between any two cities. An optimal solution to this problem consists of one or more tours resulting in minimal total fixed plus travel costs. There are three cases of the problem that are of particular interest:

- **Case 1.** $M = 1$; this is the well-known Traveling Salesman Problem [2,4,10], where it is desired to identify the order in which the cities should be visited to minimize total travel cost.
- **Case 2.** $M > 1$ and all salesmen are used; in this case, it is desired to identify M tours, each one including the base city, such that the combined travel and fixed costs are minimized.
- **Case 3.** $M > 1$ and at most M salesmen are used; in this case, the number of tours minimizing the travel cost plus the fixed charges is to be determined, in addition to the cities contained in each tour. It is noted that for travel cost data satisfying the triangular inequality the optimal solution consists of exactly one tour; hence, Case 3 would reduce to Case 1.

Existing solution methods for the MTSP are, for the most part, extensions of algorithms developed for the TSP. Most existing methods for the TSP approach the problem similarly by relaxing one or more of the three major sets of constraints:

(a) **Degree constraints.** The effect of these constraints is to prevent more than one arc from being used to enter or exit any city. When the cost is the same for any of the two directions in which two cities can be connected, the relaxation of these constraints yields a minimal spanning tree problem.
(b) **Subtour elimination constraints.** The purpose of these constraints is to eliminate any solution containing a subtour that does not include all cities. The relaxation of these constraints yields an assignment problem.
(c) **Integrality constraints.** These constraints require that all variables in the problem be integer. The relaxation of these constraints yields a linear programming problem.

5.3.1 Transformation Technique

The transformation of a multiple traveling salesman problem into a single traveling salesman problem was suggested first by Christofides and Eilon [3]. They applied the procedure to solve a version of the vehicle scheduling problem as a traveling salesman problem. This section summarizes the fundamental characteristics of the transformation procedure developed by Bellmore and Hong [2] to convert an MTSP into an equivalent TSP. The transformation procedure begins by labeling the M salesmen 1, 2, …, M, such that $F_1 \leq F_2 \leq F_3 \leq … \leq F_M$, where F_i is the

[1] *NETWORKS*, Vol. 15 (1985) 455-467, Alberto Garcia-Diaz.

fixed cost associated with salesman i. The problem is equivalent to finding r tours containing the base city, $r \leq M$, such that $\sum_{k=1}^{r} Z_k + \sum_{k=1}^{r} F_k$ is minimized over all possible values of r, where Z_k is the *length* (or travel cost) of the kth tour.

Network $G = (\mathbf{N}, \mathbf{A})$ with base city as node 1 can be expanded into network $G^* = (\mathbf{N} \cup \mathbf{N}^*, \mathbf{A} \cup \mathbf{A}^*)$ by including the following elements: (a) M-1 nodes 1, 2, …, M-1 in set $\mathbf{N}^*$ to represent salesmen 2, 3, …, M, respectively; (b) one arc $(i,j) \in \mathbf{A}^*$ for each node $i \in \mathbf{N}^*$ and each node $j \in \mathbf{N}$ such that $(1,j) \in \mathbf{A}$; (c) one arc $(j,i) \in \mathbf{A}^*$ for each node $i \in \mathbf{N}^*$ and each node $j \in \mathbf{N}$ such that $(j,1) \in \mathbf{A}$; and (d) one arc $(i+1,i)$ for each node $i \in \mathbf{N}^* \cup \{1\}$. For each arc (i,j) of G^*, a cost parameter b_{ij} is defined as follows [4]:

$b_{ij} = c_{ij}$ for $(i,j) \in \mathbf{A}$
$b_{ij} = c_{1j} + \frac{1}{2}F_i$ for (i,j) such that $i \in \mathbf{N}^*, j \in \mathbf{N}$
$b_{ji} = c_{j1} + \frac{1}{2}F_i$ for (i,j) such that $i \in \mathbf{N}^*, j \in \mathbf{N}$
$b_{i+1,i} = \frac{1}{2}F_i - \frac{1}{2}F_{i+1}$ for (i,j) such that $i \in \mathbf{N}^*, j \in \mathbf{N}$

The above transformation corresponds to Case 3. Case 2 requires a similar transformation that does not generate arcs $(i+1,i)$ such that $i \in \mathbf{N}^*$. An illustration of this transformation for Case 3 is shown in Figure 5.8. Here nodes 0, –1 and –2 represent three salesmen.

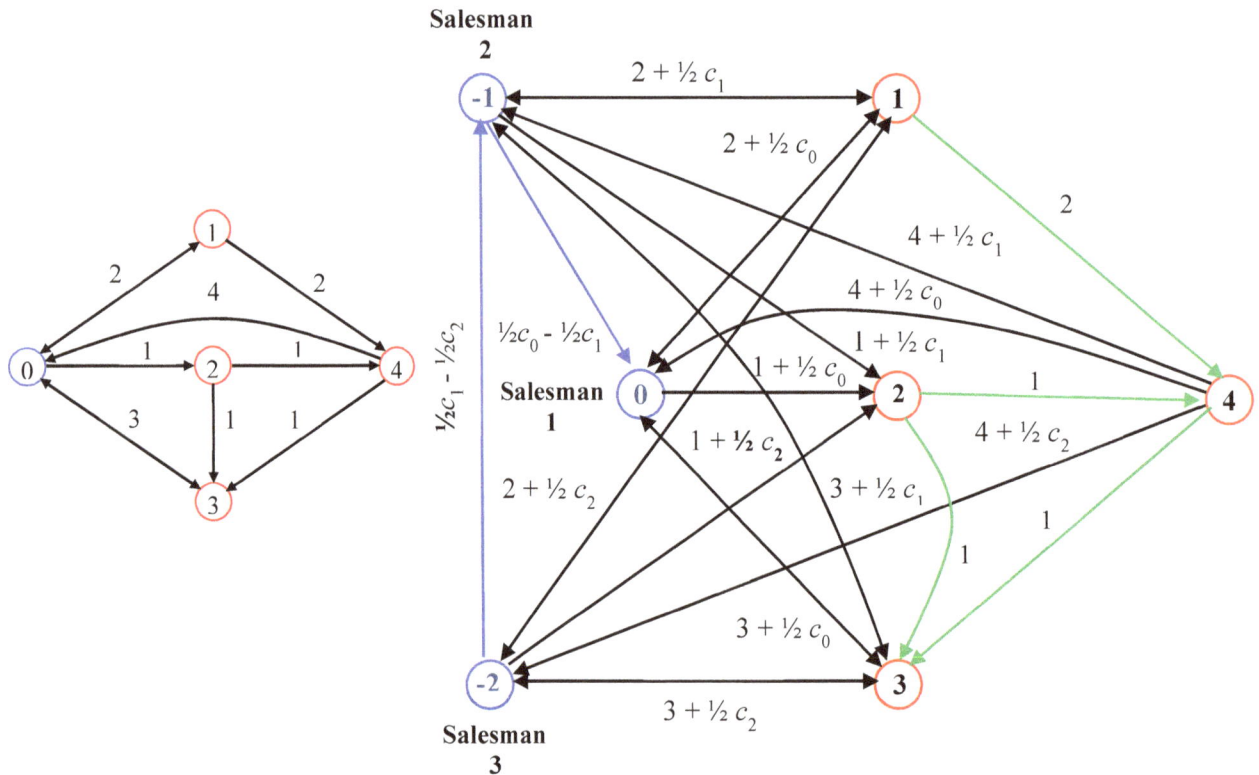

Figure 5.8. Transformation for Case 3.

5.3.2 A Heuristic Procedure

As indicated in Section 5.3.1, Bellmore and Hong [2] showed that an N-city, M-salesman problem could be transformed to an equivalent TSP problem with $N+M$-1 cities by creating M-1 additional nodes and linking these nodes to all nodes of the original network. A number of algorithms have been developed to solve the MTSP model. The objective of this section is to describe a *heuristic* circulation-network algorithm developed by Garcia-Diaz [5] to solve the Multi-Traveling Salesman Problem without transforming it into an equivalent single TSP. When the number of salesmen is equal to 1 this heuristic procedure can be used to solve the TSP, as well.

Step 1. Representation of the MTSP as a Capacitated Circulation Network. This representation is performed as follows for each of the three cases previously described.

Case 1. Let G'(**N'**,**A'**) be the capacitated circulation-network representation of the original network G(**N**,**A**). Each node $i \in$ **N**, $i \neq 1$, corresponds to two nodes i_a, $i_b \in$ **N'** and a directed arc $(i_a, i_b) \in$ **A'**. Node $1 \in$ **N**, which represents the base city, corresponds to a super source i_b, a super terminal $i_a \in$ **N'** and a directed arc $(1_a, 1_b) \in$ **A'**, which serves as the return arc of the circulation network G'(**N'**,**A'**). Each directed arc $(1_a, 1_b)$, including the return arc, has a capacity-cost triplet $[U^i_{ab}, L^i_{ab}, c^i_{ab}]$ set equal to [1,1,0] where

U^i_{ab} = upper bound on flow on arc (i_a, i_b)

L^i_{ab} = lowe bound on flow on arc (i_a, i_b)

c^i_{ab} = per-unit flow cost for arc (i_a, i_b)

Additionally, for each arc $(i, j) \in$ **A** there is directed arc $(i_b, i_a) \in$ **A'** with its capacity-cost triplet $[U^{ij}_{ba}, L^{ij}_{ba}, c^{ij}_{ba}]$ set equal to $[1, 0, c_{ij}]$.

Case 2. The construction of the capacitated circulation network G'(**N'**,**A'**) for Case 2 essentially follows the same procedure described for Case 1, with the exception that the parameters in the capacity-cost triplet for the return arc are [M,M,0], where M represents an arbitrarily large number.

Case 3. In this case $M + 1$ *additional* nodes and $2M$ *additional* arcs are inserted in the transformed network obtained in Case 1 or Case 2. These two networks have the same structure although different parameters for their return arcs. The original super source i_b is relabeled as D_0, which now acts as a dummy node. The additional nodes are denoted by S_1, S_2, …, S_M, and i_b, where node S_i represents the ith salesman and 1_b is the new super source node of the network. The additional arcs are used to connect node 1_b to each salesman node S_i, and each node S_i to node D_0. The return arc $(1_a, 1_b) \in$ **N'** has a capacity-cost triplet set equal to [M, 1, 0]. Each additional arc $(S_i, D_0) \in$ **A'**, $i \leq M$, has a capacity-cost triplet set equal to [1,0,0]. Each additional arc $(1_b, S_i) \in$ **A'**, $i \leq M$, has a capacity-cost triplet set equal to $[1, 0, F_i]$.

Step 2. Identify Minimum-Cost Flows in the Circulation Network. Each city in the original network is represented as a *node*. In the transformed circulation network, it is represented as an *arc* with its flow forced to be equal to one. This guarantees the condition that each city must be visited exactly once. An optimal solution to the circulation problem, therefore, corresponds to either a complete tour when all cities are visited, or a subtour otherwise. In the minimum-cost flow solution associated with G'(**N'**,**A'**), the flow along arc (i_a, i_b) will be represented by f_{ab}^i, and that along arc (i_b, j_a) by f_{ba}^{ij}. If a minimum-cost solution exists for the circulation network, all arcs (i_a, i_b) will have $f_{ab}^i = 1$; in this case, Step 3 is performed next. Otherwise, the procedure is stopped with an infeasible solution for the MTSP.

Step 3. Subtour Elimination. A subtour is defined as a closed chain in G'(**N'**,**A'**) which does not contain both nodes 1_a and 1_b. The subtour elimination procedure consists of three phases:

- Identification of subtours to determine the feasibility of the solution (set of tours).
- Link elimination procedure to break unacceptable subtours.
- Preservation of flow-conservation conditions for the nodes of the network.

Details about these procedures are given below.

Phase 3.1. *Identification of subtours* (*feasibility test*). If there are no subtours, the current solution is feasible. If a feasible solution was obtained in Step 2, it is also optimal; if it was obtained in Phase 3.3 it can be suboptimal. A feasible solution can be identified by tracing all arcs in **A'** having positive flow. This procedure yields exactly one tour for each salesman used; in particular, if the flow in arc $(i_b, j_a) \in$ **A'** is equal to one, city j must be visited immediately after visiting city i. When the current solution is not feasible because of the existence of subtours, Phase 3.2 is performed next. Otherwise, the procedure stops with a feasible solution to the MTSP.

Phase 3.2. *Link elimination.* Arcs (i_a, i_b) have $f_{ab}^i = L_{ab}^i = U_{ab}^i = 1$ and hence are not considered for possible elimination. In this phase of the algorithm, all arcs $(i_b, j_a) \in$ **A'** forming subtours are inspected to determine the arc with the highest value of c_{ba}^{ij}. Note that these arcs have lower bounds $L_{ba}^{ij} = 0$. The elimination of the corresponding arc is performed by setting its cost $c_{ba}^{ij} = \infty$. The eliminated arc will be relabeled as (i_b^*, j_a^*). After this, Phase 3.3 is conducted to reroute the flow along the eliminated arc. An alternative arc elimination procedure is to remove the arc which causes the minimal change in the objective function value, instead of selecting the one associated with the maximal arc cost. However, it is not possible to predict if this alternative represents an improvement over the proposed procedure since the removal of the minimal-cost arc may cause more expensive arcs to be used in rerouting the flow in the eliminated arc.

Phase 3.3. *Preservation of flow conservation conditions.* The optimality condition for any arc $(i_b, j_a) \in$ **A'** can be summarized as follows:

(a) $\overline{c}_{ba}^{ij} < 0 \Rightarrow f_{ba}^{ij} = U_{ba}^{ij}$

(b) $\overline{c}_{ba}^{ij} > 0 \Rightarrow f_{ba}^{ij} = L_{ba}^{ij}$

(c) $\overline{c}_{ba}^{ij} = 0 \Rightarrow L_{ba}^{ij} \le f_{ba}^{ij} \le U_{ba}^{ij}$

(d) $\sum_{a \in N'} f_{ba}^{ij} - \sum_{a \in N'} f_{ab}^{ij} = 0$ for all $b \in$ **N'**

208

where $\bar{c}_{ba}^{ij} = c_{ba}^{ij} + \pi_b - \pi_a$. Here π_a and π_b are the dual values associated with the flow-conservation constraints for nodes a and b, respectively. As a result of setting $\bar{c}_{ba}^{ij} = \infty$ for arc (i_b^*, j_a^*), the optimality condition (in-kilter condition) for this arc becomes (a) $f_{ba}^{ij} = 0$ and (b) $\bar{c}_{ba}^{ij} = \infty$. To reduce the flow from $f_{ba}^{ij} = 1$ to $f_{ba}^{ij} = 0$, it is necessary to reroute the flow through the arcs of the network while preserving the optimality conditions of all arcs in the modified network after removing arc (i_b^*, j_a^*). This can be done following the labeling procedure outlined in Chapter 3 for the Out-of-Kilter algorithm. Once this procedure is finished a new solution is obtained and Phase 3.1 is repeated.

5.3.3 Example with $M = 2$

Assume the following costs for the trips shown in the network of Figure 5.8: $c_{12} = 2$, $c_{13} = 4$, $c_{14} = 1$, $c_{21} = 9$, $c_{25} = 7$, $c_{34} = 9$, $c_{35} = 5$, $c_{41} = 1$, $c_{53} = 20$ and $c_{54} = 4$. It is desired to visit each of the cities 2, 3, 4 and 5 exactly once, using two salesmen stationed at city 1 (home city). The problem will be solved first using the algorithmic steps described in Section 5.3.2 and then using the NOP software.

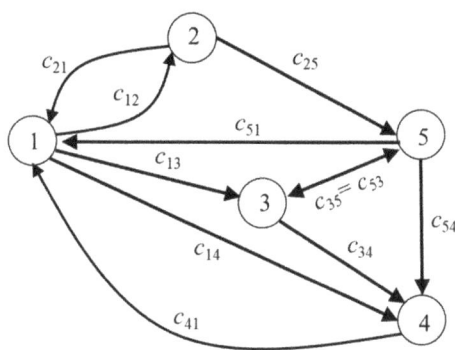

Figure 5.8. Network for MTSP.

A. Algorithmic Procedure

Step 1. Representation of the MTSP as a Capacitated Circulation Network. The transformed circulation network for Case 2 is shown in Figure 5.9, where $M = 2$.

Step 2. Identification of Optimal Flows in Circulation Network. Using the Out-of-Kilter algorithm the minimal-cost flows obtained for the circulation network of Figure 5.9 with $M=2$ are summarized in Table 5.16.

Step 3. Subtour Elimination. It can be verified that Phase 3.1 indicates that the minimum-cost circulation flows correspond to an optimal solution to the MTSP. Inspecting Table 5.16 we conclude that the tours for the two salesmen are: (1,2), (2,1) for the first salesman, and (1,3), (3,5), (5,4), (4,1). The total travel cost by both salesmen is equal to 25.

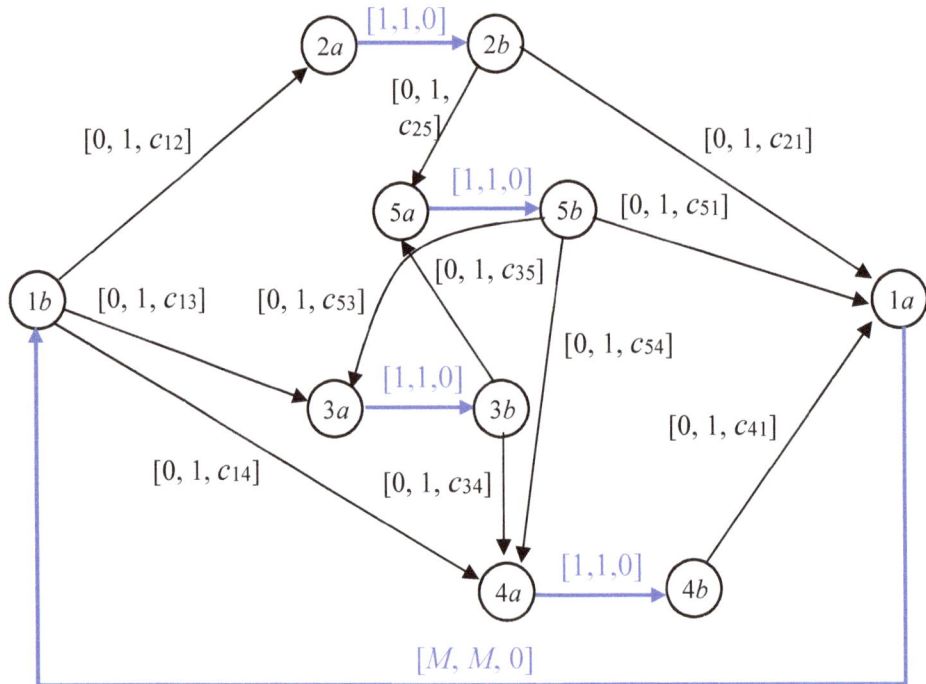

Figure 5.9. Circulation Network for Case 2, $M=2$.

Table 5.16. OPTIMAL FLOWS FOR CIRCULATION NETWORK, CASE 2

Start Node	End Node	Meaning of Arc In Original Network	Upper Bound	Lower Bound	Cost Per Unit	Arc Flow
1	2	Trip from City 1 to City 2	1	0	2	1
1	3	Trip from City 1 to City 3	1	0	4	1
1	4	Trip from City 1 to City 4	1	0	1	0
2	6	City 2	1	1	0	1
3	7	City 3	1	1	0	1
4	8	City 4	1	1	0	1
5	9	City 5	1	1	0	1
6	5	Trip from City 2 to City 5	1	0	7	0
6	10	Trip from City 2 to City 1	1	0	9	1
7	4	Trip from City 3 to City 4	1	0	9	0
7	5	Trip from City 3 to City 5	1	0	5	1
8	10	Trip from City 4 to City 1	1	0	1	1
9	3	Trip from City 5 to City 3	1	0	20	0
9	4	Trip from City 5 to City 4	1	0	4	1
10	1	City 1 (home city)	2	2	0	2

B. NOP Solution

Figure 5.10 shows the NOP screens for this example: (1) Input data file, (2) Additional Input Screens, and (3) NOP Output results. The additional screens show the number of salesmen

210

with the fixed cost for each salesman. Also, a screen shows if all salesmen must be used or not. In this example all salesmen are used (Case 2). When all salesmen may not be used, the given number of salesmen is regarded as an upper bound on the number actually used (Case 3). The NOP output results include the tour for each salesman used and the total cost (salesman costs plus travel costs). The NOP output results indicate that the tours for the two salesmen are: (1,2), (2,1) for the first salesman, and (1,3), (3,5), (5,4), (4,1). The total travel cost by both salesmen is 25.

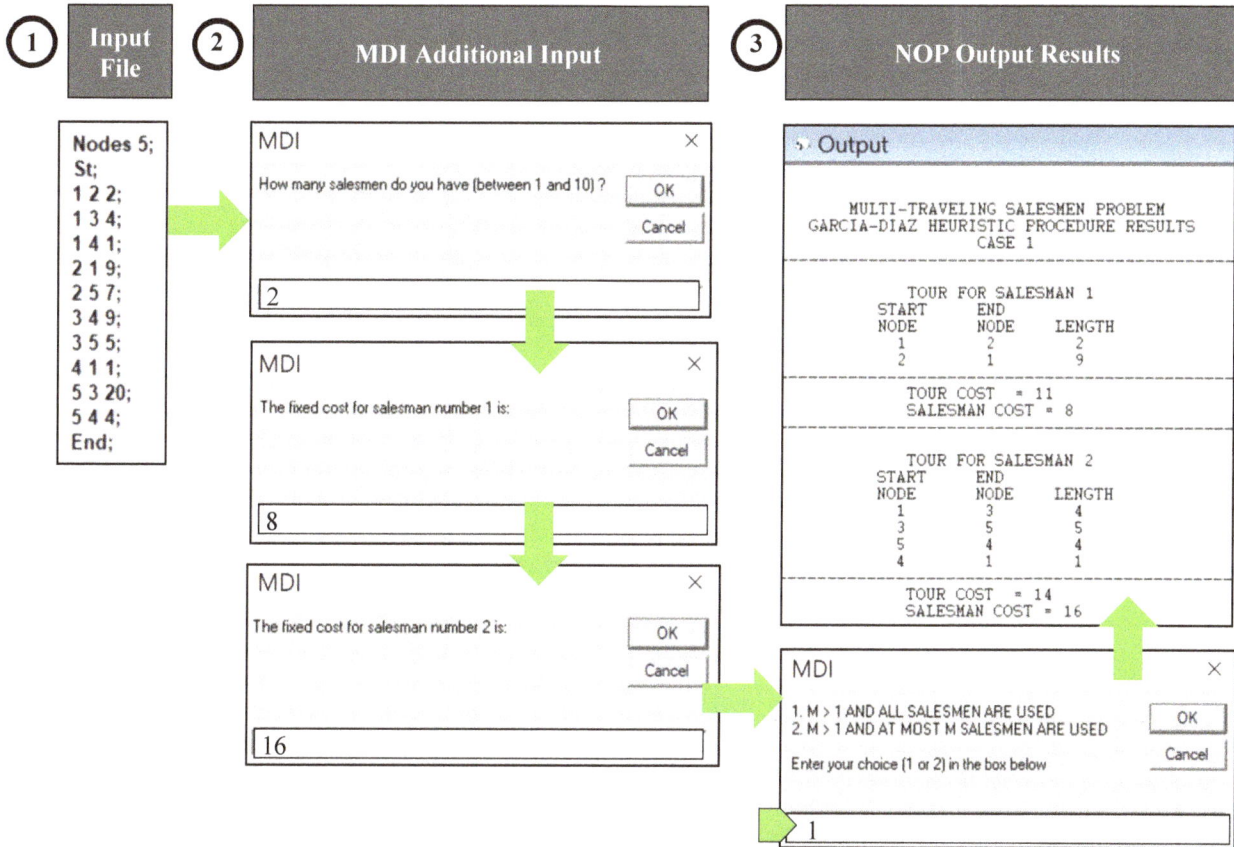

Figure 5.10. Sequence of Computer Screens for MTSP Example, Case 2.

5.3.4 Example with $1 \leq M \leq 2$

In this example the network and data given in Section 5.3.3 are considered again and each city must be visited exactly once using at most two salesmen, instead of using exactly two.

A. Algorithmic Procedure

Step 1. Representation of the MTSP as a Capacitated Circulation Network. The transformed circulation network for Case 3, assuming up to two salesmen, is shown in Figure 5.11.

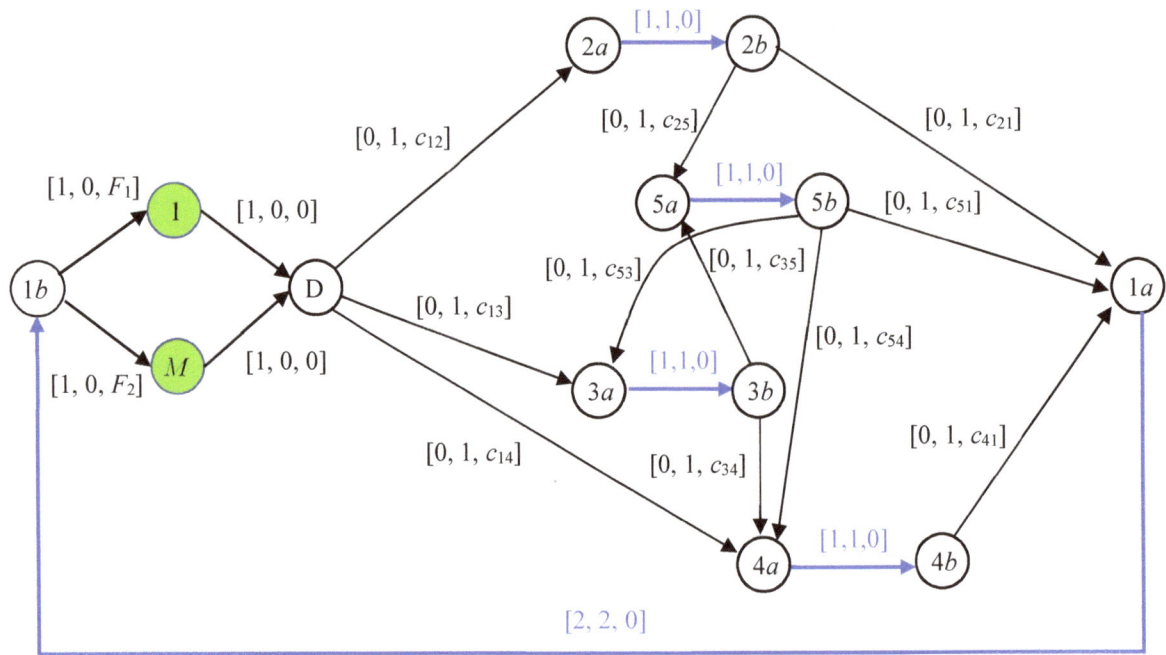

Figure 5.11. Circulation Network for Case 3, $1 \leq M \leq 2$.

Step 2. Identification of Optimal Flows in Circulation Network. Using the Out-of-Kilter algorithm, the minimal-cost flows obtained for the circulation network of Figure 5.11 with $1 \leq M \leq 2$ are summarized in Table 5.17.

Table 5.17. OPTIMAL FLOWS FOR CIRCULATION NETWORK, CASE 3

Start Node	End Node	Meaning of Arc In Original Network	Upper Bound	Lower Bound	Cost Per Unit	Arc Flow
1	2	Trip from City 1 to City 2	1	0	2	1
1	3	Trip from City 1 to City 3	1	0	4	0
1	4	Trip from City 1 to City 4	1	0	1	0
2	6	City 2	1	1	0	1
3	7	City 3	1	1	0	1
4	8	City 4	1	1	0	1
5	9	City 5	1	1	0	1
6	5	Trip from City 2 to City 5	1	0	7	1
6	10	Trip from City 2 to City 1	1	0	9	0
7	4	Trip from City 3 to City 4	1	0	9	1
7	5	Trip from City 3 to City 5	1	0	5	0
8	10	Trip from City 4 to City 1	1	0	1	1
9	3	Trip from City 5 to City 3	1	0	20	1
9	4	Trip from City 5 to City 4	1	0	4	0
10	13	Number of salesmen used	2	1	0	1
11	1	Salesman 1	1	0	0	1
12	1	Salesman 2	1	0	0	0
13	11	Cost of Salesman 1	1	0	8	1
13	12	Cost of Salesman 2	1	0	16	0

Step 3. Subtour Elimination.   Once again, it can be verified that Phase 3.1 indicates that the minimum-cost circulation flows correspond to an optimal solution to the MTSP. From Table 5.17 we conclude that the optimal solution consists of one tour (one salesman) including the following trips: (1,2), (2,5), (5,3), (3,4), (4,1). The total cost is equal to 47, which includes the travel cost of 39 plus a fixed cost of 8 due to salesman 1.

A. NOP Solution

Figure 5.12 shows the NOP screens for this example. The input data file, and the first three additional screens are the same as those shown for Case 2. The tours for the two salesmen are: (1,2), (2,1) for the first salesman, and (1,3), (3,5), (5,4), (4,1). The total travel cost by both salesmen is equal to 25.

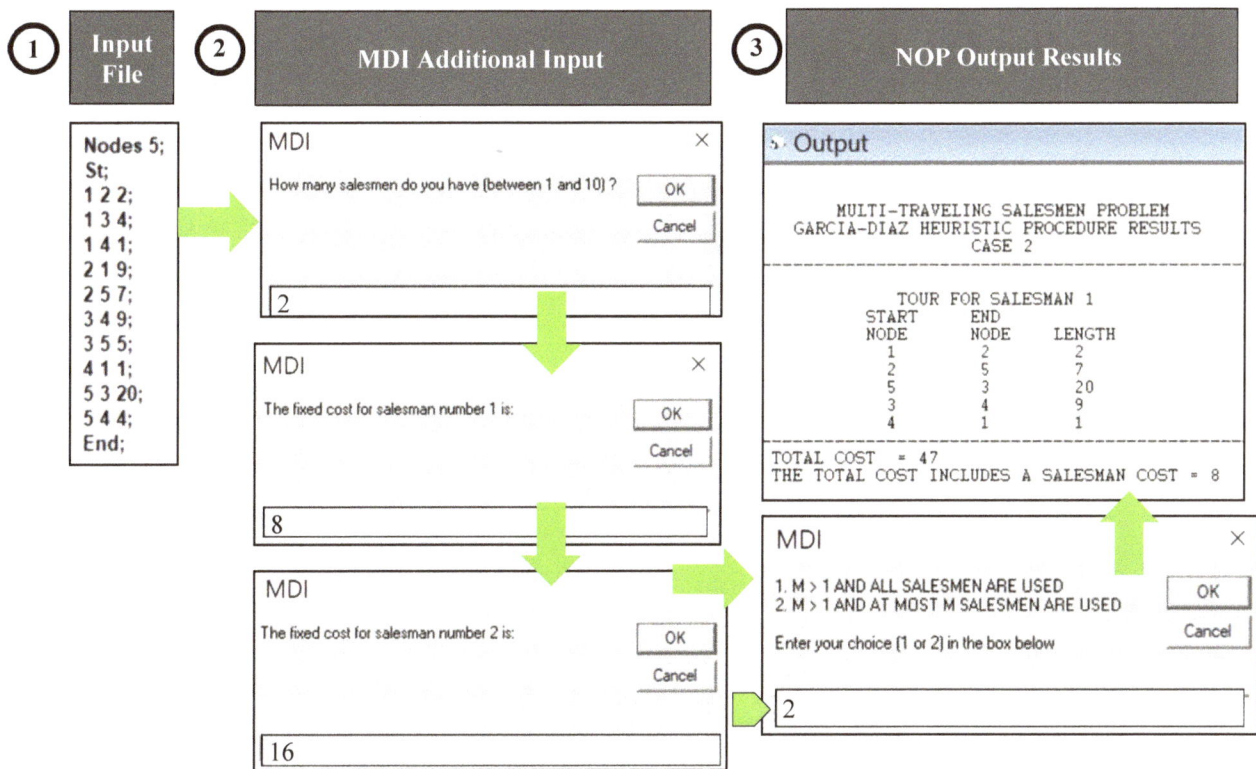

Figure 5.12. Sequence of Computer Screens for MTSP Example, Case 3.

5.3.5 Comparison of Procedures

In order to appreciate how well the heuristic procedure works, a good strategy would be to solve a number of problems having network size manageable by the optimal procedure based on the transformation into a single-traveling salesman model and compare the heuristic and optimal solutions.   As an illustration of this analysis, Case 2 and Case 3 will be considered for the example under discussion. Additional computational results are reported in the original article by Garcia-Diaz [5].

213

Case 2

Applying the transformation technique by Bellmore and Hong [2] described in Section 5.3.1, the following arc costs are calculated, ignoring the fixed costs of the salesmen: $c_{12} = 2$, $c_{13} = 4$, $c_{14} = 1$, $c_{21} = 9$, $c_{25} = 7$, $c_{34} = 9$, $c_{41} = 1$, $c_{35} = 5$, $c_{53} = 20$ and $c_{54} = 4$. Figure 5.13 shows the equivalent network and the optimal solution. From the interpretation of this solution the following tours are deducted for the two salesmen:

Salesman 1: (1,2), (2,1)   Cost = 2+9 +8 = 19

Salesman 2: (1,3), (3,5), (5,4), (4,1)   Cost = 4+5+4+1+16 = 30

The total travel cost for both salesmen is equal to 25, which is the same value obtained by the heuristic procedure. The fixed costs of the salesmen are equal to 8+16=24. Adding this to the travel cost will result in a total cost of 49.

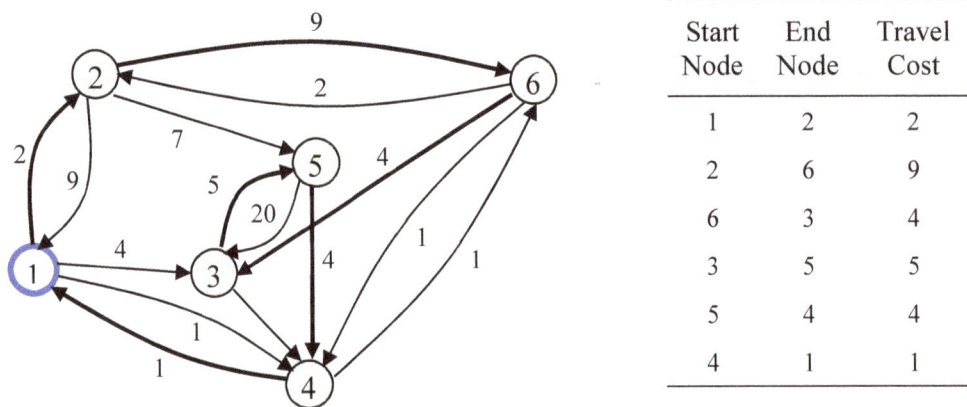

Start Node	End Node	Travel Cost
1	2	2
2	6	9
6	3	4
3	5	5
5	4	4
4	1	1

Figure 5.13. TSP Solution for Case 2.

Case 3

Assuming that the costs of the first salesman is 8 and the cost of the second one is 16, the following arc costs are calculated for the single TSP model: $c_{12} = 2$, $c_{13} = 4$, $c_{14} = 1$, $c_{21} = 9$, $c_{25} = 7$, $c_{34} = 9$, $c_{41} = 1$, $c_{35} = 5$, $c_{53} = 20$ and $c_{54} = 4$. Figure 5.14 shows the equivalent single TSP network and the corresponding optimal solution (which can be obtained by using NOP).

From the optimal solution it is concluded that only one salesman is used. The salesman associated with the lowest fixed cost should selected, that is, salesman 1. The following tour is obtained for Salesman 1: (1,2), (2,5), (5,3), (3,4), (4,1). The total cost is equal to 47. This cost has two components, 39 for travel plus a fixed cost of 8.

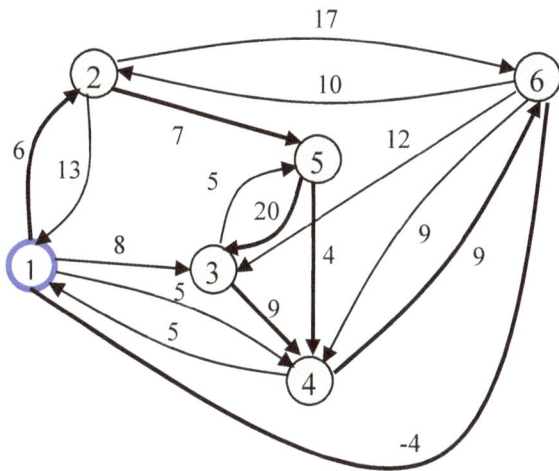

Start Node	End Node	Arc Length
1	2	6
2	5	7
5	3	20
3	4	9
4	6	9
6	1	-4

Figure 5.14. TSP Solution for Case 3.

EXERCISES

1. Explain the purpose of branching, lower bounds, backtracking and fathoming in the TSP algorithm discussed in this chapter. Indicate how to select a trip for branching and how to update lower bounds.

2. Consider the following travel cost matrix, for a network consisting of cities 1, 2, 3, and 4:

	1	2	3	4	5
1	-	38	83	37	21
2	26	-	93	10	56
3	55	40	-	48	49
4	43	77	29	-	48
5	28	61	86	73	-

A salesman wishes to determine the minimum-cost route starting at city 1 and ending at the same city after visiting each other city exactly once.

3. Solve Problem 2 assuming that the salesman is given the following instructions: (A) Go to city 4 after visiting both city 5 and city 2. (B) Do not go from city 1 to city 2.

4. Some manufacturing operations need to be performed on a given piece at locations A, B, and C by an automatic NC machine. Each type of operation requires a different tool (two tools in this problem). The following information is provided: (a) Table 1 shows the type of operations and order in which they must be performed at each location (at B and C milling must be performed before drilling); (b) Table 2 shows the time required by the machine to automatically change tools; (c) Table 3 shows the times by the tool holder to travel from one location to another. Traveling and tool changing are done simultaneously when both are needed between operations. The tool holder starts the cycle at position 1 (the origin) with the drill on and ends at the same position with the tool used in the last operation. It is desired to

215

model and solve the problem of finding the sequence of operations resulting in minimal total setup time (due to tool changing and travel) as a TSP.

Table 1

	A	B	C
2	Mill		
3		Mill	
4		Drill	
5			Mill
6			Drill

Table 2

	Mill	Drill
Mill	0	25
Drill	35	0

Table 3

1	A	B	C
0	10	20	40
A	10	0	30	35
B	20	30	0	40
C	40	35	40	0

5. Find the optimal tour for the TSP with the given distance matrix. City 1 is the home city.

6. Find the recommended tour in the network given for Problem 5 applying the heuristic procedure to solve the MTSP. Solve the circulation network problems using a standard LP software.

	1	2	3	4	5	6
1	-	27	43	16	30	26
2	7	-	16	1	30	30
3	20	13	-	35	5	0
4	21	16	25	-	18	18
5	12	46	27	48	-	5
6	23	5	5	9	5	-

7. Consider the data given in Problem 5. Assume that two salesmen must be used. Their fixed costs are equal to 15 and 18. (a) Transform this problem into an equivalent TSP and find the optimal tours. (b) Use the heuristic procedure to find the recommended tours. Solve network circulation problems using standard LP software.

8. For the TSP problem given in Problem 5, there are two salesmen available. The fixed costs associated with these salesmen are 5 and 7. (a) Find a heuristic solution for Case 2 and Case 3 of the MTSP. Solve circulation network problems using standard LP software. (b) Use the transformation technique given in Section 5.3.1 and solve the transformed network using Little's procedure.

9. Consider a production schedule problem at a machine that produces a single product with different colors. There is only one setup cost involved, the cost of changing the color in the machine. The objective is to obtain a production schedule that minimizes the sum of the setup costs in the machine. The cost of a color change is dependent upon an immediately preceding color. The cost of changing from color i to color j is denoted by C_{ij} and can be expressed in the form of a matrix (dollars), as indicated below:

	1	2	3	4	5	6	7	8	9	10
1	–	1	2	3	4	5	6	7	8	9
2	2	–	3	4	5	6	7	8	9	1
3	3	4	–	4	5	6	7	8	9	1
4	4	5	6	–	7	9	8	1	2	3
5	5	6	7	8	–	9	1	2	3	4
6	6	7	8	9	1	–	2	3	4	5
7	7	8	9	1	2	3	–	4	5	6
8	8	9	1	2	3	4	5	–	6	7
9	9	1	2	3	4	5	6	7	–	8
10	1	2	3	4	5	6	7	8	9	–

$\mathbf{C} =$ (the matrix above)

10. A tourist is planning a visit to Puerto Rico. She wants to rent a car and spend one day in each of the following cities: San Juan, Caguas, Ponce, Mayaguez and Arecibo. Determine a tour that would minimize total traveled distance, starting and ending at San Juan. (a) Use Little's algorithm; (b) use the heuristic procedure studied in this chapter.

	SJ	C	P	A	M
SJ	-	17	64	48	94
C	17	-	47	58	94
P	64	47	-	52	47
A	48	58	52	-	47
M	94	94	47	47	-

11. For the given cost matrix find the minimum cost route for a salesman starting at city 1 and returning to the same city after visiting each other city exactly once. He has been given the following instructions: (a) be sure to go to city 4 after visiting both city 2 and city 5; (b) If you go to city 4 from city 2, then do not go from city 1 to city 5.

	1	2	3	4	5
1	-	83	38	37	21
2	26	-	93	10	56
3	55	40	-	48	49
4	43	77	29	-	48
5	28	61	86	73	-

12. (a) Apply the heuristic procedure for the MTSP to the data given in Exercise 2, assuming costs equal to 21 and 25 for two salesmen; solve for the problem for Case 2 and Case 3 and interpret the solution. (b) Use Bellmore and Hong's transformation procedure for part (a) and find the optimal solution for each case.

REFERENCES

[1]   BELLMORE, M. AND G. L. NEMHAUSER, The traveling Salesman Problem: A Survey. *Operations Res.* 16 (1968) 538-558.

[2]   BELLMORE, M. AND S. HONG, Transformation of Multi-Salesman Problem to the Standard Traveling Salesman Problem. J. *ACM.* 21 (1974) 500-504.

[3]   CHRISTOFIDES, N. AND S. EILON, An Algorithm for The Vehicle Dispatching Problem. *Operational Res. Quart.* 20 (1969) 309-318.

[4]   CHRISTOFIDES, N. AND S. EILON, Algorithm for Large-Scale Traveling Salesman Problems. *Operational Res. Quart.* 23 (1972) 511-518.

[5]   GARCIA-DIAZ, A., A Heuristic Circulation-Network Approach to Solve the Multi-Traveling Salesman Problem," *Networks*, Vol. 15, pp. 455-467, 1985.

[6]   GAREY, M. R. AND D. S. JOHNSON, Computers and Intractability: A Guide to the Theory of NP-Completeness, 1979.

[7]   GAVISH, B. AND K. SRIKANTH, An Optimal Solution Method for the Multiple Traveling Salesman Problem. *Operations Research,* 34 (1986), 6, 698-710, 1986.

[8]   LAWLER, E. L., J. K. LENSTRA, AH.G. RINNOOY KAN, AND D. B. SHMOYS, The Traveling Salesman Problem, John Wiley & Sons, New York, 1985.

[9]   LAWLER, E. L., A Solvable Case of The Traveling Salesman Problem, *Math. Programming* 1, (1971) 267-269.

[10] LITTLE, J.D.C., K. G. MURTY, D. W. SWEENEY, AND C. KAREL, An Algorithm for the Traveling Salesman Problem, *Operations Research*, 11 (1963) 972-978.

Chapter 6

NETWORK-BASED PROJECT MANAGEMENT PROCEDURES

And indeed, which of you here, intending to build a tower, would not first
sit down and work out the cost to see if he had enough to complete it?
Otherwise, if he laid the foundation and then found himself unable to finish the work,
the onlookers would all start making fun of him saying,
"Here is a man who started to build and was unable to finish."

Luke, 14: 28-31
The Jerusalem Bible, 1968

A project can be defined as a collection of activities and events linked by precedence relationships and activity durations. In Chapter 2 a brief introduction to activity networks was presented to support the applicability of network analysis in the area of project management. The purpose of Chapter 5 is to study in detail specific procedures for the analysis of an activity network. Throughout this discussion the activity-on-arc representation will be used. Furthermore, for the activity network $G = (\mathbf{N}, \mathbf{A})$, we will assume that $\mathbf{N} = \{1,2,\ldots,n\}$, with node 1 representing the start of the project and node n its finish. It is possible to number the nodes of the network in such a way that for each arc (i,j) we have that $i < j$ since there are no circuits.

Network representations used for CPM and PERT project control and activity scheduling have a very special structure. In particular, such representations contain directed arcs and form *circuitless* networks. Because of this special structure, very efficient and simple computational procedures have been developed to recover relevant project information. These specialized techniques, along with related problems of cost and resource control, form the basis for this chapter. This chapter is divided into two parts. Part I concerns itself with the diagrammatical structure, network construction, and computational procedures used in CPM/PERT analysis, and Part II addresses resource and cost control in CPM/PERT networks.

PART I

Project Management with CPM and PERT

With contributions by
Warren H. Thomas
Distinguished Professor Emeritus
State University of New York

Management of complex projects that consist of a large number of interrelated activities demands significant effort in planning, scheduling, and control, especially when the project activities have to be performed in a specified sequence. With the help of CPM (critical path method) and PERT (program evaluation and review technique), project managers can:

1. Plan the project ahead of time and foresee possible sources of complications and delays in completion.
2. Schedule project activities at appropriate times to conform with proper job sequences so that the project is completed as soon as possible.
3. Coordinate and control the project activities so as to stay on schedule in completing the project.

Closely associated with the philosophies of CPM and PERT are the problems of resource balancing/utilization, compression of selected project activities to reduce total project duration, and the analysis of allowable delays known as *slack* or *float* in order to further coordinate the project. Such considerations allow project managers to perform the following tasks:

1. Sequence and schedule the use of scarce resources throughout the life of the project.
2. Dynamically control the start dates of each activity.
3. Optimally allocate further funds to the project in an effort to reduce the duration of the entire project.
4. Perform cost/time trade-off analysis among various activities using slack and float measures.

6.1 ORIGIN AND USE OF PERT

PERT (Program Evaluation and Review Technique) was developed by the U.S. Navy during the late 1950s to accelerate the development of the Polaris Fleet Ballistic Missile. The development of this weapon involved the coordination of the work of thousands of private contractors and other government agencies. The coordination by PERT was so successful that the entire project was completed 2 years ahead of schedule. This has resulted in further applications of PERT in other weapons development programs in the Navy, Air Force, and Army. The technique has been and continue to be extensively used in industry and service organizations.

The time required to accomplish the various activities in a research and development project is generally not known a priori. Thus, in its analysis PERT incorporates uncertainties in activity times. It determines the probabilities of completing various stages of the project by specified deadlines, and also calculates the expected time to complete the project. An important and extremely useful by-product of PERT analysis is its identification of various *bottlenecks* in the project. In other words, it identifies the activities that have high potential for causing delays in completing the project on schedule. Thus, even before the project has started, the project manager knows where to expect delays. He or she can take the necessary preventive measures to reduce possible delays so that the project schedule is maintained. Because of its ability to handle uncertainties in job times, PERT is extensively used in research and development projects.

CPM (Critical Path Method) closely resembles PERT in many aspects but was developed independently by *E. I. du Pont de Nemours Company*. As a matter of fact, the two techniques, PERT and CPM, were developed almost simultaneously. The major difference between them is that CPM does not incorporate uncertainties in job times. Instead, it assumes that activity times are proportional to the amount of resources allocated to them, and that by changing the level of resources the activity times and the project completion time can be varied. Thus, CPM assumes prior experience with similar projects from which relationships between resources and job times are available. CPM then evaluates the trade-off between project costs and project completion time.

CPM is mostly used in construction projects where one has prior experience in handling similar projects. In current practice, the words PERT and CPM are sometimes used interchangeably. The definitions and procedures that follow apply to both PERT and CPM.

CPM/PERT methods are applicable to projects where a structured sequence of tasks must be completed to achieve a single objective. Illustrations of such projects would be the construction of a building or other structure, a major repair activity, the design and implementation of a complex weapon system, the production of a large piece of equipment, and R&D activities. These projects have several common characteristics:

1. They are comprised of a well-defined collection of tasks which, when completed, mark the end of the project.

2. The tasks are ordered such that they must be performed in a given sequence.

3. The time to complete each task is known in advance or can be reasonably estimated. In the CPM formulation of the problem, the time to complete a task is assumed to be a single value known in advance (or can be estimated closely). In PERT, greater uncertainty is incorporated by permitting the planner to submit an upper and lower bound on the completion time of each task. The method of task completion time specification represents the only real difference between CPM and PERT.

4. A task once started is allowed to continue without interruption until it is completed.

5. A succeeding task does not need to start immediately upon completion of an immediate predecessor task, although it cannot commence until the prior one has been completed. This characteristic causes some difficulty in the application of CPM or PERT in continuous-processing industries, in which interruption between processing steps is not permitted.

6.2 NETWORK CONSTRUCTION

An activity network is a pictorial representation of a project showing the durations and interrelationships (precedence relationships) among all of its required tasks. The graph of the network is composed of directed arcs connecting pairs of nodes. The time-consuming tasks of a network (arcs) are known as activities. The nodes represent events. Events are well-defined points in time. For example, an event might represent the time at which all parts are on hand to permit the assembly of an item. The assembly itself, being a time-consuming element, is an activity.

In principle, the duration of a CPM/PERT project network can be determined by methods previously presented in this book. Indeed, the minimum duration of the project is given by the sequence of activities that yields the *longest path* through the network. This path is called a *critical path* and the activities that comprise the path are called *critical activities*. The activities are called critical because any increase or delay in their completion times will lengthen the entire project. The critical activities will play a major role in the methodologies of CPM/PERT and will not be further explored at this time. However, it is important to note that due to the specialized structure that each CPM/PERT network exhibits, there are easier ways to characterize the project duration than that of conventional maximum-flow algorithms. At this time, note the following structure of a CPM/PERT network:

1. Each arc has an orientation.

2. The project network is constrained to have no circuits.

3. Probabilistic branching is not allowed.

4. An activity cannot be started until its originating node is realized.

5. A node is realized when all activities terminating at the node are completed.

6. Assuming that the network has n nodes assigned the numerical labels 1, 2, ..., n, it is always possible to label each activity as (i,j) such that $i < j$. Because of this special structure, a very efficient algorithm can be employed to analyze project networks and determine the critical path. Such a procedure will be subsequently presented.

To clarify the discussions that follow, consider the network shown in Figure 6.1, which represents an activity network consisting of six activities, each represented by an arc. The number on each arc is the time required to perform each activity. Note that activities (2,3) and (2,4) cannot be started until node 2 is realized; also, node 5 will not be realized until activities (3,5) and (4,5) are both completed.

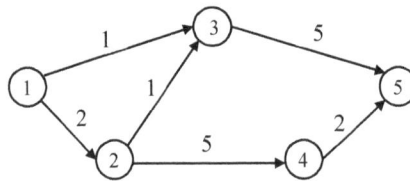

Figure 6.1. An activity network.

There are three distinct paths, described in terms of a sequence of nodes, involved in project completion:

    I. 1-3-5
    II. 1-2-4-5
    III. 1-2-3-5

Path I has a total duration of 6 units; path II has a duration of 9 units; and path III has a duration of 8 units. Path II, the longest path, is called the *critical path* since the timely scheduling of those activities on that path is necessary to prevent delay of project completion. Since activities (1,2), (2,4), and (4,5) are on the critical path, no delay in either their start or finish should be allowed. These are the *critical activities*. In other words, there is no *slack* in either the earliest possible start date or the latest allowable start date for those activities. Each activity must start on time. Hence, the earliest possible start time and the latest allowable start time coincide. In like manner, any activity could be characterized according to its earliest possible start (*ES*) and latest allowable start (*LS*). For example, the ES for activities (1,3) and (3,5) are 0 and 3, respectively. Since activity (3,5) need not be completed until time $T = 9$, the *LS* for activity (3, 5) is 4. Note that, correspondingly, the latest allowable start (*LS*) for activity (*1,3*) will be time 3. Hence, the slacks for activities (1,3) and (3,5), defined as *LS - ES*, are 3 units and 1 unit, respectively. Following this same logic, the reader can verify that activity (2,3) has slack of 1 unit.

6.2.1 Precedence Relationships

The direction of the arc represents a precedence relationship. In the network segment shown below event i must occur before activity A can be started. Similarly, event j cannot occur (be realized) until activity A has been completed.

The precedence relationships are *transitive* among the nodes (and as a result, among the activities). If node *i* precedes node *j* and node *j* precedes node *k*, then node *i* precedes node *k*, and activity *A* precedes activity *B*, as shown below:

Occasionally, a precedence relationship between activities cannot be represented correctly with conventional activity and event structure. For instance, assume that the network diagram in Figure 6.2 is intended to represent the following two sequences: (1) activity *G* follows both activities *B* and *C*, and (2) activity *E* follows activity *B* (but not activity *C*). This diagram is incorrect, for it implies that both *G* and *E* follow *C* and *B*.

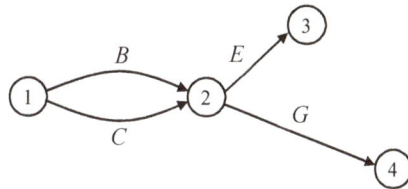

Figure 6. 2. Incorrect representation of a precedence relationship.

To achieve a correct representation, a dummy activity *X* (with *zero* duration) can be used, as shown by a dashed arrow in Figure 6.3. Dummy activities can be used whenever necessary to represent a relationship that cannot otherwise be depicted. They are merely a device for forcing a desired relationship without affecting the actual project duration. To illustrate the construction of a CPM/PERT diagram, we use the example provided in Section 6.4.2.

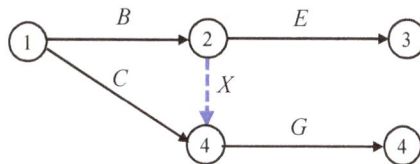

Figure 6.3. Network with a dummy activity.

6.2.2 A Manufacturing Problem

Consider the development of a project network to describe the relationships of the activities involved in the manufacture of a large machine in which subassemblies 1 and 2 are combined into subassembly 4, which is then joined with subassembly 3 to make the final product. Because of the need to match certain items in subassembly 3 with corresponding items in subassembly 2, the third subassembly cannot be built until the parts for subassembly 2 are on hand.

Let us assume that the principal activities required to build the machine are those identified in Table 6.1. This table provides information on the duration (in days) of the activities and the immediate predecessors for each activity.

The network representing the project is shown in Figure 6.4. Note that there is a single origin event and a single terminal event. With the exception of these, every event has at least one activity leading into it and at least one leading from it. Every activity provides a unique connection between two nodes, thus making it possible to identify an activity by the events that it connects. The event nodes have been numbered to provide for their identification. By convention they are numbered so that all activities lead from lower-numbered nodes to higher-numbered nodes.

Table 6.1. DATA FOR MANUFACTURING EXAMPLE

Activity		Duration (days)	Immediate Predecessors
Name	Description		
A	Procure parts for subassembly 1	5	None
B	Procure parts for subassembly 2	3	None
C	Procure parts for subassembly 3	10	None
D	Build subassembly 1	7	A
E	Build subassembly 2	10	B
F	Build subassembly 4	5	D and E
G	Build subassembly 3	9	B and C
H	Final assembly	4	F and G
I	Final inspection and test	2	H

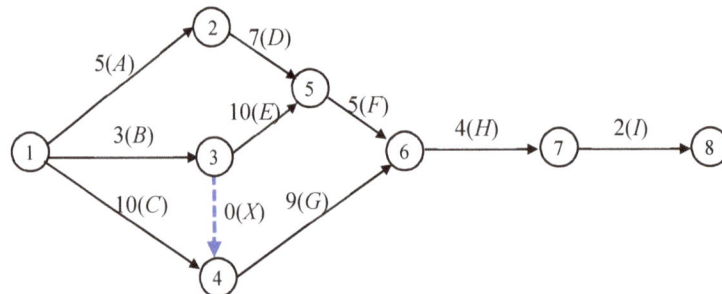

Figure 6.4. Activity-on-Arc network for manufacturing example.

The ultimate purpose of the network is to provide guidance for the scheduling of individual activities. To do this, one needs first to know the earliest possible and latest permissible times at which each event can occur.

6.3 EVENT EARLIEST AND LATEST REALIZATION TIMES

As mentioned in the previous section, an event (node) is realized when all activities ending at the node are finished. Additionally, an activity can be started as soon as its initial node is realized. Therefore, the earliest time a node can be realized is equal to the *duration* or *length* of the longest path from node 1 to the node itself.

Now we use the symbols $\prod_i^{(k)}$ and $T(\prod_i^{(k)})$ to represent the k*th* path from node 1 to node *i*, and the duration of this path, respectively. Furthermore, K_{1i} represents the number of paths between node 1 and node *i*. Using this notation, we can formally define the earliest realization time of node *i* as

$$T_i(E) = \max_{k \in \mathbf{K}} \{T(\prod_i^{(k)})\}, \ i = 2, 3, \ldots, n \qquad (6\text{-}1)$$

224

where $\mathbf{K} = \{1, 2, \ldots, K_{1i}\}$, and it is understood that $t_1(E) \equiv 0$. Instead of using Eq. (6-1), the *forward recursive relationship* given in Eq. (6-2) is computationally more efficient to calculate the *earliest realization times* of the nodes or events of a CPM project:

$$T_j(E) = \max_{i \in \beta_j} \{t_i(E) + d_{ij}\}, \; j = 2, 3, \ldots, n \qquad (6\text{-}2)$$

where $T_1(E) \equiv 0$.

The *latest realization time* for any event (node) is equal to the time at which the project is finished *minus* the duration or length of the longest path from the node representing the event to node n. Symbolically, if $\hat{\Pi}_i^{(k)}$ represents the *kth* path from node i to node n, and $T(\hat{\Pi}_i^{(k)})$ represents its duration, the latest realization time for node i can be expressed as in Eq. (6-3):

$$T_i(L) = \min_{k \in \mathbf{K}} \{\tau - T(\hat{\Pi}_i^{(k)})\}, \; i = n\text{-}1, \ldots, 1 \qquad (6\text{-}3)$$

where $\mathbf{K} = \{1, 2, \ldots, K_{in}\}$, K_{in} is the number of paths from node 1 to node n, $T_n(L) = \tau$, and, usually, $\tau = T_n(E)$. Instead of using Eq. (6-3), the following *backward recursive relationship* can be used more efficiently to compute the latest realization times of the events of a CPM project:

$$T_i(L) = \min_{j \in \alpha_i} \{T_j(L) - d_{ij}\}, \; i = n\text{-}1, \ldots, 1 \qquad (6\text{-}4)$$

where $T_n(L) = \tau$.

6.4 EVENT SLACKS, EARLIEST/LATEST ACTIVITY COMPLETION TIMES AND ACTIVITY FLOATS

The maximum time an event can be delayed without causing a delay in the total project completion time is known as the *slack* of that event. Thus, the slack for event i is defined as

$$S_i = T_i(L) - T_i(E) \qquad (6\text{-}5)$$

If the latest allowable start time and earliest possible start time for an event are the same its slack is equal to zero, and no slippage in the event is permissible. Those events that have zero slacks are events on the critical path. Any activity connecting events on this path is a critical activity.

From a management point of view, control of a project should devote primary attention to those activities on the critical path. If reduction in total project completion time is desired, efforts must be applied to the reduction of only those activities on the critical path. Such a reduction of time for any activity on the critical path is called *crashing*. Optimal crashing procedures will be discussed in Sections 6.10 and 6.11. It should be noted that as the duration of an activity is reduced, it is possible for the critical path to change. For example, the reader can verify that if either activity C or G is reduced more than 1 unit of time, the critical path will change to events 1, 3, 5, 6, 7 8.

It is necessary to distinguish between project *control* and project *time reduction*. Project time reduction is accomplished by allocating more resources to individual activities on the critical path, resulting in shorter activity durations. Project control is the manipulation of activity start/stop dates in order to balance available resources. Optimal project time reduction will be discussed in Section 6.12. At this point we will discuss project control. As previously stated, control of any project is affected by regulating the starting and stopping of activities, not their durations, since it is the accomplishment of all activities (goals) which comprises the total project. It is therefore desirable to provide specific activity scheduling information in terms of the earliest and latest times that each activity can be started and finished. As we have been, the critical path passes through those activities for which the earliest and latest start (or latest finish) times are identical, thus allowing no delays.

6.4.1 Earliest and Lates Activity Completion Times

The *earliest start* time for an activity is the earliest time the activity can be started assuming that all previous activities are completed as early as possible. Since an *activity* cannot commence until its predecessor *event* has occurred, the earliest start of activity (i, j) is

$$ES_{ij} = T_i(E)$$

Consequently, the *earliest finish* of activity (i, j) is $EF_{ij} = ES_{ij} + d_{ij}$; therefore,

$$EF_{ij} = T_i(E) + d_{ij}$$

Furthermore, since an activity can be completed no later than the latest permissible time for its successor event to occur, the *latest finish* of activity (i, j) is

$$LF_{ij} = T_j(L)$$

Finally, the *latest start* of the activity can then be calculated as

$$LS_{ij} = T_j(L) - d_{ij}$$

6.4.2 Activity Floats

The *float* of any activity is the maximum time the activity can be delayed without delaying the termination of the project. This maximum time actually depends on when the two nodes of the activity are realized. There are four cases to be considered for activity (i,j):

    (a) Node j is realized as late as possible and node i as early as possible.
    (b) Both nodes are realized as late as possible
    (c) Both nodes are realized as early as possible
    (d) Node j is realized as early as possible and node i as late as possible.

Based on the above cases, the following activity floats are defined [16]:

Total Float
$$TF_{ij} = T_j(L) - T_i(E) - d_{ij} \tag{6-6}$$
Safety Float
$$SF_{ij} = T_j(L) - T_i(L) - d_{ij} \tag{6-7}$$
Free Float
$$FF_{ij} = T_j(E) - T_i(E) - d_{ij} \tag{6-8}$$
Independent Float
$$IF_{ij} = max\ \{0;\ T_j(E) - T_i(L) - d_{ij}\} \tag{6-9}$$

226

It is noted that for any activity (i,j) we know that $T_i(L) \geq T_i(E)$, $T_j(L) \geq T_i(L) + d_{ij}$, and $T_j(E) \geq T_j(E) + d_{ij}$. Therefore, $T_j(L) - T_i(L) - d_{ij} \geq 0$, $T_j(L) - T_i(E) - d_{ij} \geq 0$, and $T_j(E) - T_j(E) - d_{ij} \geq 0$. Thus, all floats defined in Eqs. (6-6), (6-7) and (6-8) are non-negative. A graphical illustration of the floats is shown in Figure 6.5. In this figure the letters E and L represent earliest and latest realization times, respectively.

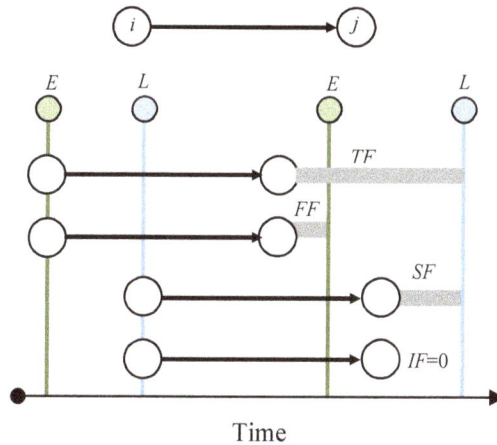

Figure 6.5. Activity floats.

The total float of an activity is the maximum time that the activity can be delayed without causing delay in the final project completion. An activity with zero total float is on the critical path. A fact too often overlooked is the implicit assumption in the total float computation that all predecessor activities must be completed as early as possible, and that all successor activities are forced to be completed as late as possible. Hence, it is generally impossible for all activities to realize their total floats, since the total float of an activity is closely coupled with the total float of others. For this reason, although the total float is vital for identifying the critical path and it is useful for overall project scheduling, its value when scheduling individual activities can be questioned.

The free float is a measure of the maximum time an activity may be delayed without affecting the start of the successor activities. Planning for the accomplishment of a particular activity could be severely disrupted if prior activities were not completed as early as possible, as it is assumed in both the total and free floats.

The independent float of an activity is the maximum time the activity can be delayed without delaying successor activities, if all prior activities are finished as late as possible. It gives a measure of time available if the worst possible condition prevail in the predecessor activities. It also gives an indication of the possible degree of decoupling the activity from the others in the project, and for this reason it is called *independent*.

The safety float is the maximum time an activity can be delayed without affecting the final project completion if all predecessor activities are completed as late as possible. This new measure appears to be one of the more useful float measures when scheduling a particular activity. Note that it permits only successor activities to be delayed, not the entire project.

One might ask, "*Of what value are the various float times?*" The first obvious answer is that zero total float identifies an activity on the critical path as one to be closely monitored and controlled. All the floats are useful to determine exactly when an activity is to be accomplished. In many cases, the labor or equipment needed to accomplish an activity must compete with other activities for available supplies of these resources. The various float times provide decision-making support when actually assigning these resources to each activity. Float times also permit some freedom in attempting to balance resource demands.

6.5 ILLUSTRATIVE CPM EXAMPLE

The activity network of Figure 6.4 constructed for the manufacturing example of section 6.2.2 will be considered to illustrate the application of the recursive relationships developed for the calculation of earliest and latest realization times. In addition, the earliest and latest completion times and the float measures for all activities will be computed.

A. Algorithmic Steps

The following results for earliest realization times and latest realization times are obtained using Eq. (6-2) and Eq. (6-4):

$$T_1(E)=0$$
$$T_2(E)=T_1(E)+d_{12}=0+5=5$$
$$T_3(E)=T_1(E)+d_{13}=0+3=3$$
$$T_4(E)=max\begin{cases}T_3(E)+d_{34} = 3+0=3\\T_1(E)+d_{14}=0+10=10\end{cases}=10$$
$$T_5(E)=13$$
$$T_6(E)=19$$
$$T_7(E)=23$$
$$T_8(E)=25$$

$$T_8(L)=T_8(E)=25$$
$$T_7(L)=T_8(L)-d_{78}=25-2=23$$
$$T_6(L)=T_7(L)-d_{67}=23-4=19$$
$$T_5(L)=T_6(L)-d_{56}=19-5=14$$
$$T_4(L)=T_6(L)-d_{46}=19-9=10$$
$$T_3(L)=min\begin{cases}T_5(L)-d_{35} = 14 - 10=4\\T_4(L)-d_{34} = 10-0=10\end{cases}=4$$
$$T_2(L)=7$$
$$T_1(L)=0$$

Figure 6.6 shows the critical path for the activity network shown in Figure 6.4. The critical path consists of activities *C, G, H,* and *I.* In this figure the slack of each node or event is shown in square brackets. Note that events (nodes) 1, 4, 6, 7, and 8 are on the critical path. The duration of the project is equal to the sum of the durations along the critical path, 10+9+4+2=25.

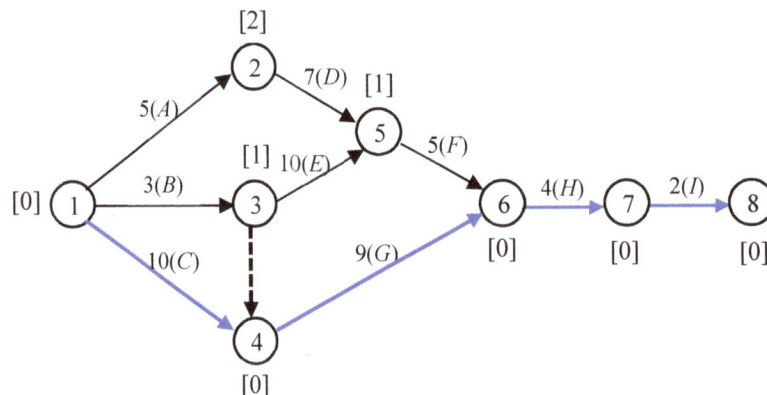

Figure 6.6. Critical Path and Node Slacks for Manufacturing Example.

228

Alternatively, those activities and events not on the critical path can be delayed without affecting the total completion time. From Figure 6.6, it can be seen, for instance, that the earliest possible start time for event 2 can be delayed up to 2 time units without affecting the project completion time. It should be noted that as the duration of an activity is reduced, it is possible for the critical path to change. For example, the reader should verify that if either activity C or G is reduced more than 1 unit of time, the critical path will change to events 1, 3, 5, 6, 7, 8.

As an illustration of the procedure to calculate the earliest and latest start and finish times of the activities of a CPM network, activity $D = (2,5)$ of the manufacturing problem example is considered:

$ES_{25} = T_2(E) = 5$
$EF_{25} = T_2(E) + d_{25} = 5 + 7 = 12$
$LS_{25} = T_5(L) - d_{25} = 14 - 7 = 7$
$LF_{25} = T_5(L) = 14$

Eqs. (6-6)-(6-9) can be used to calculate activity floats. As an illustration of these calculations, activity $D = (2,5)$ is considered again.

$TF_{25} = T_5(L) - T_2(E) - d_{25} = 14 - 5 - 7 = 2$

$FF_{25} = T_5(E) - T_2(E) - d_{25} = 13 - 5 - 7 = 1$

$IF_{25} = max \{0; T_5(E) - T_2(L) - d_{25}\} = max \{0; 13 - 7 - 7\} = 0$

$SF_{25} = T_5(L) - T_2(L) - d_{25} = 14 - 7 - 7 = 0$

Table 6.2 summarizes the above calculations for all activities in the activity network of the manufacturing example. The critical path consists of those activities identified with an asterisk (*). These activities have total float equal to zero.

Table 6.2. ACTIVITY EARLIEST/LATEST TIMES AND FLOATS
FOR MANUFACTURING EXAMPLE

Activity (i,j)	Node i	Node j	Duration	ES	EF	LS	LF	Activity Floats				Critical Path
								TF	FF	IF	SF	
A	1	2	5	0	5	2	7	2	0	0	2	
B	1	3	3	0	3	1	4	1	0	0	1	
C	1	4	10	0	10	0	10	0	0	0	0	*
D	2	5	7	5	12	7	14	2	1	0	0	
X	3	4	0	3	3	10	10	7	7	6	6	
E	3	5	10	3	13	4	14	1	0	0	0	
F	5	6	5	13	18	14	19	1	1	0	0	
G	4	6	9	10	19	10	19	0	0	0	0	*
H	6	7	4	19	23	19	23	0	0	0	0	*
I	7	8	2	23	25	23	25	0	0	0	0	*

B. NOP Solution

Figure 6.7 shows the NOP screens for this example: (1) Input data file and (2) NOP Output results. The critical path consists of arcs (1,4), (4,6), (6,7) and (7,8) corresponding to a project duration equal to 25. The output also shows earliest and latest completion times of the activities, the four floats of all activities, and the activities on the critical path of the project.

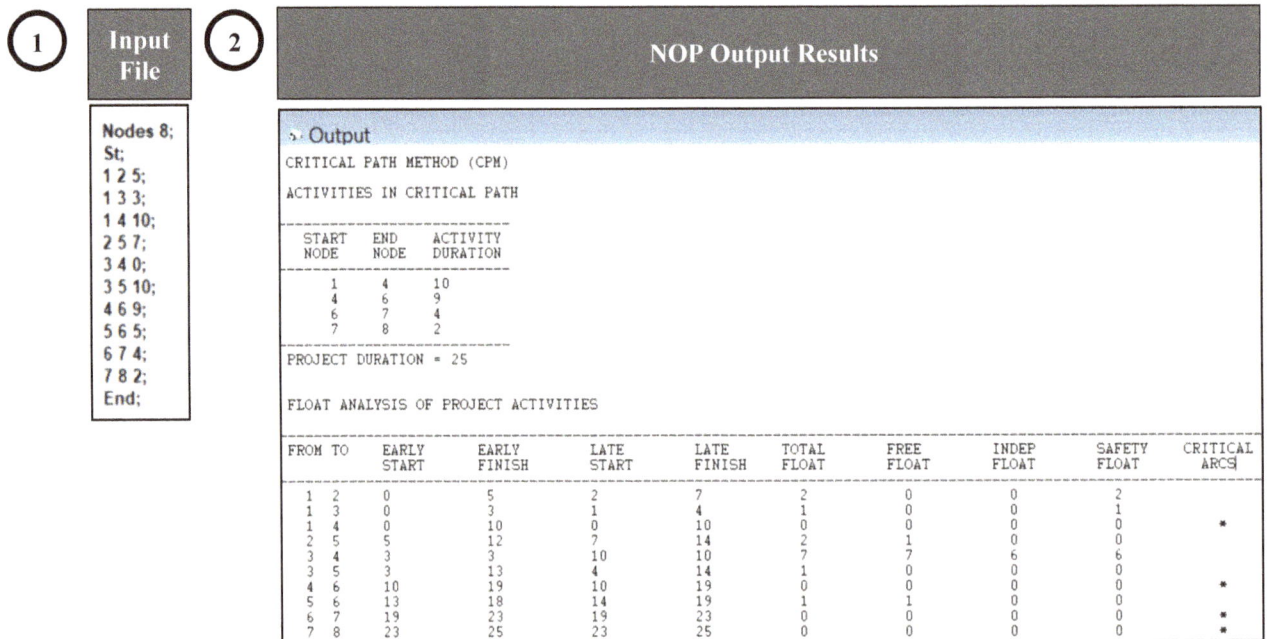

Figure 6.7. Sequence of Computer Screens for CPM Example.

6.6 PERT METHODOLOGY

For each activity in a project network, PERT assumes three estimates for its completion time:

1. A most likely time denoted by m
2. An optimistic time denoted by a
3. A pessimistic time denoted by b

The most likely time is the time required to complete the activity under normal conditions. To include uncertainties, a range of variation in job time is provided by the optimistic and pessimistic times. The optimistic estimate is a good guess on the minimum time required when everything goes according to plans; while the pessimistic estimate is a guess on the maximum time required under adverse conditions such as mechanical breakdowns, minor labor troubles, and shortage of materials or delays in their scheduled delivery. It should be remarked here that the pessimistic estimate does not take into consideration unusual or prolonged delays or other catastrophes. Because both of these estimates are only qualified guesses, the actual time for an activity could lie outside this range. (From a probabilistic viewpoint we can only say that the probability of a job time falling outside this range is very small.) PERT assumes a Beta distribution for the job times.

230

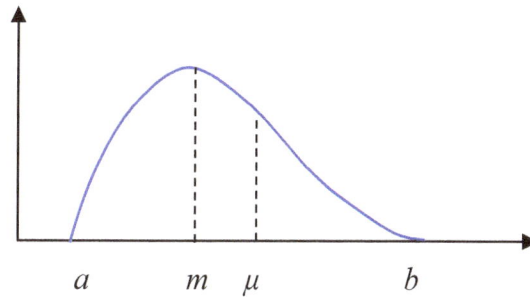

Figure 6.8. Beta distribution.

A typical distribution is shown in Figure 6.8. The mean value of μ depends on how close the values of a and b are relative to m. The mean time to complete an activity is approximated as

$$\mu = \frac{a + 4m + b}{6}$$

(6-10)

Since the actual time may vary from its mean value, we need the variance of the job time. For most *unimodal* distributions (with single peak values), the end values lie within three standard deviations from the mean value. Thus, the spread of the distribution is equal to six times the standard deviation value, σ. According to this, we can assume that $6\sigma = b - a$, or $\sigma = (b - a)/6$. Equivalently, the variance of the job time equals

$$\sigma^2 = \frac{(b-a)^2}{36}$$

(6-11)

The PERT methodology consists of four sequential steps:

Step 1. With the three time estimates on all jobs (activities), PERT calculates the mean and variance of the duration of each activity (i, j) using Eqs. (6-10) and (6-11):

$$\mu_{ij} = \frac{a_{ij} + 4m_{ij} + b_{ij}}{6}$$

$$\sigma^2_{ij} = \frac{(b_{ij} - a_{ij})^2}{36}$$

Step 2. Using the mean values for all activities, the critical path **CP** is found following the CPM procedure.

Step 3. Once the critical activities (activities in **CP**) are identified, the sum of their duration means is set equal to

$$\mu_T = \sum_{(i,j)\in\mathbf{CP}} \mu_{ij}$$

and the sum of their duration variances is set equal to

$$\sigma^2_T = \sum_{(i,j)\in\mathbf{CP}} \sigma^2_{ij}$$

231

Step 4. Assuming that the project duration is normally distributed with mean μ_T and variance σ_T^2 calculate the probability that the project completion time falls within a given range.

The originators of PERT claim three particularly attractive features: flexibility, simplicity, and potency. These are briefly described below:

Flexibility: The beta distribution can be used to fit any desired shape and location due to its four parameters.

Simplicity: Although the beta distribution function itself is a complicated expression, its application is simplified through the use of simple approximations. From that point on, calculations involve simple additions.

Potency: Not only do we possess all the information that the deterministic CPM model gives, but also we can find confidence limits (probabilistic statements) on the duration of the project by using the Normal distribution tables.

In a critical evaluation of PERT, Elmaghraby [3] concludes that:

1. The simplifying assumptions of PERT actually restrict the shape of the probability distribution of activity duration to only distributions with skewness coefficient equal to $\pm\sqrt{2}/2$ or 0.

2. Even if we are willing to approximate the probability density function of the project duration by a normal distribution, it would be a normal density function with mean and variance values different from those used by PERT.

6.7 ILLUSTRATIVE PERT EXAMPLE

For the manufacturing example of Section 6.2.2 each activity (i,j) listed in Table 6.1 is now assumed to have the m_{ij}, b_{ij}, a_{ij} estimates given in the second column of Table 6.2. It is desired to determine the probability that the project will be finished between 20 and 30 time units.

Table 6.2. PERT EXAMPLE

Activity	m_{ij}, b_{ij}, a_{ij}	μ_{ij}	σ_{ij}
A=(1,2)	5,8,2	5	1
B=(1,3)	3,4,2	3	1/3
C=(1,4)	10,15,5	10	5/3
D=(2,5)	8,9.1	7	4/3
E=(3,5)	10,15,5	10	5/3
F=(5,6)	5,8,2	5	1
G=(4,6)	10,11,3	9	4/3
H=(6,7)	3,10,2	4	4/3
I=(7,8)	2,3,1	2	1/3

A. Algorithmic Steps

Step 1. The values of $\mu_{ij} = (a_{ij} + 4m_{ij} + b_{ij})/6$ and $\sigma_{ij} = (b_{ij} - a_{ij})/6$ are shown in the third and fourth columns of Table 6.2.

Step 2. Using the mean values for all activities, the critical path **CP** is found following the CPM procedure. First, it is noted that the mean values of Table 6.2 are identical to the durations given in Table 6.1. Therefore, the critical path **CP** continues to be the same path shown in Figure 6.6. This path includes activities *C, G, H,* and *I*.

Step 3. The project duration mean and variance are set equal to $\mu_T = \sum_{(i,j) \in \mathbf{CP}} \mu_{ij} = 10 + 9 + 4 + 2$

$= 25$, and $\sigma_T^2 = \sum_{(i,j) \in \mathbf{CP}} \sigma_{ij}^2 = (5/3)^2 + (4/3)^2 + (4/3)^2 + (1/3)^2 = 58/9$.

Step 4. Assuming that the project duration is normally distributed with mean $\mu_T = 25$ and variance $\sigma_T^2 = 58/9$, calculate the probability that the project completion time falls within the given range. The *standardized normal variable* $Z = (T - \mu_T)/\sigma_T = (T-25)/2.54$ takes on the values -1.97 and 1.97 when $T = 20$ and $T = 30$, respectively. Using the probability tables of the standardized normal distribution, we obtain $P\{20 \leq T \leq 30\} = P\{-1.97 \leq Z \leq +1.97\} = 0.95$. This probability can be represented as the area under the normal curve between the values of 20 and 30 in Figure 6.9.

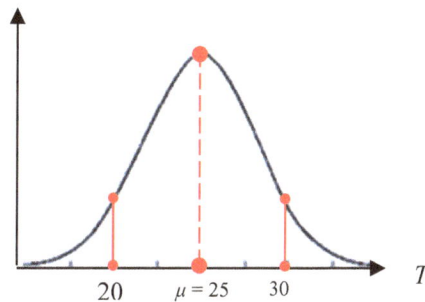

Figure 6.9. Normal distribution for project duration.

A. NOP Solution

Figure 6.10 shows the NOP screens for this example: (1) Input data file and (2) NOP Output results. The output results include:

   (a) Critical path of the project
   (b) Float analysis for all activities of the project

The critical path consists of arcs (1,4), (4,6), (6,7) and (7,8) corresponding to a mean project duration equal to 25 and standard deviation equal to 2.54. The output also shows earliest and latest completion times of the activities, the four floats of all activities, and the activities on the critical path of the project.

Input File:

```
Nodes 8;
St;
1 2 5 8 2;
1 3 3 4 2;
1 4 10 15 5;
2 5 8 9 1;
3 4 0 0 0;
3 5 10 15 5;
4 6 10 11 3;
5 6 5 8 2;
6 7 3 10 2;
7 8 2 3 1;
End;
```

Output

PROGRAM EVALUATION AND REVIEW TECHNIQUE (PERT)

ACTIVITIES IN PROJECT CRITICAL PATH

START NODE	END NODE	EXPECTED DURATION	STANDARD DEVIATION
1	4	10.00	1.67
4	6	9.00	1.33
6	7	4.00	1.33
7	8	2.00	0.33

MEAN PROJECT DURATION = 25.00

STANDARD DEVIATION = 2.54

FLOAT ANALYSIS OF PROJECT ACTIVITIES

FROM	TO	EARLY	EARLY START	LATE FINISH	LATE START	TOTAL FLOAT	FREE FLOAT	INDEP FLOAT	SAFETY FLOAT	CRITICAL ARCS
1	2	0.00	05.00	2.00	7.00	2.00	0.00	0.00	2.00	
1	3	0.00	03.00	1.00	4.00	1.00	0.00	0.00	1.00	
1	4	0.00	10.00	0.00	10.00	0.00	0.00	0.00	0.00	*
2	5	5.00	12.00	7.00	14.00	2.00	1.00	0.00	0.00	
3	4	3.00	03.00	10.00	10.00	7.00	7.00	6.00	6.00	
3	5	3.00	13.00	4.00	14.00	1.00	0.00	0.00	0.00	
4	6	10.00	19.00	10.00	19.00	0.00	0.00	0.00	0.00	*
5	6	13.00	18.00	14.00	19.00	1.00	1.00	0.00	0.00	
6	7	19.00	23.00	19.00	23.00	0.00	0.00	0.00	0.00	*
7	8	23.00	25.00	23.00	25.00	0.00	0.00	0.00	0.00	*

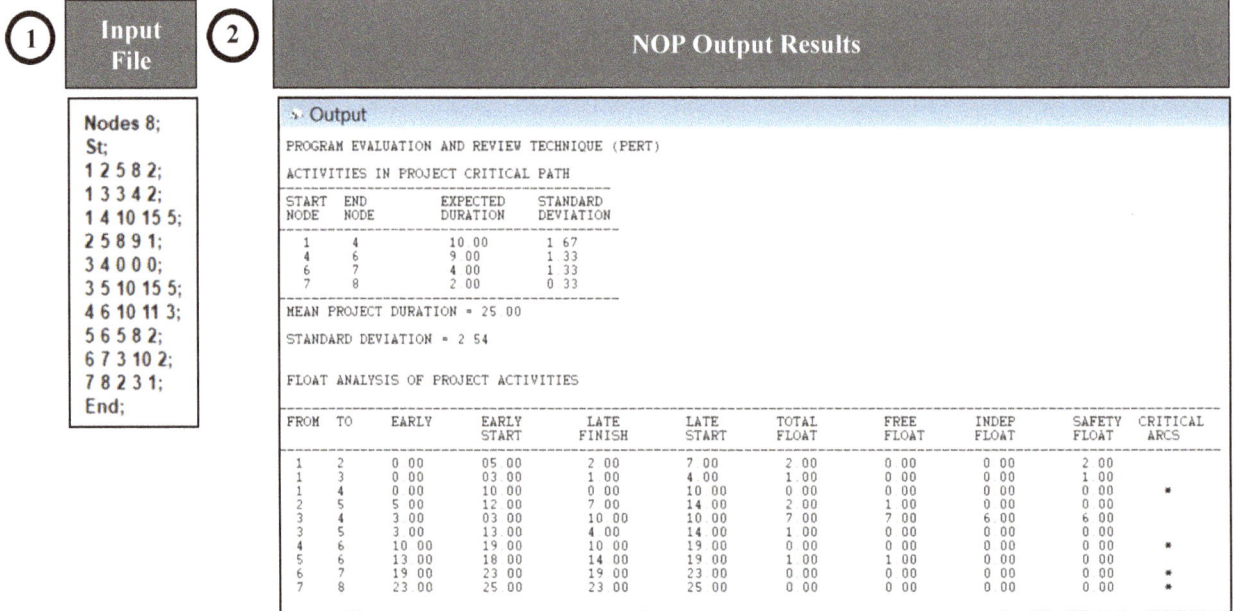

Figure 6.10. Sequence of Computer Screens for PERT Example.

PART II

RESOURCE ALLOCATION IN PROJECT NETWORKS

With contributions by
Edward W. Davis
Oliver Wight Professor Emeritus
Darden School of Business
University of Virginia

The general appearance of program evaluation and review technique (PERT) and critical path method (CPM) technique in the early 1960s saw an explosion of interest in project network methods. The ensuing popularity of these techniques demonstrated that network models are a useful means of formulating a wide variety of activity planning/scheduling problems. But it was also recognized that the basic PERT/CPM procedures are somewhat naïve models of most real-life situations in that they focus on *time* to the exclusion of *resource* requirements and availabilities.

As a result, there has been increasing attention given in recent years to the problems of resource allocation associated with project planning and scheduling. Many of the new developments that have occurs in the general area of network methods have taken place in this field. The proliferation of techniques has, in fact, progressed to the point where the persons unfamiliar with these procedures are often confused over exactly what can and cannot be done with regard to resource management. This section is an attempt to allay such confusion by categorizing the types of procedures that have been developed and summarizing the capabilities and limitations of each approach.

6.8 TIME VS. COST: DOLLAR ALLOCATIONS

It is often true that the performance of some or all project activities can be accelerated by the allocation of more resources at the expense of the higher activity direct cost. When this is so, there are many different combinations of activity durations that will yield a different value of total project cost. Time/cost trade-off procedures are directed at determining the least-cost schedule for any given project duration.

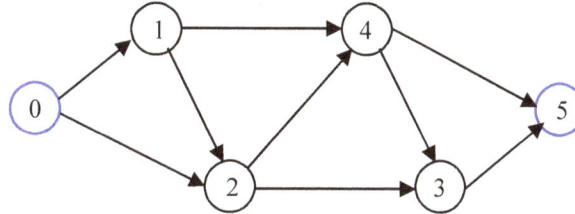

Figure 6.11. Activity network for cost crashing problem.

For example, consider the simple eight-activity project shown in Figure 6.11 with activity time/cost data given in Table 6.3. Each activity can be performed at different durations ranging from an upper "normal" value, at some associated *normal* cost, down to a lower, *crash* value, with an associated higher cost. Note that if time/cost trade-off values for each activity are assumed to be linear, the cost of intermediate activity durations between the normal and crash durations is easily determined from the single cost *slope* value for each activity. For example, the cost of performing activity (0,2) in 7 days instead of 8 is equal to $400 + $80 = $480.

Table 6.3. ILLUSTRATIVE NETWORK WITH ACTIVITY TIME/COST DATA

Activity	Normal		"Crash"		Cost Slope ($)
	Time (days)	Cost ($)	Time (days)	Cost ($)	
(0,1)	4	210	3	280	70
(0,2)	8	400	6	560	80
(1,2)	6	500	4	600	50
(1,4)	9	540	7	600	30
(2,3)	4	500	1	1100	200
(2,4)	5	150	4	240	90
(3,5)	3	150	3	150	0
(4,5)	7	600	6	750	150
		3050		3050	

If all activity durations are set at *normal* values, the project duration is 22 days, as determined by the critical path consisting of activities (0,1), (1,2), (2,4), (4,5) in Figure 6.11. The calculations are given in Table 6.4. The associated cost of project performance is $3050, as indicated in Figure 6.12. Note that this cost could be increased to $3870 through unintelligent decision-making by "crashing" all activities not on the critical path, with no decrease in project duration. Between these upper and lower cost values for a project duration of 22 days there are several other possible values, depending upon the number of noncritical activities crashed.

235

Table 6.4. CRITICAL PATH ANALYSIS

Activity	Durations	Normal Time (days)	Normal Cost ($)	"Crash" Time (days)	"Crash" Cost ($)	Cost Slope ($)
(0,1)	4	4	210	3	280	70
(0,2)	8	8	400	6	560	80
(1,2)	6	6	500	4	600	50
(1,4)	9	9	540	7	600	30
(2,3)	4	4	500	1	1100	200
(2,4)	5	5	150	4	240	90
(3,5)	3	3	150	3	150	0
(4,5)	7	7	600	6	750	150
			3050		3050	

If all activity durations are set at *crash* values, the project duration can be decreased to 17 days, with a total cost of $4280, as shown by the extreme upper left point of Figure 6.10. However, a duration of 17 days can also be achieved at lower cost by not crashing activities unnecessarily. Thus activity (0,2) can be set at 7 instead of 6, activity (1,4) at 8 instead of 7, and activity (2,3) at 4 instead of 1. With all other activities set at crash value, the associated cost of performance for 17-day project duration is reduced to $3520. This value is the lowest possible value for 17-day project duration, as may easily be determined by experimentation.

Between 22 and 17 days there are several possible values of project duration, as shown in Figure 6.12. For each such duration there is a range of possible cost values, depending upon the duration of individual activities and whether activities are crashed unnecessarily or not. Figure 6.12 shows the curve of both maximum and minimum costs and the region of possible costs for each duration between these curves.

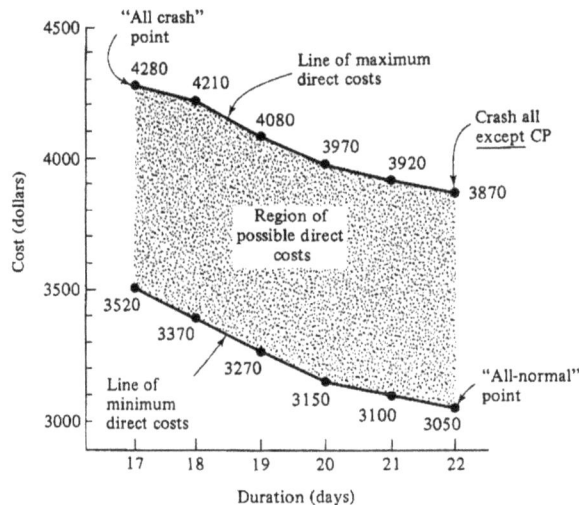

Figure 6.12. Project duration versus direct cost.

In this simple example the minimum-direct-cost curve is easily determined by trial and error. But in more realistic cases consisting of dozens or even hundreds of activities, such trial-

and-error determination becomes extremely tedious, if not impossible. Thus, various systematic computation schemes, including mathematical programming, have been developed to aid in the quick determination of the minimum-cost curve for every possible value of project duration. Some of these routines are designed to handle cases in which the activity time/cost trade-off relationships are nonlinear; many will also produce the lowest total cost curve (i.e., sum of direct plus indirect costs), as illustrated in Figure 6.13. Part III will consider the optimization of time-cost trade-offs in project network with linear and piece-wise linear cost functions.

To illustrate the fundamentals of total cost crashing, consider once again the previous example with a fixed indirect cost of $130 per day. In this case, the curve of Figure 6.13 becomes that of Figure 6.14. The calculations are summarized in Table 6.5.

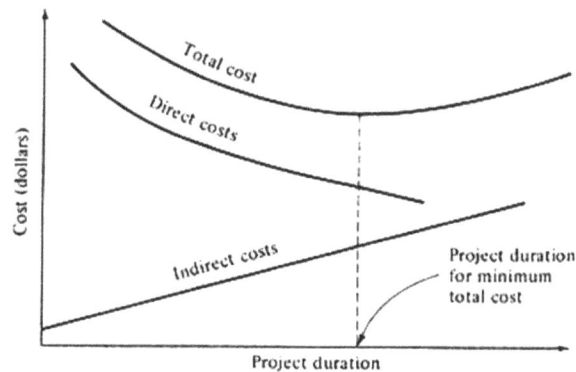

Figure 6.13. Determining project schedule for minimum total cost.

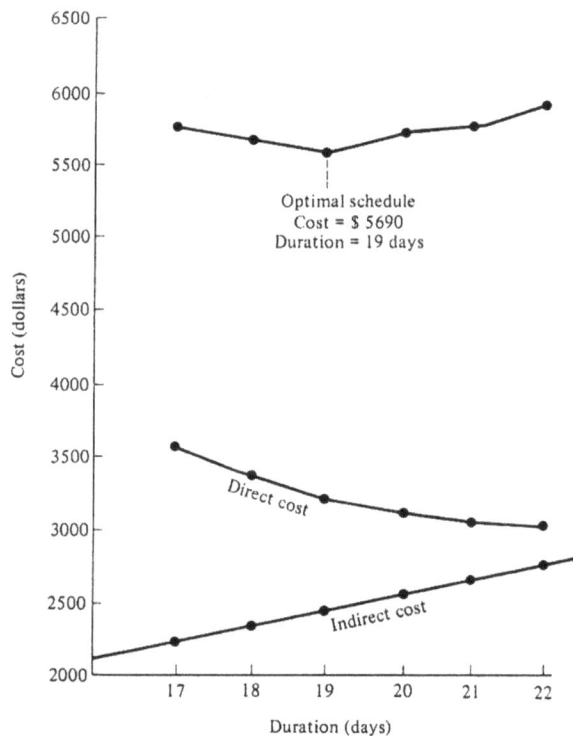

Figure 6.14. Optimal crashing with both direct and indirect costs.

237

Table 6.5. COST CRASHING

Crash Activity	Duration (days)	Direct Cost	Indirect Cost	Total Cost
(Critical path)	22	$3050	$2860	$5910
(1,2)	21	3100	2730	5830
(1,2)	20	3150	2600	5750
(2,4) and (1,4)	19	3220	2470	5690 (optimal)
(4,5)	18	3370	2340	5710
(0,1) and (0,2)	17	3520	2210	5730

6.9 RESOURCE LOADING AND RESOURCE LEVELING

One of the major advantages of the network representation of activity projects is the ability to easily generate information on the time-phased requirements of such project resources as manpower, equipment and money. This information is a direct by-product of the standard critical path method. The only assumption made is that demand for each resource associated with each activity is identified separately.

Figure 6.15, for example, shows the activity-on-node representation for a simple hypothetical project consisting of eight activities. In this figure two quantities are shown for each node. The value at the *top* is the *duration* in months and the value at the *bottom* is the *cash requirement* in hundreds of dollars per month.

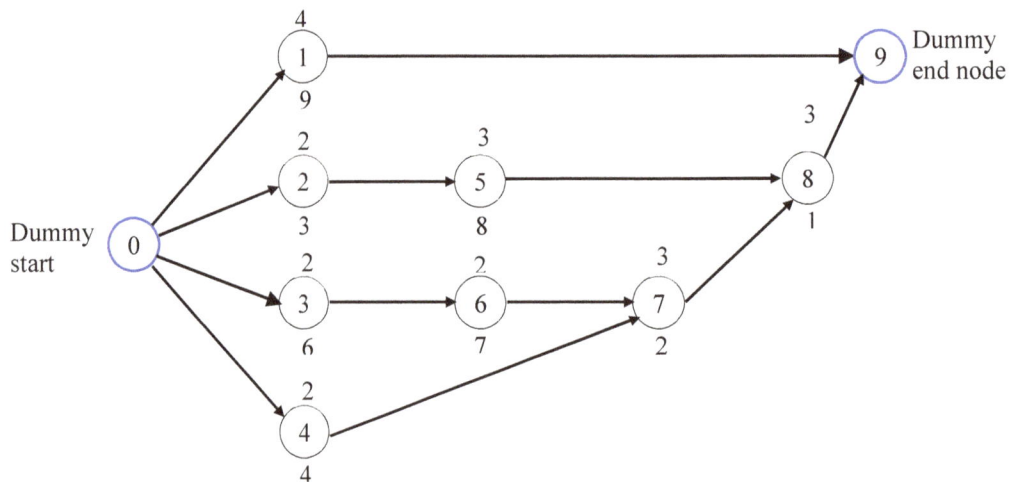

Figure 6.15. Project network with durations and cash requirements.

From the CPM calculations the *early start time* (*ES*) and the *late finish time* (*LF*) can be determined for each activity. From this information, a bar-chart diagram as the one shown in Figure 6.16(a) can be drawn, with all activities beginning at their corresponding *ES* times. The bar-chart diagram, in turn, is used to generate the profile of cash flow requirements over time, as shown in Figure 6.16(b). This kind of analysis is known as ***resource loading***.

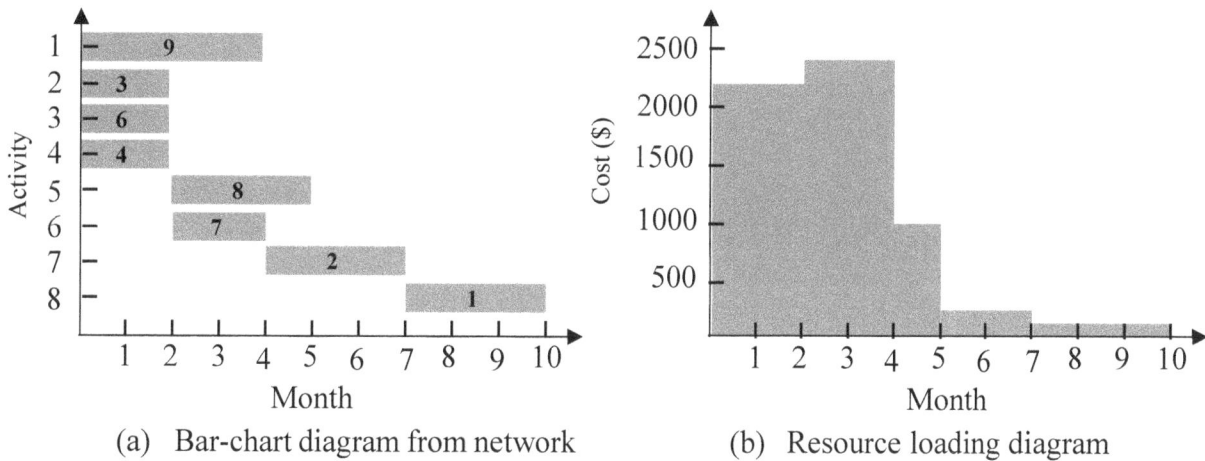

(a) Bar-chart diagram from network   (b) Resource loading diagram

Figure 6.16. Bar-chart and resource loading diagram for illustrative network.

In this example cash requirements have been used for illustration. Obviously, data on other resources, such as manpower, could be used instead; in such cases the profile of resource usage over time is commonly referred to as a *resource-loading diagram*. In case of multiple resources, a separate diagram can be obtained for each resource. Such loading diagrams are extremely useful in *project management* since they highlight the resource implications of a particular project schedule and provide the basis for more rational project planning.

In many project scheduling situations, the total levels of resource demands projected by the resource loading diagrams for a particular schedule may not be of major concern when there are sufficient amounts available of the required resources.   However, it may be that the pattern of resource usage has undesirable occurrences, such as frequent changes in the amount of a particular manpower skill category. In this situation resource-leveling techniques may be useful.

Resource leveling is an approach to provide means of distributing resource usage over time to minimize the period-by-period variations in manpower, equipment, or money expended. It can also be used to determine whether peak resource requirements can be reduced with no increase in project duration.

The concepts of resource leveling are easily grasped through a simple example. Consider again the network of Figure 6.15 but now assume that the durations are given in days instead of moths. Also, let the cash requirements be substituted for the number of workers of a particular skill required per day. Then the resource-loading diagram of Figure 6.17(a) shows an undesirable pattern of manpower requirements over time because of the extreme fluctuation from 24 workers on days 3 and 4 down to 1 worker on days 8, 9 and 10. Resource leveling can be used to achieve a more even pattern of manpower utilization without increasing project duration beyond 10 days.

As the bar chart of Figure 6.17(a) shows, jobs 1, 2, 4 and 5 have slack and can be delayed within the limits of their slack without delaying other jobs. Job 1, for example, can obviously be delayed until the last 4 days, as shown in Figure 6.17(b).   This reduces peak manpower requirements from 24 to 14, and the resulting manpower profile is more even than before. By subsequently rescheduling jobs 2 and 5, a more even pattern is obtained, as shown in Figure

239

6.17(c). As this example shows, the essential idea of resource leveling centers about *juggling* or rescheduling of jobs within the limits of available slack (float) to achieve a better distribution of resource usage. CPM computerized procedures with resource-leveling capability have been available since the early 1960s [2, 11, 12].

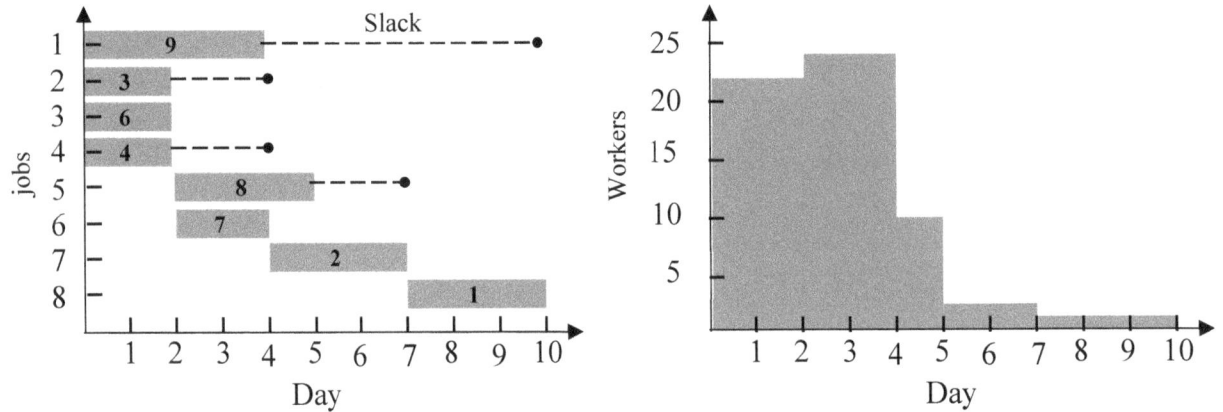

(a) Bar chart and manpower profile with all jobs started at their *ES* times

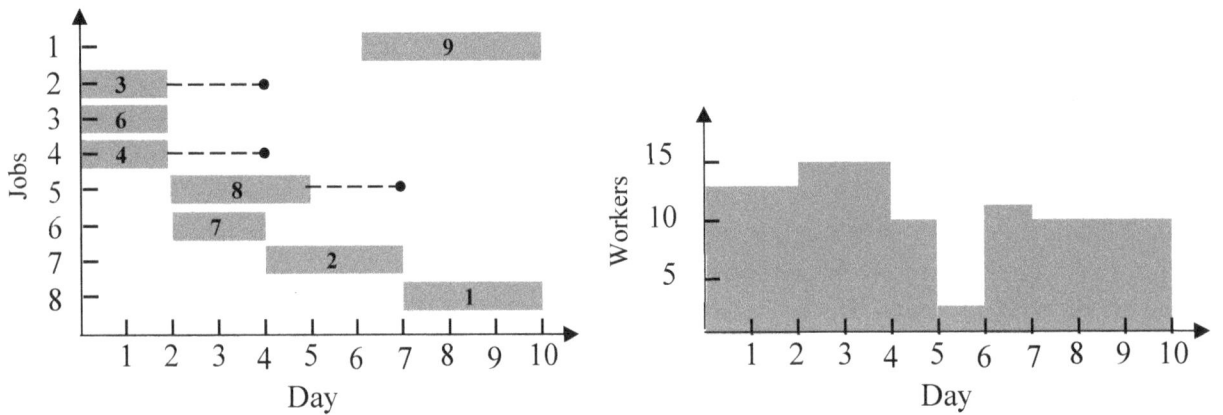

(b) Bar chart and manpower profile with job 1 rescheduled

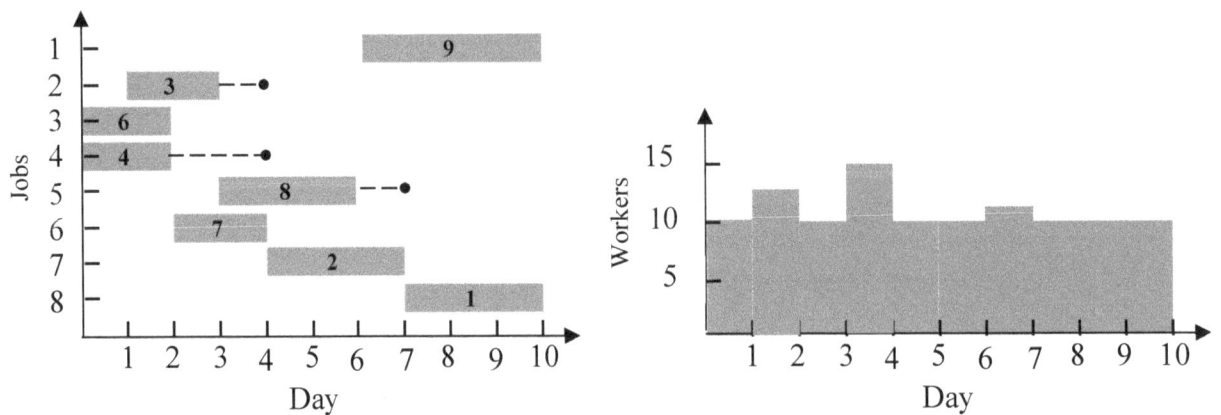

(c) Bar chart and manpower profile with jobs 1, 2 and 5 rescheduled

Figure 6.17. Illustration of resource leveling.

Through the process of resource leveling the profiles of resource usage can be smoothed to the extent allowed by available job slack. But this process does not always result in satisfactory schedules when the amounts of available resources are tightly constrained. The manpower profile of Figure 6.17(c), for example, is constant at 10 workers per day except for days 2, 4 and 7 which have peaks above that level. If the number of workers cannot exceed the level of 10, and it is not possible to get this by resource leveling, then the only way to eliminate the higher peaks would be through rescheduling, with a resultant increase in project duration. Looking at Figure 6.17(c) it is readily apparent that the manpower peaks can be eliminated by delaying the start of job 1 from day 7 until day 8. This permits job 5 to be delayed 1 day, making it concurrent with job 7, and also permits delaying job 2 for 1 day to make it concurrent with job 6. Since the resulting manpower requirements total exactly 10 or less for each day, the schedule is resource-feasible. However, the duration of the project has increased by 1 day, as shown in Figure 6.18.

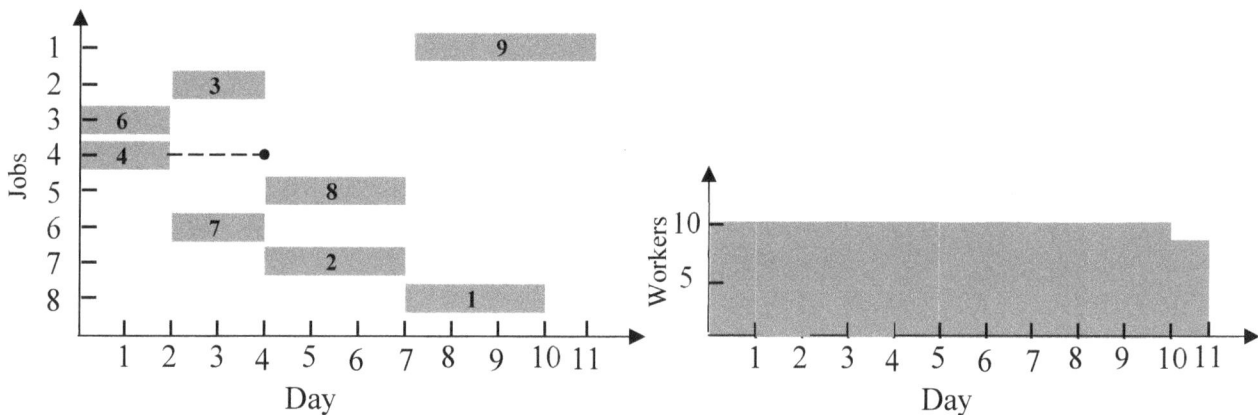

Figure 6.18. Illustration of constrained-resource scheduling.

PART III

OPTIMIZATION OF TIME/COST TRADE-OFFS IN PROJECT NETWORKS

6.10 TIME/COST TRADE-OFFS IN CPM NETWORKS

The duration of any activity of a project can be controlled by limiting the amount of resources allocated to the activity. In general, it is possible to assume that the project management office can estimate the duration of the activities as functions of the money spent on each. Under such an assumption, one can formulate a mathematical model whose purpose will be the minimization of the total cost of the project. The model will find optimal values for the event realization times and the activity durations, given the duration of the project, the precedence relationships, and upper and lower bounds on the duration of each activity.

Before we formally present the algorithm to be studied in this section, we will consider the network shown in Figure 6.19. Each activity (i,j) has a lower bound L_{ij} and an upper bound U_{ij} on its duration. Additionally, there is a constant cost a_{ij} for reducing the duration of the activity by one unit of time. Table 6.6 summarizes this information for the sample network being considered.

241

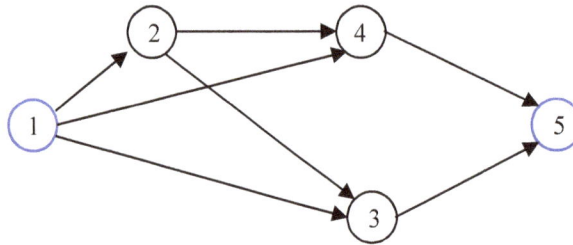

Figure 6.19. Network for sample problem.

Table 6.6. DATA FOR SAMPLE PROBLEM

Arc (i, j)	L_{ij}	U_{ij}	a_{ij}
(1,2)	2	5	3
(1,3)	1	3	2
(1,4)	4	6	2
(2,3)	2	4	5
(2,4)	3	7	1
(3,5)	0	3	4
(4,5)	3	3	-

Primal Model

The following notation will be used in the formulation of the primal model for minimizing the total cost of an activity project:

y_{ij}	duration of activity (i, j)
t_i	realization of event i
c_{ij}	cost of activity (i, j); this is a function of the duration of the activity, $c_{ij} = f(y_{ij})$
L_{ij}	lower bound on the duration of activity (i, j)
U_{ij}	upper bound on the duration of activity (i, j)
T	specified duration of the project

Since the total value of the ordinates of the activity cost lines at the origin is a constant, the objective function of the primal model can be formulated as

$$\text{Minimize } -3y_{12} - 2y_{13} - 2y_{14} - 5y_{23} - y_{24} - 4y_{35} - My_{45}$$

This objective function can be transformed into a maximization case by multiplying it by -1 to obtain the result

$$\text{Maximize } 3y_{12} + 2y_{13} + 2y_{14} + 5y_{23} + y_{24} + 4y_{35} + My_{45}$$

Therefore, the primal model for the maximization case is formulated as follows, where the dual variable corresponding to each constraint is shown to the right of the constraint.

Maximize $3y_{12} + 2y_{13} + 2y_{14} + 5y_{23} + y_{24} + 4y_{35} + My_{45}$

Subject to

t_1	t_2	t_3	t_4	t_5	y_{12}	y_{13}	y_{14}	y_{23}	y_{24}	y_{35}	y_{45}		RHS	Dual variables
-1			+1									=	15	v
+1	-1				+1							≤	0	f_{12}
+1		-1				+1						≤	0	f_{13}
+1			-1				+1					≤	0	f_{14}
	+1	-1						+1				≤	0	f_{23}
	+1		-1						+1			≤	0	f_{24}
		+1		-1						+1		≤	0	f_{35}
			+1	-1							+1	≤	0	f_{45}
					+1							≤	5	γ_{12}
						+1						≤	3	γ_{13}
							+1					≤	6	γ_{14}
								+1				≤	4	γ_{23}
									+1			≤	7	γ_{24}
										+1		≤	3	γ_{35}
											+1	≤	3	γ_{45}
					-1							≤	-2	δ_{12}
						-1						≤	-1	δ_{13}
							-1					≤	-4	δ_{14}
								-1				≤	-2	δ_{23}
									-1			≤	-3	δ_{24}
										-1		≤	0	δ_{35}
											-1	≤	-3	δ_{45}

Dual Formulation

It is noted that since the *primal variables* do not have the non-negativity conditions explicitly specified in the formulation, their corresponding dual constraints are *equality constraints*. Using the dual variables listed next to the right-hand constants of the constraints of the primal formulation, the dual model can be now formulated as follows. In this formulation we observe that the variables v, f_{12}, f_{13}, f_{14}, f_{23}, f_{24}, f_{35}, and f_{45} behave as network flows satisfying the flow conservation conditions. This allows us to view the dual formulation as a network model with additional constraints. This insight is key to the formulation of the algorithm discussed in Section 6.10.1.

Minimize $15v + 5\gamma_{12} + 3\gamma_{13} + 6\gamma_{14} + 4\gamma_{23} + 7\gamma_{24} + 3\gamma_{35} + 3\gamma_{45} - 2\delta_{12} - 2\delta_{13} - 4\delta_{14} - 2\delta_{23} - 3\delta_{24} - 3\delta_{45}$

subject to

v	f_{12}	f_{13}	f_{14}	f_{23}	f_{24}	f_{35}	f_{45}	γ_{12}	γ_{13}	γ_{14}	γ_{23}	γ_{24}	γ_{35}	γ_{45}	δ_{12}	δ_{12}	δ_{12}	δ_{12}	δ_{12}	δ_{12}	δ_{12}		
-1	+1	+1	+1																			=	0
	-1			+1	+1																	=	0
		-1		-1		+1																=	0
			-1		-1		+1															=	0
+1						-1	-1															=	0
	+1							+1							-1							=	3
		+1							+1							-1						=	2
			+1							+1							-1					=	2
				+1							+1							-1				=	5
					+1							+1							-1			=	1
						+1							+1							-1		=	4
							+1							+1							-1	=	M

$$f_{ij},\ \gamma_{ij},\ \delta_{ij} \geq 0$$

6.10.1 A Network Flow Algorithm

The cost-duration function formulated in (6-12) will be considered:

$$c_{ij} = f(y_{ij}) = b_{ij} - a_{ij}\, y_{ij} \qquad (6\text{-}12)$$

where $L_{ij} \leq y_{ij} \leq U_{ij}$. This linear relationship is shown in Figure 6.20. The lower bound is usually referred to as the *crash* duration of the activity and the upper bound as the *normal* duration. Given the specified duration of the project, and the linear cost-duration functions for a project with network representation $G = (\mathbf{N}, \mathbf{A})$, we want to know which activities in set $\mathbf{A}$ must be crashed and which ones must be left at their normal durations.

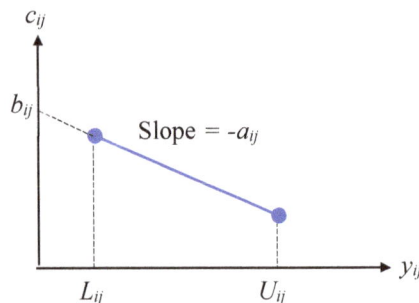

Figure 6.20. Linear cost-duration function.

The problem under consideration is formulated as a linear programming model with objective function (6-13) and Constraints (6-14)-(6-17). It is assumed that $\mathbf{N} = \{1, 2, \ldots, n\}$, where node 1 represents the start of the project and node n the end of it.

244

$$\text{Minimize} \quad \sum_{(i,j)\in\mathbf{A}}(b_{ij}-a_{ij}y_{ij}) \tag{6-13}$$

subject to

$$t_i - t_j + y_{ij} \leq 0, \text{ all } (i,j) \in \mathbf{A} \tag{6-14}$$

$$-t_1 + t_n \quad = T \tag{6-15}$$

$$y_{ij} \leq U_{ij}, \text{ all } (i,j) \in \mathbf{A} \tag{6-16}$$

$$y_{ij} \geq L_{ij}, \text{ all } (i,j) \in \mathbf{A} \tag{6-17}$$

This model is reformulated as a maximization problem with objective function (6-18) and Constraints (6-19)-(6-22):

$$\text{Maximize} \quad \sum_{(i,j)\in\mathbf{A}} a_{ij}y_{ij} \tag{6-18}$$

subject to

$$t_i - t_j + y_{ij} \leq 0, \text{ all } (i,j) \in \mathbf{A} \tag{6-19}$$

$$-t_1 + t_n = T \tag{6-20}$$

$$y_{ij} \leq U_{ij}, \text{ all } (i,j) \in \mathbf{A} \tag{6-21}$$

$$-y_{ij} \leq -L_{ij}, \text{ all } (i,j) \in \mathbf{A} \tag{6-22}$$

Since the non-negativity constraints for t_i and y_{ij} are not explicitly stated in the model, all constraints of the dual program will be of the equality type. Let f_{ij}, v, γ_{ij}, and δ_{ij} be the dual variables corresponding to Constraints (6-19), (6-20), (6-21), and (6-22), respectively.

The dual model of the formulation given in (6-18)-(6-22) has objective function (6-23) and Constraints (6-24)-(6-28):

$$\text{Minimize } Tv + \sum_{ij} U_{ij}\gamma_{ij} - \sum_{ij} L_{ij}\delta_{ij} \tag{6-23}$$

subject to

$$f_{ij} + \gamma_{ij} - \delta_{ij} = a_{ij}, \text{ all } (i,j) \in \mathbf{A} \tag{6-24}$$

$$\sum_j f_{1j} - v = 0 \tag{6-25}$$

$$\sum_j f_{ij} - \sum_j f_{ji} = 0, \ i = 2,...,n-1 \tag{6-26}$$

$$-\sum_j f_{jn} + v = 0 \tag{6-27}$$

$$f_{ij}, \gamma_{ij}, \delta_{ij} \geq 0, \text{ all } (i,j) \in \mathbf{A} \tag{6-28}$$

From the formulation of the dual model shown in (6-23)-(6-28), as was indicated in the discussion of the numerical example at the beginning of Section 6.10, it can be concluded that the dual variables f_{ij} may be viewed as *flows* in a capacitated network. Constraints (6-25), (6-26) and (6-27) correspond to flow constraints for the source, the intermediate, and the terminal nodes, respectively. In particular, Constraints (6-26) correspond to the flow-conservation conditions.

Complementary Slackness Conditions

Using the linear programming complementary slackness conditions for the model formulated in (6-18)-(6-22) the following optimality conditions are obtained:

$$\gamma_{ij} > 0 \;\Rightarrow\; y_{ij} = U_{ij}$$
$$\delta_{ij} > 0 \;\Rightarrow\; y_{ij} = L_{ij}$$
$$f_{ij} > 0 \;\Rightarrow\; t_i - t_j + y_{ij} = 0$$

The graphical representation of these optimality conditions is provided in Figure 6.21. In this figure three cases are identified: Case 1 when $\gamma_{ij} > 0$ and $\delta_{ij} = 0$; Case 2 when $\gamma_{ij} = 0$ and $\delta_{ij} = 0$; and Case 3 when $\gamma_{ij} = 0$ and $\delta_{ij} > 0$.

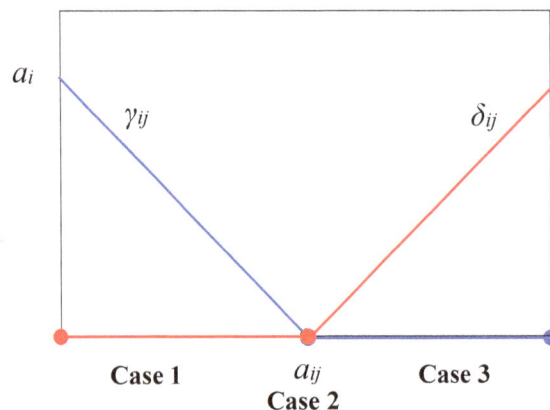

Figure 6.21. Dual variables γ_{ij} as δ_{ij} functions of f_{ij}.

In order to simplify the presentation of some fundamental results, the following definitions (transformations) are now introduced:

$$\overline{y}_{ij} = t_i - t_j + y_{ij}$$
$$\overline{U}_{ij} = t_i - t_j + U_{ij}$$
$$\overline{L}_{ij} = t_i - t_j + L_{ij}$$

Using these definitions along with the results from the complementary slackness conditions we can formulate the *optimality condition* for each case, as follows:

Case 1: $\overline{U}_{ij} = 0$ (6-29)

Case 2: $\overline{y}_{ij} = 0$ (6-30)

Case 3: $\overline{L}_{ij} = 0$ (6-31)

6.10.2 Description of the Algorithm

Briefly, this algorithm starts with the maximum project duration and iteratively identifies the points where the slope of the project curve changes, finishing at the minimum duration allowed. The final output of the procedure is the project curve.

The procedure we are going to discuss in this section was developed by Fulkerson [5] and can be viewed as a dual approach because its derivation relies on the structure of the dual program and the complementary slackness conditions of linear programs. Before describing the algorithm, some simple but helpful considerations are presented. The most typical interpretation of an arc in a flow network is that of a conduit or pipe through which a given type of commodity is transported.

A slightly more sophisticated but meaningful interpretation in the context of the algorithm to be discussed consists of viewing arc (i,j) as a two-compartment conduit, such as the one shown in Figure 6.22. The upper compartment of the representation given in Figure 6.22 has a limited flow capacity while the lower compartment has unlimited capacity. Under the current interpretation, it is assumed that flow goes through the bottom compartment only when the capacity of the top compartment is exceeded. Otherwise, the bottom compartment will remain unused.

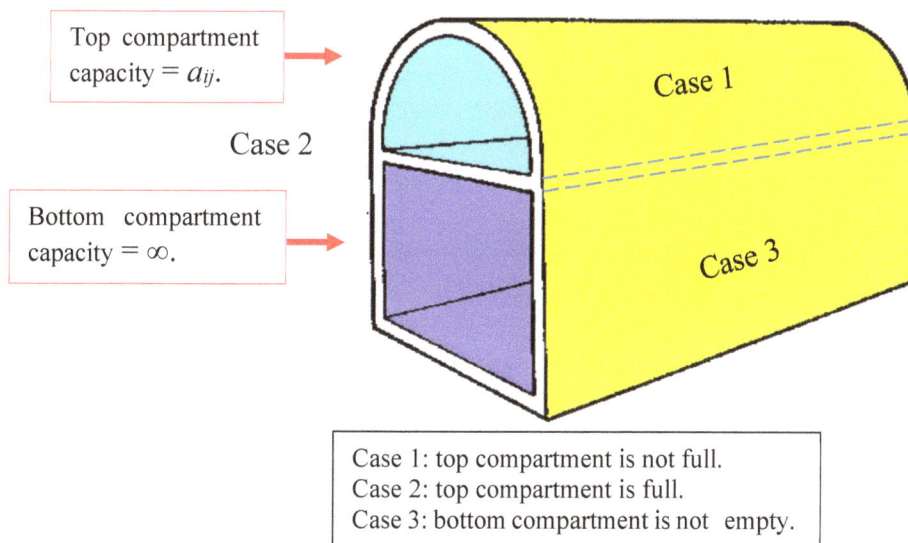

Top compartment capacity = a_{ij}.

Case 2

Bottom compartment capacity = ∞.

Case 1

Case 3

Case 1: top compartment is not full.
Case 2: top compartment is full.
Case 3: bottom compartment is not empty.

Figure 6.22. Arc (i, j) as a two-compartment conduit.

From the graphical representation of the gamma and delta versus flow functions given in Figure 6.21 it can be verified that case 1 occurs when the top compartment is not yet saturated, case 2 occurs when the top compartment is exactly full, and case 3 when the flow is forced through the bottom compartment, whose capacity, as mentioned before, is assumed unlimited. Therefore, we can interpret the *absolute value* of the slope of the linear cost-duration function of a given activity as the capacity of the top compartment of the arc representing that activity in the flow network model.

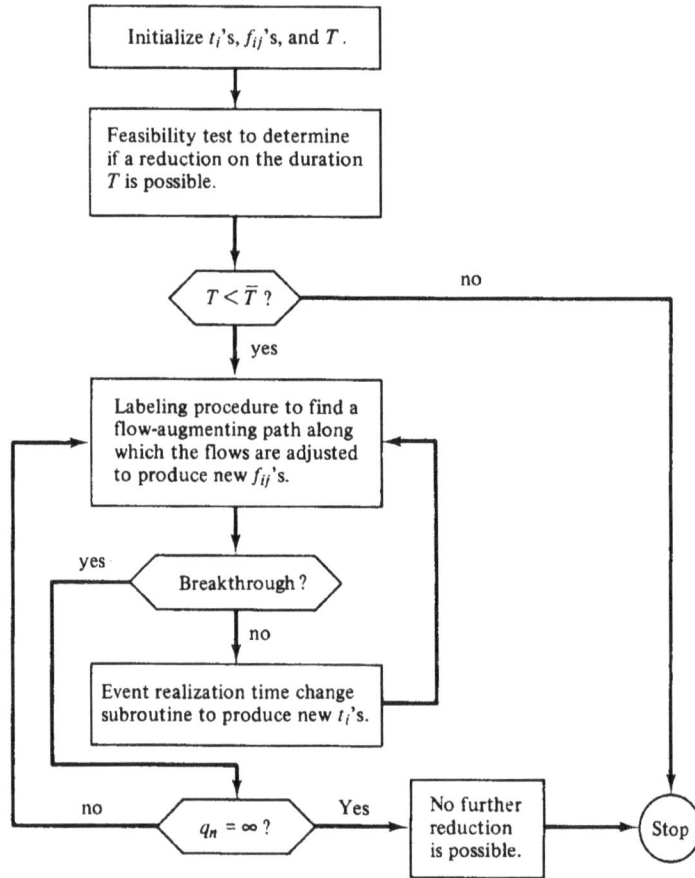

Figure 6.23. Flowchart for the project duration reduction algorithm

The algorithm consists of three fundamental steps. The first step is a feasibility test to verify if it is possible to reduce a specified duration of the project. The second step is a labeling procedure to modify the flows through the network corresponding to the dual program. The third step performs the reductions of the duration of the project, when a non-breakthrough results from the second step. A simplified description of the steps of the algorithm is given in Figure 6.23.

Initial Conditions

$f_{ij} = 0$, for all (i,j)

$t_1 = 0$, $\quad t_j = \max_{i \in \beta_j} \{t_i + U_{ij}\}$, $j = 2, 3, \cdots, n$

$\bar{T} = t_n$ (maximum project duration)

Feasibility Test

Check if a path with $\bar{L}_{ij} = 0$ for all arcs (i,j) exists. If yes, $T < \bar{T}$ is not feasible. The explanation for this is that when an infinite *flow* ($v = q_n = \Sigma_i f_{in}$) reaches the terminal, then the cost of reducing the duration of the project by one unit of time would be infinite. Otherwise, $T < \bar{T}$ is feasible.

248

The Labeling Procedure

Case 1

    Forward Arcs: $\overline{U}_{ij} = 0, f_{ij} < a_{ij}$; label node j with $q_j = min\{q_i; a_{ij} - f_{ij}\}$

    Reverse Arcs: $\overline{U}_{ij} = 0, f_{ij} > 0$; label node i with $q_i = min\{q_j; f_{ij}\}$

Case 3

    Forward Arcs: $\overline{L}_{ij} = 0$; label node j with $q_j = q_i$

    Reverse Arcs: $\overline{L}_{ij} = 0, f_{ij} > a_{ij}$; label node i with $q_i = min\{q_j; f_{ij} - a_{ij}\}$

Event realization Time Change Procedure

Let us first classify arcs into two types I and II. Arcs of type I are directed from labeled nodes to unlabeled nodes. Arcs of type II are directed from unlabeled nodes to labeled nodes. Let $\mathbf{E}$ be the set of labeled nodes, and $\overline{\mathbf{E}}$ be the set of unlabeled nodes.

For type-I arcs (i,j)

$$\mathbf{B} = \{(i,j)|\ i\in\mathbf{E},\ j\in\overline{\mathbf{E}},\ \overline{L}_{ij} < 0\ \text{ or }\ \overline{U}_{ij} < 0\}$$

$$\zeta_1 = \min_{\mathbf{B}}\{-\overline{L}_{ij},\ -\overline{U}_{ij}\}\ (\text{only }negative\ \overline{L}_{ij}\ or\ \overline{U}_{ij}\ \text{values allowed})$$

For type-II arcs (j,i)

$$\overline{\mathbf{B}} = \{(i,j)|\ i\in\overline{\mathbf{E}}, j\in\mathbf{E},\ \overline{L}_{ij} > 0\ \text{ or }\ \overline{U}_{ij} > 0\}$$

$$\zeta_2 = \min_{\mathbf{B}}\{\overline{L}_{ij},\ \overline{U}_{ij}\}\ (\text{only }positive\ \overline{L}_{ij}\ or\ \overline{U}_{ij}\ \text{values allowed})$$

$$\zeta = min\{\zeta_1, \zeta_2\}$$

The node numbers (dual values) are adjusted by subtracting ζ from the current values of the *unlabeled* nodes.

The algorithm can be used to construct a project cost-duration curve. In this case, the algorithm is started with a maximal project duration and additional budgets are estimated at each step of the method where a reduction in project duration is achieved. An illustration of such a curve is given in Figure 6.24.

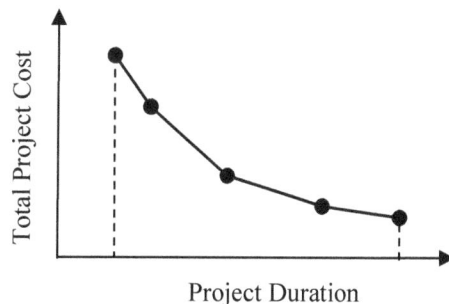

Figure 6.24. Project cost-duration curve.

6.10.3 Illustrative Example

Consider the network shown in Figure 6.19 with the data given in Table 6.6. Figure 6.25 shows the network with the initial values for the f_{ij} *dual* variables for arcs and the t_i *primal* variables for nodes. The figure also includes the table with the calculations needed to find the initial values of each t_i. Each arc (i,j) has a set of parameters $[L_{ij}, U_{ij}, a_{ij}, f_{ij}]$, and each node i has an event realization time $[t_i]$.

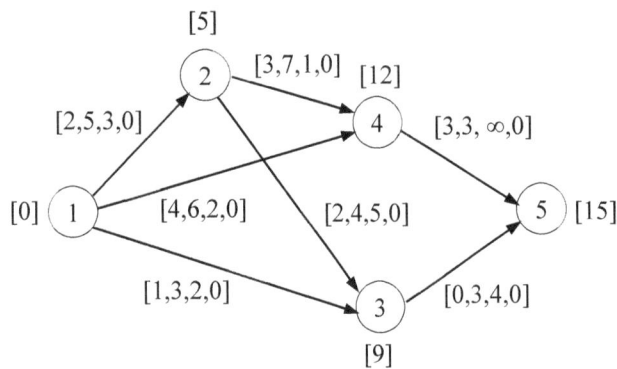

Node	Incident Arcs	Value of t_i
1	-	0
2	(1,2)	$0 + 5 = 5$
3	(1,3), (2,3)	$max \{0+3, 5+4\} = 9$
4	(1,4), (2,4)	$max \{0+6, 5+7\} = 12$
5	(3,5), (4,5)	$max \{9+3, 12+3\} = 15$

Figure 6.25. Network and initial solution.

A. Algorithmic Steps

Feasibility test: There is no path having $\overline{L}_{ij} = 0$ for each arc (i,j). Therefore, a reduction of the maximum value $T = 15$ is possible.

Iteration 1

The path consisting of arcs (1,2), (2,4), and (4,5) is selected for the labeling procedure. For these arcs, the corresponding values of $\overline{U}_{ij}$ and $\overline{L}_{ij}$ are shown below along with the results of the labeling procedure in the graph of the network. The result of this iteration is a breakthrough with $q_5 = 1$. The flow on each arc of the flow-augmenting path (1,2)-(2,4)-(4,5) is increased by 1.

Arc	$\overline{U}_{ij}$	$\overline{L}_{ij}$
(1,2)	0	-3
(2,4)	0	-4
(4,5)	0	0

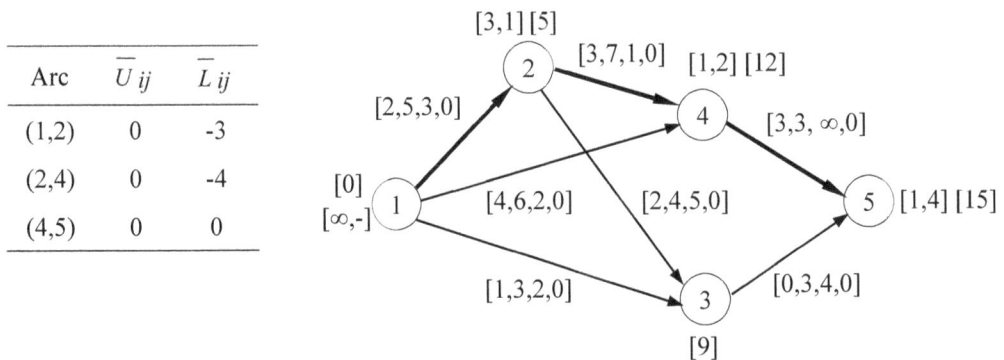

250

Iteration 2

From the results shown below, it is noted that this iteration does not result in a breakthrough because it is not possible to label node 5. The only nodes that could be labeled are nodes 1, 2, and 3. This will be referred to as a *non-breakthrough*.

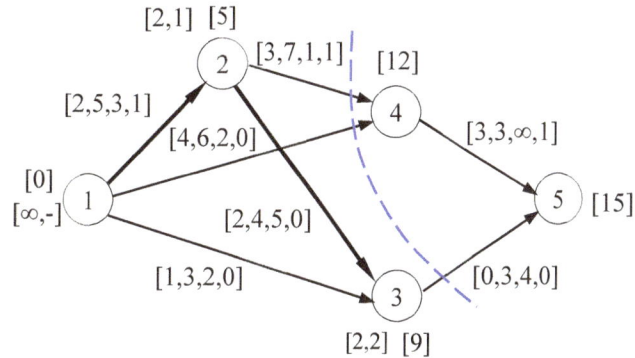

Arc	$\overline{U}_{ij}$	$\overline{L}_{ij}$
(1,2)	0	-3
(2,3)	0	-2
(3,5)	-3	-6
(2,4)	0	-4
(1,4)	-6	-8

The sets containing type-I and type-II arcs as defined in the *event realization time change procedure* are:

$\mathbf{B} = \{(3,5), (2,4), (1,4)\}$, $\zeta_1 = 3$

$\overline{\mathbf{B}} = \varnothing$, $\zeta_2 = \infty$

Therefore, $\zeta = 3$ and both t_4 and t_5 must be decreased by 3. That is, the new values for these event realization times are $t_4 = 9$ and $t_5 = 12$.

Iteration 3

The path consisting of arcs (1,2), (2,3), and (3,5) is selected for the labeling procedure. For these arcs, the corresponding values of $\overline{U}_{ij}$ and $\overline{L}_{ij}$ are shown below along with the results of the labeling procedure in the graph of the network. The result of this iteration is a breakthrough with $q_5 = 2$. The flow on each arc of the flow-augmenting path (1,2)-(2,3)-(3,5) is increased by 2.

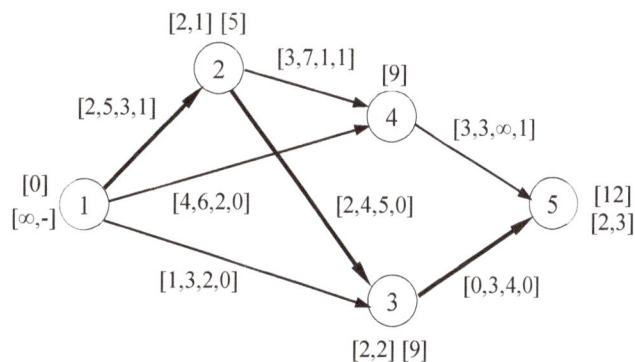

Arc	$\overline{U}_{ij}$	$\overline{L}_{ij}$
(1,2)	0	-3
(2,3)	0	-2
(3,5)	0	-3

Iteration 4

As can be seen in the analysis shown below, only node 1 could be labeled. Therefore, this iteration results in a *non-breakthrough*. This implies the realization times of some events can be reduced.

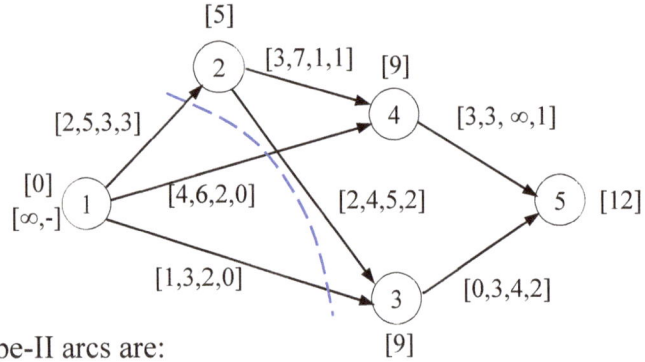

Arc	$\overline{U}_{ij}$	$\overline{L}_{ij}$
(1,2)	0	-3
(1,3)	-6	-8
(1,4)	-3	-5

The sets containing type-I and type-II arcs are:

$$\mathbf{B} = \{(1.2), (1.3), (1,4)\}, \zeta_1 = 3$$

$$\overline{\mathbf{B}} = \varnothing, \zeta_2 = \infty$$

Therefore, $\zeta = 3$ and t_2, t_3, t_4 and t_5 must be decreased by 3. That is, the new values for these event realization times are $t_2 = 2$, $t_3 = 6$, $t_4 = 6$ and $t_5 = 9$.

Iteration 5

The path consisting of arcs (1,2), (2,3), and (3,5) is selected for the labeling procedure. For these arcs, the corresponding values of $\overline{U}_{ij}$ and $\overline{L}_{ij}$ are shown below along with the results of the labeling procedure in the graph of the network. The result of this iteration is a breakthrough with $q_5 = 2$. The flow on each arc of the flow-augmenting path (1,2)-(2,3)-(3,5) is increased by 2.

Arc	$\overline{U}_{ij}$	$\overline{L}_{ij}$
(1,2)	3	0
(2,3)	0	-2
(3,5)	0	-3

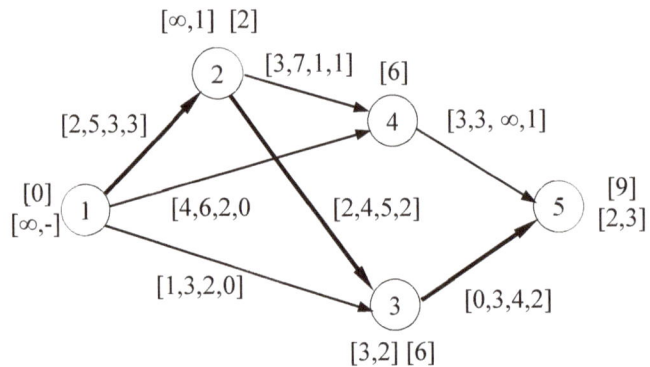

Iteration 6

The path consisting of arcs (1,4), and (4,5) is selected for the labeling procedure. The values of $\overline{U}_{ij}$ and $\overline{L}_{ij}$ are shown below along with the results of the labeling procedure in the graph of the network. The result of this iteration is a breakthrough with $q_5 = 2$. The flow on each arc of the flow-augmenting path (1,4)-(4,5) is increased by 2.

252

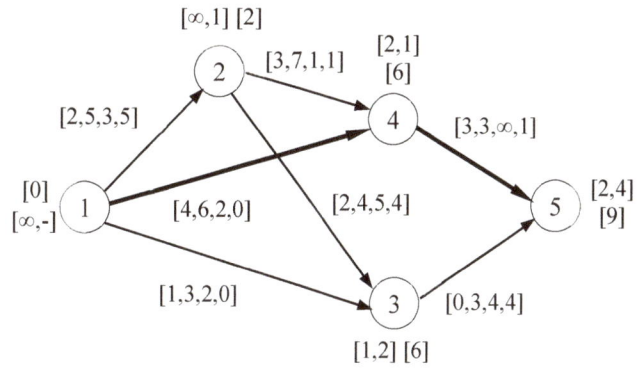

Arc	$\overline{U}_{ij}$	$\overline{L}_{ij}$
(1,2)	3	0
(2,3)	0	-2
(3,5)	0	-3
(1,4)	0	-2
(4,5)	0	0

Iteration 7

From the results shown below, it is noted that this iteration results in a *non-breakthrough* since node 5 could not be labeled. Only nodes 1, 2 and 3 could be labeled.

Arc	$\overline{U}_{ij}$	$\overline{L}_{ij}$
(1,2)	3	0
(2,3)	0	-2
(3,5)	0	-3
(2,4)	3	-1
(1,4)	0	-2

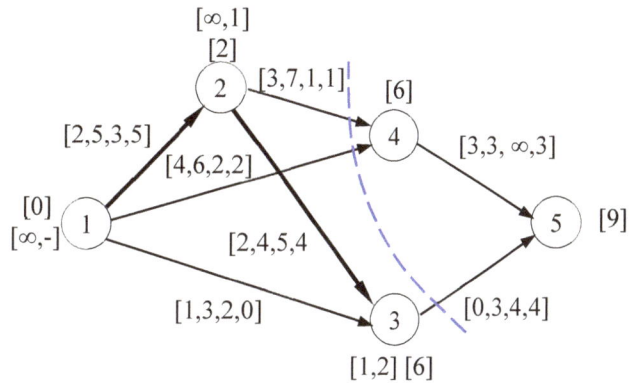

The sets containing type-I and type-II arcs are:

$$\mathbf{B} = \{(1.4), (2,4), (3,5)\}, \zeta_1 = 1$$
$$\overline{\mathbf{B}} = \varnothing, \zeta_2 = \infty$$

Therefore, $\zeta = 1$ and t_4 and t_5 must be decreased by 1. That is, the new values for these event realization times are $t_4 = 5$ and $t_5 = 8$.

Iteration 8

The path consisting of arcs (1,2), (2,4), and (4,5) is selected for the labeling procedure. The values of $\overline{U}_{ij}$ and $\overline{L}_{ij}$ are shown below along with the results of the labeling procedure in the graph of the network. The result of this iteration is a breakthrough with $q_5 = \infty$. Therefore, the project duration cannot be reduced any more, and the algorithm is terminated. The results are summarized in Table 6.7. The corresponding project cost-duration curve is given in Figure 2.

Arc	$\bar{U}_{ij}$	$\bar{L}_{ij}$
(1,2)	3	0
(2,3)	0	-2
(3,5)	1	-2
(2,4)	4	0
(4,5)	0	0

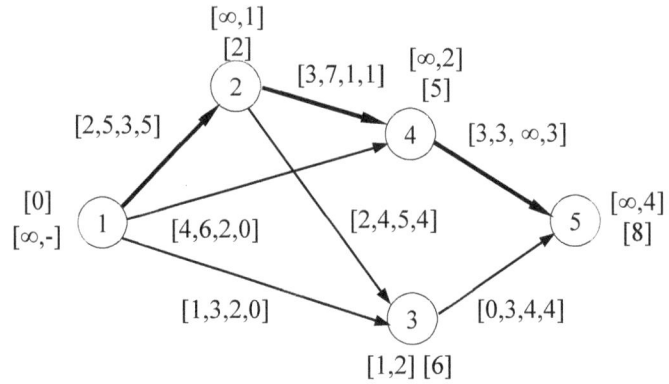

Network diagram labels: $[\infty,1]$ $[2]$ at node 2; $[\infty,2]$ $[5]$ at node 4; $[\infty,4]$ $[8]$ at node 5; $[0]$ $[\infty,-]$ at node 1; $[1,2]$ $[6]$ at node 3. Arc labels: $[2,5,3,5]$; $[3,7,1,1]$; $[3,3,\infty,3]$; $[4,6,2,0]$; $[2,4,5,4]$; $[1,3,2,0]$; $[0,3,4,4]$.

Table 6.7. OPTIMAL RESULTS

Iteration	t_5	$F = \Sigma f_{i5}$	Reduction in Duration, R	RF	Cumulative Increase
2	15	1	3	3	3
4	12	3	3	9	12
7	9	7	1	7	19
8	8	-	(terminate)	-	-

The activity durations corresponding to each of the break points of the project cost-duration curve given in Figure 6.26 can be obtained using the complementary slackness conditions. Since the flow values f_{ij} and the parameters a_{ij} are known at each iteration of the algorithm, we can compute the values of γ_{ij} and δ_{ij} from the results

$$\gamma_{ij} = max\{0; a_{ij} - f_{ij}\}$$
$$\delta_{ij} = max\{0; f_{ij} - a_{ij}\}$$

Once these dual values are determined we can calculate the duration of each activity. By the complementary slackness conditions of linear programs,

If $\gamma_{ij} > 0$, then $y_{ij} = U_{ij}$,

If $\delta_{ij} > 0$ then $y_{ij} = L_{ij}$.

Otherwise, if both dual values are zero, we can only conclude that $L_{ij} \leq y_{ij} \leq U_{ij}$. However, from the event realization times, in this case, we can use the relationship $y_{ij} = t_j - t_i$. A summary of results for the numerical example is shown in Table 6.8.

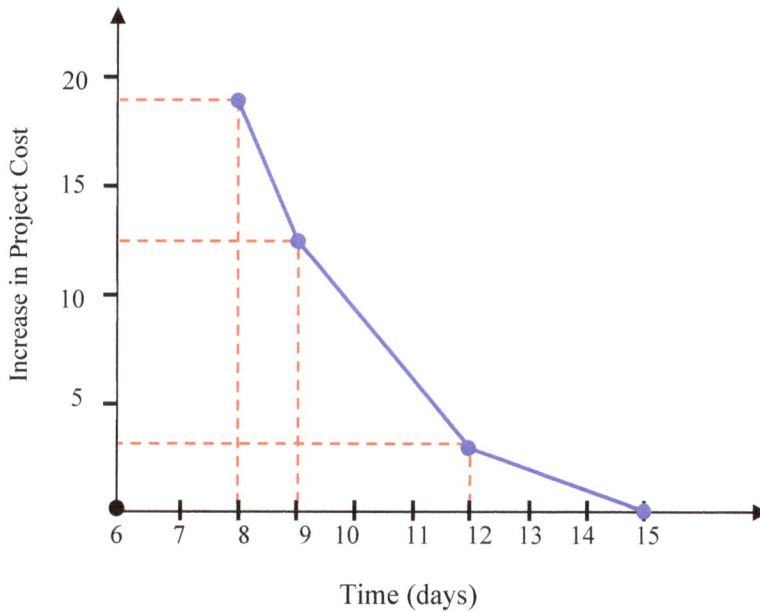

Figure 6.26. Cost-duration curve for numerical example.

Table 6.8. SUMMARY OF RESULTS [b]

$Z= \sum_{(i,j) \in A_1} a_{ij}y_{ij}$	T=8 (*Iteration 8*) Z=53				T=9 (*Iteration 5*) Z=60				T=12 (*Iteration 3*) Z=69				T= 15 (*Iteration 1*) Z=72			
Arc (i,j)	f_{ij}	γ_{ij}	δ_{ij}	y_{ij}	f_{ij}	γ_{ij}	δ_{ij}	y_{ij}	f_{ij}	γ_{ij}	δ_{ij}	y_{ij}	f_{ij}	γ_{ij}	δ_{ij}	y_{ij}
(1, 2)	5	0	2	2	3	0	0	$2^{(a)}$	1	2	0	5	0	3	0	5
(1, 3)	0	2	0	3	0	2	0	3	0	2	0	3	0	2	0	3
(1, 4)	2	0	0	$5^{(a)}$	0	2	0	6	0	2	0	6	0	2	0	6
(2, 3)	4	1	0	4	2	3	0	4	0	5	0	4	0	5	0	4
(2, 4)	1	0	0	$3^{(a)}$	1	0	0	$4^{(a)}$	1	0	0	$4^{(a)}$	0	1	0	7
(3, 5)	4	0	0	$2^{(a)}$	2	2	0	3	0	4	0	3	0	4	0	3
(4, 5)	3	∞	0	3	1	∞	0	3	1	∞	0	3	0	∞	0	3

(a) Since both dual values γ_{ij} and δ_{ij} are zero, the activity should be set at its maximal *feasible* duration, $y_{ij} = t_j - t_i$.

(b) Set A_1 includes all arcs except arc (4,5)

B. NOP Solution

Figure 6.27 shows the NOP screens: (1) Input Data and (2) NOP Output Results. The NOP output results screen consists of three tables: (a) input data, (b) log messages, and (c) output results for each iteration resulting in a reduction in project duration.

255

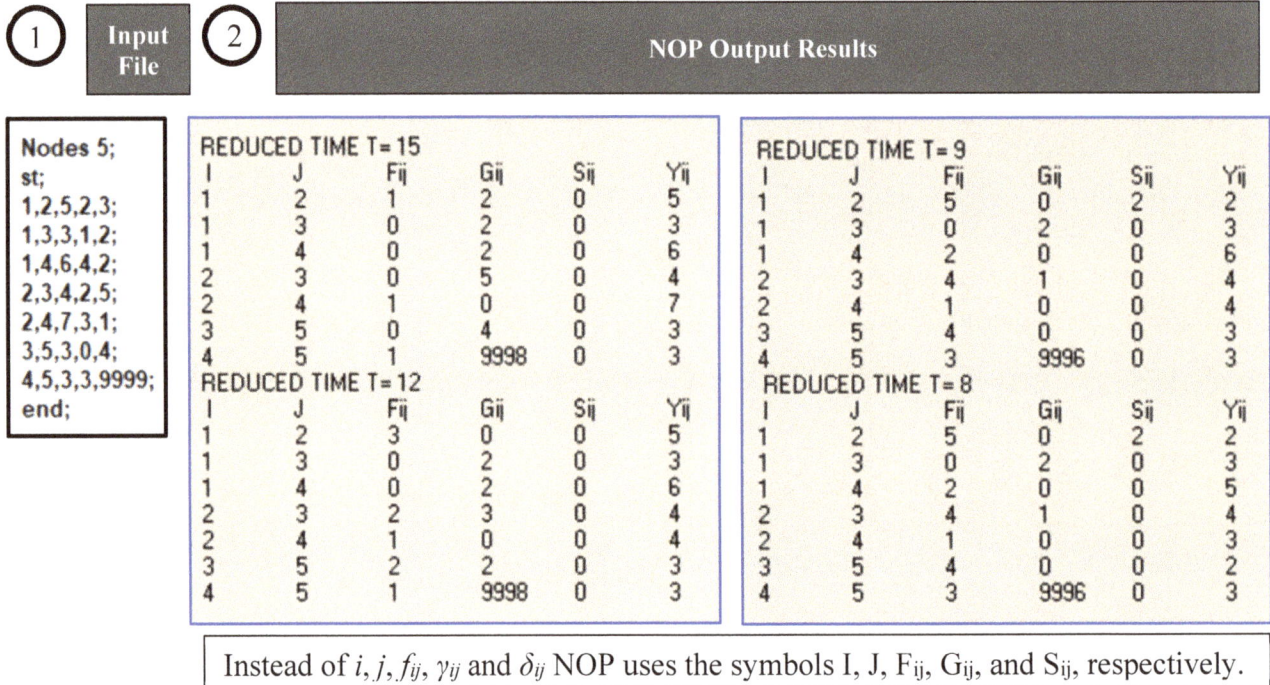

| ① Input File | ② | NOP Output Results | |

Input File:
```
Nodes 5;
st;
1,2,5,2,3;
1,3,3,1,2;
1,4,6,4,2;
2,3,4,2,5;
2,4,7,3,1;
3,5,3,0,4;
4,5,3,3,9999;
end;
```

REDUCED TIME T= 15

I	J	F_{ij}	G_{ij}	S_{ij}	Y_{ij}
1	2	1	2	0	5
1	3	0	2	0	3
1	4	0	2	0	6
2	3	0	5	0	4
2	4	1	0	0	7
3	5	0	4	0	3
4	5	1	9998	0	3

REDUCED TIME T= 12

I	J	F_{ij}	G_{ij}	S_{ij}	Y_{ij}
1	2	3	0	0	5
1	3	0	2	0	3
1	4	0	2	0	6
2	3	2	3	0	4
2	4	1	0	0	4
3	5	2	2	0	3
4	5	1	9998	0	3

REDUCED TIME T= 9

I	J	F_{ij}	G_{ij}	S_{ij}	Y_{ij}
1	2	5	0	2	2
1	3	0	2	0	3
1	4	2	0	0	6
2	3	4	1	0	4
2	4	1	0	0	4
3	5	4	0	0	3
4	5	3	9996	0	3

REDUCED TIME T= 8

I	J	F_{ij}	G_{ij}	S_{ij}	Y_{ij}
1	2	5	0	2	2
1	3	0	2	0	3
1	4	2	0	0	5
2	3	4	1	0	4
2	4	1	0	0	3
3	5	4	0	0	2
4	5	3	9996	0	3

Instead of $i, j, f_{ij}, \gamma_{ij}$ and δ_{ij} NOP uses the symbols I, J, F_{ij}, G_{ij}, and S_{ij}, respectively.

Figure 6.27. Sequence of Computer Screens for Time-Cost Trade Offs Example.

6.11 NON-LINEAR ACTIVITY COST FUNCTIONS

The content of this section is based on the following article published in the IIE Transactions, Industrial Engineering Research & Development, Volume 26, Number 6, 1994. The Institute of Industrial Engineering (IIE) is now the Institute of Industrial and System Engineering (ISE).

A Decomposition Approach to Project Compression with Concave Activity Cost Functions
Ahmet Kuyumcu and Alberto Garcia-Diaz

Given a CPM/PERT activity network with known precedence relationships, it is desired to find those durations that will result in minimal total project cost for a specified project completion time. It is assumed that each activity has a concave or convex piece-wise linear cost-duration function. Falk and Horowitz [4] developed one of the first algorithms to perform project compression in the presence of concave or convex activity cost-duration functions. A branch-and-bound procedure forms the basis of this algorithm. Subsequently, an algorithm capable of performing the time project compression when the activity cost functions are of arbitrary shape was proposed by Panagiotakopoulos [13]. The algorithm reduces the acceptable time range iteratively and produces optimal or, in some cases, sub-optimal solutions.

Phillips and Dessouky [14] formulated the problem as a network flow model and solved it using a cut search algorithm. Additionally, Tufekci [17] developed a procedure which makes use of a labeling algorithm for identifying a minimal cut in the network provided by Phillips and

Dessouky. Kuyumcu and Garcia-Diaz [10] formulate the problem under consideration as a mixed integer linear program (MILP) model and propose a solution procedure based on Benders Decomposition. This technique will be summarized and illustrated on a sample numerical problem in this section.

6.11.1 Mathematical Model Formulation

Let $G(\mathbf{N},\mathbf{A})$ be the network representation of an activity network, where $\mathbf{N} = \{1, 2, \ldots, n\}$ is the set of events (nodes) and $\mathbf{A}$ is the set of activities (arcs). It is desired to determine the duration of each activity in such a way that the total project cost is minimized, for a specified project completion time. It is assumed that each arc (i,j) has a known *piece-wise linear* cost function $f_{ij}(y_{ij})$ of the activity duration y_{ij}. Furthermore, let t_i be realization time of event i, and L_{ij} and U_{ij} denote the *crashed duration* and *normal duration*, respectively, for any activity (i,j). In general, if T represents the specified duration of the project, the problem under consideration can be formulated as minimizing $\sum_{(i,j) \in \mathbf{A}} f_{ij}(y_{ij})$ subject to constraints (6-14) through (6-17).

The following notation will be used in the formulation of the model. This notation is illustrated in Figure 6.28 for the case where $K = 2$.

K = number of linear segments for activity (i,j)

y_{ij}^k = time portion of the duration y_{ij} of activity (i,j) corresponding to the kth linear segment

a_{ij}^k = slope of the kth linear segment of activity (i,j), $k = 1, \ldots, K$

δ_{ij}^k = zero-one variable used to indicate if the kth linear segment is or is not included in the duration of activity (i,j)

L_{ij}^k = lower bound on the $(k+1)^{st}$ linear segment for activity (i,j), $k = 1, \ldots, K-1$

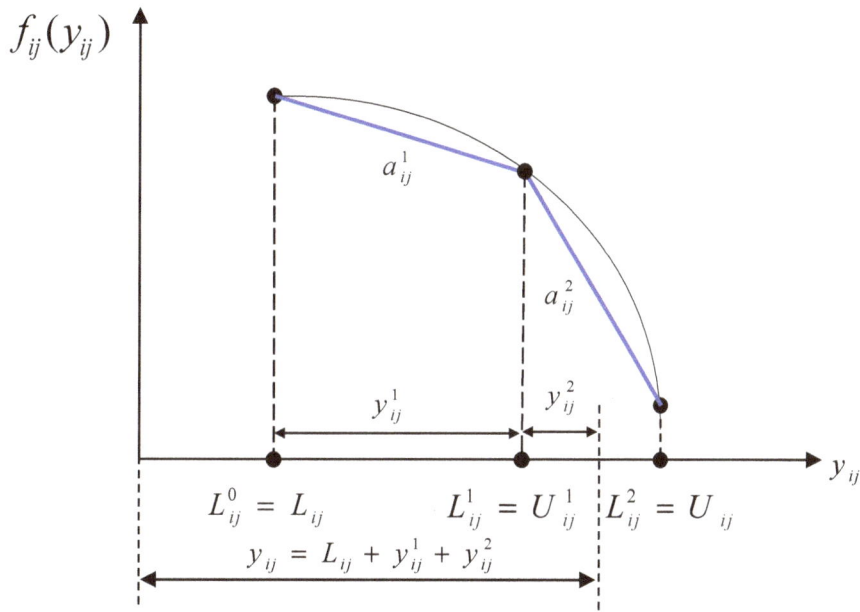

Figure 6.28. Linear segments of concave activity time-cost trade-off function for $K=2$.

257

A Benders decomposition procedure will be developed for solving the model. Using the previously introduced notation, along with the definition $L_{ij}^0 = L_{ij}$ and $L_{ij}^K = U_{ij}$, the model is reformulated as maximizing the objective function (6-32) subject to constraints (6-33) - (6-39).

$$\text{Maximize} \sum_{(ij) \in A} \sum_{k=1}^{K} a_{ij}^k y_{ij}^k \qquad (6\text{-}32)$$

subject to

$$t_i + \sum_{k=1}^{K} y_{ij}^k - t_j \leq L_{ij}^0, \text{ for all } (i,j) \in A \qquad (6\text{-}33)$$

$$-t_1 + t_n = T \qquad (6\text{-}34)$$

$$y_{ij}^k \leq L_{ij}^k - L_{ij}^{k-1}, \text{ for all } (i,j) \in A, \text{ and } k = 1, \ldots, K \qquad (6\text{-}35)$$

$$\delta_{ij}^k (L_{ij}^k - L_{ij}^{k-1}) \leq y_{ij}^k, \text{ for all } (i,j) \in A, \text{ and } k = 1, \ldots, K\text{-}1 \qquad (6\text{-}36)$$

$$\delta_{ij}^k (L_{ij}^{k+1} - L_{ij}^k) \geq y_{ij}^{k+1}, \text{ for all } (i,j) \in A, \text{ and } k = 1, \ldots, K\text{-}1 \qquad (6\text{-}37)$$

$$\delta_{ij}^k = 0 \text{ or } 1, \text{ for all } (i,j) \in A, \text{ and } k = 1, \ldots, K\text{-}1 \qquad (6\text{-}38)$$

$$-y_{ij}^k \leq 0, \text{ for all } (i,j) \in A, \text{ and } k = 1, \ldots, K \qquad (6\text{-}39)$$

The model formulation shown in (6-32) through (6-39) is a mixed integer linear program (MILP). The zero-one δ_{ij}^k variables force segments of the piece-wise linear function to come into the solution in their correct order. That is, y_{ij}^1 will be increased to its maximum value while y_{ij}^2 is kept at zero; once the first variable reaches its maximum value, the second one can start increasing. When the second variable reaches its maximum value, y_{ij}^3 can start increasing, and so on. When the activities cost functions are linear, the zero-one δ_{ij}^k variables are automatically set equal to zero, whereby (6-36) and (6-37) become redundant. When the piece-wise linear cost functions are *convex*, the above model becomes an LP model, and its solution will readily consider the δ_{ij}^k values in their proper order.

6.11.2 Solution Approach

The Benders Decomposition approach will be used assuming that the piece-wise functions are *concave*. The mixed integer program given in (6-32)-(6-39) can be partitioned into an **integer program IP** and a **linear program LP**. It is noted that the objective function coefficients of all integer variables are equal to zero, and that (6-33), (6-34), and (6-35) do not have any integer variables. The mixed integer program MILP in canonical form is formulated in (6-40)-(6-42).

$$\text{Maximize } \mathbf{cy} \qquad (6\text{-}40)$$
subject to
$$\mathbf{D\delta + Cy \leq b} \qquad (6\text{-}41)$$
$$\mathbf{\delta \in S, \ y \geq 0} \qquad (6\text{-}42)$$

In this formulation, **c** denotes the cost coefficient vector, **D** represents the constraint coefficient matrix of the zero-one vector δ, **C** is the constraint coefficient matrix for vector y and **b** is the right-hand side vector of the original problem. Furthermore, **S** is the set of all zero-one vectors of dimension equal to $|\mathbf{A}|(K\text{-}1)$, where $|\mathbf{A}|$ is the cardinality of **A**.

As the first step, let us assume that the elements of vector δ are set to be equal to fixed values. Then, the **linear program LP** can be reformulated as in (6-43)-(6-45).

$$\textit{Maximize } \mathbf{C}y \tag{6-43}$$
subject to
$$\mathbf{C}\,y \leq \mathbf{b} - \mathbf{D}\,\delta \tag{6-44}$$
$$\mathbf{y} \geq \mathbf{0} \tag{6-45}$$

The formulation of **dual model DL** corresponding to problem **LP** is given in (6-43)-(6-45).

$$\textit{Minimize } u(\mathbf{b} - \mathbf{D}\,\delta) \tag{6-46}$$
subject to
$$u\mathbf{C} \geq \mathbf{c} \tag{6-47}$$
$$u \geq 0 \tag{6-48}$$

Although the mixed integer problem given in (6-43)-(6-45) may be infeasible for some δ, it cannot be unbounded. Therefore, it is assumed that the constraint set of the model formulation in (6-46)-(6-48) is not empty and thus the dual problem **DL** reveals either an extreme point u, or an extreme ray r, which results in a single inequality for the integer problem **IP**.

If the dual problem **DL** is maximized for all δ values, its optimal solution is equivalent to the optimal solution of the mixed integer problem given in (6-32)-(6-39) or (6-46)-(6-48). Hence, at any stage, the optimal objective function value of the mixed integer problem under consideration has to be greater than or equal to the value associated with the optimal dual solution. This implies that the optimal solution of the dual problem **DL** is a lower bound z^L on the optimal mixed integer solution. It is noted that the lower bound on the objective function value of the mixed integer problem at any iteration is equal to the maximum of all the optimal dual solutions obtained until that iteration.

Let vector u^f, where $f = 1, \dots, F$, denote the extreme points, and vector r^g, where $g = 1, \dots, G$, denote the extreme rays. If for some δ exists a r^g such that $r^g(\mathbf{b} - \mathbf{D}\delta) < 0$, then the dual problem **DL** is unbounded, which causes the **LP** problem to become infeasible. This in turn would cause the original mixed integer program to be also infeasible. Hence, the necessary and sufficient condition on δ for a feasible mixed integer problem solution is $r^g(\mathbf{b} - \mathbf{D}\delta) \geq 0$.

Next, the following problem is considered:

$$\textit{Maximize } \{\textit{Minimal } u^f(\mathbf{b} - \mathbf{D}\,\delta)\} \tag{6-49}$$
subject to

$$r^g(\mathbf{b}-\mathbf{D}\boldsymbol{\delta}) \geq \mathbf{0}, \ g = 1, ..., G \qquad\qquad (6\text{-}50)$$

$$\boldsymbol{\delta} \in \mathbf{S} \qquad\qquad (6\text{-}51)$$

The formulation shown in (6.49)-(6.51) yields the Benders representation of the mixed integer problem, which can be reformulated as integer program **IP** in (6-52)-(6-55).

$$\textit{Maximize } z \qquad\qquad (6\text{-}52)$$

subject to

$$z \leq \boldsymbol{u}^f(\mathbf{b} - \mathbf{D}\boldsymbol{\delta}), \ f = 1, ..., F \qquad\qquad (6\text{-}53)$$

$$r^g(\mathbf{b}-\mathbf{D}\boldsymbol{\delta}) \geq \mathbf{0}, \ g = 1, ..., G \qquad\qquad (6\text{-}54)$$

$$\boldsymbol{\delta} \in \mathbf{S} \qquad\qquad (6\text{-}55)$$

It is noted that this formulation is equivalent to the original formulation given in (6-32)-(6-39).

Let the dual problem **DL** be solved for all entries of vector $\boldsymbol{\delta}$ being set equal to zero. The optimal solution of the dual problem **DL** reveals an extreme point $\boldsymbol{u}$, an a lower bound z^L on the optimal solution to the original mixed integer program. It is noted that if the dual problem **DL** is unbounded at the first iteration, then z^L could be set equal to zero. The solution of the dual problem provides a single inequality for the integer program **IP**, which is then solved subject to this constraint. Its optimal value yields an upper bound z^1 on the best solution of the **MILP** because the problem with one constraint is more relaxed than the original mixed integer problem. Besides, problem **IP** yields an optimal zero-one solution $\boldsymbol{\delta}$ for the dual problem **DL**, which will produce another new extreme point or an extreme ray and thus a second inequality for the integer program **IP**. This integer program, now subject to two constraints, is then solved and another upper bound, which is at least as good as the preceding one, is obtained. This process achieves optimality when the lower bound and upper bound coincide.

Finally, the **LP** program is solved with the optimal zero-one variables δ_{ij}^k obtained from the last iteration. This yields the optimal solution $\mathbf{y}$, which is the optimal solution of the original mixed integer-programming problem. The procedure previously described is summarized below.

Step 1

- Set the iteration counter equal to 1.
- Set all δ_{ij}^k variables equal to zero.
- Set lower bound z^L equal to zero.
- Set upper bound z^U equal to ∞.

Step 2

- Solve the **DL** problem with current value for each δ_{ij}^k.
- Obtain a new lower bound if possible.
- Obtain a new extreme point or a new extreme ray.

Step 3

- Identify a new inequality constraint for problem **IP**.
- Update the current upper bound.

- Obtain new δ_{ij}^k values.

Step 4

- If lower bound z^L is equal to upper bound z^U go to Step 5. Otherwise, increase the iteration counter by 1 and go to Step 2.

Step 5

- Solve the LP problem with the optimal values for the δ_{ij}^k variables.
- Obtain optimal y_{ij}^k values.

6.11.3 Illustrative Example

This example is based on the seven-activity project network shown in Figure 6.25 with the data given in Table 6.9. For convenience, the figure of the graph of the network is shown again as Figure 6.29. In this example each activity cost function has two linear segments. A computerized procedure developed by Kuyumcu [9] was used to solve the project compression problems associated with this example. Intermediate results are shown at each iteration.

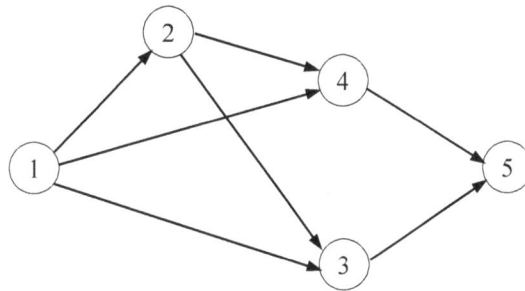

Figure 6.29. Activity Network for Example.

Table 6.9. LOWER/UPPER BOUNDS AND ABSOLUTE VALUE
OF COST SLOPES FOR TWO LINEAR SEGMENTS

(i,j)	$L_{ij}^0 = L_{ij}$	L_{ij}^1	$L_{ij}^2 = U_{ij}$	a_{ij}^1	a_{ij}^2
(1,2)	2	3	9	3	4
(1,3)	1	3	6	2	3
(1,4)	4	5	9	2	3
(2,3)	2	3	7	5	6
(2,4)	3	4	8	1	2
(3,5)	0	2	7	4	5
(4,5)	3	4	9	3	4

The feasible range for this problem is between 8 and 26. By setting all activity durations equal to their lower bounds and finding the critical path, the shortest possible duration for the project is 8. Similarly, setting all activity durations equal to their upper bounds and finding the critical path, the longest possible duration for the project is 26. By using Bender's Decomposition, we can find the optimal schedule for any desired project completion time between 8 and 26. It is desired to find the optimal schedule for a desired project completion time equal to 23 so that we can compare this result with the result obtained in Section 6.11.3.

The **MILP** problem is formulated below. To the right of each constraint the corresponding dual variable is listed within parentheses, assuming that the δ_{ij}^k variables are constants.

Objective Function

$Maximize\ \{3y_{12}^1 + 4y_{12}^2 + 2y_{13}^1 + 3y_{13}^2 + 2y_{14}^1 + 3y_{14}^2 + 5y_{23}^1 + 6y_{23}^2 + y_{24}^1 + 2y_{24}^2 + 4y_{35}^1 + 5y_{35}^2 + 3y_{45}^1 + 4y_{45}^2\}$

subject to

Precedence Relationship Constraints and their Dual Values

$t_1 + y_{12}^1 + y_{12}^2 - t_2 \leq -2 \qquad (f_{12})$

$t_1 + y_{13}^1 + y_{13}^2 - t_3 \leq -1 \qquad (f_{13})$

$t_1 + y_{14}^1 + y_{14}^2 - t_4 \leq -4 \qquad (f_{14})$

$t_2 + y_{23}^1 + y_{23}^2 - t_3 \leq -2 \qquad (f_{23})$

$t_2 + y_{24}^1 + y_{24}^2 - t_4 \leq -3 \qquad (f_{24})$

$t_3 + y_{35}^1 + y_{35}^2 - t_5 \leq 0 \qquad (f_{35})$

$t_4 + y_{45}^1 + y_{45}^2 - t_5 \leq -3 \qquad (f_{45})$

Project Duration Constraint and its Dual Value

$-t_1 + t_5 = 23 \qquad\qquad (v)$

Upper Bound Constraints and their Dual Values

$y_{12}^1 \leq 1\ (\gamma_{12}^1) \quad y_{12}^2 \leq 6\ (\gamma_{12}^2) \quad y_{13}^1 \leq 2\ (\gamma_{13}^1) \quad y_{13}^2 \leq 3\ (\gamma_{13}^2) \quad y_{14}^1 \leq 3\ (\gamma_{14}^1) \quad y_{14}^2 \leq 4\ (\gamma_{14}^2)$

$y_{23}^1 \leq 1\ (\gamma_{23}^1) \quad y_{23}^2 \leq 4\ (\gamma_{23}^2) \quad y_{24}^1 \leq 1\ (\gamma_{24}^1) \quad y_{24}^2 \leq 4\ (\gamma_{24}^2) \quad y_{35}^1 \leq 2\ (\gamma_{35}^1) \quad y_{35}^2 \leq 5 \qquad (\gamma_{35}^2$

$) \qquad y_{45}^1 \leq 1\ (\gamma_{45}^1) \quad y_{45}^2 \leq 5\ (\gamma_{45}^2)$

Ordering Constraints and their Dual Values

$\delta_{12}^1 - y_{12}^1 \leq 0 \quad (\lambda_{12}^1) \qquad 2\delta_{13}^1 - y_{13}^1 \leq 0 \quad (\lambda_{13}^1) \qquad \delta_{14}^1 - y_{14}^1 \leq 0 \quad (\lambda_{14}^1) \qquad \delta_{23}^1 - y_{23}^1 \leq 0 \quad (\lambda_{23}^1)$

$\delta_{24}^1 - y_{24}^1 \leq 0 \quad (\lambda_{24}^1) \qquad 2\delta_{35}^1 - y_{35}^1 \leq 0 \quad (\lambda_{35}^1) \qquad \delta_{45}^2 - y_{45}^2 \leq 0 \quad (\lambda_{45}^2) \qquad 6\delta_{12}^1 - y_{12}^2 \geq 0 \quad (\pi_{12}^2)$

$3\delta_{13}^1 - y_{13}^2 \geq 0 \quad (\pi_{13}^2) \qquad 4\delta_{14}^1 - y_{14}^2 \geq 0 \quad (\pi_{14}^2) \qquad 4\delta_{23}^1 - y_{23}^2 \geq 0 \quad (\pi_{23}^2) \qquad 4\delta_{24}^1 - y_{24}^2 \geq 0 \quad (\pi_{24}^2)$

$5\delta_{35}^1 - y_{35}^2 \geq 0 \quad (\pi_{35}^2) \qquad 5\delta_{45}^1 - y_{45}^2 \geq 0 \quad (\pi_{45}^2)$

Constraints for Zero-One Variables and Lower Bounds

$\delta_{ij}^k = 0\ or\ 1$, for all $(i,j) \in \mathbf{A}$, and $k = 1, 2, ..., K$

$-y_{ij}^k \leq 0$, for all $(i,j) \in \mathbf{A}$, and $k = 1, 2, ..., K$

After considering the δ_{ij}^k variables as constants, the dual problem **DL** corresponding to the above **MILP** problem is formulated below.

Minimize

$$23v - 2f_{12} - f_{13} - 4f_{14} - 2f_{23} - 3f_{24} - 0f_{35} - 3f_{45} + \gamma_{12}^1 + 6\gamma_{12}^2 + 2\gamma_{13}^1 + 3\gamma_{13}^2 +$$

$$\gamma_{14}^1 + 4\gamma_{14}^2 + \gamma_{23}^1 + 4\gamma_{23}^2 + \gamma_{24}^1 + 4\gamma_{24}^2 + 2\gamma_{35}^1 + 5\gamma_{35}^2 + \gamma_{45}^1 + 5\gamma_{45}^2 - \delta_{12}^1\lambda_{12}^1 - 2\delta_{13}^1\lambda_{13}^1 -$$

$$\delta_{14}^1\lambda_{14}^1 - \delta_{23}^1\lambda_{23}^1 - \delta_{24}^1\lambda_{24}^1 - 2\delta_{35}^1\lambda_{35}^1 - \delta_{45}^1\lambda_{45}^1 + 6\delta_{12}^1\pi_{12}^2 + 3\delta_{13}^1\pi_{13}^2 + 4\delta_{14}^1\pi_{14}^2 +$$

$$4\delta_{23}^1\pi_{23}^2 + 4\delta_{24}^1\pi_{24}^2 + 5\delta_{35}^1\pi_{35}^2 + 5\delta_{45}^1\pi_{45}^2$$

Subject to

$f_{12} + \gamma_{12}^1 - \lambda_{12}^1 \geq 3$	$f_{23} + \gamma_{23}^1 - \lambda_{23}^1 \geq 5$	$f_{23} + f_{24} - f_{12} \geq 0$	$-f_{35} - f_{45} + v \geq 0$
$f_{12} + \gamma_{12}^2 + \pi_{12}^2 \geq 4$	$f_{23} + \gamma_{23}^2 + \pi_{23}^2 \geq 6$	$f_{35} - f_{13} - f_{23} \geq 0$	$f_{ij} \geq 0$
$f_{13} + \gamma_{13}^1 - \lambda_{13}^1 \geq 2$	$f_{24} + \gamma_{24}^1 - \lambda_{24}^1 \geq 1$	$f_{45} - f_{14} - f_{24} \geq 0$	$\lambda_{ij}^k \geq 0$
$f_{13} + \gamma_{13}^2 + \pi_{13}^2 \geq 3$	$f_{24} + \gamma_{24}^2 + \pi_{24}^2 \geq 2$	$f_{23} + f_{24} - f_{12} \geq 0$	$\pi_{ij}^k \geq 0$
$f_{14} + \gamma_{14}^1 - \lambda_{14}^1 \geq 2$	$f_{35} + \gamma_{35}^1 - \lambda_{35}^1 \geq 4$	$f_{35} - f_{13} - f_{23} \geq 0$	$\gamma_{ij}^k \geq 0$
$f_{14} + \gamma_{14}^2 + \pi_{14}^2 \geq 3$	$f_{35} + \gamma_{35}^2 + \pi_{35}^2 \geq 5$	$f_{45} - f_{14} - f_{24} \geq 0$	$\delta_{ij}^k \geq 0, v$ unrestricted

The decomposition approach is now used to solve the example under consideration. Details for the procedure are shown below:

Iteration 1

Step 1

Set all $\delta_{ij}^k = 0$, $z^L = 0$, $z^U = \infty$.

Minimize

$$23v - 2f_{12} - f_{13} - 4f_{14} - 2f_{23} - 3f_{24} - 0f_{35} - 3f_{45} + \gamma_{12}^1 + 6\gamma_{12}^2 + 2\gamma_{13}^1 + 3\gamma_{13}^2 +$$

$$\gamma_{14}^1 + 4\gamma_{14}^2 + \gamma_{23}^1 + 4\gamma_{23}^2 + \gamma_{24}^1 + 4\gamma_{24}^2 + 2\gamma_{35}^1 + 5\gamma_{35}^2 + \gamma_{45}^1 + 5\gamma_{45}^2$$

Subject to

$f_{12} + \gamma_{12}^1 - \lambda_{12}^1 \geq 3$	$f_{23} + \gamma_{23}^1 - \lambda_{23}^1 \geq 5$	$f_{23} + f_{24} - f_{12} \geq 0$	$-f_{35} - f_{45} + v \geq 0$
$f_{12} + \gamma_{12}^2 + \pi_{12}^2 \geq 4$	$f_{23} + \gamma_{23}^2 + \pi_{23}^2 \geq 6$	$f_{35} - f_{13} - f_{23} \geq 0$	$f_{ij} \geq 0$
$f_{13} + \gamma_{13}^1 - \lambda_{13}^1 \geq 2$	$f_{24} + \gamma_{24}^1 - \lambda_{24}^1 \geq 1$	$f_{45} - f_{14} - f_{24} \geq 0$	$\lambda_{ij}^k \geq 0$
$f_{13} + \gamma_{13}^2 + \pi_{13}^2 \geq 3$	$f_{24} + \gamma_{24}^2 + \pi_{24}^2 \geq 2$	$f_{23} + f_{24} - f_{12} \geq 0$	$\pi_{ij}^k \geq 0$
$f_{14} + \gamma_{14}^1 - \lambda_{14}^1 \geq 2$	$f_{35} + \gamma_{35}^1 - \lambda_{35}^1 \geq 4$	$f_{35} - f_{13} - f_{23} \geq 0$	$\gamma_{ij}^k \geq 0$
$f_{14} + \gamma_{14}^2 + \pi_{14}^2 \geq 3$	$f_{35} + \gamma_{35}^2 + \pi_{35}^2 \geq 5$	$f_{45} - f_{14} - f_{24} \geq 0$	v unrestricted

Step 2

Solve the **DL** problem. The solution is shown in Table 6.11. The objective function value is equal to 26. Therefore, set $z^L = 26$.

Table 6.11. SOLUTION OF DL AFTER FIRST ITERATION

Variable	Value	Variable	Value	Variable	Value
v	0.00	γ^1_{14}	2.00	λ^1_{14}	0.00
f_{12}	0.00	γ^2_{14}	0.00	λ^1_{23}	0.00
f_{13}	0.00	γ^1_{23}	5.00	λ^1_{24}	0.00
f_{14}	0.00	γ^2_{23}	0.00	λ^1_{35}	0.00
f_{23}	0.00	γ^1_{24}	1.00	λ^1_{45}	0.00
f_{24}	0.00	γ^2_{24}	0.00	π^2_{12}	4.00
f_{35}	0.00	γ^1_{35}	4.00	π^2_{13}	3.00
f_{45}	0.00	γ^2_{35}	0.00	π^2_{14}	3.00
γ^1_{12}	3.00	γ^1_{45}	3.00	π^2_{23}	6.00
γ^2_{12}	0.00	γ^2_{45}	0.00	π^2_{24}	2.00
γ^1_{13}	2.00	λ^1_{12}	0.00	π^2_{35}	5.00
γ^2_{13}	0.00	λ^1_{13}	0.00	π^2_{45}	4.00

Step 3

Formulate the zero-one integer programming problem defined in (6.52)-(6.55) as follows:

Maximize z

subject to

$$24\delta^1_{12} + 9\delta^1_{13} + 12\delta^1_{14} + 24\delta^1_{23} + 8\delta^1_{24} + 25\delta^1_{35} + 20\delta^1_{45} - z \geq -26$$
$$\delta^k_{ij} = 0 \text{ or } 1, \text{ for all } (i,j) \in \mathbf{A}, \text{ and } k = 1.$$

The optimal solution of this model is $\delta^1_{12} = 1$, $\delta^1_{13} = 1$, $\delta^1_{14} = 1$, $\delta^1_{23} = 1$, $\delta^2_{24} = 1$, $\delta^1_{35} = 1$, $\delta^1_{45} = 1$ with an objective function equal to 148. Therefore, $z^u = 148$.

Step 4

Since $z^L = 26$ is not equal to $z^u = 148$ we proceed with iteration 2.

Iteration 2

Step 2

Solve the **DL** problem with each δ^k_{ij} set equal to the value obtained from Step 3 in iteration 1. The solution is shown in Table 6.12. The objective function value 142, which is equal to z^L.

Table 6.12. SOLUTION OF DL AFTER SECOND ITERATION

Variable	Value	Variable	Value
v	2.00	γ_{35}^1	4.00
f_{12}	2.00	γ_{35}^2	0.00
f_{13}	0.00	γ_{45}^1	1.00
f_{14}	0.00	γ_{45}^2	0.00
f_{23}	0.00	λ_{12}^1	0.00
f_{24}	2.00	λ_{13}^1	0.00
f_{35}	0.00	λ_{14}^1	0.00
f_{45}	2.00	λ_{23}^1	0.00
γ_{12}^1	2.00	λ_{24}^1	1.00
γ_{12}^2	0.00	λ_{35}^1	0.00
γ_{13}^1	2.00	λ_{45}^1	0.00
γ_{13}^2	0.00	π_{12}^2	2.00
γ_{14}^1	2.00	π_{13}^2	3.00
γ_{14}^2	0.00	π_{14}^2	3.00
γ_{23}^1	5.00	π_{23}^2	6.00
γ_{23}^2	0.00	π_{24}^2	0.00
γ_{24}^1	0.00	π_{35}^2	5.00
γ_{24}^2	0.00	π_{45}^2	2.00

Step 3

Formulate the zero-one integer programming problem by using (6.56)-(6.55) as follows:

Maximize z

subject to

$$24\delta_{12}^1 + 9\delta_{13}^1 + 12\delta_{14}^1 + 24\delta_{23}^1 + 8\delta_{24}^1 + 25\delta_{35}^1 + 20\delta_{45}^1 - z \leq 226$$
$$-1\delta_{24}^1 + 12\delta_{12}^1 + 9\delta_{13}^1 + 12\delta_{14}^1 + 24\delta_{23}^1 + 25\delta_{35}^1 + 10\delta_{45}^1 - z \leq 251$$
$$\delta_{ij}^k = 0 \text{ or } 1, \text{ for all } (i,j) \in \mathbf{A}, \text{ and } k = 1, 2, \ldots, K\text{-}1$$

The solution of this zero-one linear programming problem is $\delta_{12}^1 = 1$, $\delta_{13}^1 = 1$, $\delta_{14}^1 = 1$, $\delta_{23}^1 = 1$, $\delta_{24}^2 = 1$, $\delta_{35}^1 = 1$, $\delta_{45}^1 = 1$ with an objective function equal to 142.

Step 4

Since $z^L = 142$ is equal to $z^U = 142$, the optimum δ_{ij}^k values are obtained. Go to step 5.

265

Step 5

Solve the MILP problem with the optimum δ_{ij}^k values. The optimum schedule is given in Table 6.13 which corresponds to an optimal value of the objective function equal to 142.

Table 6.13. OPTIMAL SOLUTION

Variable	Value	Variable	Value	Variable	Value	Variable	Value
t_1	0	y_{12}^1	1	y_{14}^2	4	y_{35}^1	2
t_2	9	y_{12}^2	6	y_{23}^1	1	y_{35}^2	5
t_3	16	y_{13}^1	2	y_{23}^2	4	y_{45}^1	1
t_4	14	y_{13}^2	3	y_{24}^1	1	y_{45}^2	5
t_5	23	y_{14}^1	1	y_{24}^2	1		

EXERCISES

1. A building contractor is planning construction activities for a custom home. The description of the corresponding sequence of activities required for the execution of this project, along with the precedence relationships and the time durations, are given in the attached table. The building contractor is especially interested in constructing the appropriate CPM activity network for the sequence of activities. Use both (a) activity-on-arc and (b) activity-on-node representations. Discuss the advantages and disadvantages of both representations.

Activity	Description	Immediate Predecessors	Time Duration (days)
a	Start		0
b	Excavate and pour footings	a	4
c	Pour concrete foundation	b	2
d	Erect wooden frame, including rough roof	c	4
e	Lay brickwork	d	6
f	Install basement drains and plumbing	c	1
g	Pour basement floor	f	2
h	Install rough plumbing	f	3
i	Install rough wiring	d	2
j	Install heating and ventilating	d,g	4
k	Fasten plaster board and plaster (including	i,j,h	10
l	drying)	k	3
m	Lay finish flooring	l	1
n	Install kitchen fixtures	l	2
o	Install finish plumbing	l	3
p	Finish carpentry	e	2
q	Finish roofing and flashing	p	1
r	Fasten gutters and downspouts	c	1
s	Lay storm drains for rain water	o,t	2
t	Sand and varnish flooring	m,n	3
u	Paint	t	1
v	Finish electrical work	q,r	2
w	Finish grading	v	5
x	Pour walks and complete landscaping	s,u,w	0
	Finish		

2. Consider a project to promote a new product. The activity durations to complete the project are given in the table. Find the minimum total time to complete the project.

No.	Activity	Time Duration (weeks)	Precedence Activities
0	Lead-time planning	3	-
1	Develop training plan	6	0
2	Select trainees	4	0
3	Draft brochure	3	0
4	Conduct training course	1	1,2,3
5	Deliver sample products	4	0
6	Print brochure	5	3
7	Prepare advertising	5	0
8	Release advertising	1	7
9	Distribute brochure	2	6

3. A project consists of the activities shown in the table. The three given duration estimates represent the optimistic (a), most likely (m), and pessimistic (b) times in days for each activity. Draw the activity network and find the critical path for using PERT. From the result for the critical path, determine both the mean and the standard deviation of the project completion time.

Activity	Duration (days)

	a	m	b
(1,2)	5	8	10
(1,3)	18	20	22
(1,4)	26	33	40
(2,5)	16	18	20
(2,6)	15	20	25
(3,6)	6	9	12
(4,7)	7	10	12
(5,7)	5	7	8
(6,7)	3	4	5

4. The optimistic (a), most likely (m), and pessimistic (b) times for a certain project are given in the table. The scheduled completion time is 17.5 days. Find the probability of finishing the project by the scheduled time.

Activity	Duration (days)		
	a	m	b
(1,2)	6	8	10
(1,3)	4	6	7
(1,4)	4	8	12
(2,5)	5	6	8
(3,5)	7	8	9
(4,6)	7	10	14
(5,6)	3	4	5

5. What are three uses of dummy activities and dummy events in the activity-on-arc representation? Illustrate use by examples.

6. A hospital building foundation consists of four consecutive sections. The activities for each section need excavation, reinforcement, and filling concrete. The excavation of one section cannot start until the preceding one is completed. The same applies to filling concrete. After excavating all sections, the plumbing jobs can be started, but only 15% of the job can be completed before any concrete is poured. After completion of each section of the foundation, an additional 10% of the plumbing can be started provided that the preceding 10% portion is complete. Develop an activity network for this project.

7. Consider the activity network shown. Assume that the latest allowable project completion date is set at 49 days. $[T_8(L) \neq T_8(E)]$. Find the critical path for this network.

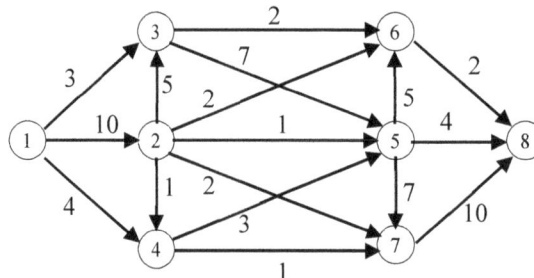

8. Calculate the total float, free float, safety float, and independent float for each of the activities in Exercise 2. This problem statement is due to The Methods and Standards group has furnished the data shown in the table. (a) Draw the activity network and find the critical path. (b) Find the values of the activity floats.

9. This problem statement is due to Thomas [16].

Activity	Description	Immediate Predecessors	Duration (days)	Immediate Successors
Start	-	-	-	Q, R
Q	Lead time	*Start*	10	A
R	Line available	*Start*	30	F
A	Measure and sketch	Q	2	B
B	Develop material list	A	1	C, D, F, G
C	Procure pipe	B	30	E
D	Procure valves.	B	45	K
E	Prefabricate sections	C	5	I
F	Deactivate line	B, R	1	H, K
G	Erect scaffold	B	2	H, K
H	Remove old pipe and valves	F. G	6	I
I	Place new pipe	E, H	6	J
J	Weld pipe	I	2	L, N
K	Place valves	D, F, G	1	L, N
L	Fit up pipe and valves	J, K	1	M, O
M	Pressure test	L	1	P
N	Insulate	J, K	4	O
O	Remove scaffold	L, N	1	P
P	Cleanup	M, O	1	*Finish*
Finish	-	P	-	-

A reactor and storage tank are interconnected by a 3-inch insulated process line. Because of erosion, the line needs to be periodically replaced. Interspersed along the line and at the terminals are valves that also need to be replaced. Pipe and valves must be ordered. Accurate drawings do not exist and must be made. The line is overhead, so that scaffolding is necessary for the replacement operation, but adequate craft labor is available. It is assumed that work on the project will proceed around the clock. The line cannot be shut down without a 30-day notice. Furthermore, a lead time of 10 days is required before the engineering work can be started. The Methods and Standards group has furnished the data shown in the table. (a) Draw the activity network and find the critical path. (b) Find the values of the activity floats.

10. Consider a project to promote a new product. The probabilistic time durations to complete the activities are given in the table.

Precedence Activities	No.	Activities	Duration (days)		
			Optimistic a	*Most Likely, m*	*Pessimistic, b*

269

None	0	Lead-time planning	2	3	5
0	1	Develop training plan	2	6	10
1	2	Select trainees	3	4	5
0	3	Draft brochures	1	3	4
9,10,5, and 8	4	Field test material	1	1	1
13	5	Deliver sample products	3	4	4
3	6	Print brochure	4	5	6
3	7	Prepare advertising	2	5	7
7	8	Release advertising	1	1	1
6	9	Distribute brochure	2	2	3
6 and 2	10	Train sales force	3	5	6
4	11	Review market survey	2	4	5
0	12	Develop prototype product	5	7	8
12	13	Manufacture sample products	2	3	4

(a) Compute the mean and variance for each activity.
(b) Compute the critical path using the activity-on-node representation.
(c) Compute the critical path using the activity-on-arc representation.
(d) What is the probability that the network will be realized in less than 20 days? 25 days? 30 days?
(e) Determine the number of project days for which the probability of exceeding this duration is only 10%.

11. Determine the following for Exercise 1: (a) the critical path; (b) total floats; (c) free floats; (d) safety floats; (e) independent floats.

12. An activity project network consists of eleven tasks with activity duration data given in days in the table shown below.

Task	Activity	Duration (days)
A	(1,2)	8
B	(2,3)	10
C	(2,4)	2
D	(3,4)	16
E	(3,5)	4
F	(4,5)	8
G	(3,6)	7
H	(4,6)	12
I	(5,7)	3
J	(6,7)	8
K	(7,8)	2

(a) Draw the CPM network.
(b) Determine the project duration

(c) Assume that each task requires exactly one person to perform the associated activity. Draw a bar chart and a resource-loading diagram that profiles the resource usage through time (assume that all jobs start at the earliest possible time).

(d) Use the resource-leveling procedures given in Section 6.11 to minimize the total manpower usage per day over the entire project duration.

13. The table shows data for a particular activity network.

	Time (days)		Cost		Preceding
Job	Normal	Crash	Normal	Crash	Activities
A	7	4	$ 95	$100	None
B	6	3	90	97	None
C	5	4	86	104	None
D	7	7	92	98	A and C
E	6	5	87	93	A, B, and C
F	7	5	112	120	B
G	8	5	101	113	F
H	9	6	97	109	K, E, and D
I	12	10	95	100	A and C
J	10	7	100	110	None
K	9	8	105	114	F

(a) Construct the CPM activity network using both the activity-on-arc and activity-on-node representations.

(b) Find the critical path using the activity-on-node representation.

(c) Find the critical path using the activity-on-arc representation.

(d) Crash the network to minimum-time duration along a minimum-cost schedule. Plot the results.

14. Consider the following CPM network (activity-on-node) with the activity durations (weeks) above the node and the crew sizes required to perform the activity below the node.

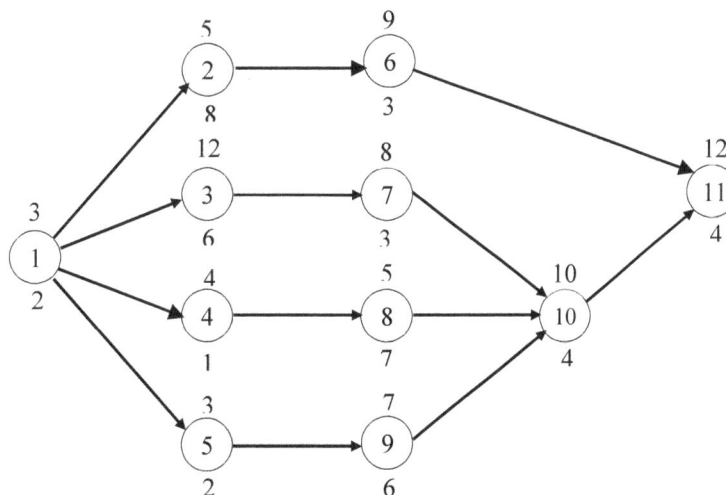

(a) Find the critical path for this project network.

(b) Compute the total float, free float, safety float, and independent float for all activities.

(c) Draw a bar chart and a resource-loading diagram showing the resource requirements profile for this project over the total time period (assume all jobs at earliest possible start).

(d) Use resource-leveling techniques to "smooth out" the manpower requirements.

15. Consider the following activity-on-arc CPM network with associated cost data. Resources requires are dollars per day to run the project.

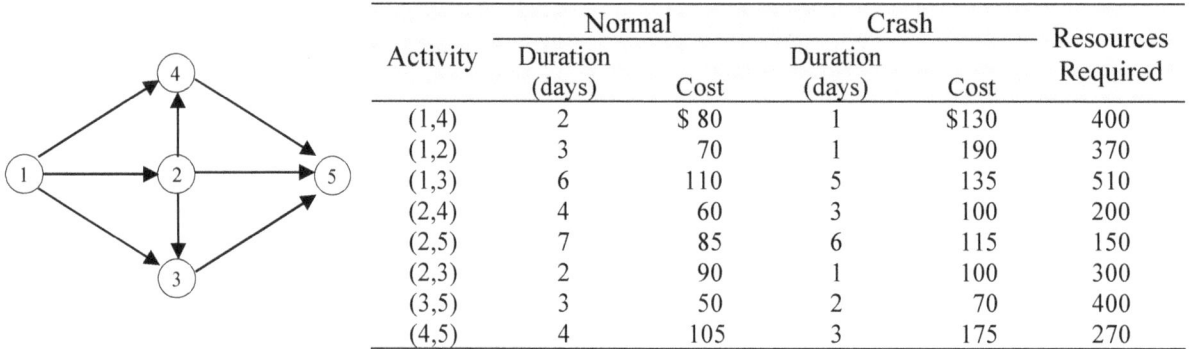

	Normal		Crash		Resources Required
Activity	Duration (days)	Cost	Duration (days)	Cost	
(1,4)	2	$ 80	1	$130	400
(1,2)	3	70	1	190	370
(1,3)	6	110	5	135	510
(2,4)	4	60	3	100	200
(2,5)	7	85	6	115	150
(2,3)	2	90	1	100	300
(3,5)	3	50	2	70	400
(4,5)	4	105	3	175	270

(a) Compute the critical path.

(b) Reduce project duration to 9 days at minimum cost.

(c) Reduce the project to minimum duration and calculate the cost.

(d) Smooth the flow of capital (create level cash expenditures) into the project using the results of part (a) by using resource-leveling techniques.

16. Using the flow algorithm of Section 6.12.2, find the cost-duration curve for the project represented by the network shown. Show all relevant results for each iteration of the algorithm. Check your results solving the problem with NOP.

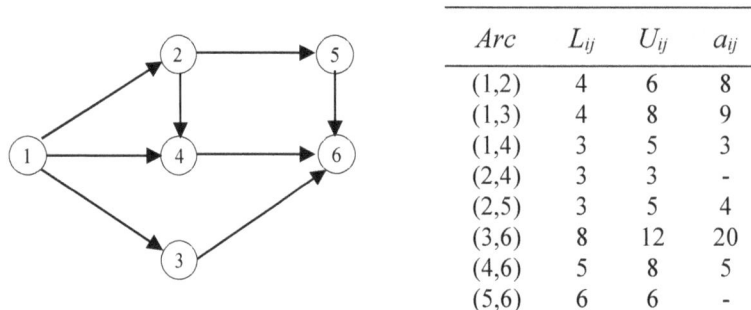

Arc	L_{ij}	U_{ij}	a_{ij}
(1,2)	4	6	8
(1,3)	4	8	9
(1,4)	3	5	3
(2,4)	3	3	-
(2,5)	3	5	4
(3,6)	8	12	20
(4,6)	5	8	5
(5,6)	6	6	-

17. Given the following network with activities having normally distributed durations, determine the numbers w_i such that the probability of realization of each node by time w_i is at least 0.95. The numbers assigned to the arcs are the means, and all variances are equal to 4. Formulate the chance-constrained event realization model.

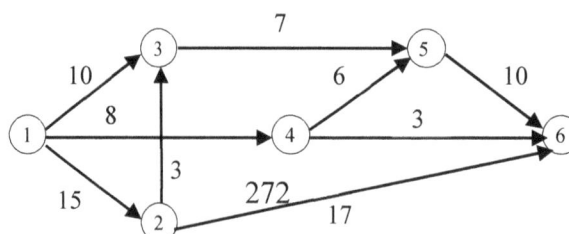

18. Consider a PERT network with p independent paths exponentially distributed with mean duration (path length) equal to $1/b$. Write an expression in terms of b and p for the probability that the project is finished between 10 and 12 time units.

19. Find the optimal activity durations for the activity network shown below considering the data given in the table for two linear segments for the cost functions of the five activities.

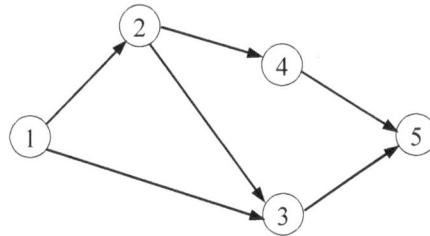

Lower/Upper Bounds and Absolute Value of Cost Slopes for Two Linear Segments

(i,j)	$L_{ij}^0 = L_{ij}$	L_{ij}^1	$L_{ij}^2 = U_{ij}$	a_{ij}^1	a_{ij}^2
(1,2)	2	4	9	3	4
(1,3)	1	2	6	2	3
(2,3)	2	4	7	5	6
(2,4)	3	6	8	1	2
(3,4)	3	5	7	4	5

REFERENCES

[1] DAVIS, E. W., AND G. E. HEIDORN, "An Algorithm for Optimal Project Scheduling under Multiple Resource Constraints," *Management Science* (August 1971).

[2] DAVIS, E. W., "Resource Allocation in Project Network Models-A Survey," *Journal of Industrial Engineering* (April 1966).

[3] ELMAGHRABY, S.E., *Activity Networks: Project Planning and Control by Network Models*, New York: John Wiley & Sons, 1977.

[4] FALK, J.E., AND J.L. HAROWITZ, "Critical Path Problems with Concave Cost-Time Curves," *Management Science*, 19(4), 446-455, 1972.

[5] FULKERSON, D. R., "A Network Flow Computation for Project Cost Curves," *Management Science* (January 1961).

[6] KELLEY, J. E., AND M. R. WALKER, "Scheduling Activities to Satisfy Resource Constraints," in *Industrial Scheduling*, by J. Muth and G. Thompson. Englewood Cliffs, N.J.: Prentice-Hall, Inc., 1963.

[7] KELLEY, J. E., AND M. R. WALKER, "Critical Path Planning and Scheduling," *Proceedings of the Eastern Joint Computer Conference*, 1959.

[8] KELLEY, J. E., AND M. R. WALKER, "Critical Path Planning and Scheduling: Mathematical Basis," *Operations Research* (May-June 1961).

[9] KUYUMCU, A, "A Decomposition Approach to Project Compression in CPM/PERT Networks with Concave Activity Cost Functions," *Master of Science Thesis*, Department of Industrial Engineering, Texas A&M University, August 1991.

[10] KUYUMCU, A. AND A. GARCIA-DIAZ, "A Decomposition Approach to Project Compression with Concave Activity Cost Functions," *IIE Transactions*, Industrial Engineering Research & Development, Volume 26, No. 6, pp.63-73, 1994.

[11] LEVY, F. K., AND J. D. WIEST, A Management Guide to PERT/CPM, Englewood Cliffs, N.J.: Prentice-Hall, Inc., 1969.

[12] MODER, J. J., AND C. R, PHILLIPS, *Project Management with CPM and PERT*. New York: Van Nostrand Reinhold Company, 1964; 2nd ed., 1970.

[13] PANAGIOTAKOPOULOS, D., "Cost-Time Model for Large CPM Project Networks," *Journal of the Construction Division*, ASCE, Vol. 103, No. CO2, pp. 201-211, 1977.

[14] PHILLIPS, S. JR, AND M.I. DESSOUKY, "The Cut Search Algorithm with Arc Capacities and Lower Bounds," *Management Science*, 25(4), 396-404, 1979.

[15] SHAFFER, L.R., J.B. RITTER, AND W.L. MEYER, *The Critical Path Method*. New York: McGraw-Hill Book Company, 1964.

[16] THOMAS, W.H., "Four Float Measures for Critical Path Scheduling," *Industrial Engineering* (October 1969).

[17] TUFEKCI, S., "A Flow Preserving Algorithm for Time-Cost Tradeoff Problem," *AIIE Transactions*, Vol. 12, No. 3, 1982.

Chapter 7

GRAPHICAL EVALUATION AND REVIEW TECHNIQUE (GERT)

"You may have trouble believing it, but every time we've tested the old saying,
it has paid off for us in spades: the more you give, the more you get.
It's all a matter of attitude and the capacity to constantly study and question
the management of the business."

Sam Walton
SAM WALTON Made in America
Doubleday 1992

In all previous discussions we have considered only systems with deterministic network representations. In typical project networks, all arcs of the network must be realized for a realization of the overall network. This condition implies that feedback operations cannot be included in the model, since they are represented by branches that close loops, and the existence of these loops in turn implies that the final node of an activity must be realized before the starting node. Two situations have been widely studied in the field of deterministic networks. When there is *exactly* one realization time for each arc we have the critical path model, and when there are *several* possible realization times for each arc we have a PERT model.

Often, in modeling industrial systems, a more flexible and powerful representation is provided by a network with a stochastic structure. A stochastic network is defined as one which is realized when a *proper* subset of its arcs is realized, assuming that arcs are selected for realization according to known probabilities. In particular, the nature of a stochastic network is such that for the realization of a node, it is not necessary to realize all the arcs incident to the node. For this reason, cycles and self-loops are also allowed in the representation of the system.

7.1 NETWORK REPRESENTATION

The nodes of a stochastic network can be interpreted as the states of the system. The arcs represent transitions from one state to another. Such transitions can be viewed as generalized activities, characterized by a unique probability density or mass function and a probability of realization. Each intermediate node in a stochastic network performs two functions, one on the receiving side and another on the emitting side. Usually, these two functions are classified as input and output functions:

1. *Input function*: It indicates the condition under which the node can be realized.

2. *Output function*: It indicates the branching conditions following the node realization. In other words, the output function defines if only one or all the activities emanating from the node are undertaken.

Notice that the initial node of the network performs only an output function, while the ending node performs an input function. There are three types of input functions and two types of output functions, as defined below.

7.1.1 Input Functions

The three types of input functions of a node in a stochastic network are defined as follows. In this chapter we are especially interested in nodes with the third type of input function.

Type 1: The node is realized when all arcs leading into it are realized.

Type 2: The node is realized when any arc leading into it is realized.

Type 3: The node is realized when any arc leading into it is realized, under the condition that only one arc can be realized at a given time.

7.1.2 Output Functions

Type 1: All arcs emanating from the node are undertaken if the node is realized. This function is referred to as a deterministic output function.

Type 2: Exactly one arc emanating from the node is undertaken if the node is realized. The selection of the arc can be described by means of a unique probability. Hence, this output is called a probabilistic function.

In this chapter we consider only two types of nodes: (a) nodes with type-3 input function and *deterministic* output function followed by at most one single arc, and (b) nodes with type-3 input function and *probabilistic* output function. Either type of node will be referred to as a *GERT node*. The two symbols shown in Figure 7.1 are used to represent GERT nodes [4,5]. A network consisting of only GERT nodes is called a GERT network.

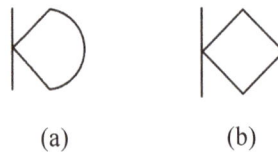

(a)   (b)

Figure 7.1. Gert nodes: (a) deterministic output; (b) probabilistic output

As an illustration, let us consider a simple quality control system, where after an inspection operation it is decided that parts should be sold for scrap, reworked, or sent to the assembly line. If a part is reworked, it can be sold to a secondary retail outlet, scrapped, or sent to assembly for internal use. The GERT network for this process is shown in Figure 7.2. Each arc has a probability of being chosen, given the *state* corresponding to the beginning node of the arc.

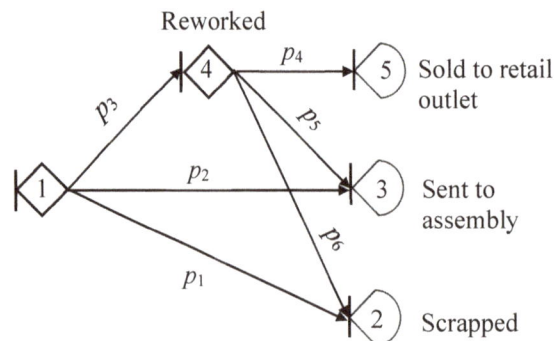

Figure 7.2. Stochastic network.

Note that $p_1 + p_2 + p_3 = 1$ and $p_4 + p_5 + p_6 = 1$. Also, arcs (4,2), (1,4), (1,2), (1,3), (4,3), and (4,5) represent physical processes which might be described by probability density functions. For example, activities (1,4) and (4,5) might be normally distributed, activities (1,2) and (4,3) exponentially distributed, and activities (1,3) and (4,2) uniformly distributed. The GERT methodology is directed toward answering the following typical questions:

1. What is the probability that a part will be scrapped?
2. What is the probability that a part will be used in assembly?
3. What is the probability that a part will be sold at retail?
4. What is the mean and variance of the time needed to produce a part for assembly?
5. How much time will be lost if a part is scrapped?

In the following sections, we will develop an analytical approach to answer these and other similar questions.

7.2 GERT BASIC PROCEDURES

Consider a network $G = (\mathbf{N}, \mathbf{A})$ with only GERT nodes in set $\mathbf{N}$. Let the random variable Y_{ij} be the *duration* (or any other quantity that is *pathwise additive*) of activity (i,j). By definition, activity (i,j) can be undertaken only if node i is realized. Therefore, we must know the conditional probability (discrete case) or density (continuous case) function of Y_{ij}, given that node i is realized, in order to study the realization of the activity. This, in turn, would allow us to investigate the realization of the overall network. In particular, we would be able to find the moments of the distribution of the realization time of the network, from which we can determine the mean and variance of the network realization time.

Let $f_{ij}(y)$ be the probability or density function of the duration of activity (i,j). The *moment generating function* (MGF) of the random variable Y_{ij} is defined as $M_{ij}(s) = E[e^{sY_{ij}}]$. That is,

$$M_{ij}(s) = \begin{cases} \int e^{sy_{ij}} f(y_{ij}) dy_{ij} & (continous\ random\ variables) \\ \\ \sum e^{sy_{ij}} f(y_{ij}) & (discrete\ random\ variables) \end{cases}$$

A particular case is $y_{ij} = a$, where a is a constant. In this case, $M_{ij}(s) = E[e^{sa}] = e^{sa}$. Furthermore, when $a = 0$, then $M_{ij}(s) = 1$.

Table 7.1 contains several important probability distributions with their respective moment generating functions and first (mean) and second moments about the origin. The following list includes some typical discrete distributions with the definition of the corresponding random variables;

Bernoulli distribution

Random variable: $X=1$ if event $\mathbf{U}$ happens and $X=0$ if event $\mathbf{V}$ happens.

Probability mass function: $f(1) = p$, $f(0) = 1-p$.

Binomial distribution

Random variable: number of times event **U** occurs in n independent Bernoulli trials. This implies that $Y = X_1 + X_2 + \ldots + X_n$.

Probability mass function: $f(y) = \binom{n}{y} p^y (1-p)^{n-y}, y = 0, 1, \ldots, n.$

Geometric distribution

Random variable: number of trials until event **U** occurs for the first time.

Probability mass function: $f(y) = p(1-p)^{y-1}, y = 1, 2, 3, \ldots$

Negative binomial distribution

Random variable: number of times event **V** occurs before event **U** happens r times.

Probability mass function: $f(y) = \binom{y+r-1}{r-1} p^r (1-p)^y, y = 0, 1, 2, \ldots$

Table 7.1. MGFS, MEAN AND SECOND MOMENT

Type of Distribution	MGF $M_E(s)$	Mean	Second Moment
Binomial (B)	$\left(pe^s + 1 - p\right)^n$	np	$np(np+1-p)$
Discrete (D)	$\sum_i p_i e^{sT_i} / \sum_i p_i$	$\sum_i p_i T_i / \sum_i p_i$	$\sum_i p_i T_i^2 / \sum_i p_i$
Exponential (E)	$\left(1 - \frac{s}{a}\right)^{-1}$	$1/a$	$2/a^2$
Gamma (GA)	$\left(1 - \frac{s}{a}\right)^{-b}$	b/a	$b(b+1)/a^2$
Geometric (GE)	$pe^s / \left(1 - e^s + pe^s\right)$	$1/p$	$(2-p)/p^2$
Negative Binomial (NB)	$\left(p/\left(1 - e^s + pe^s\right)\right)^r$	$r(1-p)/p$	$r(1-p)(1+r-rp)/p^2$
Normal (NO)	$e^{sm + \frac{1}{2}s^2\sigma^2}$	m	$m^2 + \sigma^2$
Poisson (P)	$e^{\lambda(e^s - 1)}$	λ	$\lambda(1+\lambda)$
Uniform (U)	$(e^{sa} - e^{sb})/((a-b)s)$	$(a+b)/2$	$(a^2 + b^2 + ab)/3$

Let p_{ij} be the conditional probability that activity (i,j) will be undertaken given that node i is realized. The *W-function* [4,5] for the random variable Y_{ij} is defined as

$$W_{ij}(s) = p_{ij} M_{ij}(s) \tag{7-1}$$

By using the transformation of Eq. (7-1) it is possible to construct a network G′ with the same nodes and arcs as network G, except that a single parameter $W_{ij}(s)$ is defined for each arc (i,j) of network G′, instead of the two quantities p_{ij} and y_{ij} defined for the arcs of network G. Figure 7.3 shows a generic arc for each of the networks G and G′.

Figure 7.3. Generic arcs for: (a) network G; (b) network G′

Network G′ has several attractive computational properties under the assumption that the durations of the activities of network G are statistically independent random variables. To show these properties, we will consider three special cases: (1) network G′ consists of two arcs in series; (2) network G′ consists of two branches in parallel; and (3) network G′ consists of one branch and one self-loop.

7.2.1 Branches in Series

Consider the two branches in series shown in Figure 7.4. As indicated in this figure, these two branches can be substituted by an equivalent branch, for which $Y_{ik} = Y_{ij} + Y_{jk}$.

Figure 7.4. Arcs in series and equivalent one-branch representation.

In order to find the *W*-function of the equivalent one-branch representation, we proceed as follows. The *W*-functions of the individual arcs are:

$$W_{ij}(s) = p_{ij} M_{ij}(s)$$
$$W_{jk}(s) = p_{jk} M_{jk}(s)$$

Furthermore, the *W*-function for the equivalent one-branch representation is

$$W_{ik}(s) = p_{ik} M_{ik}(s)$$

By the assumption of independence, we can conclude that $p_{ik} = p_{ij}\, p_{jk}$. Moreover, the moment-generating function of the *sum* of two independent random variables is equal to the product of the individual moment-generating functions. Hence, $M_{ik}(s) = M_{ij}(s)\, M_{jk}(s)$. Therefore, we conclude that

$$W_{ik}(s) = W_{ij}(s)\ W_{jk}(s) \tag{7-2}$$

The fundamental result of Eq. (7-2) can be extended to a case with three or more branches. The equivalent branch has a W-function equal to the product of the -functions of the branches in series.

7.2.2 Branches in Parallel

Consider the two branches in parallel shown in Figure 7.5. These branches can be substituted by one equivalent arc for which $Y_{ij} = Y_a$ with probability p_a and $Y_{ij} = Y_b$ with probability p_b.

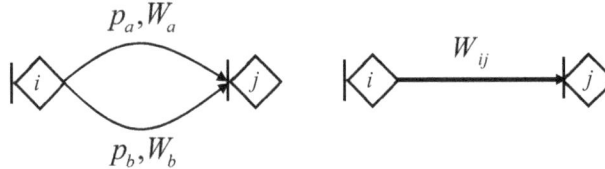

Figure 7.5. Branches in parallel and their one-branch representation.

By definition, the moment-generating function of the one-branch representation is formulated as $M_{ij}(s) = p_{ij}\ W_{ij}(s)$, where

$$p_{ij} = p_a + p_b$$

and

$$M_{ij}(s) = \frac{p_a M_a(s) + p_b M_b(s)}{p_a + p_b}$$

Therefore,

$$W_{ij}(s) = W_a(s) + W_b(s) \tag{7-3}$$

Again, the result given in Eq. (7-3) can be generalized for three or more parallel branches. The equivalent branch has its W-function equal to the sum of the individual W-functions of the arcs in parallel.

7.2.3 Single Branch with Self-Loop

Let us consider the simple network component shown in Figure 7.6. This component consists of one self-loop and one arc and can also be reduced to an equivalent one-branch representation.

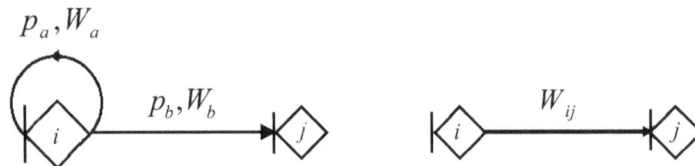

Figure 7.6. Arc with self-loop and the one-branch representation.

Note that the network under consideration can be transformed into that given in Figure 7.7. This new network consists of an infinite sequence of parallel chains, each chain being a sequence of branches in series. Therefore, we can first reduce the chains to equivalent single arcs, and then these arcs can be reduced to the one-branch network equivalent to the original system.

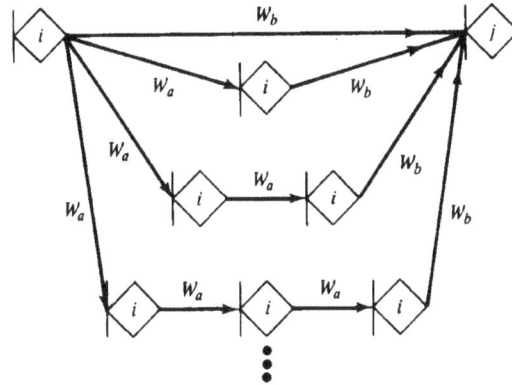

Figure 7.7. Series-parallel representation of an arc with self-loop.

Let (i,j) be the equivalent branch for the network of Figure 7.7. Using Eqs. (7-2) and (7-3), its weight is given by

$$W_{ij} = W_b + W_a W_b + W_a^2 W_b + \ldots = W_b (1 + \sum_{m=1}^{\infty} W_a^m)$$

where we have temporarily omitted the argument s of the W-functions. The expression above can be further simplified after noting that the binomial series $(1-W_a)^{-1}$ can be expanded as

$$(1-W_a)^{-1} = 1 + W_a + W_a^2 + W_a^3 + \ldots = 1 + \sum_{m=1}^{\infty} W_a^m$$

Therefore, now we can re-write the expression for W_{ij} as

$$W_{ij}(s) = \frac{W_b(s)}{1-W_a(s)} \tag{7-4}$$

Hence, the network of Figure 7.7 reduces to a single equivalent branch with W-function given by Eq. (7-4). Note that this procedure can be also used in the case where there is a maximum finite number of times allowed for the self-loop to be traversed. In such a case the network of Figure 7.7 has a finite number of parallel chains.

Following the basic procedures for reducing parallel chains, series chains, and self-loops, a GERT network can be reduced to an equivalent *single-branch* network. In summary the GERT methodology in essence consist of these steps:

1. Represent the system as a stochastic network with GERT nodes.
2. Determine probabilities and moment-generating functions for all the arcs of the network.
3. Compute the W-functions for all the arcs of the network.
4. Reduce the network to an equivalent one-branch network.

Before we proceed, we would like to provide the reader with fundamental background information on flowgraphs that is needed for further development.

7.3 BASIC CONCEPTS OF LINEAR FLOWGRAPHS

A system can be defined as a collection of active and interactive elements that perform a function. In our present discussion we will only consider systems whose elements and relationships among elements can be represented by a set of linear equations. One of the most popular diagrammatic representations of such systems is a *flowgraph*. In a flowgraph, the elements of the system are represented by nodes and the relationships or transfer functions by arcs.

The fundamental element of a linear flowgraph is a branch (i,j) directed from node i to node j with arc parameter equal to t_{ij}. Node i corresponds to an independent variable x_i, and node j to a dependent variable x_j. The direction of the branch indicates the input/output relationship between the two variables represented by the nodes of the branch. Additionally, the parameter t_{ij}, known as the *arc transmittance*, is equal to the factor used to transform the value of x_i, before it is considered as part of the value of x_j. The value of the variable x_j corresponding to a given node j is equal to the sum $\sum_{i \in \beta_j} t_{ij} x_i$, where set β_j is the set of all nodes directly connected to node i.

An illustration of this is given in Figure 7.8, where nodes 1 and 2 are directly connected to node 3.

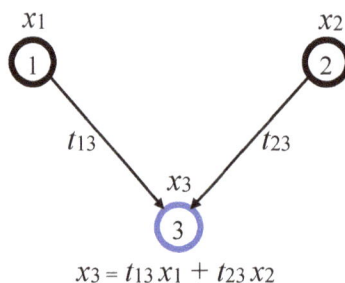

$$x_3 = t_{13} x_1 + t_{23} x_2$$

Figure 7.8. Basic property of linear flowgraphs.

7.4 DEFINITIONS

Mason's *topological equation* [1,2] is a fundamental result that can be used to (a) represent a linear flowgraph as an equivalent single branch; (b) determine the transmittance of the one-branch representation. Before presenting this equation, some definitions are necessary. This equation actually allows the representation of a GERT network, with specified source and terminal nodes, as a single branch, after recognizing that a GERT network can be viewed as a linear flowgraph with transmittances equal to the *W*-functions.

> *Loop:* a connected sequence of directed branches with every node being common to exactly two branches. A loop is usually referred to as a first-order loop to indicate that it does not contain another loop, and that each node can be reached from every other node. A self-loop can be viewed as a degenerate first-order loop.
>
> *Loop of order n*: a set of *n* disjoint first-order loops.
>
> *Closed flowgraph:* a graph in which each branch belongs to at least one loop.

As an illustration, the closed flowgraph shown in Figure 7.9 will be examined to identify all the loops and classify them according to their order. It can be seen that this flowgraph is closed since it is composed entirely of loops.

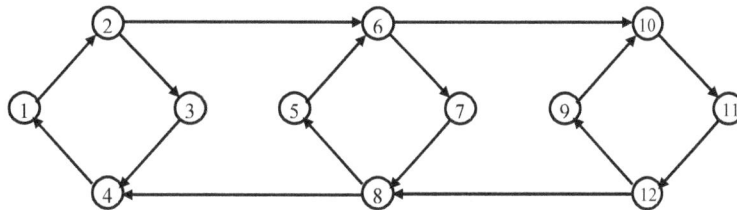

Figure 7.9. Closed flowgraph.

Figures 7.10, 7.11 and 7.12 contain the first- order, second-order, and third-order loops of the graph given in Figure 7.9. Note that a second (third)-order loop is a set of two (three) disjoint first-order loops. For example, L_1 and L_2 form a second-order loop, while L_1, L_2, and L_3, form a third-order loop.

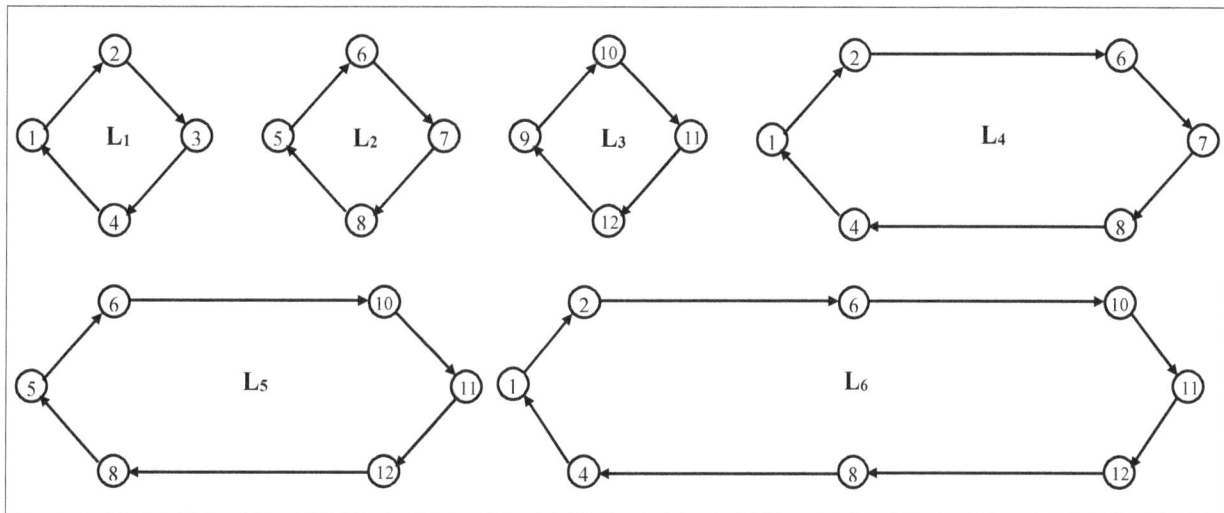

Figure 7.10. First-order loops.

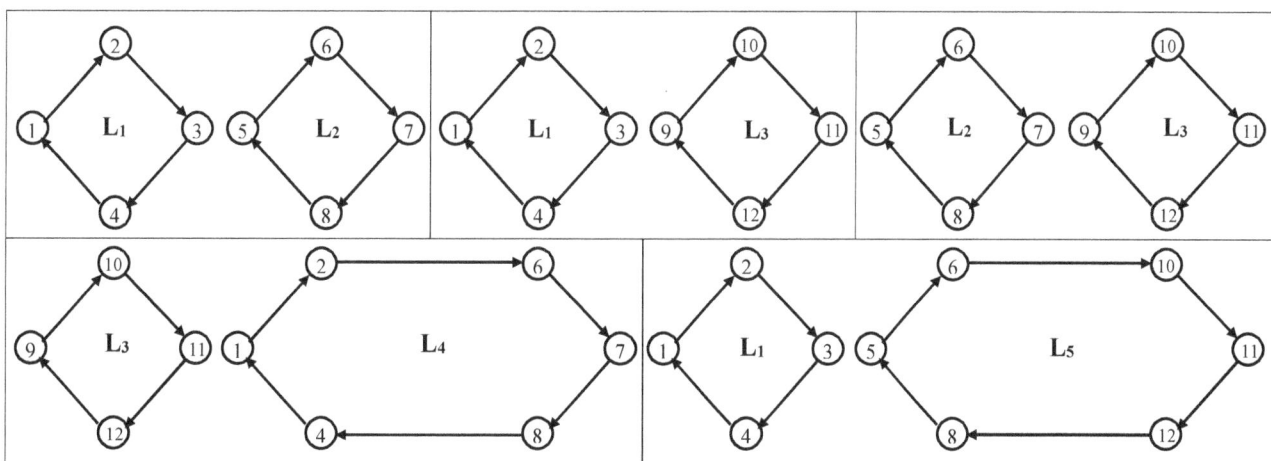

Figure 7.11. Second-order loops.

282

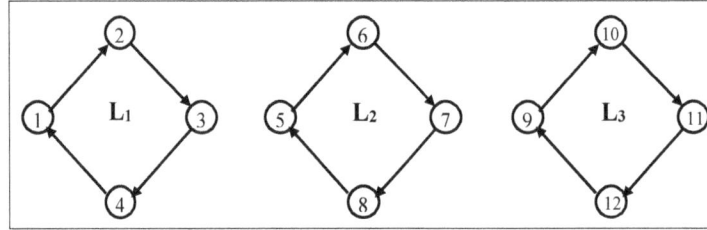

Figure 7.12. Third-order loop.

7.5 MASON'S RULE FOR CLOSED FLOWGRAPHS

Let t_{ij} be the transmittance of arc (i,j) in the kth first-order loop $\mathbf{L}_{k1}$ of a closed flowgraph. The transmittance T_k of this loop is given by Eq. (7-5).

$$T_k = \prod_{(i,j)\in \mathbf{L}_{k1}} t_{ij} \qquad (7\text{-}5)$$

Let $\mathbf{L}_n$ be an n-order loop consisting of $\mathbf{L}_{11}, \mathbf{L}_{21}, \dots, \mathbf{L}_{n1}$. The equivalent transmittance of $\mathbf{L}_n$ is given by

$$T(\mathbf{L}_n) = \prod_{k=1}^{n} T_k \qquad (7\text{-}6)$$

The following fundamental result is known as *Mason's topological equation* for closed signal flowgraphs [2]:

$$1 - \sum_{all\,\mathbf{L}_1} T(\mathbf{L}_1) + \sum_{all\,\mathbf{L}_2} T(\mathbf{L}_2) + \dots + (-1)^m \sum_{all\,\mathbf{L}_m} T(\mathbf{L}_m) + \dots = 0 \qquad (7\text{-}7)$$

where m is the *highest* order associated with the loops of a given closed flowgraph. The following is a summary of the entire GERT procedure.

7.6 GERT PROCEDURAL STEPS

At this time of our analysis, we can use the view of a GERT network as a flowgraph with the transmittance of each arc defined as its *W*-function. A realization of the GERT network must have a specific source node and a specific terminal node. Once this is done, the realization of the network is the result of the realization of individual components that allow the system to transfer from the source to the terminal through the arcs of the network. In order to transform the GERT network into a closed flowgraph a return arc is added.

In Figure 7.13, the original network is represented by a *block box* with *W*-function equal to $W_E(s)$. The main purpose of the GERT methodology is to find this *W*-function, since once it is available, it can be used to determine the moment-generating function for the realization of the network. This moment generation function completely characterizes the distribution on the random variable (duration, cost, number of trials, etc) associated with the realization of the network defined by the pair of nodes s and t.

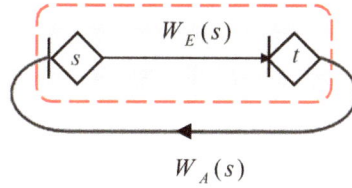

Figure 7.13. GERT network with return arc.

The closed flowgraph shown in Figure 7.13 has one first-order loop, with equivalent transmittance equal to $W_E(s) W_A(s)$. From Eq. (7-7) with $m = 1$, we can conclude that

$$1 - W_A(s)W_E(s) = 0 \Rightarrow W_A(s) = \frac{1}{W_E(s)}$$

From the above expression, we can find the W-function of the equivalent one-branch representation, $W_E(s) = 1/W_A(s)$. Since $W_E(s) = p_E M_E(s)$ we can write $W_E(0) = p_E M_E(0) = p_E$, since $M_E(0) = 1$. Therefore, we conclude that

$$M_E(s) = \frac{W_E(s)}{W_E(0)} \tag{7-8}$$

From the moment-generating function, we can determine the jth moments using the result

$$\mu_{jE} = \frac{\partial^j}{\partial s^j} M_E(s)\big|_{s=0} \tag{7-9}$$

Eq. (7-9) can be used to find the mean, μ_{1E}, and μ_{2E}. From these values, we can determine the mean and the variance for the network realization:

$$\mu = \mu_{1E} \tag{7-10}$$
$$\sigma^2 = \mu_{2E} - \mu^2 \tag{7-11}$$

Before illustrating the methodology with a numerical example, a summary of the steps followed by the GERT methodology is shown below:

- Add return arc (t,s) to close the network, and let $W_{ts}(s) = W_A(s)$.
- Find $W_{ij}(s) = p_{ij} M_{ij}(s)$ for each original arc (i,j).
- Identify all loops of order $n = 1,2, …, m$. Here m is the highest order among all possible loops.
- Apply Mason's equation, Eq. (7-7), and find $W_A(s)$.
- Find $W_E(s) = 1/W_A(s)$.
- Find $M_E(s)$ from Eq. (7-8).
- Find the mean and variance for the network realization, using Eqs. (7-9), (7-10) and (7-11).

284

We now illustrate the overall GERT approach using the network shown in Figure 7.14. It is desired to find the equivalent *W*-function for the one-branch representation of the network realization having node 1 as *source* and node 4 as a *terminal*.

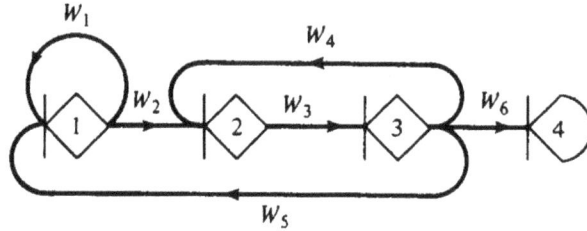

Figure 7.14. Open GERT network.

Figure 7.15 shows the closed GERT network for this example.

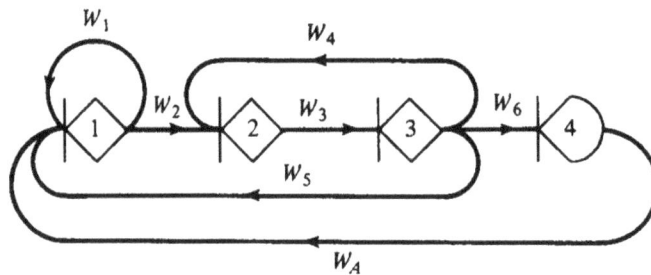

Figure 7. 15. Closed GERT network.

If $W_A(s)$ is substituted with $1/W_E(s)$, the following loop transmittances (*W*-functions) are obtained using Eqs. (7-5) and (7-6):

$$\text{Loops of order 1:} \quad W_1, W_3 W_4, W_2 W_3 W_5, W_2 W_3 W_6 \frac{1}{W_E}$$

$$\text{Loop of order 2:} \quad W_1 W_3 W_4$$

By using Eq. (7-7) with $m = 2$, we obtain

$$H = 1 - W_1 - W_3 W_4 - W_2 W_3 W_5 - W_2 W_3 W_6 \frac{1}{W_E} + W_1 W_3 W_4 = 0$$

Therefore,

$$W_E(s) = \frac{W_2 W_3 W_6}{1 - W_1 - W_3 W_4 - W_2 W_3 W_5 + W_1 W_3 W_4}$$

which is the equivalent *W*-function for the one-branch representation of the GERT network of Figure 7.15.

7.7 GERT APPLICATIONS

In this section we will consider tree applications of GERT to illustrate the wide scope of the GERT approach. Specifically, the following applications will be discussed:

285

(a) Production of a high-risk item.
(b) Semiconductor material processing.
(c) Determination of time standards.

7.7.1 Production of a High-Risk Item

A government contractor has agreed to produce a high-risk item. Because of stringent inspections, there is only a 20 % probability that a successful item will be produced from raw material. What is the expected number of trials necessary to achieve two good items?

In this problem it is assumed that the time needed to produce an item is constant; recall that for the case in which the random variable Y_{ij} is equal to a constant a, then

$$M_{ij}(s) = E[e^{sa}] = e^{sa}$$

By letting $a = 1$, we obtain $M_{ij}(s) = e^s$, which is the moment-generating function for the *duration* of each arc (i,j) in the original network. The GERT network for this application is shown in Figure 7.16. Since there is a 20% probability of success,

$$W_1(s) = W_3(s) = 0.80\ e^s$$
$$W_2(s) = W_4(s) = 0.20\ e^s$$

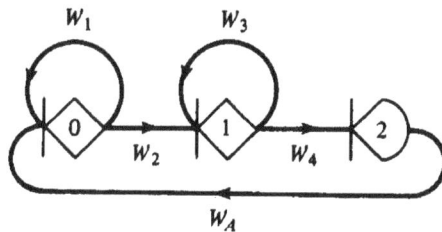

Figure 7.16. GERT network for high-risk item.

Using the topology equation, we obtain (omitting the argument s of the W-functions):

$$H = 1 - W_1 - W_3 - W_2 W_4 W_A + W_1 W_3 = 0 \Rightarrow 1 - W_1 - W_3 - W_2 W_4 \left(\frac{1}{W_E}\right) + W_1 W_3 = 0$$

Therefore,

$$W_E(s) = \frac{W_2 W_4}{1 - W_1 - W_3 - W_1 W_3}$$

Using the expressions for each of the W-functions W_1, W_2, W_3, and W_4, the following result is obtained:

$$W_E(s) = \frac{0.04 e^{2s}}{1 - 1.6 e^s + 0.64 e^{2s}}$$

286

Since $p_E = 1$, $W_E(s) = M_E(s)$. Thus, the expected number of trials necessary to produce two good items is:

$$\mu_{1E} = \left. \frac{\partial M_E(s)}{\partial s} \right|_{s=0} = 10$$

Similarly, the variance of the total number of trials is found to be equal to 40.

Example

In the problem considered in Section 7.7.1, find the expected number of trials needed to make two *consecutive* good items.

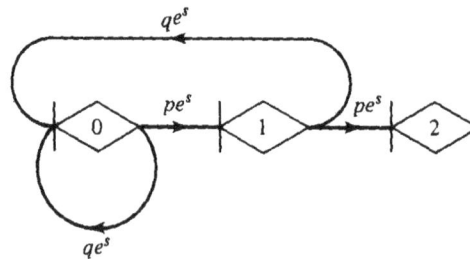

Figure 7.17. GERT network for two good consecutive items.

Figure 7.17 shows the GERT network. In this figure, it is assumed that the probability of success is p and the probability of failure is $q = 1$-p. Using Mason's equation, the equivalent W-function for this network is found to be equal to

$$W_E(s) = \frac{(pe^s)^2}{1 - pq(e^s)^2 - qe^s}$$

Setting $p=0.20$ and $q=0.80$, we obtain the result

$$W_E(s) = \frac{(0.20e^s)^2}{1 - 0.16(e^s)^2 - 0.80e^s}$$

It can be verified that $W_E(0)=1$. Therefore, $M_E(s) = W_E(s)$. Using this result, it is possible to show that the expected number of trials to produce two consecutive good items is

$$\mu_{1E} = \left. \frac{\partial M_E(s)}{\partial s} \right|_{s=0} = 30$$

Note that the expected number of trials is equal to 10 when the two good items do not have to be consecutive.

7.7.2 Material Processing [6]

We will now consider the processing of semiconductor material. As shown in Figure 7.18, the material is first inserted into the furnace to alter its impurities. The output of the furnace is either retreated in the furnace, declared inferior, or declared acceptable and cut (sliced) into wafers. After the slicing operation, the material is either in the form of acceptable wafers, lost due to the slicing process, or captured and returned as raw material. In this application, it is desired to calculate the average time and variance for obtaining an acceptable wafer (i.e., for arriving at terminal A). Table 7.2 contains the information related to each branch in the network.

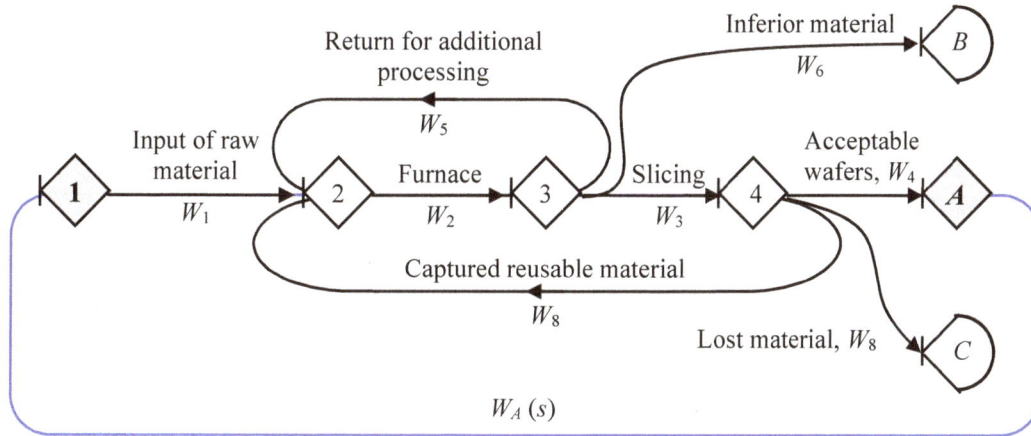

Figure 7.18. Material processing application.

Table 7.2. MATERIAL-PROCESSING ACTIVITIES

Branch	p_i	Type of Distribution	Parameters (in hours)	MGF
$(1, 2)$	1	Constant	$a = 1$	$exp\,(s)$
$(2, 1)$	1	Nirmal	$m = 0.5$ $\sigma = 0.1$	$exp\,(0.5s + \tfrac{1}{2}\,(0.01)\,s^2$
$(3, 2)$	0.12	Normal	$m = 0.1$ $\sigma = 0.1$	$exp\,(0.1s + \tfrac{1}{2}\,(0.01)\,s^2$
$(3, B)$	0.03	Constant	$a = 0.25$	$exp\,(s)$
$(3, 4)$	0.85	Normal	$m = 0.25$ $\sigma = 0.20$	$exp\,(0.25s + \tfrac{1}{2}\,(0.04)\,s^2$
$(4, A)$	0.75	Constant	$a = 0.20$	$exp\,(0.20s)$
$(4, C)$	0.05	Constant	$a = 0.05$	$exp\,(0.05s)$
$(4, 2)$	0.20	Constant	$a = 0.10$	$exp\,(0.10s)$

Since we are interested in the mean and variance for the realization time of an acceptable wafer, we must introduce arc $(A, 1)$ as in Figure 7.18, where $W_A(s) = 1/W_E(s)$. Note that arcs $(3, B)$ and $(4, C)$ are not considered in the analysis, since they are excluded when node A is chosen as the terminal. The topology equation for this application is

$$H = 1 - W_1 W_2 W_3 W_4 \left(\frac{1}{W_E} \right) - W_2 W_5 - W_2 W_3 W_7 = 0$$

Solving the above equation for $W_E(s)$, we can write

$$W_E(s) = \frac{W_1 W_2 W_3 W_4}{1 - W_2 W_5 - W_2 W_3 W_7}$$

After substituting each W-function with its respective probability and MGF, we obtain:

$$W_E(s) = \frac{0.6375\, e^{1.95s + 0.025s^2}}{1 - 0.12\, e^{0.6s + 0.01s^2} - 0.17\, e^{0.85s + 0.025s^2}}$$

It can be verified that for this problem, $W_E(0) = 0.8979$. This is interpreted as the probability that an acceptable wafer will be produced.

$$M_E(s) = \frac{W_E(s)}{W_E(0)} = \frac{0.71\, e^{1.95s + 0.025s^2}}{1 - 0.12\, e^{0.6s + 0.01s^2} - 0.17\, e^{0.85s + 0.025s^2}}$$

Now, by obtaining the first and second partial derivatives of M(s) with respect to s and setting s = 0, we obtain

$$\mu_{1E} = \left. \frac{\partial M_E(s)}{\partial s} \right|_{s=0} = 2.255\, hr$$

$$\mu_{2E} = \left. \frac{\partial^2 M_E(s)}{\partial s} \right|_{s=0} = 5.477\, hr^2$$

$$\sigma^2 = \mu_{2E} - (\mu_{1E})^2 = 0.392\, hr^2$$

Thus, the mean realization time for an acceptable wafer is 2.255 hours, with a variance of 0.392 hour^2. The same analysis can be conducted to obtain similar information for terminals B and C.

7.7.3. Determination of Time Standards [3]

In this application we compute the mean and standard deviation of the standard time required by an operator for assembling a printed circuit board. Whenever the worker is involved with composite tasks, such as the ones in this problem, it is reasonable to view the standard time as a random variable, possessing a finite mean and variance, and described by a probability distribution function. Figure 7.19(a) contains a description of the assembly operation under study.

In order to investigate variance estimates, assumptions must be made regarding the stochastic behavior of each element (task) conducted in the standard. Although this is a more complex description of a task, it is clearly more accurate than a single time. Table 7.3 contains the mean and variance for each element of the standard. For this example, the stochastic behavior of each element can be described by a normal density function.

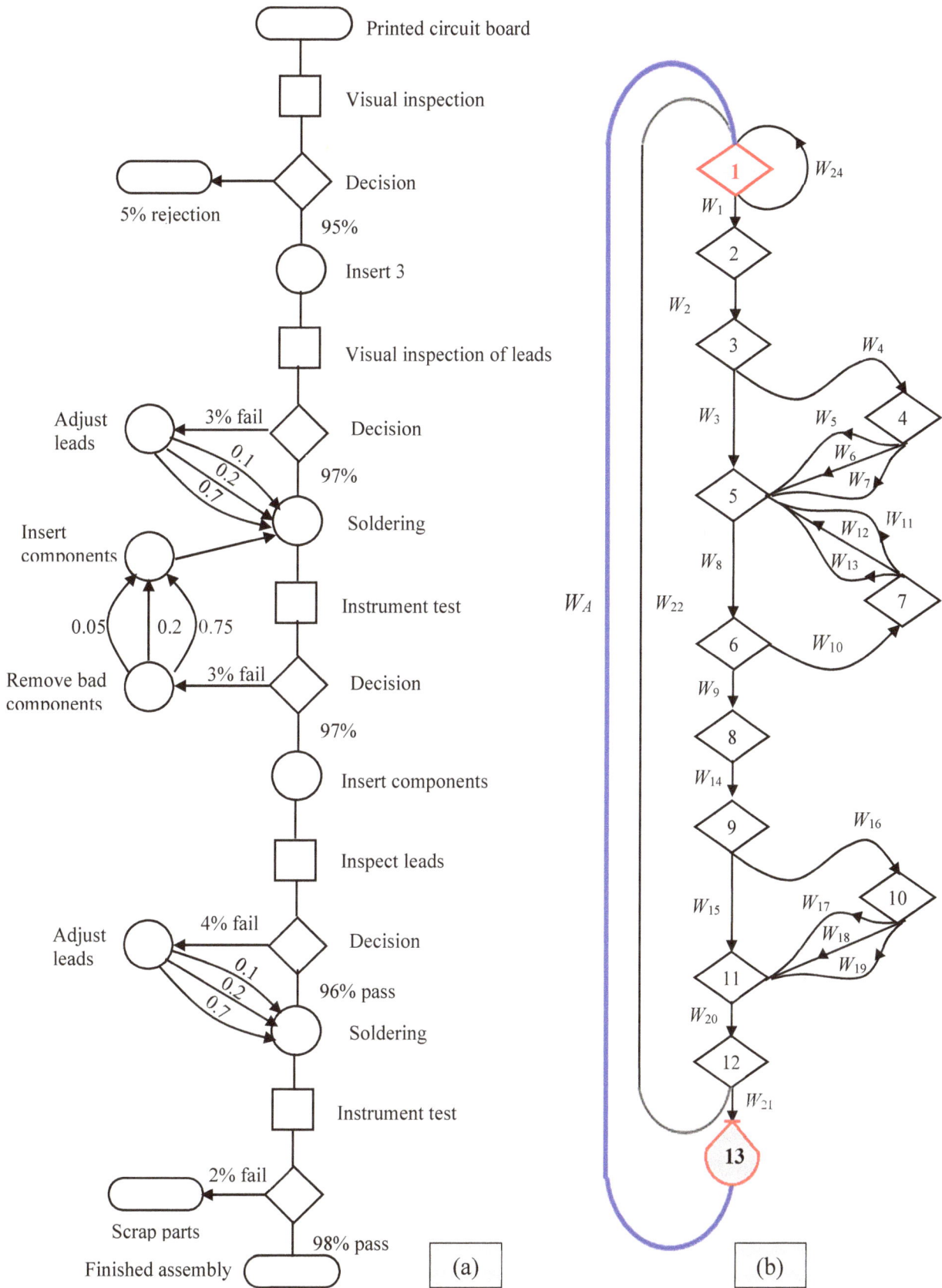

Figure 7.19. Stochastic assembly chart and GERT network.

Table 7.3. BRANCH PARAMETERS FOR ASSEMBLY GERT NETWORK

From/To	W_{ij}	Operation Description	p_{ij}	Distribution	Mean	Variance
½	W_1	Success on inspecting circuit board	0.95	Normal	1.40	0.35
1/1	W_{24}	Fail on inspecting circuit board	0.05	Normal	1.40	0.35
2/3	W_2	Inserting components	1.00	Normal	2.65	0.30
3/5	W_3	Success on inspecting leads	0.97	Normal	0.75	0.20
¾	W_4	Fail on inspecting leads	0.03	Normal	0.85	0.25
4/5	W_5	Adjust component; one component's leads unqualified	0.70	Normal	1.65	0.35
4/5	W_6	Adjust components; two components' leads unqualified	0.20	Normal	2.15	0.40
4/5	W_7	Adjusted components; three components' leads unqualified	0.10	Normal	2.65	0.55
5/6	W_8	Soldering operation	1.00	Normal	1.30	0.10
6/8	W_9	Pass on instrumental test	0.97	Normal	0.90	0.25
6/7	W_{10}	Fail on instrumental test	0.03	Normal	1.10	0.30
7/5	W_{11}	Remove and replace component; one component failed	0.75	Normal	2.75	0.55
7/5	W_{12}	Remove and replace two components	0.20	Normal	3.65	0.68
7/5	W_{13}	Remove and replace three components	0.05	Normal	4.25	0.79
8/9	W_{14}	Inserting components	1.00	Normal	2.65	0.30
9/11	W_{15}	Success on inspecting leads	0.96	Normal	0.80	0.25
9/10	W_{16}	Fail on inspecting leads	0.04	Normal	0.95	0.30
10/11	W_{17}	Adjust component; one components' leads unqualified	0.70	Normal	2.00	0.50
10/I1	W_{18}	Adjust components; two components' leads unqualified	0.20	Normal	2.65	0.60
10/11	W_{19}	Adjust components, three components' leads unqualified	0.10	Normal	3.15	0.75
11/12	W_{20}	Soldering operation	1.00	Normal	2.10	0.40
12/13	W_{21}	Passing instrumental test	0.98	Normal	0.80	0.15
12/1	W_{22}	Failing instrumental test	0.02	Normal	0.85	0.15

A. Algorithmic Procedure

The GERT network shown in Figure 7.19(b) is formed by nodes that represent the beginning and ending points of all individual tasks, as well as arcs that represent the actual time duration of each task. Table 7.4 contains the W-functions associated with the arcs of the GERT network of Figure 7.19(b).

Table 7.4. GERT W-FUNCTIONS FOR ASSEMBLY APPLICATION

Branch Number	Branch (i,j) Node i	Node j	p_{ij}	$W = pe^{\mu t+\frac{1}{2}(\sigma^2 t^2)}$	Branch Number	Branch (i,j) Node i	Node j	p_{ij}	$W = pe^{\mu t+\frac{1}{2}(\sigma^2 t^2)}$
1	1	2	0.95	$0.95e^{1.4t+0.175t^2}$	13	7	8	0.05	$0.05e^{4.25t+0.395t^2}$
2	2	3	1.0	$1.0e^{2.65t+0.15t^2}$	14	8	9	1.0	$1.0e^{2.65t+0.15t^2}$
3	3	5	0.97	$0.97e^{0.75t+0.10t^2}$	15	9	11	0.96	$0.96e^{0.8t+0.125t^2}$
4	3	4	0.03	$0.03e^{0.85t+0.125t^2}$	16	9	10	0.04	$0.04e^{0.95t+0.15t^2}$
5	4	5	0.70	$0.70e^{1.65t+0.175t^2}$	17	10	11	0.70	$0.70e^{2t+0.25t^2}$
6	4	5	0.20	$0.20e^{2.1t+0.20t^2}$	18	10	11	0.20	$0.20e^{2.65t+0.325t^2}$
7	4	5	0.10	$0.10e^{2.6t+0.275t^2}$	19	10	11	0.10	$0.10e^{3.15t+0.378t^2}$
8	5	6	1.0	$1.0e^{1.3t+0.05t^2}$	20	11	12	1.0	$1.0e^{2.1t+0.20t^2}$
9	6	8	0.97	$0.97e^{0.9t+0.125t^2}$	21	12	13	0.98	$0.98e^{0.8t+0.075t^2}$
10	6	7	0.03	$0.03e^{1.1t+0.15t^2}$	22	12	1	0.02	$0.02e^{0.85t+0.075t^2}$
11	7	8	0.75	$0.75e^{2.75t+0.275t^2}$	23	13	1	1.9	$1.0/W_E$
12	7	8	0.20	$0.20e^{3.65t+0.34t^2}$	24	1	1	0.05	$0.05e^{1.4t+0.175t^2}$

Before continuing with the GERT analysis, the network shown in Figure 7.19(b) is modified by performing three simplifications resulting in the network shown in Figure 7.20. These simplifications provide a computationally more convenient equivalent GERT network with fewer loops than the original network, and for which it is much easier to identify the loops. Using Figure 7.20, the equivalent transmittances for loops of first and second order are calculated below.

Loops of order 1

W_{24}

$W_8W_{10}(W_{11}+W_{12}+W_{13})$

$W_1W_2W_8W_9W_{14}W_{20}W_{22}[W_3+W_4(W_5+W_6+W_7)][W_{15}+W_{16}(W_{17}+W_{18}+W_{19})]$

$W_1W_2W_8W_9W_{14}W_{20}W_{22}W_{21}(1/W_E)[W_3+W_4(W_5+W_6+W_7)][W_{15}+W_{16}(W_{17}+W_{18}+W_{19})]$

Loop of order 2

$W_8W_{10}W_{24}(W_{11}+W_{12}+W_{13})$

From Mason's equation, we obtain the following equivalent W-function:

$$W_E(t) = \frac{W_1W_2W_8W_9W_{14}W_{20}W_{21}\left[W_3+W_4\left(W_5+W_6+W_7\right)\right]\left[W_{15}+W_{16}\right]\left(W_{17}+W_{18}+W_{19}\right)}{1-W_{24}-W_8W_{10}\left(W_{11}+W_{12}+W_{13}\right)-W_1W_2W_8W_9W_{14}W_{20}W_{22}\left[W_3+W_4\left(W_5+W_6+W_7\right)\right]\left[W_{15}+W_{16}\left(W_{17}+W_{18}+W_{19}\right)\right]}$$

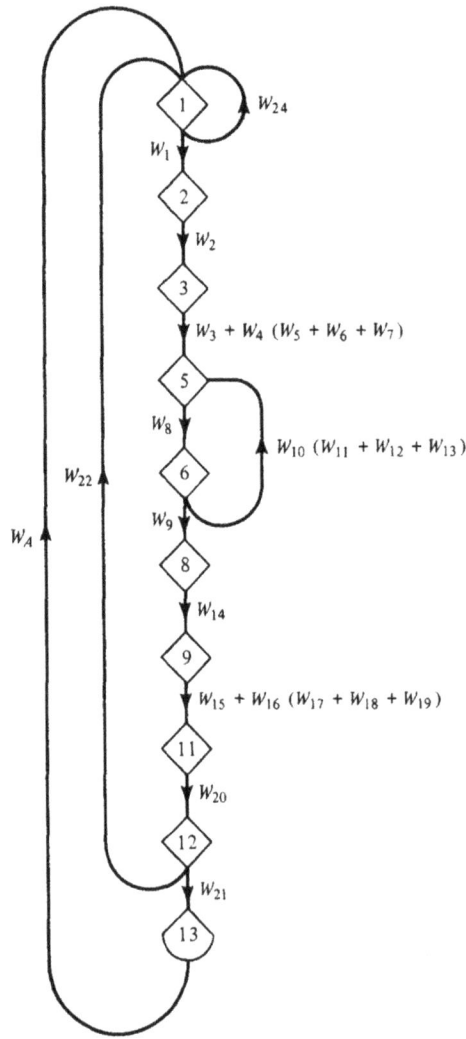

Figure 7.20. Simplified GERT network.

In order to simplify the presentation of the analysis required to get the mean and the variance of the operation time, we will write $W_E(t)$ as $W_E(t) = A(t)/B(t)$, where:

$$A(t) = (0.90307e^{11.8t+0.925t^2}) \times (0.97e^{0.75t+0.1t^2} + 0.021e^{2.5t+0.3t^2} + 0.006e^{3t+0.325t^2} + 0.003e^{2.5t+0.4t^2})$$

$$\times (0.96e^{0.8t+0.125t^2} + 0.028e^{2.95t+0.4t^2} + 0.008e^{3.6t+0.45t^2} + 0.004e^{4.1t+0.525t^2})$$

$$B(t) = 1 - (0.01843e^{11.85t+0.925t^2}) \times (0.97e^{0.75t+0.1t^2} + 0.021e^{2.5t+0.3t^2} + 0.006e^{3t+0.325t^2} + 0.003e^{2.5t+0.4t^2})$$

$$\times (0.96e^{0.8t+0.125t^2} + 0.028e^{2.95t+0.4t^2} + 0.008e^{3.6t+0.45t^2} + 0.004e^{4.1t+0.525t^2}) - (0.05e^{1.4t+0.175t^2})$$

$$- (0.0225e^{5.15t+0.475t^2} + 0.006e^{6.05t+0.58t^2} + 0.0015e^{6.65t+0.595t^2})$$

$$+ (0.001125e^{6.55t+0.65t^2} + 0.0003e^{7.45t+0.715t^2} + 0.000075e^{8.05t+0.77t^2}).$$

Continuing with the analysis, we set $t = 0$ in the expressions for both A(t) and B(t) to get:

(a) $A(0) = 0.90307 \, (0.97 + 0.021 + 0.006 + 0.003) \, (0.96 + 0.028 + 0.008 + 0.004) = 0.9037$

(b) $B(0) = 1 - 0.01843 \, (0.97 + 0.021 + 0.006 + 0.003) \, (0.96 + 0.028 + 0.008 + 0.004) - 0.05$

$- \, (0.0225 + 0.006 + 0.0015) \quad + (0.001125 + 0.0003 + 0.000075) = 0.9037$.

(c) $W_E(0) = 0.9037/0.9037 = 1$, and $M_E(t) = \dfrac{W_E(t)}{W_E(0)} = W_E(t)$.

Now, we get the first and second derivatives of the moment generating function with respect to t and get the first moment (mean) and the second moment by setting $t = 0$ in the corresponding derivatives. Final results are shown below without intermediate calculations:

$$\mu_{1E} = \left.\frac{\partial M_E(t)}{\partial t}\right|_{t=0} = 14.08, \, \mu_{2E} = \left.\frac{\partial^2 M_E(t)}{\partial t^2}\right|_{t=0} = 204.561 \text{ and } \sigma^2 = \mu_{2E} - \mu_{1E}^2 = 204.561 - 14.026^2 = 7.78.$$

Since all component times are normally distributed, the standard time is also normally distributed and exact probability statements concerning the time to completion can be made. In addition, if a standard is set against this result, accurate hypothesis tests governing changes in standards can be run to maintain standards in the future.

It is also instructive to note that the GERT approach presents a unique alternative to the traditional processes of setting time standards. Using conventional methods, the individual elements of time are treated as constants, and after summation, an adjustment is made to simulate random fluctuations or absorb imbalances in the actual processing times.

On the other hand, the GERT approach incorporates random variation and uncertainties directly into each time element. Hence, the resultant standard already includes all random imbalances and needs no further adjustment except for the appropriate personal and fatigue allowances. Finally, since an estimate of the variance is obtained, confidence intervals on the standard time can be constructed for specified levels of significance.

B. NOP Solution

Figure 7.21 shows the NOP screens for this example: (1) Input data file, (2) NOP Output results. The output results include the following:

(a) Probability, mean and variance for each arc of the GERT network.
(b) All paths from the specified source node to the specified terminal node.
(c) Mean and variance for the specified network realization.

The mean and variance of the assembly operation under consideration are equal to 14.08 hr and 7.78 hr^2, respectively.

Input File

```
Nodes 20;
St;
14 1 1.00 con 0 0;
1 2 0.95 nor 1.4 0.5916;
1 1 0.05 nor 1.4 0.5916;
2 3 1.0 nor 2.65 0.5477;
3 5 0.97 nor 0.75 0.4472;
3 4 0.03 nor 0.85 0.50;
4 5 0.70 nor 1.65 0.5916;
4 15 0.20 nor 2.15 0.6324;
15 5 1.0 con 0 0;
4 16 0.10 nor 2.65 0.7416;
16 5 1.0 con 0 0;
5 6 1.0 nor 1.30 0.3162;
6 8 0.97 nor 0.90 0.50;
6 7 0.03 nor 1.10 0.5477;
7 5 0.75 nor 2.75 0.7416;
7 17 0.20 nor 3.65 0.8246;
17 5 1.0 con 0 0;
7 18 0.05 nor 4.25 0.8888;
18 5 1.0 con 0 0;
8 9 1.0 nor 2.65 0.5477;
9 11 0.96 nor 0.80 0.50;
9 10 0.04 nor 0.95 0.5477;
10 11 0.70 nor 2.0 0.7071;
10 19 0.20 nor 2.65 0.7746;
19 11 1.0 con 0 0;
10 20 0.10 nor 3.15 0.8660;
20 11 1.0 con 0 0;
11 12 1.0 nor 2.10 0.6324;
12 1 0.02 nor 0.80 0.3873;
12 13 0.98 nor 0.85 0.3873;
End;
```

NOP Output Results

Output

GRAPHICAL EVALUATION AND REVIEW TECHNIQUE (GERT)

START	END	PROB	MEAN	VARIANCE
14	01	1.000	00.000	00.000
01	02	0.950	01.400	00.350
01	01	0.050	01.400	00.350
02	03	1.000	02.650	00.300
03	05	0.970	00.750	00.200
03	04	0.030	00.850	00.250
04	05	0.700	01.650	00.350
04	15	0.200	02.150	00.400
15	05	1.000	00.000	00.000
04	16	0.100	02.650	00.550
16	05	1.000	00.000	00.000
05	06	1.000	01.300	00.100
06	08	0.970	00.900	00.250
06	07	0.030	01.100	00.300
07	05	0.750	02.750	00.550
07	17	0.200	03.650	00.680
17	05	1.000	00.000	00.000
07	18	0.050	04.250	00.790
18	05	1.000	00.000	00.000
08	09	1.000	02.650	00.300
09	11	0.960	00.800	00.250
09	10	0.040	00.950	00.300
10	11	0.700	02.000	00.500
10	19	0.200	02.650	00.600
19	11	1.000	00.000	00.000
10	20	0.100	03.150	00.750
20	11	1.000	00.000	00.000
11	12	1.000	02.100	00.400
12	01	0.020	00.800	00.150
12	13	0.980	00.850	00.150

RESULTS

GERT EXCLUSIVE-OR PATHS

START NODE	END NODE	PROBABILITY	MEAN	VARIANCE	PATH
14	13	0.931	0013.92	0007.41	14 -1 -2 -3 -5 -6 -8 -9 -11 -12 -13
14	13	0.027	0016.07	0007.96	14 -1 -2 -3 -5 -6 -8 -9 -10 -11 -12 -13
14	13	0.008	0016.72	0008.06	14 -1 -2 -3 -5 -6 -8 -9 -10 -19 -11 -12 -13
14	13	0.004	0017.22	0008.21	14 -1 -2 -3 -5 -6 -8 -9 -10 -20 -11 -12 -13
14	13	0.020	0015.67	0007.81	14 -1 -2 -3 -4 -5 -6 -8 -9 -11 -12 -13
14	13	0.001	0017.82	0008.36	14 -1 -2 -3 -4 -5 -6 -8 -9 -10 -11 -12 -13
14	13	0.000	0018.47	0008.46	14 -1 -2 -3 -4 -5 -6 -8 -9 -10 -19 -11 -12 -13
14	13	0.000	0018.97	0008.61	14 -1 -2 -3 -4 -5 -6 -8 -9 -10 -20 -11 -12 -13
14	13	0.006	0016.17	0007.86	14 -1 -2 -3 -4 -15 -5 -6 -8 -9 -11 -12 -13
14	13	0.000	0018.32	0008.41	14 -1 -2 -3 -4 -15 -5 -6 -8 -9 -10 -11 -12 -13
14	13	0.000	0018.97	0008.51	14 -1 -2 -3 -4 -15 -5 -6 -8 -9 -10 -19 -11 -12 -13
14	13	0.000	0019.47	0008.66	14 -1 -2 -3 -4 -15 -5 -6 -8 -9 -10 -20 -11 -12 -13
14	13	0.003	0016.67	0008.01	14 -1 -2 -3 -4 -16 -5 -6 -8 -9 -11 -12 -13
14	13	0.000	0018.82	0008.56	14 -1 -2 -3 -4 -16 -5 -6 -8 -9 -10 -11 -12 -13
14	13	0.000	0019.47	0008.66	14 -1 -2 -3 -4 -16 -5 -6 -8 -9 -10 -19 -11 -12 -13
14	13	0.000	0019.97	0008.81	14 -1 -2 -3 -4 -16 -5 -6 -8 -9 -10 -20 -11 -12 -13

GERT EXCLUSIVE-OR FINAL RESULTS

ENTRY	EXIT	PROB	MEAN TIME	VARIANCE
14	13	1.000	0014.08	0007.78

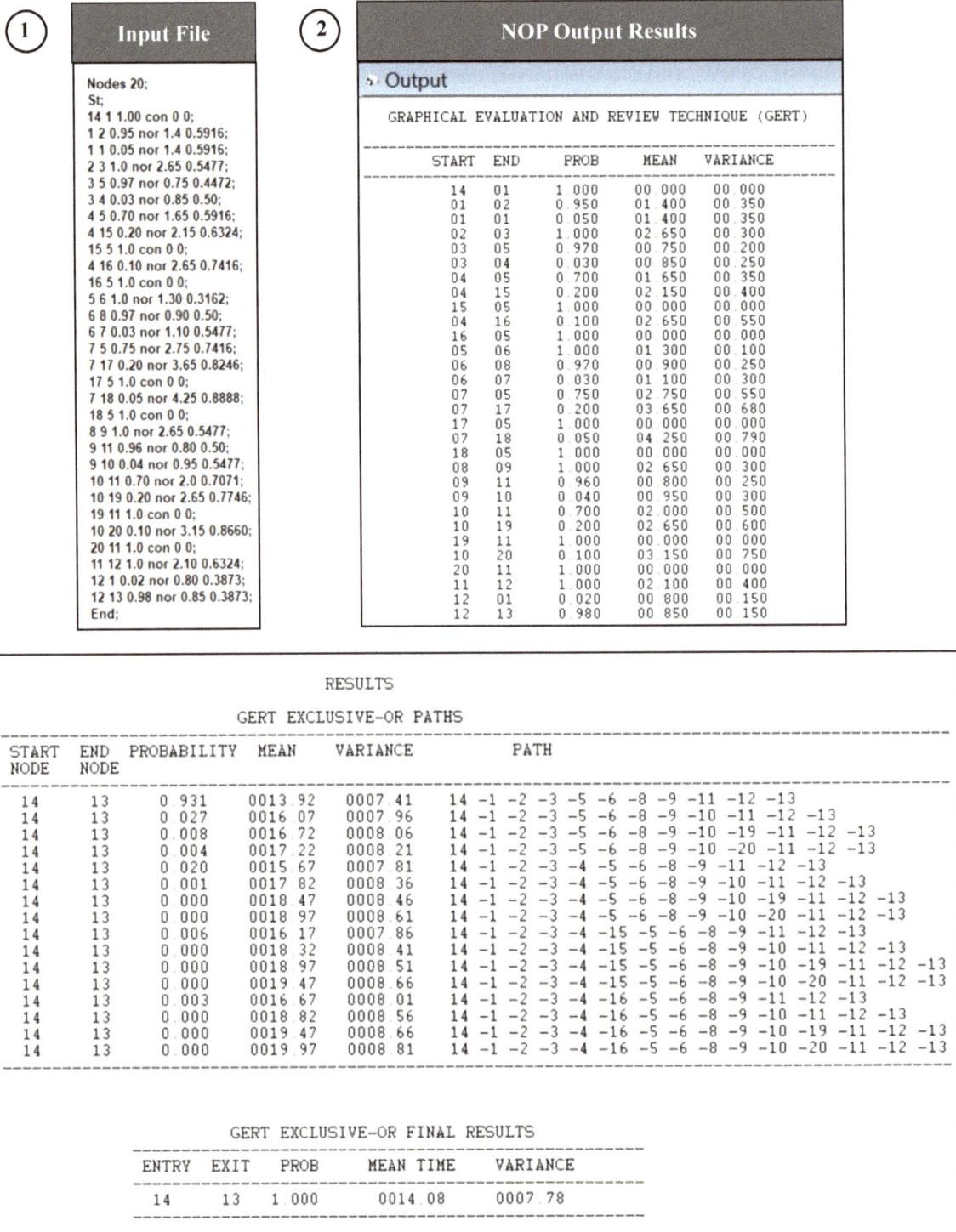

Figure 7.21. Computer Screens for GERT Example.

295

EXERCISES

1. The probability of winning a game is 1/6. (a) Use the GERT approach to find the moment generating function for the number of trials to win the game three *consecutive* times. (b) Find the mean and variance for the number of trials needed, using the result obtained in (a).

2. Develop a GERT approach to find the moment generating function of the binomial, geometric, negative binomial, and Pascal distributions.

3. Consider a shop that produces electromechanical equipment characterized by the following sequence. There is a series of operations that takes 25 minutes and terminates at an inspection station. The inspection activity takes 3 minutes. As a result of the inspection, 70% of the units are dispatched to a testing area for an additional test that takes 2 minutes, and 30% to an adjustment operation that requires 1 minute. The transit time from the first inspection to the testing or adjustment areas is 3 minutes. After the testing, 70% of the units are accepted and sent to the 1-minute adjustment operation, and 30% are rejected and sent to a 4-minute repair operation. The time for the movement from test to repair is 1 minute, while the time for the movement to adjustment follows a normal distribution with mean 2 and variance 6. From the repair area, material flows back to be tested (this transit time is practically negligible). After the adjustment, the units are finally packed. The packing time is assumed to be equal to 1 minute. (a) Draw the GERT network for the entire process. (b) Find the expected time per unit. (c) Find the variance of the time per unit. [Adapted from "*An Algebra for the Analysis of Generalized Activity Networks,*" *Management Science,* Vol. 10, No. 3, (1964), pp. 494-514, by S. F. Elmaghraby].

4. Assume that a dart thrown at a board has a probability of hitting the target of 0.20. Develop the GERT network to find the expected number of throws to achieve three hits. Repeat this exercise, assuming that the three hits must be consecutive.

5. Donald-Do-Right has recently opened up a TV fix-it shop in College Station. Being an IE graduate from Texas A&M, he wishes to analyze his business. The procedure is as follows. Donald receives three basic types of jobs: (1) black-and-white portables, (2) black-and-white consoles, and (3) color sets. Recent experience indicates 42 %, 31 %, and 27 % arrivals, respectively. The repair time for a B&W portable is exponential with mean 3 hours. A B&W console repair is exponential with mean 5 hours. He is not experienced in servicing color sets, and so he must send them to another shop to be repaired until he becomes more experienced. However, he does find that approximately one in three color sets have minor troubles, and so he repairs them immediately at a rate of 40 minutes 30 % of the time, 55 minutes 30 % of the time and 75 minutes 40 % of the time. The outside contractor repairs sets according to a gamma probability law with a mean repair rate of 7 hours and a variance of 4 hours^2. Mr. Do-Right puts all his repairs on temporary test, and he has found that 10 % of the outside contract jobs have to be returned for rework, and 5 % of his jobs must be redone. The rework jobs take approximately I of their original repair times and are also put on temporary test. Can you analyze this business?

6. A manufacturing operation takes place in Workstation A and an inspection is performed after the operation. The inspection results in either of three equally likely outcomes. If the item is found acceptable, it is immediately sent to Workstation B for an assembly operation. Otherwise, the item is reprocessed in Workstation A. However, there are two alternatives for

this, depending on the specific result of the inspection. One alternative requires one third of the time (1 minute) of the other alternative (3 minutes).

  (a) Draw the GERT network with all W-functions.
  (b) Find the moment-generating function of the time in Workstation A.
  (c) Find the expected time in Workstation A.
  (d) Find the variance of the time in Workstation A.

7. Suppose that a poor merchant is cutting a diamond for King Richrocks and he requires a perfect cut for his crown. Any diamonds not perfectly cut will result in industrial tools. Suppose that the time to cut a diamond (in hours) is Poisson-distributed with mean equal to 18 hours. What is the expected time to cut and present to the King a diamond if the probability of success is 0.15 and delivery takes 1 hour?

8. In Exercise 7, suppose that the King is off fighting a war (as good kings do) and the diamond cutter is instructed to deliver the diamond personally. Suppose that the search time is exponential with mean equal to 9 hours. How long will total delivery take?

9. A steel ingot is being prepared for final processing at the Heavy Steel Company in Black Smoke, Texas. The raw steel comes from three different sources A, B, and C according to known arrival percentages 50 %, 25 %, and 25 %, respectively. The raw steel is melted in a cupola, and the processing time is normally distributed according to a normal density with mean equal to 25 hours and variance equal to 9 hours^2. A hot steel ingot is then sent to a cooling operation. The cooling operation takes 2.5 days for 30 % of the ingots, 1.8 days for 15 % of the ingots, and 2.7 days for 55 % of the ingots. After cooling, the ingots are heat-treated. Heat-treatment times are exponentially distributed with mean of 46 hours. After heat treating, an ingot is pulled and tested. 85 % of the ingots pass testing and proceed to the cutting station. 15 % are reprocessed with the second heat-treatment time being exponentially distributed with mean of 14.6 hours. After the second heat treatment, the ingot is again re-inspected. 95 % of these ingots proceed to cutting, but 5 % are rejected and sold as second-grade steel. All inspections take exactly 5 minutes by a computer scanner. After a cutting operation that requires an exponential time with mean of 6.2 minutes, the new slabs (cut up ingots) are inspected: 85 % are acceptable in present form; 10 % are recut to enforce standards; and 5 % are rejected as scrap. (a) Draw the GERT network for this example. (b) Construct the W-functions for each arc. (c) What percentage of items starts as raw material A and B and end up being unacceptable slabs? (d) What is the mean and variance of the time to produce a good slab?

10. This problem is taken from Reference [7].

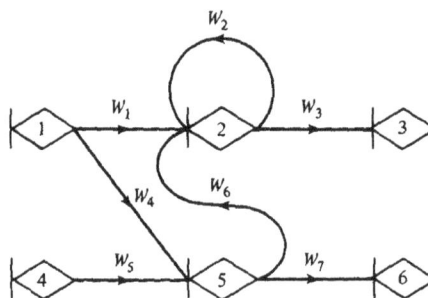

The GERT network has two source nodes with probabilities p_a and p_b and two terminal nodes. We want the equivalent W-function for each terminal node.

11. The probability of making a good item is $p = 0.80$. We want to find the probability of and expected number of trials required for manufacturing three consecutive good items. (a) Draw the GERT network (with W-functions). (b) Do the GERT analysis. (c) Solve using NOP.

12. A machine produces parts continuously throughout the day. The machine gets out of adjustment due to wear and tear. When it becomes inoperative it must be repaired. Assume that there are two successive stages of deterioration. In stage 1 the machine can still do its job, but it is apparent that it will soon need repair. Stage 2 is the stage in which the machine is inoperative and must be repaired to become operative again. If the machine is repaired when it is discovered in stage 1 of deterioration, the repair task is not difficult; but if it is allowed to run until it breaks down, then the repair takes longer. Suppose that if the machine is in good condition at the beginning of one of its runs, the probability is a that it will be in good condition after the run and $1-a$ that it will be in stage 1 after the run. Similarly, if the machine is in stage 1 at the beginning of the run, the probability is b that it will remain there and $1-b$ that it will fail. Assume that the moment generating functions of the time of a production run, the time to adjust the machine, and the time to repair the machine are $M_p(s)$, $M_a(s)$, and $M_r(s)$, respectively. Consider two policies. Policy A is to adjust the machine as soon as it gets into stage 1. Policy B is to wait until the machine fails to repair it. For each policy draw a GERT network.

REFERENCES

[1] MASON, S.J., "Feedback Theory: Some Properties of Signal Flow Graphs," *Proc. Inst. Radio Engrs.,* 41, 1144-1156, September 1953.

[2] MASON, S.J., "Feedback Theory: Further Properties of Signal Flow Graphs," *Proc. Inst. Radio Engrs.,* 44, 920-926, July 1956.

[3] PHILLIPS, D. T. AND D. R. SMITH, "Determination of Probabilistic Time Standards for Tasks Performed Under Uncertainty," Proceedings of the AIIE National Conference, San Francisco, Ca., 1979.

[4] PRITSKER, A. A. B., AND W. HAPP, "GERT: Graphical Evaluation and Review Technique, Part I. Fundamentals," *Journal of Industrial Engineering*, pp. 267-274, May 1966.

[5] PRITSKER, A. A. B., AND G. E. WHITEHOUSE, "GERT: Graphical Evaluation and Review Technique, Part II. Probabilistic and Industrial Engineering Applications," *Journal of Industrial Engineering,* pp. 293-301, June 1966.

[6] PRITSKER, A. A. B., The Production Engineer, pp. 499-506, October 1968.

[7] WHITEHOUSE, GARY E., System Analysis and Design Using Network Techniques. Englewood Cliffs, N.J.: Prentice-Hall, Inc., 1973.

[8] WHITEHOUSE G. E, AND A. B. PRITSKER, "GERT-Generating Functions, Conditional Distributions, Counters, Renewal Times and Correlations," *AIIE Transaction* 1969, 1(1): 45-50.

Chapter 8

AN INTRODUCTION TO MULTICOMMODITY NETWORK FLOW OPTIMIZATION

"Anyone who stops learning is old, whether at twenty or eighty. Anyone who keeps learning stays young. The greatest thing in life is to keep your mind young."

— Henry Ford

The aim of this chapter is to present and discuss fundamental results about multicommodity networks and simplifying solution procedures for special classes of these networks. For the solution of the more general linear minimal flow cost multicommodity network, we have selected the well-known *decomposition method* and the *Lagrangian relaxation method* combined with *subgradient optimization.* The material presented in this chapter has been divided into two parts. Most of the material covered in Part I contributed by Professor James R. Evans. Part II consists of a brief description of the decomposition method and the Lagrangian relaxation method and detailed numerical examples to illustrate the solution procedures. Among the bibliographical references we have included four survey articles [2, 16, 20, 21] to provide the reader with supplementary material regarding both applications and solution procedures.

PART I

MULTICOMMODITY NETWORK FLOWS

Contributed by
James R. Evans
Professor Emeritus, Department of Operations, Business Analytics, and Information Systems
University of Cincinnati

In Chapter 3 the out-of-kilter algorithm was shown to be a powerful tool in modeling and solving a variety of deterministic network-flow problems. All the examples and applications had one important aspect in common: the flow along any arc consisted of a single commodity. There are many applications which involve the shipment of many distinct commodities whose identities must be preserved along the arcs of a network. One must be able to distinguish between commodities. For example, consider the simplified problem of transporting three varieties of fruit from California orchards to wholesale grocers. Suppose that orchards owned by a particular company are located in Santa Barbara, Bakersfield, and Sacramento. During the peak of harvesting, the orchards are capable of producing various amounts of oranges, lemons, and limes, as given in Table 8.1.

Contractual agreements with wholesale grocers in Denver, Seattle, and Kansas City require amounts of these commodities as given in Table 8.2. The fruit is shipped by boxcar to these cities. However, only a limited amount of space on outgoing trains is available for the fruit. The capacities between the various origins and destinations are given in Table 8.3. The capacities are given in terms of crates per week, irrespective of the type of fruit (we assume that all commodities are packed in crates of similar size). Because of differences in packaging, the weight of the crates vary by commodity; so also then do the transportation costs. Unit shipping costs are given in Table 8.4.

Table 8.1. PRODUCTION RATE (CRATES/WEEK)

	Oranges	Lemons	Limes
Santa Barbara	800	700	700
Bakersfield	1000	800	500
Sacramento	500	1000	500

Table 8.2. DEMAND RATE (CRATES/WEEK)

	Oranges	Lemons	Limes
Denver	900	900	700
Seattle	700	1000	500
Kansas City	700	600	500

Table 8.3. Train Capacity (crates/week)

From	To		
	Denver	Seattle	Kansas
Santa Barbara	1200	900	1000
Bakersfield	1200	1000	1100
Sacramento	950	1300	1000

Table 8.4. UNIT TRANSPORTATION COSTS (CENTS/CRATE)

From	To		
	Denver	Seattle	Kansas City
Santa Barbara			
Oranges	3.2	2.5	5.6
Lemons	2.4	1.9	4.3
Limes	2.3	1.7	4.0
Bakersfield			
Oranges	3.0	2.1	5.5
Lemons	2.3	1.7	4.4
Limes	2.0	1.4	4.3
Sacramento			
Oranges	3.1	2.0	5.7
Lemons	2.2	1.6	4.8
Limes	2.0	1.5	4.3

The problem of determining a minimal-cost shipping policy is a multi-commodity network-flow problem. Notice that this example is similar in structure to the transportation problem developed in Section 2.2.4 in that shipments can only occur between origins and destinations. The major distinguishing features of multicommodity flow problems are that several non-homogeneous commodities share common arcs, and that the flow of all commodities on an arc is constrained by the arc capacity. Were this not so, one would be able to solve a minimal-cost transportation problem for each commodity independently of the others, using an algorithm such as the out-of-kilter method. The capacity constraints, however, make these problems much more difficult to solve.

8.1 LINEAR PROGRAMMING FORMULATIONS

As in the single-commodity case, multicommodity network-flow problems can be formulated as linear programs. The fruit distribution problem is an example of the ***multicommodity transportation problem***, hereafter denoted MCTP. Define x^k_{ij} to be the flow of commodity k from origin i to destination j having unit transportation cost c^k_{ij}. Also, let a_i^k and b_j^k be the supply at node i and demand at node j, respectively, of commodity k, and let u_{ij} represent the capacity of arc (i,j). The mathematical model for the MCTP is given in Eqs. (8-1) to (8-5).

$$minimize \sum_{k=1}^{r}\sum_{i=1}^{m}\sum_{j=1}^{n} c^k_{ij} x^k_{ij} \tag{8-1}$$

subject to

$$\sum_i x^k_{ij} = b^k_j \qquad \text{all } j,k \tag{8-2}$$

$$\sum_j x^k_{ij} = a^k_i \qquad \text{all } i,k \tag{8-3}$$

$$\sum_k x^k_{ij} \le u_{ij} \qquad \text{all } i,j \tag{8-4}$$

$$x^k_{ij} \ge 0 \qquad \text{all } i,j,k \tag{8-5}$$

It is assumed, as in the single-commodity transportation problem, that total supply equals total demand for each commodity, that is,

$$\sum_i a^k_i = \sum_j b^k_j \quad \text{for all } k \tag{8-6}$$

Many multicommodity models include transshipment nodes. At transshipment nodes there is neither a supply nor a demand; the only requirement is that conservation of flow be satisfied. An example of a multicommodity transshipment network is given in Figure 8.1. Nodes 1 and 2 are sources for commodity 1, while node 2 is the only source for commodity 2. Node 5 is the destination for both commodities; nodes 3 and 4 are transshipment nodes.

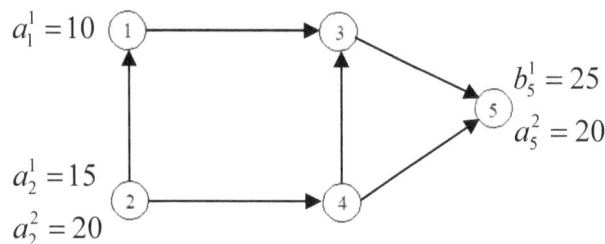

Figure 8.1. Multicommodity transshipment network.

The general linear programming formulation for the ***multicommodity cost minimization problem*** is given in Eqs. (8-7) to (8-12).

301

$$minimize \sum_{k=1}^{r} \sum_{(i,j) \in \mathbf{A}} c_{ij}^k x_{ij}^k \qquad (8\text{-}7)$$

subject to

$$\sum_j x_{ij}^k - \sum_j x_{ji}^k = a_i^k \quad \text{if node } i \text{ is a source for commodity } k \qquad (8\text{-}8)$$

$$\sum_j x_{ij}^k - \sum_j x_{ji}^k = 0 \quad \text{if node } i \text{ is a transshipment node} \qquad (8\text{-}9)$$

$$\sum_j x_{ij}^k - \sum_j x_{ji}^k = -b_j^k \quad \text{if node } j \text{ is a sink for commodity } k \qquad (8\text{-}10)$$

$$\sum_k x_{ij}^k \leq u_{ij} \qquad \text{all } i,j \qquad (8\text{-}11)$$

$$x_{ij}^k \geq 0 \qquad \text{all } i,j,k \qquad (8\text{-}12)$$

In this formulation, $\mathbf{A}$ is the set of arcs in the network.

One of the difficulties encountered in solving multicommodity network-flow problems is caused by the fact that optimal solutions may be *noninteger*. Noninteger solutions arise because linear programming formulations do not have unimodular constraint matrices in general. The total unimodularity condition, introduced in Chapter 3, provides a sufficient condition for integer optimal solutions. Since single-commodity problems have this property, it is possible to devise highly efficient algorithms whose computations are essentially composed of only additions and subtractions. Division and matrix inversion, for instance, are not required. From a computer standpoint, arithmetic operations performed on integers are much faster than using floating-point (decimal) numbers. This leads to computer implementations that can solve extremely large problems very quickly.

Special algorithms for multicommodity flow problems have been developed based on the structure and properties of this class of problems. They have been shown to be faster than general-purpose linear programming procedures but are not nearly as fast as the corresponding single-commodity algorithms. A full discussion of these methods is beyond the scope of this book. However, in the remainder of this chapter we discuss a variety of special topics and solution techniques. In Part II we will outline and illustrate the procedure followed in the the *decomposition method* for solving linear multicommodity minimal-cost flow models.

8.2 A SPECIAL CLASS OF INTEGER MULTICOMMODITY NETWORKS

There exist several classes of multicommodity networks that have totally unimodular constraint matrices and hence integer solutions when the right-hand constants of the constraints are integer. More surprisingly, many of these problems can be transformed to equivalent single-commodity network flow problems. One of the special classes is the multicommodity transportation problem introduced in Section 8.1 with the number of sources or the number of destinations being less than or equal to 2. This result states that an optimal integer solution is guaranteed to exist for the model formulated in (8-1)-(8-5) whenever $m \leq 2$ or $n \leq 2$ *regardless* of the number of commodities. For $m = 2$, the model is reformulated as shown in (8-13)-(8-19), where non-negative slack variables are added to the capacity Constraints (8-4).

$$\text{minimize} \sum_{k=1}^{r}\sum_{i=1}^{m} c_{1j}^{k} x_{1j}^{k} + c_{2j}^{k} x_{2j}^{k} \tag{8-13}$$

subject to

$$x_{1j}^{k} + x_{2j}^{k} = b_{j}^{k} \qquad j = 1, \cdots n; k = 1, \cdots, r \tag{8-14}$$

$$\sum_{j} x_{1j}^{k} = a_{1}^{k} \qquad k = 1, \cdots, r \tag{8-15}$$

$$\sum_{j} x_{2j}^{k} = a_{2}^{k} \qquad k = 1, \cdots, r \tag{8-16}$$

$$\sum_{k} x_{1j}^{k} + s_{1j} = u_{1j} \qquad j = 1, \cdots, n \tag{8-17}$$

$$\sum_{k} x_{2j}^{k} + s_{2j} = u_{2j} \qquad j = 1, \cdots, n \tag{8-18}$$

$$x_{1j}^{k}, x_{2j}^{k}, s_{1j}, s_{2j} \geq 0 \quad j = 1, \cdots, n; k = 1, \cdots, r \tag{8-19}$$

From Eq. (8-14) we can see that $x_{1j}^{k} = b_{j}^{k} - x_{2j}^{k}$. Using this in Eq. (8-15) it is concluded that $\sum_{j} b_{j}^{k} - x_{2j}^{k} + \sum_{j} s_{1j} = a_{1}^{k}$. This combined with Eq. (8-16) results in $\sum_{j} x_{2j}^{k} + \sum_{j} s_{2j} = a_{2}^{k}$. Furthermore, from Eqs. (8-17) and (8-18) along with Eq. (8-14) it is possible conclude that $\sum_{k} b_{j}^{k} + s_{1j} + s_{2j} = u_{1j} + u_{2j}$; and from Eq. (8-18) we can get the result $\sum_{j}\sum_{k} x_{2j}^{k} + \sum_{j} s_{2j} = \sum_{j} u_{2j}$. All these developments allow the reformulation of the model in (8-13)-(8-19) as

$$\text{minimize} \sum_{k=1}^{r}\sum_{j=1}^{n} \left[(c_{2j}^{k} - c_{1j}^{k}) x_{2j}^{k} + c_{1j}^{k} b_{j}^{k} \right] \tag{8-20}$$

$$\sum_{k} x_{2j}^{k} + s_{2j} = u_{2j} \qquad j = 1, \cdots, n \tag{8-21}$$

$$-\sum_{j} x_{2j}^{k} = -a_{2}^{k} \qquad k = 1, \cdots, r \tag{8-22}$$

$$-\sum_{j} s_{2j} = -\sum_{j} u_{2j} + \sum_{k} a_{2}^{k} \tag{8-23}$$

$$x_{1j}^{k} + x_{2j}^{k} = b_{j}^{k} \qquad j = 1, \cdots n; k = 1, \cdots, r \tag{8-24}$$

$$s_{1j} + s_{2j} = u_{1j} + u_{2j} - \sum_{k} b_{j}^{k} \qquad j = 1, \cdots, n \tag{8-25}$$

$$x_{1j}^{k}, x_{1j}^{k}, s_{1j}, s_{2j} \geq 0 \qquad j = 1, \cdots, n; k = 1, \cdots, r \tag{8-26}$$

By closely inspecting the model in (8-20)-(8-26) we can verify that the coefficient matrix of Constraints (8-21), (8-22) and (8-23) has the form of a node-arc incidence matrix. Notice also that x_{1j}^{k} and s_{1j} appear only in Eqs. (8-24) and (8-25) and are nonnegative. Thus, we may think of these as slack variables and eliminate them by writing Eqs. (8-24) and (8-25) as

$$x_{2j}^k \le b_j^k \qquad\qquad j = 1, \cdots n; k = 1, \cdots, r$$

$$s_{2j} \le u_{1j} + u_{2j} - \sum_k b_j^k \qquad j = 1, \cdots, n$$

The right-hand-side constants of Eqs. (8-21)-(8-23) represent either supplies if they are positive or demands otherwise. The structure of the network represented by Eqs. (8-21) to (8-23) is a transportation model as shown in Figure 8.2. The right-hand-side constants of Eqs. (8-24) and (8-25) are the capacities of the arcs with flows equal to x_{2j}^k and s_{2j}. The out-of-kilter algorithm can be used to find the optimal values for these flows. Eqs. (8-24) and (8-25) could then be used to solve for to x_{1j}^k and s_{1j}.

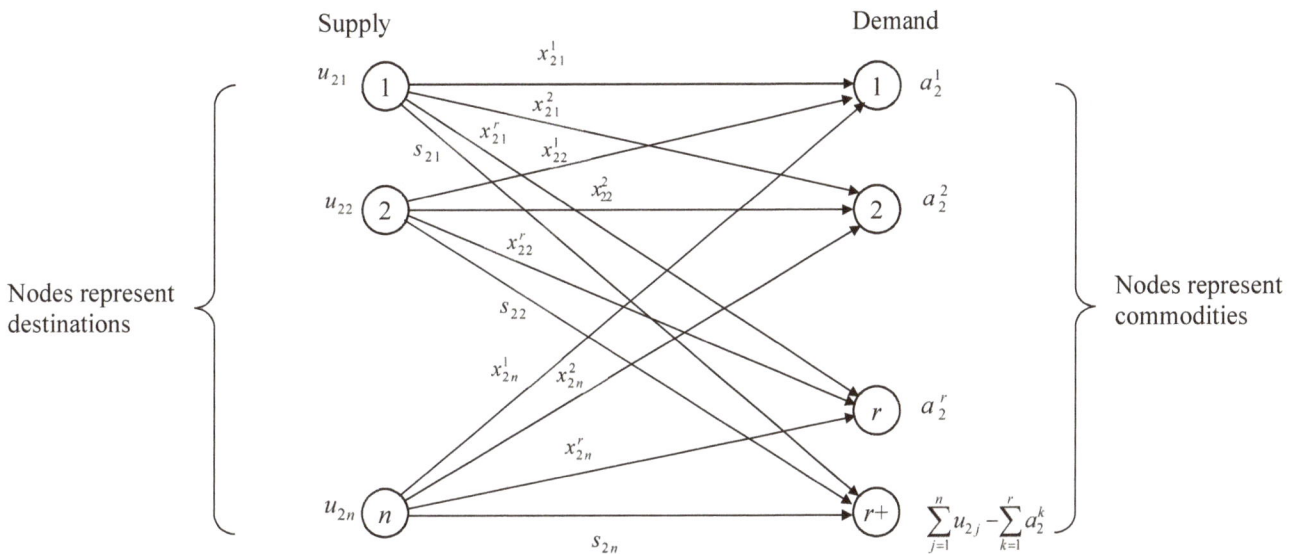

Figure 8.2. Single-commodity network representation of 2-source MCTP.

This new formulation was obtained by performing row operations on the original constraints in a manner similar to the employment scheduling example in Section 2.2.6. The optimal flows for outlet B can be obtained using the OKA and the flows for outlet A can be calculated using Eqs. (8-24) and (8-25).

Table 8.5. INPUT DATA FOR SAMPLE MCTP

Plant	Commodity	Destination C	Destination D	Destination E	Production (units/month)			
A	1	4.65	5.50	5.00	4000			
	2	4.95	5.90	5.45	5000			
	3	5.30	6.30	6.00	2000			
B	1	5.15	4.85	5.25	2500			
	2	5.70	4.95	5.90	3000			
	3	6.00	5.15	6.25	5000			
Demand (units/month)	1	3500	1500	1500	Capacity	C	D	E
	2	4000	1500	2500	A	6000	3000	4000
	3	3000	2000	2000	B	6000	4000	3000

As an illustration, let us consider two manufacturing outlets, three shipping destinations and three commodities, with the data given in Table 8.5.

The corresponding *circulation network* is shown in Figure 8.3. Nodes 1, 2 and 3 represent the three shipping destinations; nodes 4, 5 and 6 represent the three commodities; node 7 represents a dummy commodity. Nodes 8 and 9 represent the super source and super terminal needed for designing the circulation network. Furthermore, arcs out of node 8 represent the flow from outlet B to the shipping destination nodes 1, 2 and 3. Arcs from commodity nodes 4, 5 and 6 into node 9 represent the production flows, and the flow on the arc from the dummy commodity to the super terminal represent the difference between total transportation capacity for outlet *B* and total production capability of all commodities for this outlet. Arcs from nodes 1, 2 and 3 to nodes 4, 5, 6 and 7 represent the shipping strategies for outlet *B*. All these arcs have lower bounds equal to zero and upper bounds equal to the corresponding demands in units per month shown in Table 8.5.

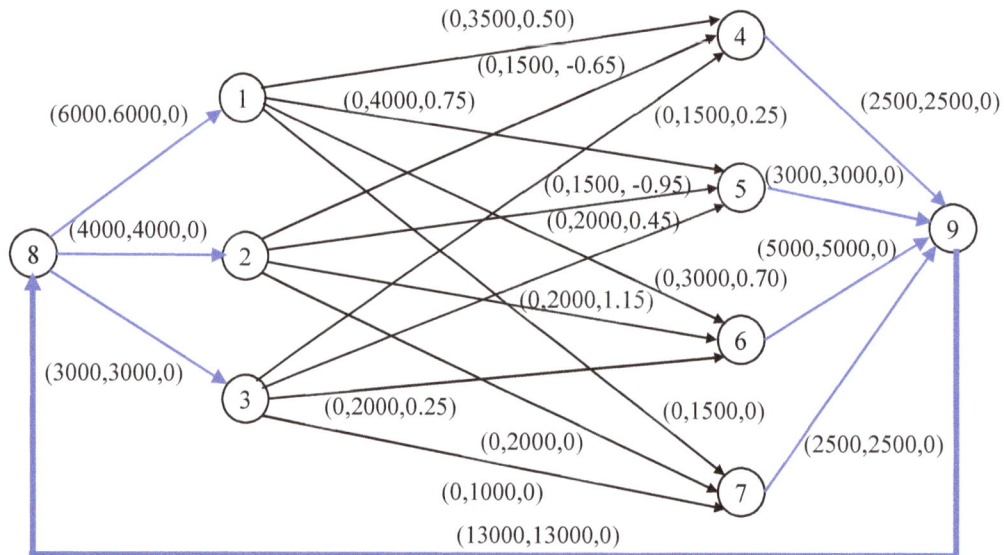

Figure 8.3. Circulation Network for Single-Commodity Two-Source MCTP.

To illustrate the calculation of costs for the arcs representing shipping strategies, let us consider arc (1,4). Node 1 represents destination 1 and node 4 represents commodity 1. Therefore, the flow along this arc represents the amount of commodity 1 shipped from outlet B to destination 1. According to (8-20), and using the data in Table 8.5, the per-unit flow cost for this arc is 5.15 – 4.65 = 0.50; all costs for the shipping strategies are shown in Figure 8.3.

Figure 8.4 shows the optimal flows for Outlet B obtained by NOP after choosing the out-of-kilter algorithm.

```
Nodes 9;
St;
1 4 3500 0 0.50;
1 5 4000 0 0.75;
1 6 3000 0 0.70;
1 7 1500 0 0;
2 4 1500 0 -0.65;
2 5 1500 0 -0.95;
2 6 2000 0 1.15;
2 7 2000 0 0;
3 4 1500 0 0.25;
3 5 2500 0 0.45;
3 6 2000 0 0.25;
3 7 1000 0 0;
4 9 2500 2500 0;
5 9 3000 3000 0;
6 9 5000 5000 0;
7 9 2500 2500 0;
8 1 6000 6000 0;
8 2 4000 4000 0;
8 3 3000 3000 0;
9 8 13000 13000 0;
End;
```

```
------------ SOLUTION BY OUT OF KILTER ALGORITHM ------------

Number of Nodes = 9
Number of Arcs = 20

----------------------------------------------------------------

Arc   I   J      UPPER       LOWER        FLOW        COST
 1    1   4     3500.00     0000.00     1000.00     0000.50
 2    1   5     4000.00     0000.00     0500.00     0000.75
 3    1   6     3000.00     0000.00     3000.00     0000.70
 4    1   7     1500.00     0000.00     1500.00     0000.00
 5    2   4     1500.00     0000.00     1500.00    -0000.65
 6    2   5     1500.00     0000.00     1500.00    -0000.95
 7    2   6     2000.00     0000.00     0000.00     0001.15
 8    2   7     2000.00     0000.00     1000.00     0000.00
 9    3   4     1500.00     0000.00     0000.00     0000.25
10    3   5     2500.00     0000.00     1000.00     0000.45
11    3   6     2000.00     0000.00     2000.00     0000.25
12    3   7     1000.00     0000.00     0000.00     0000.00
13    4   9     2500.00     2500.00     2500.00     0000.00
14    5   9     3000.00     3000.00     3000.00     0000.00
15    6   9     5000.00     5000.00     5000.00     0000.00
16    7   9     2500.00     2500.00     2500.00     0000.00
17    8   1     6000.00     6000.00     6000.00     0000.00
18    8   2     4000.00     4000.00     4000.00     0000.00
19    8   3     3000.00     3000.00     3000.00     0000.00
20    9   8    13000.00    13000.00    13000.00     0000.00

Total Cost = 1525.00
```

Figure 8.4. Optimal commodity flows from outlet B obtained by the OKA.

From Figure 8.4 we obtain the following results for outlet B:

$$
\begin{array}{ccc}
x^1_{BC} = 1000 & x^1_{BD} = 1500 & x^1_{BE} = 0 \\
x^2_{BC} = 500 & x^2_{BD} = 1500 & x^2_{BE} = 1000 \\
x^3_{BC} = 3000 & x^3_{BD} = 0 & x^3_{BE} = 2000
\end{array}
$$

Using Eqs. (8-24) for commodities 1, 2 and 3:

$$
x^1_{BC} + x^1_{AC} = b^1_C \quad x^1_{BD} + x^1_{AD} = b^1_D \quad x^1_{BE} + x^1_{AE} = b^1_E
$$

$$
x^2_{BC} + x^2_{AC} = b^2_C \quad x^2_{BD} + x^2_{AD} = b^2_D \quad x^2_{BE} + x^2_{AE} = b^2_E
$$

$$
x^3_{BC} + x^3_{AC} = b^3_C \quad x^3_{BD} + x^3_{AD} = b^3_D \quad x^3_{BE} + x^3_{AE} = b^3_E
$$

From the given demands and OKA results in Figure 8.4 for outlet B, we get for outlet A:

$$
1000 + x^1_{AC} = 3500 \quad 1500 + x^1_{AD} = 1500 \quad 0 + x^1_{AE} = 1500
$$

$$
500 + x^2_{AC} = 4000 \quad 1500 + x^2_{AD} = 1500 \quad 1000 + x^2_{AE} = 2500
$$

$$
3000 + x^3_{AC} = 3000 \quad 0 + x^3_{AD} = 2000 \quad 2000 + x^3_{AE} = 2000
$$

Therefore, the results for outlet A are:

$$
\begin{array}{|lll|}
\hline
x^1_{AC} = 2500 & x^1_{AD} = 0 & x^1_{AE} = 1500 \\[4pt]
x^2_{AC} = 3500 & x^2_{AD} = 0 & x^2_{AE} = 1500 \\[4pt]
x^3_{AC} = 0 & x^3_{AD} = 2000 & x^3_{AE} = 0 \\
\hline
\end{array}
$$

8.3 APPROXIMATE SOLUTIONS OF MULTICOMMODITY TRANSPORTATION PROBLEMS

Often a decision maker is satisfied with good, though suboptimal solutions, to complex problems. One method of simplifying a problem in order to make it easier to solve is called *aggregation*. In the case of multicommodity transportation problems we aggregate sources or destinations. More specifically, we will consider cases in which the aggregated problem has exactly two origins and solve it by means of the single-commodity transformation.

Let us suppose that a multicommodity transportation problem consists of m origins, n destinations, and r commodities. Partition the origins into two sets S_1 and S_2 and replace the sets of nodes by two *pseudo nodes* as illustrated in Figure 8.5. The two-source network is the aggregated problem. We need to relate the parameters of this problem to those of the original problem in some meaningful fashion, and also be able to recover a feasible solution to the original problem after solving the aggregated problem.

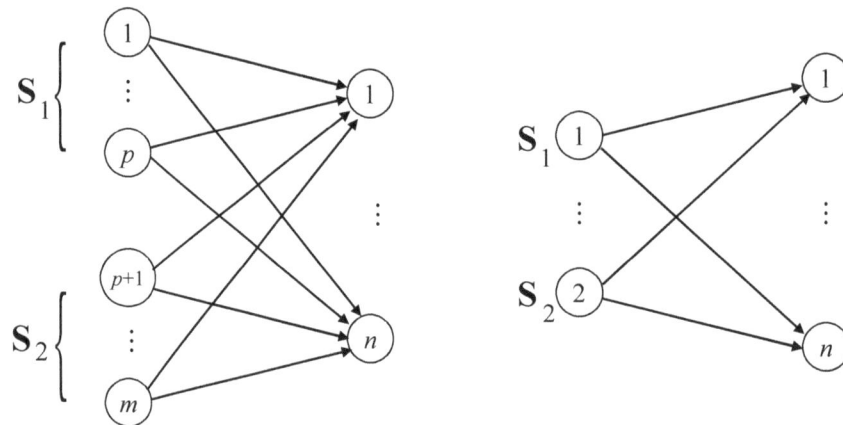

Figure 8.5. Aggregation of origins in a MCTP.

Since each of the two pseudo nodes represents a collection of original sources, it is natural to specify the supplies at these nodes as

$$
\bar{a}^k_l = \sum_{i \in S_l} a^k_i \quad l = 1, 2 \tag{8-27}
$$

Let y^k_{lj} represent the flow of commodity k from node l to node j in the aggregate problem. Since arc (l, j) is an aggregation of arcs from nodes $i \in S_l$ to destination j, we can write

307

$$y_{lj}^k = \sum_{i \in S_l} x_{ij}^k \qquad (8\text{-}28)$$

There still remains the question of specifying costs and capacities on the aggregate arcs. There are several ways to do this; the method we will use has certain advantages over others. To specify costs the concept of *weighted aggregation* will be used; that is, costs are weighted proportionally to the supplies of the origins represented by the pseudo node as follows:

$$\overline{c}_{lj}^k = \sum_{i \in S_l} c_{ij}^k \frac{a_i^k}{\overline{a}_l^k} \qquad l = 1,2 \qquad (8\text{-}29)$$

At first thought it would seem reasonable to define the capacity on the aggregate arc (l, j) as the sum of the capacities of the original arcs from $i \in S_l$ to destination j. The problem with this is that it may be difficult to recover a feasible solution to the original problem. Therefore, the following method will be used. Define

$$\delta_i = \min_k \left\{ \frac{a_i^k}{\overline{a}_l^k} \right\} \qquad i \in \mathbf{S}_l \qquad (8\text{-}30)$$

Now, let

$$\overline{u}_{lj} = \min_{i \in \mathbf{S}_l} \left\{ \delta_i u_{ij} \right\} \qquad (8\text{-}31)$$

We may now state the linear program in Eqs. (8-32)-(8-36) for the aggregate problem:

$$minimize \sum_{k=1}^{r} \sum_{l=1}^{2} \sum_{j=1}^{n} \overline{c}_{lj}^k y_{lj}^k \qquad (8\text{-}32)$$

subject to

$$\sum_{l} y_{lj}^k = b_j^k \qquad j = 1, \cdots, n; k = 1, \cdots, r \qquad (8\text{-}33)$$

$$\sum_{j} y_{lj}^k = \overline{a}_l^k \qquad l = 1, 2; k = 1, \cdots, r \qquad (8\text{-}34)$$

$$\sum_{k} y_{lj}^k \leq \overline{u}_{lj} \qquad l = 1, 2; j = 1, \cdots, n \qquad (8\text{-}35)$$

$$y_{lj}^k \geq 0 \qquad j = 1, \cdots, n; k = 1, \cdots, r; l = 1, 2 \qquad (8\text{-}36)$$

We now need to recover a solution to the original problem once we have an optimal solution to the aggregate problem defined in (8-32)-(8-36). Consider the following definition,

$$x_{ij}^k = y_{lj}^k \frac{a_i^k}{\overline{a}_l^k} \qquad i \in \mathbf{S}_l \qquad (8\text{-}37)$$

Note that Eq. (8-37) can be used to obtain $\displaystyle \sum_{i \in \mathbf{S}_l} x_{ij}^k = y_{lj}^k \sum_{i \in \mathbf{S}_l} \frac{a_i^k}{\overline{a}_l^k} = y_{lj}^k \left(\frac{\overline{a}_l^k}{\overline{a}_l^k} \right) = y_{lj}^k,$ which corresponds

top definition (8-28). The procedure defined in Eq. (8-37) is called *fixed-weight disaggregation*. It is not difficult to show that the values found by Eq. (8-37) are feasible to the original problem (specifying $\bar{u}_{ij}$ as we did was necessary in order to guarantee satisfying the original capacity constraints). Also,

$$\sum_{k=1}\sum_{i=1}\sum_{j=1}c_{ij}^{k}x_{ij}^{k}=\sum_{k=1}\sum_{l=1}\sum_{j=1}\bar{c}_{lj}^{k}y_{lj}^{k} \tag{8-38}$$

In other words, the fixed-weight disaggregation has the same objective function value as the aggregate problem. The reader should be cautioned however, that the aggregate problem may not be feasible even though the original problem is, due to the nature of $\bar{u}_{ij}$. One may have to try different aggregations in this case.

8.3.1   A Fruit Distribution Problem

Let us solve the fruit distribution problem posed in the introduction of Part I by aggregating sources 2 (Bakersfield) and 3 (Sacramento). The supplies, costs, and capacities of the aggregate problem are given in Table 8.6. The optimal solution to the original problem has a cost of $18,570 and is shown in Table 8.7. Table 8.8 gives the optimal solution to the aggregate problem having a cost of $18,799.10, an increase of 0.26% over the true optimal cost. It is left as an exercise to the reader to compute the disaggregate solution and verify that it has the same cost.

Table 8.6. AGGREGATE PROBLEM PARAMETERS

SUPPLIES			
Origin	Oranges	Lemons	Limes
Santa Barbara	800	700	700
Bakersfield/Sacramento	1500	1800	1000

COSTS	To		
From	Denver	Seattle	Kansas City
Santa Barbara			
Oranges	3.200	2.500	5.600
Lemons	2.400	1.900	4.300
Limes	2.300	1.700	4.000
Bakersfield/Sacramento			
Oranges	3.033	2.067	5.567
Lemons	2.244	1.644	4.622
Limes	2.000	1.450	4.300

CAPACITIES	To		
From	Denver	Seattle	Kansas City
Santa Barbara	1200	900	1000
Bakersfield/Sacramento	1500	1800	1000

Table 8.7. SOLUTION TO FRUIT DISTRIBUTION PROBLEM

From	To		
	Denver	Seattle	Kansas City
Santa Barbara			
Oranges	800	0	0
Lemons	200	0	500
Limes	0	200	500
Bakersfield			
Oranges	100	200	700
Lemons	500	200	100
Limes	200	300	0
Sacramento			
Oranges	0	500	0
Lemons	200	800	0
Limes	500	0	0

Table 8.8. SOLUTION TO AGGREGATED PROBLEM

From	To		
	Denver	Seattle	Kansas City
Santa Barbara			
Oranges	800	0	0
Lemons	200	0	500
Limes	0	200	500
Bakersfield/Sacrament			
Oranges	100	200	700
Lemons	500	200	100
Limes	200	300	0

8.3.2   Error Bounds for Aggregation

This section summarizes the final result developed in Section 5.21 of *Fundamentals of Network Analysis* [18].   By considering the dual of the aggregate problem formulated in (8-32) through (8-36) we can derive a bound on the error resulting from the disaggregate solution.   The bound is useful in determining the error resulting from aggregation, although it should be realized that the *actual error* is generally less than the bound.   The dual linear program to the aggregate problem is defined in Eqs. (8-39) to ((8-41).

$$maximize \sum_{k=1}^{r}\sum_{l=1}^{2}\overline{a}_l^k \alpha_l^k + \sum_{k=1}^{r}\sum_{j=1}^{n}b_j^k \beta_j^k + \sum_{j=1}^{n}\sum_{l=1}^{2}\overline{u}_{lj}\gamma_{lj} \qquad (8\text{-}39)$$

subject to

$$\alpha_l^k + \beta_j^k - \gamma_{lj} \le \overline{c}_{lj} \quad l = 1,2; j = 1,\cdots,n; k = 1,\cdots,r \qquad (8\text{-}40)$$

$$\gamma_{lj} \ge 0 \quad l = 1,2; j = 1,\cdots,n \qquad (8\text{-}41)$$

Also, let $\overline{z}$ be the value of the objective function (8-39).   This value is an upper bound on the optimum cost z^*.   It can be proved that

$$\overline{z} - z^* \leq \sum_l \sum_{i \in S_l} \max_{j,k} \left(\overline{\alpha}_l^k + \overline{\beta}_l^k - \overline{\gamma}_{ij} - c_{ij}^k \right) + \sum_k a_i^k - \sum_l \sum_j (\overline{u}_{lj} - \sum_{i \in S_l} u_{ij}) \overline{\gamma}_{lj} \ \square \ \varepsilon$$

Therefore, $\overline{z} - \varepsilon \leq z^* \leq \overline{z}$.

8.4 MAXIMAL FLOWS IN MULTICOMMODITY NETWORKS

For single-commodity problems it was shown that the value of the maximal flow is equal to that of the minimal cut set. An elegant and efficient algorithm based of labeling concepts was developed for finding the maximal flow. At first, one might believe that perhaps these results can be generalized to multicommodity flow problems. But again, except for some special cases, this is not true. One of the earliest counterexamples is shown in Figure 8.6. Nodes s^k and t^k represent the source and sink for commodity k, respectively. Each arc has capacity of 1, and the problem is to maximize the *sum* of the flows of the three commodities from their sources to their respective sinks. For instance, if a flow of 1 unit is sent from s^1 to t^1 along the only path available, no other commodity flow can be sent through the network. This is not the maximal flow since we can send 1/2 unit from s^k to t^k for each commodity, a *noninteger* flow.

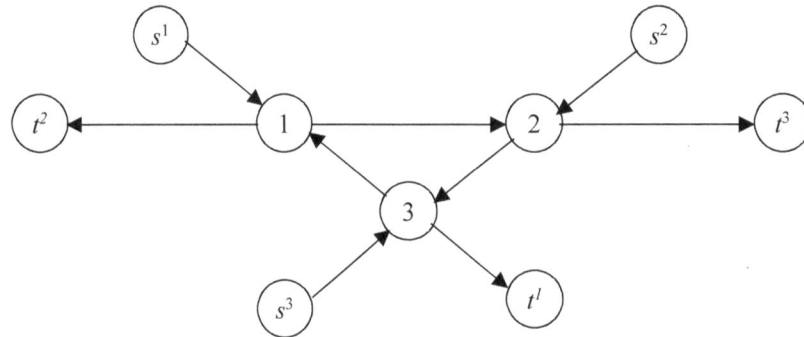

Figure 8.6. Multicommodity maximal flow problem.

In multicommodity problems, the analogous concept of a cut is called a *disconnecting set* and is defined as a set of arcs whose removal from the network destroys all paths from each source to its respective sink. In this example, the minimal disconnecting set consists of any two arcs between nodes 1, 2, and 3. The value of the minimal disconnecting set is 2; hence, we have that the maximal multicommodity flow can be *less than* the value of the minimal disconnecting set.

In general, the multicommodity maximal-flow problem, just as the minimal-cost problem, is difficult to solve. The development of most solution procedures are beyond the scope of this chapter. However, for special types of networks, it is true that maximum flow equals the value of the minimum disconnecting set and that all flow values are integer. This result can be shown to hold true for networks that are called *completely planar*. A network is *planar* if it can be drawn so that no two arcs intersect. For example, the network in Figure 8.7(a) is planar while the network in Figure 8.7(b) is not. A multicommodity network is completely planar if it can be drawn as in Figure 8.8 so that the resulting network is planar. An example with two commodities is given in Figure 8.9.

311

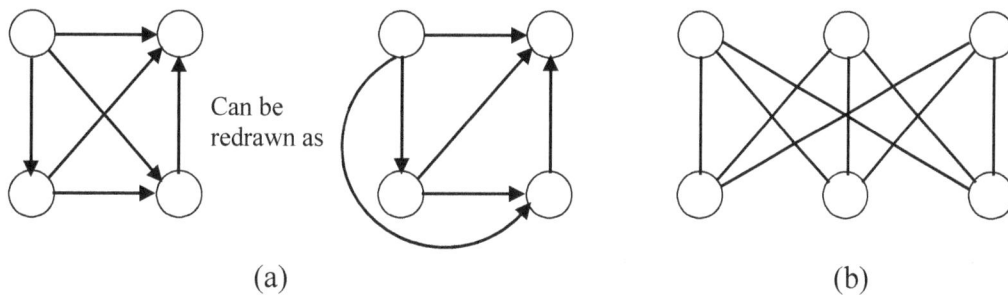

Figure 8.7. Examples of Planar and nonplanar networks.

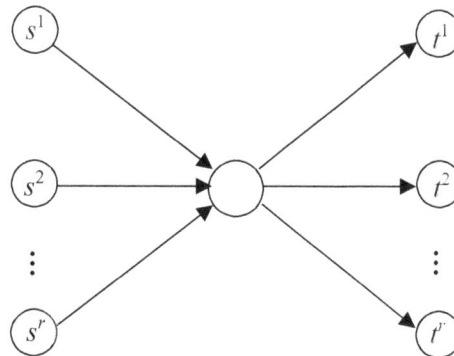

Figure 8.8. Structure of completely planar networks.

There is a very simple algorithm for solving problems of this type. Start with commodity 1. Identify the *uppermost* path from s^1 to t^1 and assign as much flow as possible. Adjust all capacities by subtracting the flow on the arcs. At least one arc will reach capacity; remove this and all other arcs whose total flow equals the capacity. When there are no flow-augmenting paths remaining from s^1 to t^1, repeat this procedure for commodity 2, 3, and so on, until it is no longer possible to send any additional flow from a source to a sink. The resulting flow pattern is a maximal flow.

As an illustrative example of the algorithm, let us find the maximum flow for the two-commodity network of Figure 8.9. The corresponding results are shown in Figure 8.10. A detailed description of this figure is provided in the discussion of this figure.

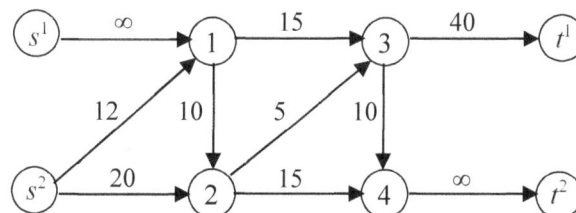

Figure 8.9. Two-commodity completely planar network.

312

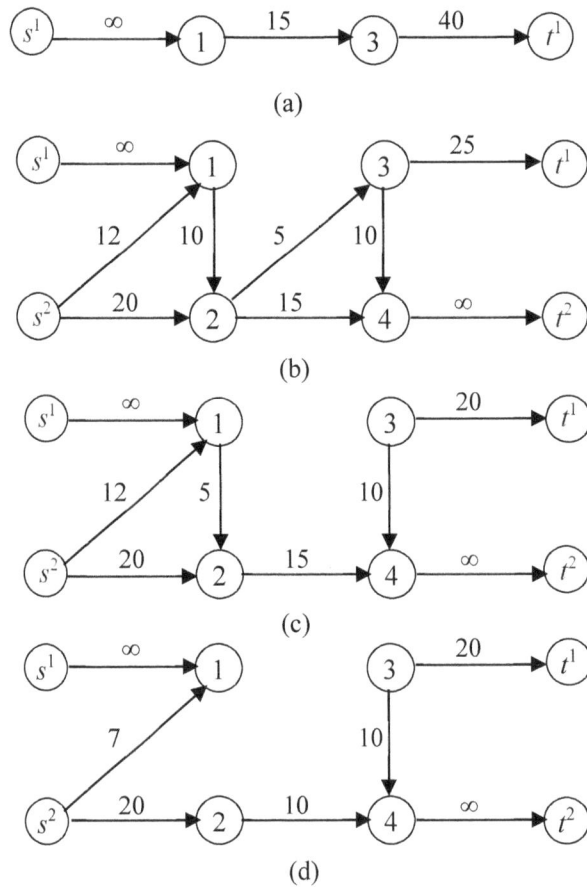

Figure 8.10. Flow-augmenting paths and modified networks for the example.

The uppermost path from s^1 to t^1 is the sequence of nodes $s^1 - 1 - 3 - t^1$, shown in Figure 8.10 (a). The maximum flow that can be sent along this path is equal to 15 units. Adjusting the residual capacities and removing saturated arcs yields the network in Figure 8.10 (b). We next identify the path s^1-1-2-3-t^1 as the uppermost path and assign a flow of 5 units. The new network is shown in Figure 8.10 (c).

No flow-augmenting paths exist from s^1 to t^1 in the modified network of Figure 8.10 (c). Considering commodity 2, we start with the path s^2-1-2-4-t^2. Five units of flow can be sent along this path. The modified network is shown in Figure 8.10 (d).

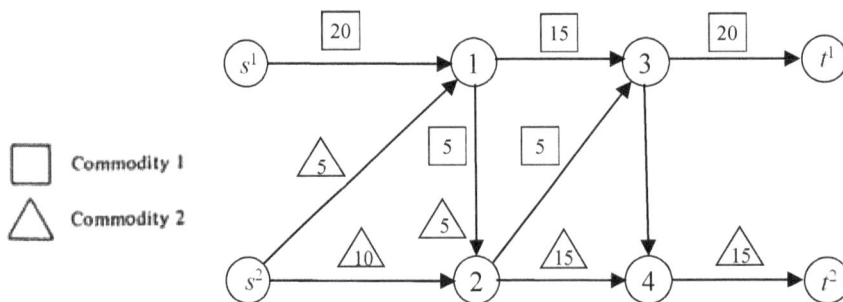

Figure 8.11. Maximal two-commodity flow.

8.5 MULTICOMMODITY FLOWS IN UNDIRECTED NETWORKS

In networks with undirected arcs, flow may traverse an arc in either direction. In the single-commodity case, one can replace an undirected arc by a pair of oppositely directed arcs. This can be done since flows in opposite directions cancel each other. However, for multicommodity problems, this cannot be done since flows *of different* commodities in opposite directions do not cancel out, but add to the total consumption of capacity of the arc. And as before, optimal flows need not be integer, nor does the maximum-flow/minimum-cut theorem always hold.

Although the general problem is difficult to solve, we shall present an algorithm for the special case of *two-commodity maximal flows*. For this class of problems, it can be shown that optimal flows are always multiples of 1/2, and that the maximum flow equals the value of the minimum disconnecting set.

To develop the algorithm, some new concepts and definitions need to be introduced. The flow of commodity k on arc (i, j) is denoted f_{ij}^k. Since flow can occur in either direction, a positive flow from i to j can be viewed as a negative flow from j to i; hence, $f_{ij}^k = -f_{ji}^k$. Regardless of the direction of flow, the net amount of commodity k on arc (i, j) will be denoted $|f_{ij}^k|$. As before, s^k and t^k are the source and sink, respectively, of commodity k, and u_{ij} is the capacity of arc (i, j), *assumed to be integer*. Let us assume that we have a network flow pattern satisfying conservation-of-flow restrictions. We define two sets of nodes relative to the given flow pattern, **F** and **B**:

1. s^2 is defined to be in **F**. If $i \in$ **F** and $f_{ji}^1 + f_{ij}^2 < u_{ij}$ then $j \in$ **F**.
2. s^2 is defined to be in **B**. If $i \in$ **B** and $f_{ij}^1 + f_{ij}^2 < u_{ij}$ then $j \in$ **B**.

Suppose that we wish to send additional flow of commodity 1 from i to j. If $f_{ij}^1 > 0$, then the *residual capacity* is defined as in Eq. (8-42),

$$u_{ij}^1 = u_{ij} - f_{ij}^1 - |f_{ij}^2| \qquad (8\text{-}42)$$

If we wish to send additional flow from j to i and $f_{ij}^1 > 0$, the residual capacity is given by

$$u_{ij}^1 = u_{ij} + f_{ij}^1 - |f_{ij}^2| \qquad (8\text{-}43)$$

since the flow of the same commodity in opposite directions cancel. Similarly, for commodity 2, if $f_{ij}^2 > 0$, the residual capacity in the direction from i to j is

$$u_{ij}^2 = u_{ij} - |f_{ij}^1| - f_{ij}^2 \qquad (8\text{-}44)$$

and in the direction from j to i,

$$u_{ij}^2 = u_{ij} - |f_{ij}^1| + f_{ij}^2 \qquad (8\text{-}45)$$

Consider any path from s^2 to t^2 with arc (i, j) on the path. As we travel from s^2 to t^2 and node i is reached before node j, then arc (i, j) is said to be in the *positive direction* with respect to the path. In this case, a flow from i to j is said to be in *the forward direction,* and a flow from j to i in the *backward direction.*

Now, if $t^2 \in$ **B,** there is a path from s^2 to t^2 such that $u_{ij} + f^1_{ji} - f^2_{ij} > 0$ for all arcs on the path and where the positive direction of arc (i, j) is from i to j. We shall call this a *backward path* from s^2 to t^2. Similarly, if $t^2 \in$ **F,** then $u_{ij} + f^1_{ij} - f^2_{ij} > 0$ for all arcs on the path. This is called a *forward path.* A pair of forward and backward paths from s^2 to t^2 is called a *double path.* If the positive direction of all arcs in a path is from i to j, the *capacity of a backward path* is

$$\alpha_B = min\left\{u_{ij} + f^1_{ji} - f^2_{ij}\right\} \tag{8-46}$$

and the *capacity of a forward path* is

$$\alpha_F = min\left\{u_{ij} + f^1_{ij} - f^2_{ij}\right\} \tag{8-47}$$

These are simply the maximum amounts that can be shipped along these paths.

Let

$$\Delta^1_F = min\left\{f^1_{ij} > 0\right\}, \ f^1_{ij} \text{ is on a forward path with } f^1_{ij} \text{ in the forward direction.}$$

$$\Delta^1_B = min\left\{f^1_{ij} > 0\right\}, \ f^1_{ij} \text{ is on a backward path with } f^1_{ij} \text{ in the backward direction.}$$

Finally, for any double path, let

$$\Delta^1 = min\left\{\Delta^1_F, \Delta^1_B\right\} \tag{8-48}$$

$$\alpha = min\left\{\alpha_B; \alpha_F\right\} \tag{8-49}$$

We may now state an algorithm for finding maximal two-commodity flows in an undirected network.

Step 0*:* Find the maximal flow for commodity 1 using any single-commodity algorithm.

Step 1*:* Compute residual capacities u^2_{ij}, and find the maximal flow for commodity 2.

Step 2*:* Locate double paths from s^2 to t^2. If none exist, we have the maximal flow. Otherwise, let $\varepsilon = min\left\{\Delta^1, \frac{1}{2}\alpha\right\}$. Send ε units of commodity 1 from s^2 to t^2 along the backward path and from t^2 to s^2 along the forward path.

Step 3*:* Increase the total flow of commodity 2 by sending ε units of commodity 2 from s^2 to t^2 along both the forward and backward paths. Return to step 2.

What the algorithm essentially does is to redistribute the flow of commodity 1 along a double path so that the flow of commodity 2 can be increased along each of these paths.

8.5.1 A Two-Commodity Flow Problem

Let us solve the two-commodity problem for the network shown in Figure 8.12. In this figure all arcs have capacity equal to 1 unit.

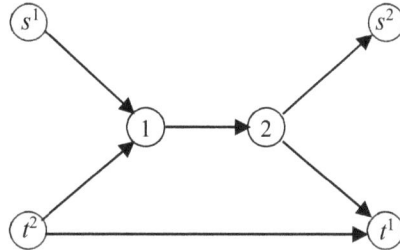

Figure 8.12. Two-Commodity Network with arc capacities equal to 1.

8.5.2 Algorithmic Steps

***Step* 0.** The maximal flow for commodity 1 is equal to 1 ($\square\!\!\rightarrow$ denotes commodity 1, $\Delta\!\!\rightarrow$ denotes commodity 2):

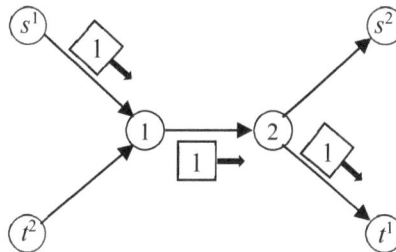

***Step* 1.** The residual capacities u_{ij}^2 are shown below. The maximal flow from s^2 to t^2 is zero.

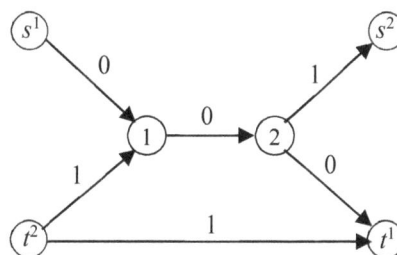

***Step* 2.** To locate double paths, we will determine the sets **F** and **B,**

$$\mathbf{F} = \{s_2, 2, t_1, t_2\}$$
$$\mathbf{B} = \{s_2, 2, t_1, t_2\}$$

Since $t_2 \in \mathbf{F}$ and $t_2 \in \mathbf{B},$ we have a pair of double paths. The forward path is $s^2\text{-}2\text{-}t^1\text{-}t^2$ with capacity equal to

$$\alpha_F = min\ [1, 2, 1] = 1$$

The backward path is $s^2\text{-}2\text{-}1\text{-}t^2$ with capacity equal to

316

$$\alpha_B = min\ [1, 2, 1] = 1$$

Thus $\alpha = min\ [\alpha_F, \alpha_B] = 1$. Also, $\Delta_F^1 = 1$, $\Delta_B^1 = 1$. Therefore, $\Delta^1 = min\{\Delta_F^1, \Delta_B^1\} = 1$. We have $\varepsilon = 1/2$. To complete step 2, send 1/2 unit of commodity 1 from s^2 to t^2 along the backward path and from t^2 to s^2 along the forward path as follows:

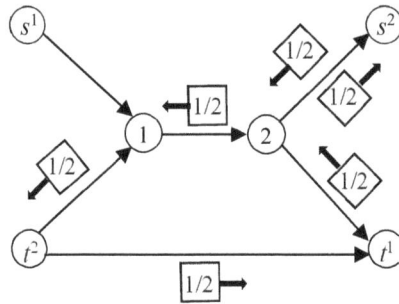

Step 3. Send 1/2 unit of commodity 2 from s^2 to t^2 along both paths:

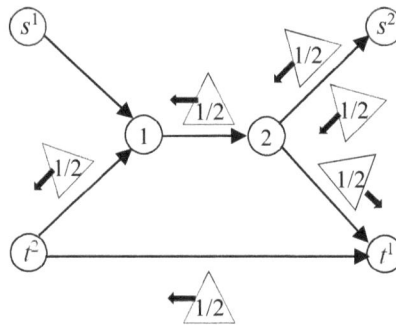

The current flow pattern is shown in Figure 8.13.

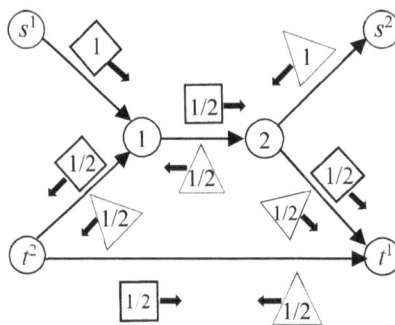

Figure 8.13. Optimal solution.

When we return to step 2, we find that no double paths exist; therefore, this is the optimal solution. Notice that ε is either an integer or a multiple of 1/2; thus, all flows will be integer or multiples of 1/2. If all capacities are even integers, the optimal flow will be all-integer.

8.6 MAXIMAL FLOWS AND FUNNEL NODES

A not uncommon logistics problem, particularly in military operations, can be described as follows. Supplies are located at one point for shipment to a destination. Vehicles that can transport the supplies are stationed at a different facility. Given a transportation network with limited capacity, one seeks an optimal routing of vehicles from their origin to the supply pickup point and then to the ultimate destination. A related problem that has a similar structure is that of establishing a central message center for an existing communication network through which all messages must pass. The maximum possible message flow must be maintained.

In both of these examples, a special node is singled out through which all flow from source to sink must pass. This node is called a *funnel node*. If we let f_{ij}^1 be the flow of the commodity from node i to j on its way to the funnel node, and f_{ij}^2, the flow from i to j after passing through the funnel node, we have the mathematical formulation in (8-69) to (8-74). In this model, s, a, and t represent the source, funnel node, and sink, respectively.

$$\text{maximize } v \tag{8-69}$$

$$\text{subject to}$$

$$\sum_j \left(f_{ij}^1 - f_{ji}^1 \right) = \begin{cases} v^1 & \text{if } i = s \\ 0 & \text{if } i \neq s, a \\ -v^1 & \text{if } i = a \end{cases} \tag{8-70}$$

$$\sum_j \left(f_{ij}^2 - f_{ji}^2 \right) = \begin{cases} v^2 & \text{if } i = a \\ 0 & \text{if } i \neq a, t \\ -v^2 & \text{if } i = t \end{cases} \tag{8-71}$$

$$\left| f_{ij}^1 \right| + \left| f_{ij}^2 \right| \leq u_{ij} \tag{8-72}$$

$$v = v^1 = v^2 \tag{8-73}$$

$$f_{ij}^k \geq 0 \tag{8-74}$$

Note that this is similar to the two-commodity flow problem on an undirected network. In this case, however, there are not really two commodities; they only represent different types of flow. In addition, the flows of "commodities" 1 and 2 are required to be equal. In the two-commodity problem we maximize the sum of the commodity flows; in this case we maximize $\frac{1}{2}(v^1 + v^2)$.

We can show that the maximal funnel node flow is equal to

$$\min \left\{ \overline{v}^1, \overline{v}^2, \frac{1}{2} \max [v^1 + v^2] \right\}$$

where $\overline{v}^1$ and $\overline{v}^2$ are the maximal flows from s to a and from a to t, respectively. This result leads to a very simple algorithm, which depends only upon solving a sequence of ordinary maximum-flow problems.

***Step* 1:** Find $\overline{v}^1$ and $\overline{v}^2$ using a single-commodity flow algorithm.

***Step* 2:** Construct a new network G' by adding an additional node s' and arcs (s', s), (s', t) having infinite capacity. Let $\bar{v}$ be the maximal flow in this network from s' to a.

***Step* 3:** Determine $v^* = min\left\{\bar{v}^1, \bar{v}^2, \frac{1}{2}\bar{v}\right\}$. If $v^* = 0$, stop. No flow is possible.

***Step* 4:** Construct a new network G'' by adding a node s'' to the original network along with arcs (s'', s) and (s'', t), each with capacity v^*. Find the maximal flow in this network from s'' to a. Decompose the flow into a flow from s'' through s to a, and a flow from a through t to s''. Remove the additional nodes and arcs. The result is a maximal funnel node flow through a.

Notice that Step 2 determines $\bar{v} = max\ \{v^1 + v^2\}$ by solving a single-commodity problem. It is easy to preserve the identities of "commodities" since any flow on a path from s' through s to a can be called "commodity 1" and from s' through t to a can be called "commodity 2". Step 4 simply finds a flow pattern having all of flow v^* passing through node a.

8.7 APPLICATIONS OF MULTICOMMODITY NETWORKS

Multicommodity network models have obvious applications in communication systems, railway transportation, production and distribution planning, and military logistics to name just a few. In this section we shall examine some practical applications of multicommodity network models and introduce some new formulations of certain multicommodity problems.

8.7.1 Tanker Scheduling

One problem that often arises in the logistics area is optimal scheduling of a fleet of vehicles. Suppose that a company owns a fixed fleet of non-homogeneous tankers, which differ in speeds, capacities, and operating costs. We associate a utility with the delivery of a shipment by a particular type of tanker on a specific delivery date that reflects the desirability and cost of making that shipment. In addition, there is a cost (negative utility) of reassigning a tanker from a delivery port to a port of origin of a new shipment. The objective is to determine a schedule and routing of the fleet with maximal utility.

The problem may be viewed as a network where each node j corresponds to a receipt of a shipment at a delivery port on one of its allowable delivery dates; each node i corresponds to an origin port at a time equal to the delivery date minus the transit time. Transit times are assumed to be deterministic. Arcs (i, j) correspond to shipments and (j, i) to reassignments from delivery points to origins. Each type of tanker originates at a source node s^k; thus, arcs (s^k, i) correspond to initial deployment of the tankers. A pseudo-sink t is included and arcs (j, t) represent removal from service of all tankers. Finally, the total flow in all arcs corresponding to alternative shipping dates and tanker types for a given shipment must be restricted to the amount being shipped.

Mathematically, we have the linear network model formulated in Eqs. (8-60) through (8-63).

$$maximize\ z = \sum_k \sum_i \sum_j c_{ij}^k x_{ij}^k \qquad (8\text{-}60)$$

subject to

$$\sum_k \sum_{A_v} r_{ij}^k x_{ij}^k \le b_v \qquad (8\text{-}61)$$

$$\sum_j \left(x_{ij}^k - x_{ji}^k \right) = \begin{cases} d^k & \text{If } i=s^k \\ 0 & \text{If } i \ne s^k, t \\ -d^k & \text{If } i=t \end{cases} \qquad (8\text{-}62)$$

$$0 \le x_{ij}^k \le u_{ij}^k \qquad (8\text{-}63)$$

In this model, c_{ij}^k represents the utility associated with a particular tanker, shipment, and route. A_v represents the set of feasible routes for a given shipment; and r_{ij}^k represents the capacity of tanker type k over route (i, j). Thus, the total amount shipped cannot exceed the demand of the shipment. The remaining constraints represent conservation of flow on the number of tankers, and u_{ij}^k will either be 0 or ∞, depending on whether a particular tanker type is restricted from use or not.

To illustrate the application of the above model, consider the problem with data given in Table 8.9. Two tankers of each type are available. The network representation is shown in Figure 8.14. Note that in this model class, nodes represent location and time combinations.

Table 8.9. DATA FOR ILLUSTRATIVE EXAMPLE

Shipment	Available at Time	Feasible Delivery Dates
1	0	4, 5
2	2	6, 7, 8
3	3	7, 8

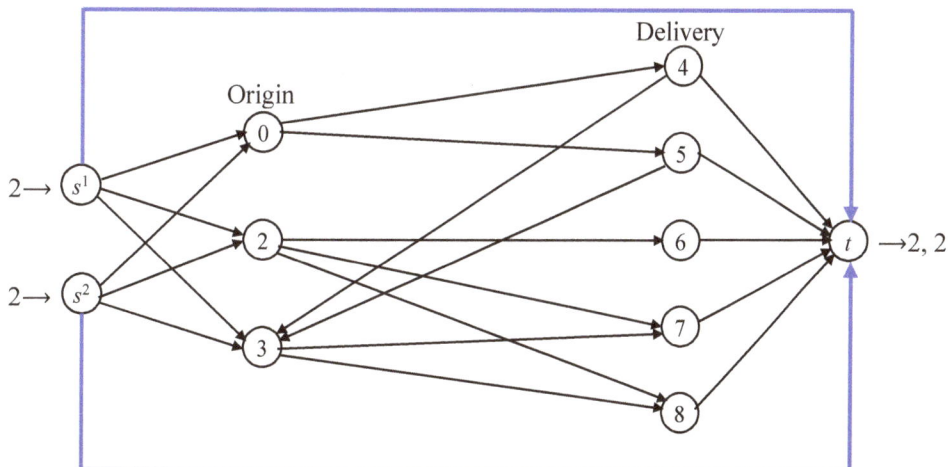

Figure 8.14. Tanker scheduling problem.

8.7.2 Urban Transportation Planning

Many multicommodity network models arise in planning urban transportation systems. Nodes represent zones or areas in a city while arcs correspond to streets or other types of roadways. Transportation demand is given by a trip matrix, $\mathbf{D}$, in which d_{ij} denotes the number of vehicles that travel from zone i to zone j during a fixed time interval. Each arc has a fixed capacity u_{ij} which may be increased (by road improvements for example) at a cost of c_{ij}. Improvements are restricted by a total budget, B. "Commodities" are the flows from each origin to all destinations. Let x_{ij}^k be the number of vehicles on arc (i, j) originating at node k, y_{ij} the improvement in capacity, and let the travel time on arc (i, j) be some function of the flow:

$$f_{ij}\left(\sum_k x_{ij}^k\right)$$

The network design problem is to determine which arcs to improve, and to find the flows on each arc so that total travel time is minimized. Mathematically, we have

$$\text{minimize} \sum_{(i,j)} f_{ij}\left(\sum_k x_{ij}^k\right) \tag{8-64}$$

subject to

$$\sum_j x_{ji}^k - \sum_j x_{ij}^k = d_{ki} \tag{8-65}$$

$$\sum_k x_{ij}^k \leq u_{ij} + y_{ij} \tag{8-66}$$

$$\sum_{(i,j)} c_{ij} y_{ij} \leq B \tag{8-67}$$

$$x_{ij}^k, y_{ij} \geq 0 \tag{8-68}$$

The model formulated in (8-64) to (8-68) makes a critical assumption regarding user behavior. This is based on a classic principle of distribution of traffic presented by Wardrop [22]: users choose routes so that the sum of journey times over the system is minimal. Typically, the objective function (8-64) is nonlinear and convex.

8.7.3 Computer Communication Models

Many applications in computer networks closely resemble the urban transportation model we have just discussed. In this case, arcs represent transmission lines and nodes represent terminals, systems, storage devices, and so on. Each arc has a fixed capacity u_{ij} which may be increased by y_{ij} units at a cost of c_{ij}, up to a maximum b_{ij}. Commodities again represent flows (messages) from an origin to all destinations. The traffic rate matrix $\mathbf{D}$ represents the transmission rate d_{ij} from node i to node j. The unit cost of transmission over arc (i, j) is e_{ij}.

The mathematical model can be formulated as follows:

$$\text{minimize} \sum_k \sum_{(i,j)} e_{ij} x_{ij}^k + \sum_{(i,j)} c_{ij} y_{ij}$$

subject to

$$\sum_j x_{ji}^k - \sum_j x_{ij}^k = d_{ij}$$

$$\sum_k x_{ij}^k \le u_{ij} + y_{ij}$$
$$y_{ij} \le b_{ij}$$
$$x_{ij}^k, y_{ij} \ge 0$$

One use of this model is to determine the optimal capacity of a network that satisfies demand at least cost. In this case, e_{ij} is usually 0, $u_{ij} = 0$, and b_{ij} is very large. A second use is to consider a network with fixed capacities as already given, and to determine the optimal routing of messages to minimize cost. In this case, $b_{ij} = 0$.

8.8 NOTES AND REMARKS

Early Surveys of multicommodity flow problems and various solution techniques have been written by Kennington [16] and Assad [2]. These also provide comprehensive bibliographies on the subject. Aspects of unimodularity in multicommodity flow problems and transformation to equivalent single-commodity problems were first proposed by Evans, Jarvis, and Duke [6], and extended by Evans and Jarvis [4] and Evans [6-8].

Aggregation results are generalizations of work performed by Geoffrion [13] and Zipkin [23] and were developed by Evans [11, 12]. Results for completely planar networks are due to Sakarovitch [19]. Hu [14] developed the algorithm for two-commodity flows in undirected networks. The funnel-node flow application is due to Jarvis and Miller [15]. Other applications in this section are discussed in Kennington's survey [16]; the tanker scheduling problem is due to Bellmore, Bennington, and Lubore [4]. References to urban design and computer networks can be found in Kennington's survey [16].

Wang [21] overviews two classes of solution methodologies. The first class consists of primal and dual based solution methods, including basis partitioning methods, resource-directive methods, price-directive methods, primal-dual methods, methods for integral MCNF problems, and heuristic procedures. The second class includes, among others, approximation methods, interior-point methods, and convex programming methods. Basis Partitioning methods exploit the fact that the multicommodity flow problem is a specially structured LP with embedded network flow problems.

The *network simplex method* is used to solve any single-commodity flow problem by generating a sequence of *spanning tree basic solutions*. Additional arcs are required to coordinate these solutions to satisfy the bundle constraints. Kennington and Helgason [17] illustrate this method on a two-commodity network problem.

PART II

MULTICOMMODITY NETWORK FLOW COST MINIMIZATION PROCEDURES

8.9 MATHEMATICAL FORMULATION

The linear programming model for the ***multicommodity cost minimization problem*** can be formulated using matrix notation as in Eqs. (8-27) through (8-31).

$$\textit{minimize} \sum_k \mathbf{c}^k \mathbf{x}^k \qquad (8\text{-}27)$$

$$\textit{subject to}$$

$$\mathbf{A}\mathbf{x}^k = \mathbf{b}^k, \quad k = 1,\ldots,K \qquad (8\text{-}28)$$

$$\sum_k \mathbf{x}^k + \mathbf{s} = \mathbf{u} \qquad (8\text{-}29)$$

$$\mathbf{0} \le \mathbf{x}^k \le \mathbf{u}^k, \, k = 1, \ldots, K \qquad (8\text{-}30)$$

$$\mathbf{s} \ge \mathbf{0} \qquad (8\text{-}31)$$

The coefficient matrix $\mathbf{A}$ in Constraint (8-28) is the node-arc incidence matrix for the network. This constraint is the *net flow condition for each commodity* for the nodes of the network. In Constraint (8-29) $\mathbf{s}$ is a vector of slack variables. The dual variable for this constraint is λ.

To illustrate the formulation of the model represented by Eqs. (8-27)-(8-31) the network shown in Figure 8.15 will be considered. This network has two source nodes, 1 and 2, and two terminal nodes, 3 and 4, for commodity 1. The second commodity has a source node 2 and a terminal node 3. The network is assumed to have zero lower bounds on their arc flows.

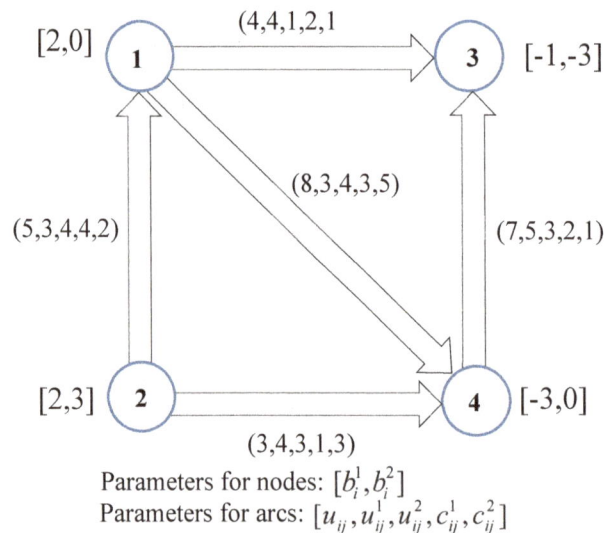

Parameters for nodes: $[b_i^1, b_i^2]$
Parameters for arcs: $[u_{ij}, u_{ij}^1, u_{ij}^2, c_{ij}^1, c_{ij}^2]$

Figure 8.15. Two commodity minimum cost flow problem.

The LP model for this problem is formulated below.

$$Minimize \quad 2x_{13}^1 + 3x_{14}^1 + 4x_{21}^1 + x_{24}^1 + 2x_{43}^1 + x_{13}^2 + 5x_{14}^2 + 2x_{21}^2 + 3x_{24}^2 + x_{43}^2$$

subject to

$$x_{13}^1 + x_{14}^1 - x_{21}^1 = 2$$

$$x_{21}^1 + x_{24}^1 = 2$$

$$-x_{13}^1 - x_{43}^1 = -1$$

$$-x_{14}^1 - x_{24}^1 + x_{43}^1 = -3$$

$$x_{131}^2 + x_{14}^2 - x_{21}^2 = 0$$

$$x_{21}^2 + x_{24}^2 = 3$$

$$-x_{13}^2 - x_{43}^2 = -3$$

$$-x_{14}^2 - x_{24}^2 + x_{43}^2 = 0$$

$$x_{13}^1 + x_{13}^2 + s_1 = 4$$

$$x_{14}^1 + x_{14}^2 + s_2 = 8$$

$$x_{21}^1 + x_{21}^2 + s_3 = 5$$

$$x_{24}^1 + x_{24}^2 + s_4 = 3$$

$$x_{43}^1 + x_{43}^2 + s_5 = 7$$

$$x_{13}^1 \le 4, x_{14}^1 \le 3, x_{21}^1 \le 3, x_{24}^1 \le 4, x_{43}^1 \le 5$$

$$x_{13}^2 \le 1, x_{14}^2 \le 4, x_{21}^2 \le 4, x_{24}^2 \le 3, x_{43}^2 \le 3$$

$$x_{13}^1, x_{14}^1, x_{21}^1, x_{24}^1, x_{43}^1 \ge 0$$

$$x_{13}^2, x_{14}^2, x_{21}^2, x_{24}^2, x_{43}^2 \ge 0$$

$$s_1, s_2, s_3, s_4, s_5 \ge 0$$

Additionally, the constraint matrix for the LP model is shown in Table 8.10. In this table, the constraints are arranged in the following order:

1. Net-flow conditions for Commodity 1.
2. Net-flow conditions for Commodity 2.
3. Bundle constraints (coupling constraints).
4. Upper-bound constraints for Commodity 1.
5. Upper-bound constraints for Commodity 2.

Table 8.10. CONSTRAINT MATRIX OF EXAMPLE

Description	x^1_{13}	x^1_{14}	x^1_{21}	x^1_{24}	x^1_{43}	x^2_{13}	x^2_{14}	x^2_{21}	x^2_{24}	x^2_{43}	s_1	s_2	s_3	s_4	s_5	RHS	Vector
Node-arc incidence matrix for subproblem 1	1	1	-1	0	0	0	0	0	0	0	0	0	0	0	0	2	$\mathbf{b}^1$
	0	0	1	1	0	0	0	0	0	0	0	0	0	0	0	2	
	-1	0	0	0	-1	0	0	0	0	0	0	0	0	0	0	-1	
	0	-1	0	-1	1	0	0	0	0	0	0	0	0	0	0	-3	
Node-arc incidence matrix for subproblem 2	0	0	0	0	0	1	1	-1	0	0	0	0	0	0	0	0	$\mathbf{b}^2$
	0	0	0	0	0	0	0	1	1	0	0	0	0	0	0	3	
	0	0	0	0	0	-1	0	0	0	-1	0	0	0	0	0	-3	
	0	0	0	0	0	0	-1	0	-1	1	0	0	0	0	0	0	
Bundle constraints	1	0	0	0	0	1	0	0	0	0	1	0	0	0	0	4	$\mathbf{u}$
	0	1	0	0	0	0	1	0	0	0	0	1	0	0	0	8	
	0	0	1	0	0	0	0	1	0	0	0	0	1	0	0	5	
	0	0	0	1	0	0	0	0	1	0	0	0	0	1	0	3	
	0	0	0	0	1	0	0	0	0	1	0	0	0	0	1	7	
Upper bound constraints for commodity 1	1	0	0	0	0	0	0	0	0	0	0	0	0	0	0	4	$\mathbf{u}^1$
	0	1	0	0	0	0	0	0	0	0	0	0	0	0	0	3	
	0	0	1	0	0	0	0	0	0	0	0	0	0	0	0	3	
	0	0	0	1	0	0	0	0	0	0	0	0	0	0	0	4	
	0	0	0	0	1	0	0	0	0	0	0	0	0	0	0	5	
Upper bound constraints for commodity 2	0	0	0	0	0	1	0	0	0	0	0	0	0	0	0	1	$\mathbf{u}^2$
	0	0	0	0	0	0	1	0	0	0	0	0	0	0	0	4	
	0	0	0	0	0	0	0	1	0	0	0	0	0	0	0	4	
	0	0	0	0	0	0	0	0	1	0	0	0	0	0	0	3	
	0	0	0	0	0	0	0	0	0	1	0	0	0	0	0	3	

Figure 8.16 shows the LINDO linear programming formulation. It is noted that LINDO assumes the non-negativity condition for all variables. Additionally, the inequality symbol < is assumed to represent the symbol $\leq$. Both of these assumptions are made in LINDO to facilitate the formulation of LP models.

Figure 8.17 shows the LINDO optimal solution. To solve this problem, we used the VB program **ex_mps** which creates a LINDO environment to solve the LP model stored in a file named **multi.ltx**. The optimal solution to this problem, as shown in Figure 8.17 is:

$$x^1_{13} = 1 \quad x^2_{13} = 1 \quad s_1 = 2$$
$$x^1_{14} = 1 \quad x^2_{14} = 1 \quad s_2 = 6$$
$$x^1_{21} = 0 \quad x^2_{21} = 2 \quad s_3 = 3$$
$$x^1_{24} = 2 \quad x^2_{24} = 1 \quad s_4 = 0$$
$$x^1_{43} = 0 \quad x^2_{43} = 2 \quad s_5 = 5$$

```
min 2x131 + 3x141 + 4x211 + x241 + 2x431 + x132 + 5x142 + 2x212 +3 x242 + x432
s.t.
x131 + x141 - x211 = 2
x211 + x241 =2
-x131 - x431 = -1
-x141 - x241 + x431 = -3
x132 + x142 - x212 = 0
x212 + x242 = 3
-x132 - x432 = -3
-x142- x242 + x432 = 0
x131 + x132 + s1 = 4
x141 + x142 + s2 = 8
x211 + x212 + s3 = 5
x241 + x242 + s4 = 3
x431 + x432 + s5 = 7
x131 < 4
x141 < 3
x211 < 3
x241 < 4
x431 < 5
x132 < 1
x142 < 4
x212 < 4
x242 < 3
x432 < 3
end
```

Figure 8.16. Model Formulation (multi.ltx).

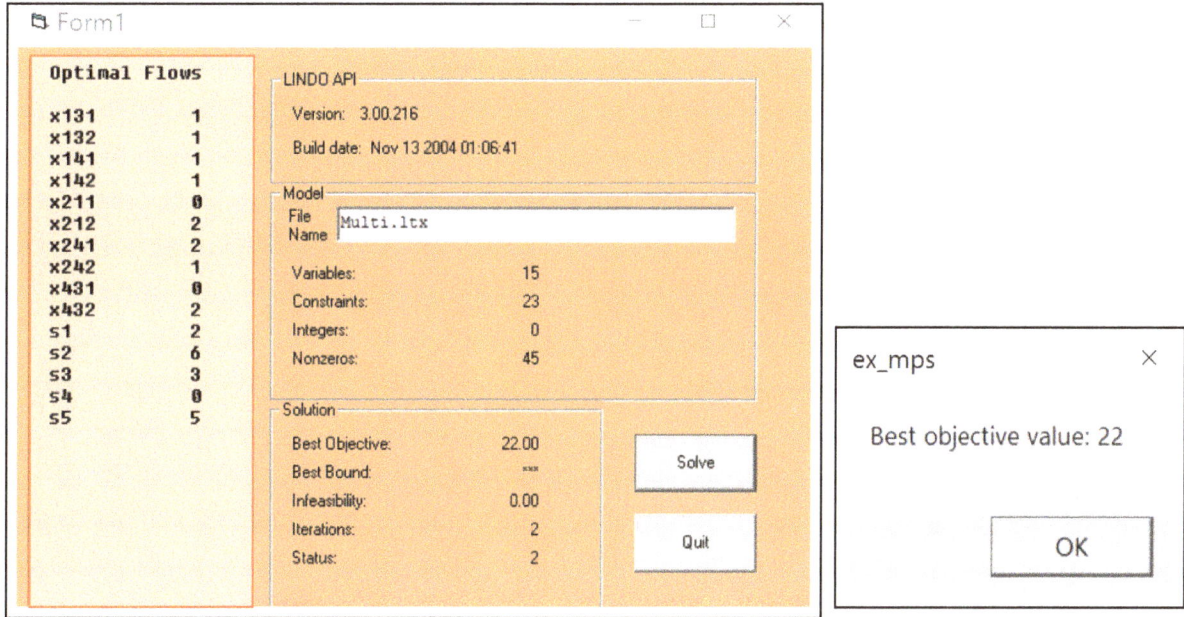

Figure 8. 17. Optimal Solution from *ex_msp* VB Code.

8.10 SOLUTION PROCEDURES

Although multicommodity flow problems do not have the same nice properties as single-commodity flow problems they still have a special structure that can be exploited to solve the problem efficiently. There are several common approaches for solving the multicommodity flow problem. The solution methods generally attempt to exploit the network flow structure of the

326

individual single commodity flow problems. In Section 8.8 we cite several bibliographical references that document some of these procedures. More recently, Wang has written two outstanding survey articles on applications, formulations, and solution procedures for multicommodity network problems [20,21]. One paper focuses on applications and the second one discusses solution procedures. In particular, the second paper addresses the conventional methods such as price-directive, resource-directive, and basis partitioning methods and then summarizes recent progress regarding the application of approximation methods, interior-point algorithms, quadratic programming algorithms, and heuristics to solve either the linear or integral network multicommodity flow problems. The computational performance of different solution methods is compared and some research directions are suggested.

Although a full discussion of available procedures for solving linear multicommodity network problems is beyond the scope of this chapter, in the following sections we will briefly explain the underlying concepts and procedures of the Dantzig-Wolfe *decomposition method* [5] and the *Lagrangian relaxation method* combined with *subgradient optimization*. The decomposition method creates a master optimization problem that coordinates the solution of a collection of minimal cost network flow subproblems. The Lagragian relaxation method decomposes the problem into a sequence of subproblems whose solution converges to the optimal solution of the original problem. Decomposition methods can be classified as *price-directive* and *resource-directive* methods.

8.10.1 Price-Directive Dantzig-Wolfe Decomposition Method

The price-directive decomposition method considers the dual variables associated with the bundle constraints as *prices* and brings theme into the objective function. It removes the complicating capacity constraint and charges each commodity for the use of the arc. The objective is to obtain a set of prices such that the combined solution for all subproblems yields an optimum for the original problem [3,16,17].

For each commodity $k = 1, \ldots, K$, let $\mathbf{X}_k = \{\mathbf{x}^k : \mathbf{A}\mathbf{x}^k = \mathbf{b}^k, \ \mathbf{0} \le \mathbf{x}^k \le \mathbf{u}^k\}$ and let $\mathbf{x}_j^k, j = 1, \ldots, q$, denote the *extreme points* of $\mathbf{X}_k$. If $\mathbf{X}_k$ is the null set for any k, then the original problem has no solution. Now, let λ be the dual variable associated with the bundle constraints (8-17) and let $\mathbf{X}_k$ be non-empty and bounded; then for any $\mathbf{x}^k \in \mathbf{X}_k$, $\mathbf{x}^k = \sum_j \lambda_j^k \mathbf{x}_j^k$, where $\sum_j \lambda_j^k = 1$, and $\lambda_j^k \ge 0$. Substituting these results in the multicommodity minimal cost flow problem, we obtain the following *master problem*.

$$minimize \ \sum_{j,k} \mathbf{c}^k \mathbf{x}_j^k \lambda_j^k \tag{8-32}$$

$$subject \ to$$

$$\sum_{j,k} \mathbf{x}_j^k \lambda_j^k + \mathbf{s} = \mathbf{u} \qquad (\mathbf{w}) \tag{8-33}$$

$$\sum_j \lambda_j^k = 1, \text{ all } k \qquad (\alpha_k) \tag{8-34}$$

$$\lambda_j^k \ge 0, \text{ all } j, k. \tag{8-35}$$

where $\mathbf{w}$ and α_k are the dual variables associated with Constraints (8-33) and (8-34), respectively. The optimality conditions are:

(a) $w_m \leq 0$ for each slack variable s_m

(b) $(\mathbf{w} - \mathbf{c}^k)\mathbf{x}_j^k + \alpha_k \leq 0$ for each λ_j^k

Any variable violating any of these conditions is a candidate to enter the master basis. In order to determine the variable associated with the *largest violation* of the optimality condition we solve the single-commodity *subproblem* formulated in Eqs. (8-36)-(8-38).

$$minimize \ (\mathbf{c}^k - \mathbf{w}) \ \mathbf{x}^k \tag{8-36}$$
$$subject \ to$$
$$\mathbf{A}\mathbf{x}^k = \mathbf{b}^k \tag{8-37}$$
$$\mathbf{0} \leq \mathbf{x}^k \leq \mathbf{u}^k \tag{8-38}$$

The master problem is solved by the *revised simplex method*. The subproblems are solved to test for optimality and select candidates for entering the master problem basis. The subproblems are single-commodity problems and can be solved efficiently by any well-known techniques such as out-of-kilter algorithm or primal simplex algorithm for network optimization. The *price-directive decomposition algorithm* can be summarized as follows [1,19].

Initialization.

Find an initial feasible basis for the master problem. The inverse basis matrix $\mathbf{B}^{-1}$ and the vector $(\mathbf{u}, \mathbf{1})$ are used to obtain $\mathbf{b'} = \mathbf{B}^{-1} (\mathbf{u}, \mathbf{1})^{\mathrm{T}}$. Calculate the flow costs $\mathbf{c}^k \mathbf{x}^k$ for all commodities $(k = 1, \ldots, K)$. The values of the corresponding dual variables are found as $(\mathbf{w}, \boldsymbol{\alpha}) = \hat{\mathbf{c}}_B \mathbf{B}^{-1}$, where $\hat{\mathbf{c}}_j^k = \mathbf{c}^k \mathbf{x}_j^k$ for each λ_j^k variable. If a feasible basis is not available, then use artificial variables and a two-phase method to find a starting feasible solution for the master problem. The revised simplex array is set up using the following table layout:

$(\mathbf{w},\boldsymbol{\alpha})$	$\hat{\mathbf{c}}_B \mathbf{B}^{-1} \mathbf{b'}$
$\mathbf{B}^{-1}$	$\hat{\mathbf{c}}_B \mathbf{B}^{-1}$

Step 1. *Pricing Operation.*

Let $\mathbf{x}^k$ denote an optimum extreme point for $z_k = min \ \{(\mathbf{c}^k - \mathbf{w}) \ \mathbf{x}^k : \mathbf{A}\mathbf{x}^k = \mathbf{b}^k, \ \mathbf{0} \leq \mathbf{x}^k \leq \mathbf{u}^k \}$. If there is no solution, then the master problem has no solution. If $(\mathbf{w} - \mathbf{c}^k) \ \mathbf{x}_j^k + \alpha_k > 0$ (or $\alpha_k - z_k > 0$), then λ^k is a candidate to enter the basis of the master. Otherwise, no extreme point of $\mathbf{X}_k$ is a candidate for basis entry. If $w_m > 0$, then the corresponding slack, s_m, is a candidate for basis entry. If there is at least one candidate for basis entry continue with step 2. If not, an optimal solution is available and the algorithm is terminated.

Step 2. Pivot Operation in Master Problem.

Select an eligible entering variable, update the chosen column, perform pivot operation in the master program, and return to step 1 with a new set of dual variables. The chosen column is updated by $\mathbf{B}^{-1}(\mathbf{e}_{pq}, \mathbf{0})^T$, where $\mathbf{e}_{pq}$ is a unit vector with the entry equal to 1 in the row associated with arc (p, q).

8.10.2 An Example of the Dantzig-Wolfe Decomposition Method

We will consider the numerical example shown in Figure 8.15 with constraint matrix given in Table 8.8 to illustrate the application of this method.

Initialization

In this example we will use as initial feasible solutions $\mathbf{x}_1^1$ and $\mathbf{x}_1^2$. The superscripts in this notation correspond to the commodities and the subscripts represent the *order* of the solutions for the corresponding commodities. The *first* solution for Commodity 1 is $\mathbf{x}_1^1 = (0, 2, 0, 2, 1)^T$ and the *first* solution for Commodity 2 is $\mathbf{x}_1^2 = (0, 2, 2, 1, 3)^T$, where $\mathbf{x}_i^k = (x_{13}^k, x_{14}^k, x_{21}^k, x_{24}^k, x_{43}^k)^T$ for $k = 1, 2$. The cost vectors for the two commodities are $\mathbf{c}^1 = (c_{13}^1, c_{14}^1, c_{21}^1, c_{24}^1, c_{43}^1) = (2, 3, 4, 1, 2)$ and $\mathbf{c}^2 = (c_{13}^2, c_{14}^2, c_{21}^2, c_{24}^2, c_{43}^2) = (1, 5, 2, 3, 1)$.

The master program is reformulated below for the specific initial feasible solution and cost data. The dual variables are listed in parentheses next to the corresponding constraints. It is noted that the *subscripts* $j = 1, 2, 3, 4, 5$ represent the arcs of the network. Arc 1 is (1,3), arc 2 is (1,4), arc 3 is (2,1), arc 4 is (2,4) and arc 5 is (4,3).

$$\text{minimize } 10\lambda_1^1 + 20\lambda_1^2$$

subject to

$$s_1 = 4 \qquad\qquad (w_1)$$
$$2\lambda_2^1 + 2\lambda_2^2 + s_2 = 8 \qquad\qquad (w_2)$$
$$2\lambda_3^2 + s_3 = 5 \qquad\qquad (w_3)$$
$$2\lambda_4^1 + \lambda_4^2 + s_4 = 3 \qquad\qquad (w_4)$$
$$\lambda_5^1 + 3\lambda_5^2 + s_5 = 7 \qquad\qquad (w_5)$$
$$\lambda_1^1 = 1 \qquad\qquad (\alpha_1)$$
$$\lambda_1^2 = 1 \qquad\qquad (\alpha_2)$$
$$\lambda_j^k \geq 0, \, j = 1, ..., 5; k = 1, 2$$

From the above master problem formulation, we can have the following basis matrix and its inverse.

$$\mathbf{B} = \begin{pmatrix} 1 & 0 & 0 & 0 & 0 & 0 & 0 \\ 0 & 1 & 0 & 0 & 0 & 2 & 2 \\ 0 & 0 & 1 & 0 & 0 & 0 & 2 \\ 0 & 0 & 0 & 1 & 0 & 2 & 1 \\ 0 & 0 & 0 & 0 & 1 & 1 & 3 \\ 0 & 0 & 0 & 0 & 0 & 1 & 0 \\ 0 & 0 & 0 & 0 & 0 & 0 & 1 \end{pmatrix} \quad \mathbf{B}^{-1} = \begin{pmatrix} 1 & 0 & 0 & 0 & 0 & 0 & 0 \\ 0 & 1 & 0 & 0 & 0 & -2 & -2 \\ 0 & 0 & 1 & 0 & 0 & 0 & -2 \\ 0 & 0 & 0 & 1 & 0 & -2 & -1 \\ 0 & 0 & 0 & 0 & 1 & -1 & -3 \\ 0 & 0 & 0 & 0 & 0 & 1 & 0 \\ 0 & 0 & 0 & 0 & 0 & 0 & 1 \end{pmatrix}$$

From the given initial basic solution, we have the following results:

$$\mathbf{c}^1 \mathbf{x}_1^1 = 10, \quad \mathbf{c}^2 \mathbf{x}_1^2 = 20$$

$$(\mathbf{w}, \boldsymbol{\alpha}) = \hat{\mathbf{c}}_B \mathbf{B}^{-1} = (0,0,0,0,0,10,20)\,\mathbf{B}^{-1} = (0,0,0,0,0,10,20)$$

$$\mathbf{b}' = \mathbf{B}^{-1}(\mathbf{u},\,1)^{\mathrm{T}} = \mathbf{B}^{-1}(4,8,5,3,7,1,1)^{\mathrm{T}} = (4,\,8,\,5,\,3,\,7,\,1,\,1)^{\mathrm{T}}$$

$$\mathbf{B}^{-1}\mathbf{b}' = (4,\,4,\,3,\,0,\,3,\,1,\,1)^{\mathrm{T}}$$

$$Z = \hat{\mathbf{c}}_B \mathbf{B}^{-1} \mathbf{b}' = (0,0,0,0,0,10,20)\,\mathbf{b}' = 30$$

The revised simplex array for the master problem is:

	w_1	w_2	w_3	w_4	w_5	α_1	α_2	RHS
Z	0	0	0	0	0	10	20	30
s_1	1	0	0	0	0	0	0	4
s_2	0	1	0	0	0	-2	-2	4
s_3	0	0	1	0	0	0	-2	3
s_4	0	0	0	1	0	-2	-1	0
s_5	0	0	0	0	1	-1	-3	3
λ_1^1	0	0	0	0	0	1	0	1
λ_1^2	0	0	0	0	0	0	1	1

Iteration 1.

In the above revised simplex tableau, all $w_j \le 0$. That is, all w_j satisfy dual feasibility. Therefore, we need to check whether a candidate from either subproblem 1 for Commodity 1 or Subproblem 2 for Commodity 2 is eligible for entering the basis of the master problem. These are single-commodity flow problems. If we find any $z_j^k - c_j^k > 0$, then λ_j^k can be a candidate to enter the basis. The following parameters for the nodes and arcs are obtained from Table 8.8 (constraint matrix).

Nodes	1	2	3	4	Arcs	(1,3)	(1,4)	(2,1)	(2,4)	(4,3)	Arcs	(1,3)	(1,4)	(2,1)	(2,4)	(4,3)
$\mathbf{b}^1$	2	2	-1	-3	$\mathbf{u}^1$	4	3	3	4	5	$\mathbf{c}^1$	2	3	4	1	2
$\mathbf{b}^2$	0	3	-3	0	$\mathbf{u}^2$	1	4	4	3	3	$\mathbf{c}^2$	1	5	2	3	1

Based on Eqs. (8-38)-(8-38), the formulations for the subproblems for Commodity 1 and

330

Commodity 2 are shown below. In these formulations, we use the following notation for the dual values associated with the slack variables: $w_1 = w_{13}, w_2 = w_{14}, w_3 = w_{21}, w_4 = w_{23}$, and $w_5 = w_{43}$. Furthermore, $x_1^k = x_{13}^k, x_2^k = x_{14}^k, x_3^k = x_{21}^k, x_4^k = x_{24}^k$ and $x_5^k = x_{43}^k$ for $k = 1, 2$.

Commodity 1

$$minimize\ (2 - w_{13})x_{13}^1 + (3 - w_{14})x_{14}^1 + (4 - w_{21})x_{21}^1 + (1 - w_{23})x_{24}^1 + (2 - w_{43})x_{43}^1$$

$$x_{13}^1 + x_{14}^1 - x_{21}^1 = 2$$

$$x_{21}^1 + x_{24}^1 = 2$$

$$-x_{13}^1 - x_{43}^1 = -1$$

$$-x_{14}^1 - x_{24}^1 + x_{43}^1 = -3$$

$$0 \le x_{13}^1 \le 4, 0 \le x_{14}^1 \le 3, 0 \le x_{21}^1 \le 3, 0 \le x_{24}^1 \le 4, 0 \le x_{43}^1 \le 5$$

Commodity 2

$$minimize\ (1 - w_{13})x_{13}^2 + (5 - w_{14})x_{14}^2 + (2 - w_{21})x_{21}^2 + (3 - w_{23})x_{24}^2 + (1 - w_{43})x_{43}^2$$

$$x_{13}^2 + x_{14}^2 - x_{21}^2 = 0$$

$$x_{21}^2 + x_{24}^2 = 3$$

$$-x_{13}^2 - x_{43}^2 = -3$$

$$-x_{14}^2 - x_{24}^2 + x_{43}^2 = 0$$

$$0 \le x_{13}^2 \le 1, 0 \le x_{14}^2 \le 4, 0 \le x_{21}^2 \le 4, 0 \le x_{24}^2 \le 3, 0 \le x_{43}^2 \le 3$$

Subproblem for *Commodity* 2

The single-commodity network flow cost minimization model is formulated below after setting $\mathbf{w} = (0, 0, 0, 0, 0)$.

$$minimize\ x_{13}^2 + 5x_{14}^2 + 2x_{21}^2 + 3x_{24}^2 + x_{43}^2$$

$$x_{13}^2 + x_{14}^2 - x_{21}^2 = 0$$

$$x_{21}^2 + x_{24}^2 = 3$$

$$-x_{13}^2 - x_{43}^2 = -3$$

$$-x_{14}^2 - x_{24}^2 + x_{43}^2 = 0$$

$$0 \le x_{13}^2 \le 1, 0 \le x_{14}^2 \le 4, 0 \le x_{21}^2 \le 4, 0 \le x_{24}^2 \le 3, 0 \le x_{43}^2 \le 3$$

The minimum cost optimal solution is $\mathbf{x}_2^2 = (1, 0, 1, 2, 2)^{\mathrm{T}}$ with cost equal to 11.

Now we proceed to calculate $\mathbf{c}_2 - \mathbf{w} = (0, 0, 0, 0, 0) - (1, 5, 2, 3, 1) = (-1, -5, -2, -3, -1)$ to check $z_2^2 - c_2^2 = (\mathbf{w} - \mathbf{c}_1)\mathbf{x}_2^2 + \alpha_2 = -11 + 20 = 9$. Since this value is greater than zero, λ_2^2 is a candidate to enter the basis. Next, we generate the column and perform the pivoting operation. The column for λ_2^2 is generated using $\mathbf{B}^{-1} (\mathbf{x}_2^2, \mathbf{e}^2)^{\mathrm{T}} = \mathbf{B}^{-1} (1, 0, 1, 2, 2, 0, 1)^{\mathrm{T}} = (1, -2, -1, 1, -1, 0, 1)^{\mathrm{T}}$. The current revised simplex array for the master problem with the new column is shown below. The result of the pivoting operation is shown to the right of the array.

	w_1	w_2	w_3	w_4	w_5	α_1	α_2	RHS	λ_2^2
Z	0	0	0	0	0	10	20	30	9
s_1	1	0	0	0	0	0	0	4	1
s_2	0	1	0	0	0	-2	-2	4	-2
s_3	0	0	1	0	0	0	-2	3	-1
s_4	0	0	0	1	0	-2	-1	0	(1)
s_5	0	0	0	0	1	-1	-3	3	-1
λ_1^1	0	0	0	0	0	1	0	1	0
λ_1^2	0	0	0	0	0	0	1	1	1

	w_1	w_2	w_3	w_4	w_5	α_1	α_2	RHS	λ_2^2
Z	0	0	0	-9	0	28	29	30	0
s_1	1	0	0	-1	0	2	1	4	0
s_2	0	1	0	2	0	-6	-4	4	0
s_3	0	0	1	1	0	-2	-3	3	0
λ_2^2	0	0	0	1	0	-2	-1	0	1
s_5	0	0	0	1	1	-3	-4	3	0
λ_1^1	0	0	0	0	0	1	0	1	0
λ_1^2	0	0	0	-1	0	0	1	1	0

Iteration 2.

After the pivoting operation, in the above revised simplex array, we obtain the result $\mathbf{w} =$ $(0,0,0,-9,0)$ and $\boldsymbol{\alpha} = (28,29)$. It is noted that $w_m \leq 0$ for each slack variable s_m, $m=1,2, \ldots, 5$. This means that we must proceed to solve another subproblem to find a candidate to enter the basis of the master problem.

Subproblem for *Commodity* 2

The single-commodity network flow cost minimization model for this subproblem is formulated below after setting $\mathbf{w} = (0, 0, 0, -9, 0)$. From the solution $\mathbf{x}_3^2$ to this subproblem we will determine $z_3^2 - c_3^2 = (\mathbf{w} - \mathbf{c}_2)\mathbf{x}_3^2 + \alpha_2$.

$$minimize \ x_{13}^2 + 5x_{14}^2 + 2x_{21}^2 + 12x_{24}^2 + x_{43}^2$$

$$x_{13}^2 + x_{14}^2 - x_{21}^2 = 0$$

$$x_{21}^2 + x_{24}^2 = 3$$

$$-x_{13}^2 - x_{43}^2 = -3$$

$$-x_{14}^2 - x_{24}^2 + x_{43}^2 = 0$$

$$0 \leq x_{13}^2 \leq 1, 0 \leq x_{14}^2 \leq 4, 0 \leq x_{21}^2 \leq 4, 0 \leq x_{24}^2 \leq 3, 0 \leq x_{43}^2 \leq 3$$

The minimum cost optimal solution is $\mathbf{x}_3^2 = (1, 2, 3, 0, 2)^T$ with cost equal to 19.

Now we proceed to calculate $\mathbf{w} - \mathbf{c}_2 = (0, 0, 0, -9, 0) - (1, 5, 2, 3, 1) = (-1, -5, -2, -12, -1)$. Based on this, we get $z_3^2 - c_3^2 = (\mathbf{w} - \mathbf{c}_2) \mathbf{x}_3^2 + \alpha_2 = -19 + 29 = 10$. Since $z_3^2 - c_3^2$ is greater than $z_3^1 - c_3^1$, we choose λ_3^2 as an entering variable to the master problem basis. Next, we generate the column and perform pivoting process. The column for λ_3^2 is generated by $\mathbf{B}^{-1} (\mathbf{x}_3^2, \mathbf{e}^2)^T = \mathbf{B}^{-1} (1, 2, 3, 0, 2, 0, 1)^T = (2, -2, 0, -1, -2, 0, 2)^T$. The revised simplex array for the master problem and the new column are shown below along with the result of the pivoting operation.

	w_1	w_2	w_3	w_4	w_5	α_1	α_2	RHS	λ_3^2
Z	0	0	0	0	0	28	29	30	10
s_1	1	0	0	-1	0	2	1	4	2
s_2	0	1	0	2	0	-6	-4	4	-2
s_3	0	0	1	0	0	0	-2	3	0
λ_2^2	0	0	0	1	0	-2	-1	0	-1
s_5	0	0	0	0	1	-3	-4	3	-2
λ_1^1	0	0	0	0	0	1	0	1	0
λ_1^2	0	0	0	-1	0	2	2	1	②︎

	w_1	w_2	w_3	w_4	w_5	α_1	α_2	RHS	λ_3^2
Z	0	0	0	-4	0	18	19	25	0
s_1	1	0	0	0	0	0	-1	3	0
s_2	0	1	0	1	0	-4	-2	5	0
s_3	0	0	1	1	0	-2	-3	3	0
λ_2^2	0	0	0	1/2	0	-1	0	1/2	0
s_5	0	0	0	0	1	-1	-2	4	0
λ_1^1	0	0	0	0	0	1	0	1	0
λ_3^2	0	0	0	-1/2	0	1	1	1/2	1

Iteration 3.

In the above revised simplex array, $w_m \leq 0$ for each slack variable s_m. Thus, we proceed to solve another subproblem to find a candidate to enter the master basis.

Subproblem for *Commodity* 2

The single-commodity network flow cost minimization model is formulated below after setting $\mathbf{w} = (0, 0, 0, -4, 0)$.

$$minimize\ x_{13}^2 + 5x_{14}^2 + 2x_{21}^2 + 7x_{24}^2 + x_{43}^2$$

$$x_{13}^2 + x_{14}^2 - x_{21}^2 = 0$$

$$x_{21}^2 + x_{24}^2 = 3$$

$$-x_{13}^2 - x_{43}^2 = -3$$

$$-x_{14}^2 - x_{24}^2 + x_{43}^2 = 0$$

$$0 \leq x_{13}^2 \leq 1, 0 \leq x_{14}^2 \leq 4, 0 \leq x_{21}^2 \leq 4, 0 \leq x_{24}^2 \leq 3, 0 \leq x_{43}^2 \leq 3$$

The minimum cost optimal solution is $\mathbf{x}_4^2 = (1, 2, 3, 0, 2)^T$ with cost equal to 19.

Next, we proceed to calculate $\mathbf{w} - \mathbf{c}_2 = (0, 0, 0, -4, 0) - (1, 5, 2, 3, 1) = (-1, -5, -2, -7, -1)$. Based on this we get $z_4^2 - c_4^2 = (\mathbf{w} - \mathbf{c}_2)\,\mathbf{x}_4^2 + \alpha_2 = -19 + 19 = 0$. Since $z_4^2 - c_4^2 = 0$ there is no candidate for this subproblem.

Iteration 4.

Since $w_m \leq 0$ for $m = 1, \ldots, 5$ we proceed to solve subproblems for Commodity 1 to find a candidate to enter the master basis.

Subproblem for *Commodity* 1

The single-commodity network flow cost minimization model is formulated below after setting $\mathbf{w} = (0, 0, 0, -4, 0)$.

$$minimize\ 2x_{13}^1 + 3x_{14}^1 + 4x_{21}^1 + 5x_{24}^1 + 2x_{43}^1$$

$$x_{13}^1 + x_{14}^1 - x_{21}^1 = 2$$

$$x_{21}^1 + x_{24}^1 = 2$$

$$-x_{13}^1 - x_{43}^1 = -1$$

$$-x_{14}^1 - x_{24}^1 + x_{43}^1 = -3$$

$$0 \le x_{13}^1 \le 4, 0 \le x_{14}^1 \le 3, 0 \le x_{21}^1 \le 3, 0 \le x_{24}^1 \le 4, 0 \le x_{43}^1 \le 5$$

The optimal solution is $\mathbf{x}_2^1 = (1,1,0,2,0)^T$ with cost equal to 15.

Next, we proceed to calculate $\mathbf{w} - \mathbf{c}^1 = (0, 0, 0, -4, 0) - (2, 3, 4, 1, 2) = (-2, -3, -8, -1, -2)$. Based on this, we find that $z_2^1 - c_2^1 = (\mathbf{w} - \mathbf{c}_1)\mathbf{x}_1^1 + \alpha_1 = -15 + 18 = 3$. Since this value is greater than zero, λ_2^1 is chosen as candidate to enter the basis, with column $\mathbf{B}^{-1}(\mathbf{x}_2^1, \mathbf{e}^1)^T = \mathbf{B}^{-1}(1, 1, 0, 2, 0, 1, 0)^T = (1, -1, 0, 0, -1, 1, 0)^T$. The revised simplex array and the new column are shown below.

	w_1	w_2	w_3	w_4	w_5	α_1	α_2	RHS	λ_2^1
Z	0	0	0	-4	0	18	19	25	3
s_1	1	0	0	0	0	0	-1	3	1
s_2	0	1	0	1	0	-4	-2	5	-1
s_3	0	0	1	1	0	-2	-3	3	0
λ_2^2	0	0	0	1/2	0	-1	0	1/2	0
s_5	0	0	0	0	1	-1	-2	4	-1
λ_1^2	0	0	0	0	0	1	0	1	(1)
λ_3^2	0	0	0	-1/2	0	1	1	1/2	0

	w_1	w_2	w_3	w_4	w_5	α_1	α_2	RHS	λ_2^1
Z	0	0	0	-4	0	15	19	22	0
s_1	1	0	0	0	0	-1	-1	2	0
s_2	0	1	0	1	0	-3	-2	6	0
s_3	0	0	1	1	0	-2	-3	3	0
λ_2^2	0	0	0	1/2	0	-1	0	1/2	0
s_5	0	0	0	0	1	0	-2	5	0
λ_2^1	0	0	0	0	0	1	0	1	1
λ_3^2	0	0	0	-1/2	0	1	1	1/2	0

Iteration 5.

Since $w_m \le 0$ for $m = 1, \ldots, 5$ we proceed to solve subproblems for Commodity 1 to find a candidate to enter the master basis.

Subproblem for Commodity 1

Since the dual values associated with the slack variables continue to be $\mathbf{w} = (0, 0, 0, -4, 0)$, single-commodity model remains the same as the model formulated in in iteration 4. The optimal solution is $\mathbf{x}_3^1 = (1,1,0,2,0)^T$ with cost equal to 15. Next, we calculate $\mathbf{w} - \mathbf{c}^1 = (0, 0, 0, -4, 0) - (2, 3, 4, 1, 2) = (-2, -3, -4, -5, -2)$ to compute $z_3^1 - c_3^1 = (\mathbf{w} - \mathbf{c}_1)\mathbf{x}_3^1 + \alpha_1 = -15 + 15 = 0$. Therefore, there is no candidate from this subproblem, and the decomposition method is terminated.

Optimal Solution for Original Problem

$$\lambda_1^2 = 1, \lambda_2^2 = 1/2, \lambda_3^2 = 1/2$$

$$z^* = 22$$

$$\mathbf{x}_1^* = \lambda_1^2 \mathbf{x}_1^2 = (1,1,0,2,0)^T$$

$$\mathbf{x}_2^* = \lambda_2^2 \mathbf{x}_2^2 + \lambda_3^2 \mathbf{x}_3^2 = 1/2(1,0,1,2,2)^T + 1/2(1,2,3,0,2)^T = (1,1,2,1,2)^T$$

The optimal cost is equal to 22. Figure 8.18 shows the optimal arc flows for both commodities.

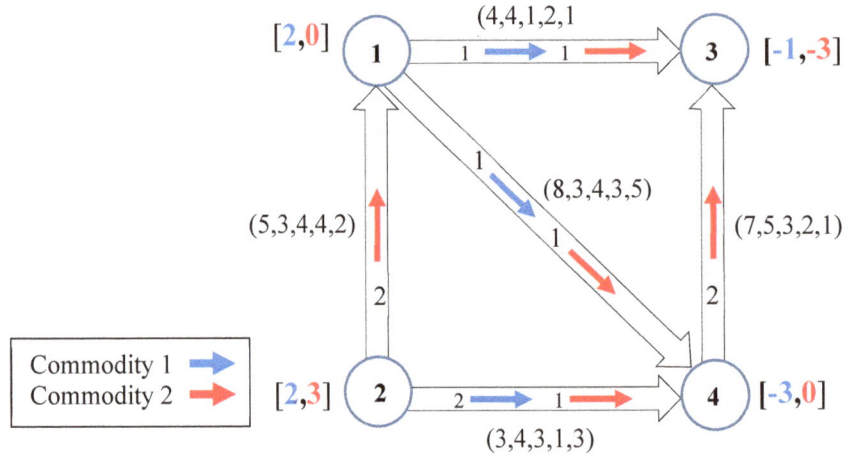

Figure 8.18. Optimal solution for sample two-commodity problem.

8.10.3 Lagrangian Relaxation Method

Another popular price-directive solution method is *Lagrangian relaxation*. This technique is often referred to as price-directive Lagrangian decomposition. The method brings the bundle or coupling constraints into the objective function and place *Lagrangian multipliers* (*prices*) on these constraints. The resulting problem can be decomposed into a separate minimum cost flow subproblem for each commodity.

To formulate the Lagrangian subproblem, we will start with the formulation in (8-39) through (8-42) for the linear multicommodity flow cost problem. This is a reformulation of the model in (8-27) through (8-31).

$$minimize \sum_{k=1}^{K} \mathbf{c}^k \mathbf{x}^k \tag{8-39}$$

$$\sum_{k=1}^{K} x_{ij}^k \le u_{ij} \quad \text{for all } (i,j) \in \mathbf{A} \tag{8-40}$$

$$\mathbf{A}\mathbf{x}^k = \mathbf{b}^k \quad \text{for } k = 1, 2, ..., K \tag{8-41}$$

$$0 \le x_{ij}^k \le u_{ij}^k \quad \text{for all } (i,j) \in \mathbf{A}; k = 1, 2, ..., K \tag{8-42}$$

If we assign a multiplier w_{ij} to the bundle constraint (8-40) of arc (i, j) the objective function of the corresponding Lagrangian subproblem is defined as

$$L(\mathbf{w}) = min \sum_{k=1}^{K} \mathbf{c}^k \mathbf{x}^k + \sum_{(i,j)\in \mathbf{A}} w_{ij} \left(\sum_{k=1}^{K} x_{ij}^k - u_{ij} \right)$$

The Lagrangian subproblem can be formulated as shown below:

$$L(\mathbf{w}) = \min \sum_{k=1}^{K} \sum_{(i,j) \in \mathbf{A}}^{K} (c_{ij}^{k} + w_{ij}) x_{ij}^{k} - \sum_{(i,j) \in \mathbf{A}}^{K} w_{ij} u_{ij}$$

subject to

$$\mathbf{Ax}^{k} = b^{k} \quad k = 1, \cdots, K$$

$$x_{ij}^{k} \geq 0 \qquad (i,j) \in \mathbf{A}; k = 1, \cdots, K$$

If we assume that the multipliers are known, the arc flows for the Commodity k can be obtained from the solution to the following Lagrangian subproblem:

$$L(\mathbf{w}) = \min \sum_{k=1}^{K} \sum_{(i,j) \in \mathbf{A}}^{K} (c_{ij}^{k} + w_{ij}) x_{ij}^{k}$$

subject to

$$\mathbf{Ax}^{k} = b^{k} \quad k = 1, \cdots, K$$

$$x_{ij}^{k} \geq 0 \qquad (i,j) \in \mathbf{A}; k = 1, \cdots, K$$

It should be noted that this problem decomposes into K single-commodity problems since no constraint has variables for more than exactly one commodity.

Subgradient optimization [1,17,21] is commonly used to determine the multipliers. Ahuja, Magnanti, and Orlin [1] present illustrative examples. To apply the subgradient optimization procedure we alternate two steps. In the first step the minimum cost flow model is solved for given values of the multipliers. The resulting subproblem is decomposed into single-commodity models. In the second step we use the solutions of the single-commodity models to generate new values for the multipliers. At the qth iteration the Lagrangian multiplier for the *capacitated* arc (i,j) is W_{ij}^{q} and the solution is x_{ij}^{k}. The total flow on the arc is equal to $y_{ij} = \sum_{k} x_{ij}^{k}$. To update the multiplier for the next iteration we use the following formula:

$$w_{ij}^{q+1} = \left[w_{ij}^{q} + \theta_q (y_{ij} - u_{ij}) \right]^{+}$$

where $[r]^{+} = \max(0, r)$. The selection of the *step size* θ_q is of critical importance. If may not be too large or too small to avoid lack of convergence. The step size θ_q should be chosen so that $\lim_{q \to \infty} \theta_q = 0$, and $\sum_{q=1}^{\infty} \theta_q = \infty$. Based on the above conditions, a simple choice is $\theta_q = \frac{1}{q}$, or $\theta_q = \frac{1}{q+1}$, or $\theta_q = \frac{a}{\sqrt{q}}$ where $a > 0$. In practice this choice does not always lead to quick convergence, hence other step size sequences are commonly used.

8.10.4 An Example of the Lagrangian Relaxation Method

The same model formulated in Section 8.8 and used to illustrate the Dantzig-Wolfe decomposition method will be used in this section to illustrate the application of the Lagrangian relaxation method combined with subgradient optimization.

$$
\begin{aligned}
&\textit{Minimize} \quad 2x_{13}^1 + 3x_{14}^1 + 4x_{21}^1 + x_{24}^1 + 2x_{43}^1 + x_{13}^2 + 5x_{14}^2 + 2x_{21}^2 + 3x_{24}^2 + x_{43}^2 \\
&\textit{subject to} \\
&x_{13}^1 + x_{14}^1 - x_{21}^1 = 2 \\
&x_{21}^1 + x_{24}^1 = 2 \\
&-x_{13}^1 - x_{43}^1 = -1 \\
&-x_{14}^1 - x_{24}^1 + x_{43}^1 = -3 \\
&x_{13}^2 + x_{14}^2 - x_{21}^2 = 0 \\
&x_{21}^2 + x_{24}^2 = 3 \\
&-x_{13}^2 - x_{43}^2 = -3 \\
&-x_{14}^2 - x_{24}^2 + x_{43}^2 = 0 \\
&x_{13}^1 \le 4, x_{14}^1 \le 3, x_{21}^1 \le 3, x_{24}^1 \le 4, x_{43}^1 \le 5 \\
&x_{13}^2 \le 1, x_{14}^2 \le 4, x_{21}^2 \le 4, x_{24}^2 \le 3, x_{43}^2 \le 3 \\
&x_{13}^1, x_{14}^1, x_{21}^1, x_{24}^1, x_{43}^1 \ge 0 \\
&x_{13}^2, x_{14}^2, x_{21}^2, x_{24}^2, x_{43}^2 \ge 0
\end{aligned}
$$

Initialization

Set all multipliers equal to zero, that is, $\mathbf{w} = \mathbf{0}$. In this example we will use $\theta_q = \frac{1}{q}$ for iterations1, …, 10; $\theta_q = \frac{2}{q}$ for iterations 11, 12, 13; $\theta_q = \frac{3}{q}$ for iteration 14; and $\theta_q = \frac{1}{q}$ for iteration 15. The original model can be decomposed into the two single-commodity problems formulated in Figure 8.18. The objective function coefficients of these models will be updated at each iteration of the subgradient optimization procedure.

$$
\begin{aligned}
&\textit{Minimize} \quad 2x_{13}^1 + 3x_{14}^1 + 4x_{21}^1 + x_{24}^1 + 2x_{43}^1 \\
&\textit{subject to} \\
&x_{13}^1 + x_{14}^1 - x_{21}^1 = 2 \\
&x_{21}^1 + x_{24}^1 = 2 \\
&-x_{13}^1 - x_{43}^1 = -1 \\
&-x_{14}^1 - x_{24}^1 + x_{43}^1 = -3 \\
&x_{13}^1 \le 4, x_{14}^1 \le 3, x_{21}^1 \le 3, x_{24}^1 \le 4, x_{43}^1 \le 5 \\
&x_{13}^1, x_{14}^1, x_{21}^1, x_{24}^1, x_{43}^1 \ge 0
\end{aligned}
$$

$$
\begin{aligned}
&\textit{Minimize} \quad x_{13}^2 + 5x_{14}^2 + 2x_{21}^2 + 3x_{24}^2 + x_{43}^2 \\
&\textit{subject to} \\
&x_{13}^2 + x_{14}^2 - x_{21}^2 = 0 \\
&x_{21}^2 + x_{24}^2 = 3 \\
&-x_{13}^2 - x_{43}^2 = -3 \\
&-x_{14}^2 - x_{24}^2 + x_{43}^2 = 0 \\
&x_{13}^2 \le 1, x_{14}^2 \le 4, x_{21}^2 \le 4, x_{24}^2 \le 3, x_{43}^2 \le 3 \\
&x_{13}^2, x_{14}^2, x_{21}^2, x_{24}^2, x_{43}^2 \ge 0
\end{aligned}
$$

Figure 8.19. Single-commodity models to be updated at each iteration.

The solutions for the two models are shown below. Only non-zero variables are listed.

$Z_1 = 7, x_{13}^1 = 1, x_{14}^1 = 1, x_{24}^1 = 2$

$Z_2 = 11, x_{13}^2 = 1, x_{21}^2 = 1, x_{24}^2 = 2, x_{43}^2 = 2$

$L(\mathbf{0}) = 7 + 11 = 18$.

Iteration 1

Set $q=1$, <mark>$\theta_1 = 1$</mark>
$y_{13} = 2, y_{14} = 1, y_{21} = 1, y_{24} = 4, y_{43} = 2$
$u_{13} = 4, u_{14} = 8, u_{21} = 5, u_{24} = 3, u_{43} = 7$

$$w_{ij}^1 = \left[w_{ij}^0 + \theta_1(y_{ij} - u_{ij}) \right]^+ = \left[(y_{ij} - u_{ij}) \right]^+$$

$w_{13}^1 = 0, w_{14}^1 = 0, w_{21}^1 = 0, w_{24}^1 = 1, w_{43}^1 = 0$.

Since only w_{24}^1 is different from zero, only the coefficients of x_{24}^1 and x_{24}^2 must be updated in the models shown in Figure 8.18. The corresponding updated terms are $2x_{24}^1$ in the model for the first commodity, and $4x_{24}^2$ in the model for the second commodity. The solutions for the two updated models and the value of the objective function of the Lagrangian subproblem are:

$Z_1 = 9, x_{13}^1 = 1, x_{14}^1 = 1, x_{24}^1 = 2$
$Z_2 = 13, x_{13}^2 = 1, x_{21}^2 = 1, x_{24}^2 = 2, x_{43}^2 = 2$

$$L(1) = min \sum_{k=1}^{K} \sum_{(i,j) \in A}^{K} (c_{ij}^k + w_{ij})x_{ij}^k - \sum_{(i,j) \in A}^{K} w_{ij}u_{ij} = 9 + 13 - 3 = 19$$

Iteration 2

Set $q=2$, <mark>$\theta_2 = 1/2$</mark>
$y_{13} = 2, y_{14} = 1, y_{21} = 1, y_{24} = 4, y_{43} = 2$
$u_{13} = 4, u_{14} = 8, u_{21} = 5, u_{24} = 3, u_{43} = 7$
$w_{13}^2 = w_{14}^2 = w_{21}^2 = w_{43}^2 = \left[0 + \frac{1}{2}(y_{ij} - u_{ij}) \right]^+ = 0, \ w_{24}^2 = 1 + \frac{1}{2} = 1.5$

The corresponding updated terms are $2.5x_{24}^1$ in the model for the first commodity, and $4.5x_{24}^2$ in the model for the second commodity. The solutions for the two updated models and the value of the objective function of the Lagrangian subproblem are:

$Z_1 = 10, x_{13}^1 = 1, x_{14}^1 = 1, x_{24}^1 = 2$
$Z_2 = 14, x_{13}^2 = 1, x_{21}^2 = 1, x_{24}^2 = 2, x_{43}^2 = 2$

$$L(2) = min \sum_{k=1}^{K} \sum_{(i,j) \in A}^{K} (c_{ij}^k + w_{ij})x_{ij}^k - \sum_{(i,j) \in A}^{K} w_{ij}u_{ij} = 10 + 14 - 4.5 = 19.5$$

Iteration 3

Set $q=3$, <mark>$\theta_3 = 1/3$</mark>
$y_{13} = 2, y_{14} = 1, y_{21} = 1, y_{24} = 4, y_{43} = 2$
$u_{13} = 4, u_{14} = 8, u_{21} = 5, u_{24} = 3, u_{43} = 7$
$w_{13}^3 = 0, w_{14}^3 = 0, w_{21}^3 = 0, w_{43}^3 = 0, w_{24}^3 = w_{24}^2 + \frac{1}{3}(y_{24} - 3) = \frac{3}{2} + \frac{1}{3} = 1\frac{5}{6}$

The corresponding updated terms are $2\frac{5}{6}x_{24}^1$ in the model for the first commodity, and $4\frac{5}{6}x_{24}^2$ in the model for the second commodity. The solutions for the two updated models and the value of the objective function of the Lagrangian subproblem are:

$$Z_1 = 10\tfrac{2}{3}, x_{13}^1 = 1, x_{14}^1 = 1, x_{24}^1 = 2$$

$$Z_2 = 12, x_{24}^2 = 3, x_{43}^2 = 3$$

$$L(\mathbf{3}) = min \sum_{k=1}^{K} \sum_{(i,j)\in A}^{K} (c_{ij}^k + w_{ij}) x_{ij}^k - \sum_{(i,j)\in A}^{K} w_{ij} u_{ij} = 22\tfrac{2}{3} - 5\tfrac{1}{2} = 17\tfrac{1}{6}$$

Iteration 4

Set q=4, $\theta_4 = 1/4$

$$y_{13} = 1, y_{14} = 1, y_{21} = 0, y_{24} = 5, y_{43} = 3$$

$$u_{13} = 4, u_{14} = 8, u_{21} = 5, u_{24} = 3, u_{43} = 7$$

$$w_{13}^4 = 0, w_{14}^4 = 0, w_{21}^4 = 0, w_{43}^4 = 0, w_{24}^4 = w_{24}^3 + \tfrac{1}{4}(y_{24} - 3) = 1\tfrac{5}{6} + \tfrac{1}{2} = 2\tfrac{1}{3}$$

The corresponding updated terms are $3\tfrac{1}{3} x_{24}^1$ in the model for the first commodity, and $5\tfrac{1}{3} x_{24}^2$ in the model for the second commodity. The solutions for the two updated models and the value of the objective function of the Lagrangian subproblem are:

$$Z_1 = 11\tfrac{2}{3}, x_{13}^1 = 1, x_{14}^1 = 1, x_{24}^1 = 2$$

$$Z_2 = 15\tfrac{2}{3}, x_{13}^2 = 1, x_{21}^2 = 1, x_{24}^2 = 2, x_{43}^2 = 2$$

$$L(\mathbf{4}) = min \sum_{k=1}^{K} \sum_{(i,j)\in A}^{K} (c_{ij}^k + w_{ij}) x_{ij}^k - \sum_{(i,j)\in A}^{K} w_{ij} u_{ij} = 11\tfrac{2}{3} + 15\tfrac{2}{3} - 7 = 20\tfrac{1}{3}$$

Iteration 5

Set q=5, $\theta_5 = 1/5$

$$y_{13} = 2, y_{14} = 1, y_{21} = 1, y_{24} = 4, y_{43} = 2$$

$$u_{13} = 4, u_{14} = 8, u_{21} = 5, u_{24} = 3, u_{43} = 7$$

$$w_{13}^5 = 0, w_{14}^5 = 0, w_{21}^5 = 0, w_{43}^5 = 0, w_{24}^5 = w_{24}^4 + \tfrac{1}{5}(y_{24} - 3) = 2\tfrac{1}{3} + \tfrac{1}{5} = 2\tfrac{8}{15}$$

The corresponding updated terms are $3\tfrac{8}{15} x_{24}^1$ in the model for the first commodity, and $5\tfrac{8}{15} x_{24}^2$ in the model for the second commodity. The solutions for the two updated models and the value of the objective function of the Lagrangian subproblem are:

$$Z_1 = 12\tfrac{1}{15}, x_{13}^1 = 1, x_{14}^1 = 1, x_{24}^1 = 2$$

$$Z_2 = 16\tfrac{1}{15}, x_{13}^2 = 1, x_{21}^2 = 1, x_{24}^2 = 2, x_{43}^2 = 2$$

$$L(\mathbf{5}) = min \sum_{k=1}^{K} \sum_{(i,j)\in A}^{K} (c_{ij}^k + w_{ij}) x_{ij}^k - \sum_{(i,j)\in A}^{K} w_{ij} u_{ij} = 12\tfrac{1}{15} + 16\tfrac{1}{15} - 7\tfrac{3}{5} = 20\tfrac{8}{15}$$

Proceeding in this fashion it will be observed that at iteration 15 the solution is converging to an optimal solution. Table 8.11 summarizes the results for the remaining iterations. It is noted again that $w_{13}^q = w_{14}^q = w_{21}^q = w_{43}^q = 0, w_{24}^q = w_{24}^{q-1} + \theta_q(y_{24} - 3)$. The first column of Table 8.11 shows the iteration number; the second one displays the step size; the third column shows the calculation for the multiplier of the bundle constraint of arc (2,4); the next to columns show the terms that must be updated in the models of Figure 8.19; the following 12 columns show the solution to the updated modes; and the last column shows the value of $L(\mathbf{w})$. It is noted that after iteration 15, this value is converging to the optimal value of 22 as the multiplier converges to its optimal value of 4. Figure 8.20 shows the optimal flows for both commodities. These results are identical to those shown in Figure 8.18 for the Dantzig-Wolfe decomposition method.

Table 8.11. SUMMARY OF RESULTS FOR ITERATIONS 6-15

Iteration q	θ_q	w^q_{24}	Updated Costs x^1_{24}	x^2_{24}	Z_1	x^1_{13}	x^1_{14}	x^1_{21}	x^1_{24}	x^1_{43}	Z_2	x^2_{13}	x^2_{14}	x^2_{21}	x^2_{24}	x^2_{43}	$L(\mathbf{w})$
6	$\frac{1}{6}$	$2\frac{8}{15}+\frac{1}{6}=2\frac{7}{10}$	3.7	5.7	12.4	1	1	0	2	0	16.4	1	0	1	2	2	21.20
7	$\frac{1}{7}$	$2\frac{7}{10}+\frac{1}{7}=2\frac{59}{70}$	3.8428	5.8428	12.685	1	1	0	2	0	16.685	1	0	1	2	2	20.84
8	$\frac{1}{8}$	$2\frac{59}{70}+\frac{1}{8}=2\frac{271}{280}$	3.9678	5.9678	12.935	1	1	0	2	0	16.935	1	0	1	2	2	21.34
9	$\frac{1}{9}$	$2\frac{271}{280}+\frac{1}{9}=3\frac{199}{2520}$	4.0789	6.0789	13.158	1	1	0	2	0	17.158	1	0	1	2	2	21.08
10	$\frac{1}{10}$	$3\frac{199}{2520}+\frac{1}{10}=3\frac{451}{2520}$	4.1789	6.1789	13.358	1	1	0	2	0	17.358	1	0	1	2	2	21.18
11	$\frac{2}{11}$	$3\frac{451}{2520}+\frac{2}{11}=3\frac{10001}{27720}$	4.3607	6.3607	13.721	1	1	0	2	0	17.721	1	0	1	2	2	21.36
12	$\frac{2}{12}=\frac{1}{6}$	$3\frac{10001}{27720}+\frac{1}{6}=3\frac{14621}{27720}$	4.5274	6.5274	14.055	1	1	0	2	0	18.055	1	0	1	2	2	21.53
13	$\frac{2}{13}$	$3\frac{14621}{27720}+\frac{2}{13}=3\frac{245513}{360360}$	4.6813	6.6813	14.363	1	1	0	2	0	18.363	1	0	1	2	2	21.68
14	$\frac{3}{14}$	$3\frac{245513}{360360}+\frac{3}{14}=3\frac{322733}{360360}$	4.8956	6.8956	14.791	1	1	0	2	0	18.791	1	0	1	2	2	21.90
15	$\frac{1}{15}$	$3\frac{322733}{360360}+\frac{1}{15}=3\frac{346757}{360360}$	4.9622	5.9622	14.924	1	1	0	2	0	18.924	1	0	1	2	2	21.96

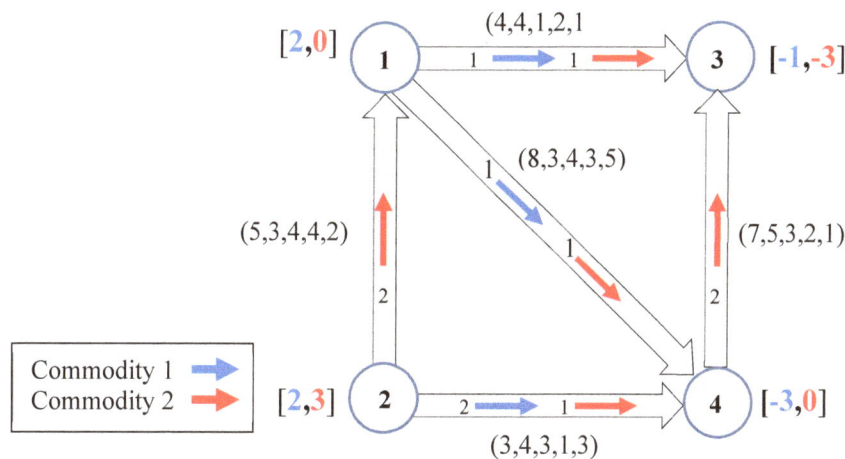

Figure 8.20. Optimal flows for both commodities.

8.10.5 Resource-Directive Decomposition Method

This method considers the multicommodity network problem as a capacity allocation problem, in which all commodities compete for the fixed capacity of every arc. Initially, the method allocates capacity to commodities and decomposes the problem into a set of K independent single-commodity problems. At each iteration, an allocation is made and the K single-commodity minimal cost flow problems are solved. The sum of the capacities allocated to an arc over all commodities is less than or equal to the arc capacity in the original problem. Hence the combined flow from the solutions of the subproblems provides a feasible flow for the original problem.

Optimality is tested and the procedure is either terminated or a new arc-capacity allocation is performed.

After the artificial variables are added, the formulation in (8-27)-(8-31) becomes:

$$minimize \ \sum_k \mathbf{c}^k \mathbf{x}^k + \gamma \sum_k \mathbf{1} a^k \qquad (8\text{-}39)$$

$$subject \ to$$
$$\mathbf{A}\mathbf{x}^k + a^k = \mathbf{b}^k, \ \ k = 1, \ ..., \ K \qquad (8\text{-}40)$$
$$\sum_k \mathbf{x}^k + \mathbf{s} = \mathbf{u} \qquad (8\text{-}41)$$
$$0 \leq \mathbf{x}^k \leq \mathbf{u}^k, \ k = 1, \ ..., \ K \qquad (8\text{-}42)$$
$$a^k, \mathbf{s} \geq 0 \qquad \ 8\text{-}43)$$

where γ is a large positive number, a^k is the vector of artificial variables, and $\mathbf{1}$ is a vector of ones.

Now let us consider the following formulation where $\boldsymbol{\mu}$ and v^k are the dual values associated with Constraints (8-45) and (8-46), respectively.

$$V(\mathbf{y}^k) = min \ \mathbf{c}^k \mathbf{x}^k + \gamma \mathbf{1} a^k \qquad (8\text{-}44)$$
$$subject \ to$$
$$\mathbf{A}\mathbf{x}^k + a^k = \mathbf{b}^k \qquad (\boldsymbol{\mu}) \qquad (8\text{-}45)$$
$$0 \leq \mathbf{x}^k \leq \mathbf{y}^k \qquad (v^k) \qquad (8\text{-}46)$$

The dual model for the above formulation is shown below:

$$V(\mathbf{y}^k) = max \ \mathbf{b}^k \boldsymbol{\mu}^k - \mathbf{y}^k v^k \qquad (8\text{-}47)$$
$$subject \ to$$
$$\boldsymbol{\mu}^k \mathbf{A} - v^k \leq \mathbf{c}^k \qquad (8\text{-}48)$$
$$\boldsymbol{\mu}^k \leq \gamma \mathbf{1} \qquad (8\text{-}49)$$
$$v^k \geq 0$$

Therefore, the model in (8-39)-(8-43) is equivalent to

$$g(y) = min \ \sum_k V_k(\mathbf{y}^k) \qquad (8\text{-}50)$$
$$subject \ to$$
$$\sum_k \mathbf{y}^k + \mathbf{s} = \mathbf{u} \qquad (8\text{-}51)$$
$$0 \leq \mathbf{y}^k \leq \mathbf{u}^k \qquad (8\text{-}52)$$
$$\mathbf{s} \geq 0 \qquad (8\text{-}53)$$

There are several ways to solve the model in (8-50)-(8-53). We will use the *tangential approximation method* [16].

Let $\mathbf{R}_k = \{(\boldsymbol{\mu}^k, \mathbf{v}^k): \boldsymbol{\mu}^k\mathbf{A} - \mathbf{v}^k \leq \mathbf{c}^k, \boldsymbol{\mu}^k \leq \gamma\mathbf{1}, \mathbf{v}^k \geq \mathbf{0}$, and $(\boldsymbol{\mu}^k, \mathbf{v}^k)$ an extreme point$\}$. Then the model in (8-50)-(8-53) may be reformulated as

$$minimize \sum_k \sigma_k \qquad (8\text{-}54)$$

subject to

$$\sigma_k \geq \mathbf{b}^k\boldsymbol{\mu}^k - \mathbf{y}^k\mathbf{v}^k, \quad \text{all } (\boldsymbol{\mu}^k, \mathbf{v}^k) \in \mathbf{R}_k \text{ and all } k \qquad (8\text{-}55)$$

$$\sum_k \mathbf{y}^k + \mathbf{s} = \mathbf{u} \qquad (8\text{-}56)$$

$$\mathbf{0} \leq \mathbf{y}^k \leq \mathbf{u}^k, \ k = 1, \ldots, K \qquad (8\text{-}57)$$

$$\mathbf{s} \geq \mathbf{0} \qquad (8\text{-}58)$$

Let $\mathbf{Q}_k \subset \mathbf{R}_k$, let Z_R denote the optimal objective value of (8-54)-(8-58) and let Z_Q denote the optimal objective value of (8-54)-(8-58) with $\mathbf{Q}_k$ substituted for $\mathbf{R}_k$ in (8-55). Then $Z_Q \leq Z_R$ provides a lower bound. The *Resource-Directive Decomposition Algorithm* using *Tangential Approximation* can be summarized as follows [16,17].

Initialization.

Let $i = 0$ and $\left(\mathbf{y}_0^1, \cdots, \mathbf{y}_0^K\right)$ be any member of the set $\left\{\left(\mathbf{y}^1, \cdots, \mathbf{y}^K\right): \sum_k \mathbf{y} \leq \mathbf{u}, \mathbf{0} \leq \mathbf{y}^k \leq \mathbf{u}^k\right\}$. Set $\mathbf{Q}_k = \varnothing$ and $\sigma_k = -\infty$ for each k.

Step 1. Solve a Subproblem to Determine an Upper Bound .

$$V_k(\ y_i^k\ ) = min \ \mathbf{c}^k\mathbf{x}^k + \gamma\mathbf{1}a^k$$

subject to

$$\mathbf{A}\mathbf{x}^k + a^k = \mathbf{b}^k \qquad (\ \boldsymbol{\mu}_i^k\ )$$

$$\mathbf{x}^k \leq \mathbf{y}^k \qquad (\ \mathbf{v}_i^k\ )$$

$$0 \leq \mathbf{x}^k \text{ for each } k{=}1, \ldots, K.$$

The dual values for the constraints are shown in parentheses.

Step 2. Check Optimality Condition.

If $\sum_k \sigma_k = \sum_k V_k(\mathbf{y}_i^k)$ the algorithm is terminated with an optimal solution $(\mathbf{x}_i^1, \cdots, \mathbf{x}_i^K)$ and $(a_i^1, \cdots, a_i^K)$. Otherwise, add $(\boldsymbol{\mu}_i^k, \mathbf{v}_i^k)$ to $\mathbf{Q}_k$ for each k and continue with step 3.

Step 3. Solve Master Program.

Set $i = i + 1$ and solve the following model. Afterwards, return to step 1.

$$\text{minimize} \sum_k \sigma_k$$

subject to

$$\sigma_k \geq \mathbf{b}^k \mathbf{\mu}^k - \mathbf{y}^k \mathbf{v}^k, \text{ for each } k \text{ and all } (\mathbf{\mu}^k, \mathbf{v}^k) \in \mathbf{Q}_k$$

$$\sum_k \mathbf{y}^k + \mathbf{s} = \mathbf{u}$$

$$\mathbf{0} \leq y^k \leq \mathbf{u}^k, k = 1, \ldots, K$$

$$\mathbf{s} \geq \mathbf{0}$$

EXERCISES

1. A classical application of a multicommodity model is the route design problem with several trip-generation zones. Given a network $G = (\mathbf{N}, \mathbf{A})$ and certain original availabilities or requirements of each of L commodities at node i, designated as Q_{ik}, $k = 1,2,\ldots,L$, find for each link (i,j) and each commodity k a flow x_{ijk}, which would result in the minimization of the total route construction and travel cost. Assume that the cost of constructing the link (i,j) is p_{ij} and the variable cost per unit of flow over the arc (i,j) is C_{ij}. For each of the following general types of problems, discuss possible interpretations of the nodes, the arcs, and the flows, in the context of a multicommodity network:

   (a) Freight distribution among cities by means of trucks, rail, air freight routes, and waterways.
   (b) Communication network synthesis with several message centers and several transmission lines.
   (c) Distribution of mail, including letters and parcels, among several post offices.
   (d) Production and inventory control with periodic supply and demand requirements.
   (e) Network design for collection, treatment, and discharge of solid waste disposal.
   (f) Regional transportation planning for passenger cars and hauling vehicles.
   (g) Urban traffic planning.
   (h) Mass transit design with several stations and several transit lines.

2. Two commodities can flow over the network shown. Each commodity may flow along any arc in either direction. An unlimited supply of commodity 1 is available at nodes 1, 2, 3, and the final destination of this commodity is node 4. Additionally, an unlimited supply of commodity 2 is available at node 5, and the final destination of this commodity is node 1. The number on each arc represents the capacity of the arc. Each unit of commodity 1 that reaches its final destination is worth 4, and each unit of commodity 2 that reaches its final destination is worth 6. Find feasible flows of commodities resulting in a maximum total value.

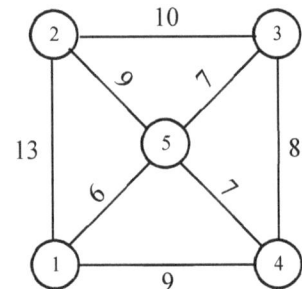

3. In meeting demand requirements, the standard transportation model does not distinguish among the supply centers. However, most multicommodity firms would determine a complete distribution system by taking account of the entire product line. Discuss how to use the regular

single-commodity transportation model to solve a simple multicommodity model with a product mix required at each demand center. List all your assumptions.

4. Write down the *constraints* to the linear programming formulation of the multi-commodity transshipment problem shown. Numbers shown for each arc are flow capacities. Additionally,

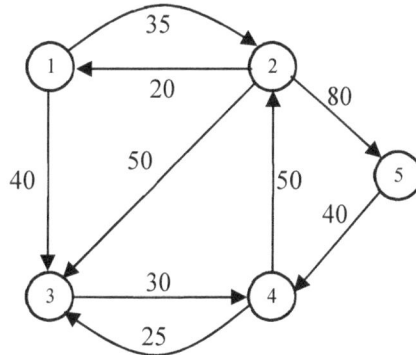

5. $a_1^1 = 50$, $b_5^1 = 30$, $b_4^1 = 20$, $a_4^2 = 35$, $b_3^2 = 35$, $a_4^3 = 30$, $b_5^3 = 30$. What type of structure does the linear program have?

6. Convert the following problem to a single-commodity equivalent network problem and solve with the out-of-kilter algorithm. Use both formulations, that is, with and without capacitated arcs.

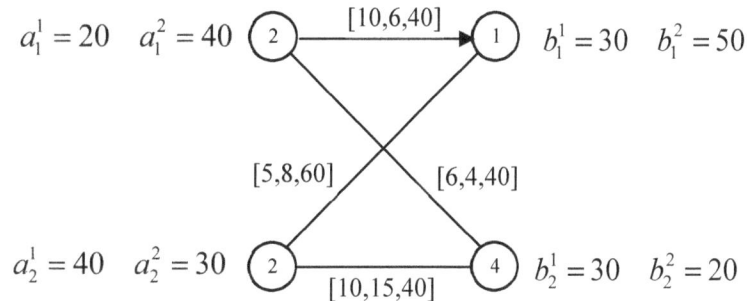

7. Resolve the fruit distribution problem by aggregating Santa Barbara and Bakers-field. Also try Santa Barbara and Sacramento. Which gives the best solution? Can you justify any reasons why?

8. Solve the multicommodity maximal-flow problem for the network shown below.

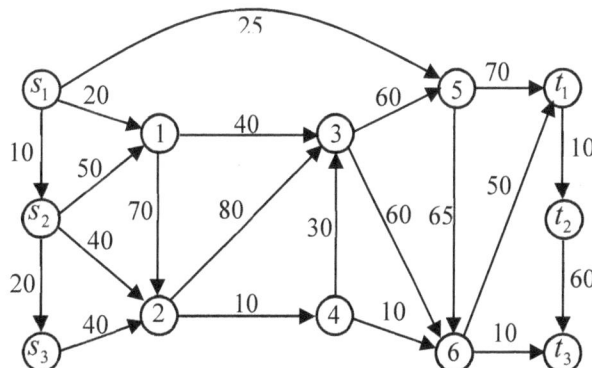

9. Find the maximal two-commodity flow for the following network.

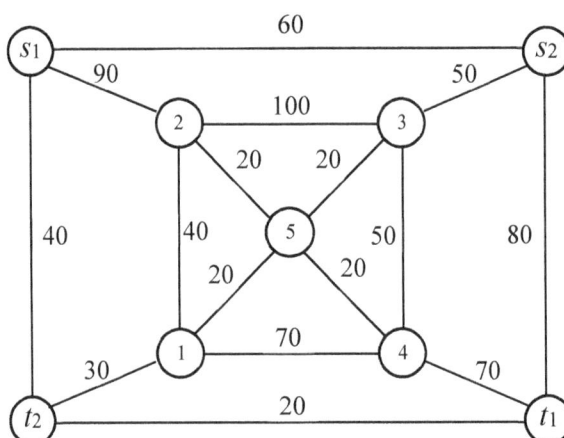

10. Show that the MCTP formulated in Eqs. (8-10 to (8-5) and the transportation problem formulated in Eqs. (8-20) to (8-26) are equivalent; that is, any feasible solution to one is also a feasible solution to the other.

11. Compute the disaggregate solution to the fruit distribution problem in Table 8.1-Table 8.4. verify that it has the same cost as the aggregated solution.

12. Consider the numerical example shown in Figure 8.15 with constraint matrix given in Table 8.8 to illustrate the application of the resource-directive decomposition method.

13. Apply the price-directive decomposition method to solve the two-commodity minimal flow problem corresponding to the given network.

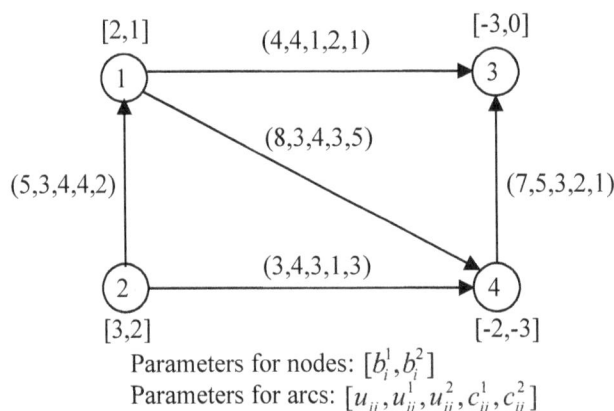

Parameters for nodes: $[b_i^1, b_i^2]$
Parameters for arcs: $[u_{ij}, u_{ij}^1, u_{ij}^2, c_{ij}^1, c_{ij}^2]$

14. Solve Exercise 13 using the Lagrangian method combined with subgradient optimization.

15. Consider the multicommodity network shown in the following figure. Nodes 1 and 2 are sources and nodes 3 and 4 are terminals. Each node has two numbers representing the supplies (for sources) or demands (for terminals) for two commodities. Each arc has three values representing the costs for each of the two commodities and the arc capacity.

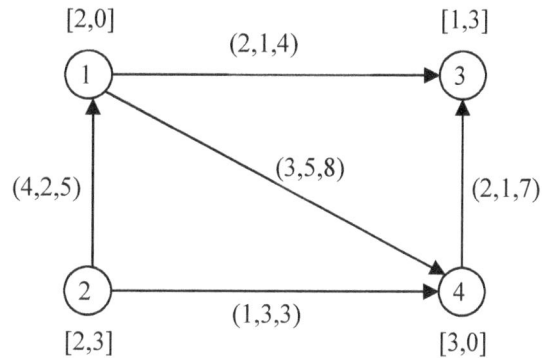

Do two iterations of the Lagrangian Relaxation method starting with values equal to 1 for all multipliers.

REFERENCES

[1] AHUJA, R., MAGNANTI, T., AND ORLIN, J., *Network Flows-Theory, Algorithms, and Applications*. Prentice-Hall, Inc., 1993.

[2] ASSAD, A. A., "Multicommodity Network Flows-A Survey," *Networks* 8, 37-91 (1978).

[3] BAZARAA, M. S., J. J. JARVIS, AND H. D. SHERALI, *Linear Programming and Network Flows,* John Wiley & Sons, Inc., 1990.

[4] BELLMORE, M., G. BENNINGTON, AND S. LUBORE, "A Multivehicle Tanker Scheduling Problem," *Transportation Sciences, 5,* 35-47 (1971).

[5] DANTZIG G. B. AND P. WOLFE, "Decomposition Principle for Linear Programs," *Operations Research*, 8, 101-111, 1960.

[6] EVANS, J. R., J.J. JARVIS, AND R. A. DUKE, "Graphic Matroids and the multicommodity Transportation Problem," *Mathematical Programming,* 13, 323-328 (1977).

[7] EVANS, J. R., AND J.J. JARVIS, "Network Topology and Integral Multicommodity Flow Problems," *Networks,* 8, 107-119 (1978).

[8] EVANS, J. R., "A Combinational Equivalence between a Class of Multicommodity Flow Problems and the Capacitated Transportation Problem," *Mathematical Programming*, 10, 401-404 (1976).

[9] EVANS, J.R., "A Single Commodity Transformation for Certain Multicommodity Networks," *Operations Research,* 26, 673-680 (1978).

[10] EVANS, J. R., "An Equivalent Formulation of Certain Multicommodity Networks as Single Commodity Flow Problems," *Mathematical Programming*, 15, 92-99 (1978).

[11] EVANS, J. R., "Solving Multicommodity Transportation Problems through Aggregation," ORSA/TIMS National Conference, Los Angeles, November 1978.

[12] EVANS, J. R., "Model Simplification in Multicommodity Distribution Systems though Aggregation," *Proceedings, American Institute of Decision Sciences,* National Meeting, New Orleans, November 1979.

[13] GEOFFRION, A., "Customer Aggregation in Distribution Modeling," Working Paper No. 259, Western Management Science Institute, University of California, Los Angeles, 1976.

[14] HU, T. C., "Multicommodity Network Flows," *Operations Research,* 11, 344-360 (1963).

[15] JARVIS, J. J., AND D. D. Miller, "Maximal Funnel-Node Flow in an Undirected Network," *Operations Research, 11*, 344-360 (1963).

[16] KENNINGTON, J. L., "A Survey of Linear Cost Multicommodity Network Flows," *Operations Research,* 26, 209-236 (1978).

[17] KENNINGTON, J. L. AND R. V. HELGASON, *Algorithms for Network Programming.* John Wiley & Sons, Inc., 1980.

[18] PHILLIPS, D.T. AND A. GARCIA-DIAZ, *Fundamentals of Network Analysis,* Prentice-Hall International Series in Industrial Engineering, 1981.

[19] SAKAROVITCH, M., "The Multicommodity Maximal Flow Problems," ORC 66-25, University of California, Berkeley, 1966.

[20] WANG, I-Lin, "Multicommodity Network Flows: A Survey, Part I: Applications and Formulations," *International Journal of Operations Research,* Vol. 15, No. 4, 145-153 2018).

[21] WANG, I-Lin, "Multicommodity Network Flows: A Survey, Part II: Solution Methods," *International Journal of Operations Research,* Vol. 15, No. 4, 155-173 (2018).

[22] WARDROP, J.G., "Some Theoretical Aspects of Road Traffic Research". *Proceedings of the Institution of Civil Engineers.* **1** (3): 325–362, 1952.

[23] ZIPKIN, P. H., "Aggregation in Linear Programming," Ph.D. dissertation, Yale University, December 1977.

Chapter 9

AN INTRODUCTION TO NETWORK RELIABILITY

*"People are the quintessential element in all technology... Once we recognize
the inescapable human nexus of all technology our attitude toward the
reliability problem is fundamentally changed."*

Garret Hardin
SKEPTIC (July-August, 1976)

The purpose of *network reliability* is to study in detail the ability of a network to carry out successfully its intended operation. Typical networks for which a reliability study is essential include systems such as computer communication, telecommunication, transportation, traffic management, and electrical power distribution systems. Network reliability can be measured in terms of the frequency of failure, time it takes to recover from failure, the network's robustness, and other means to quantify the ability of the network to function as intended. For a network to be reliable it is necessary that the connectivity and functionality of its components remain significantly unaffected by a number of incidents that may cause disruptions that can render some arcs and or nodes to become non-operational or perform in a way that is far from what is expected.

A full discussion of network reliability is beyond the scope of this introductory chapter. Most of the material presented in this chapter is based on the excellent work by D. R. Shier on network reliability [12]. We want to express our appreciation to Mr. Mike Hare, a former graduate student at the University of Tennessee, for contributing with an earlier version of this chapter. This final version is the result of some editing, reorganization and addition of some topics. In this chapter we focus on the difference between deterministic and probabilistic (or stochastic) networks, discuss the failure of nodes and arcs in a network, and consider typical ways to quantify the reliability of both deterministic and probabilistic networks illustrating the procedures with small numerical examples.

The study of network reliability began with the pioneering work by More and Shannon [8] in 1956. In their work they assumed the nodes of the network to be functional at all times with the arcs subject to independent failures according to a given probability distribution. Under these assumptions the objective of their research was to determine the probability that the network continues to be connected. Pérez-Rosés [9] reviews basic concepts and results accumulated during six decades after the article by More and Shannon, provides some recent developments, and shares relevant direction for future research efforts. Rebaiaia and Ait-Kadi [10] also review the most important methods, algorithms and software available. This work focuses on telecommunications and transport networks.

9.1 INTRODUCTION

Network models can be formulated to represent physical or organizational systems as graphs composed of nodes and arcs (directed or undirected). Different ideas concerning network reliability have been proposed by a number of researchers from industry and academia in order to investigate the capability of a network system to operate as intended by its designers

To motivate the topic to be studied in this chapter, let us consider a printed circuit board in which the arcs correspond to traces and nodes to solder joints. Each trace consists of a flat, narrow part of the copper foil that remains after etching. One of the ever-present malfunctioning occurrences in manufacturing printed circuit boards is caused by cold joints. These are joints where the solder did not melt completely, causing a disruption in the electrical connection between the external component and the traces of the printed circuit board. On the other hand, a break in a trace, from poor manufacturing or external physical damage, will cause a disruption in the flow of electricity between two components. In this example we have failure of nodes (cold solder joints) and failures of arcs (breaks in a trace), both causing disruption in the electrical flow of the network.

We will consider a *directed* network $G = (\mathbf{N}, \mathbf{A})$ with arcs in set $\mathbf{A}$ and nodes in set $\mathbf{N}$. From the point of view of reliability, the network can be classified as either *deterministic* or *probabilistic*. As we have seen in previous chapters, an *undirected network* can be converted into an equivalent directed network by substituting each undirected arc with two arcs of opposite directions. Either when the network is directed or undirected, we usually say that the arc is *incident* to both of its nodes. If the network is undirected, the number of arcs incident with a node is known as de *degree* of the node.

In a *deterministic network* an arc fails when it is removed from the network and a node fails when it *and* its adjacent arcs are removed from the network. The network may cease to perform as intended if the *connectivity* of its nodes and arcs is not preserved when a subset of its arcs and or its nodes is removed. In deterministic networks typically the emphasis is on worst-case network performance when the removal of a set of components has the greatest undesirable effect on the connectivity of the network.

Alternatively, in a *probabilistic network* each node and each arc are viewed as components with known reliabilities or probabilities of being operational. In this case, the reliability of the network may be defined as the probability that it will perform its intended function successfully over a given time period. Since node and arc failures are probabilistic the reliability of the entire network is a random variable itself, and, therefore, an important and relevant assessment is the *average reliability* of the network.

9.2 NODE AND ARC FAILURES IN DETERMINISTIC NETWORKS

In this section we focus on how network analysis can play a major role in the study of the ability of a network to effectively operate or function after one or more of its components fail to operate as intended by the network designers. As already stated in Chapter 1, a network is said to be connected if there is at least one path or chain joining any pair of arbitrary distinct nodes. Otherwise, the network is considered to be disconnected. More specifically, the main goal of this section is to determine how many nodes or arcs must be removed from the graph of a network to break all paths between two arbitrary or specified nodes.

As an illustration of this, let us consider the network shown in Figure 9.1(a). In this figure we consider the connectivity between nodes 1 and 7. Figure 9.1(b) shows three arc-disjoint paths and three arcs that if removed will cause the source and terminal nodes to become disconnected. Figure 9.1(c) shows two node-disjoint paths and two nodes that if removed will cause the source and terminal nodes to become disconnected.

(a) Undirected graph.

(b) Three arc-disjoint paths.

(c) Two node-disjoint paths.

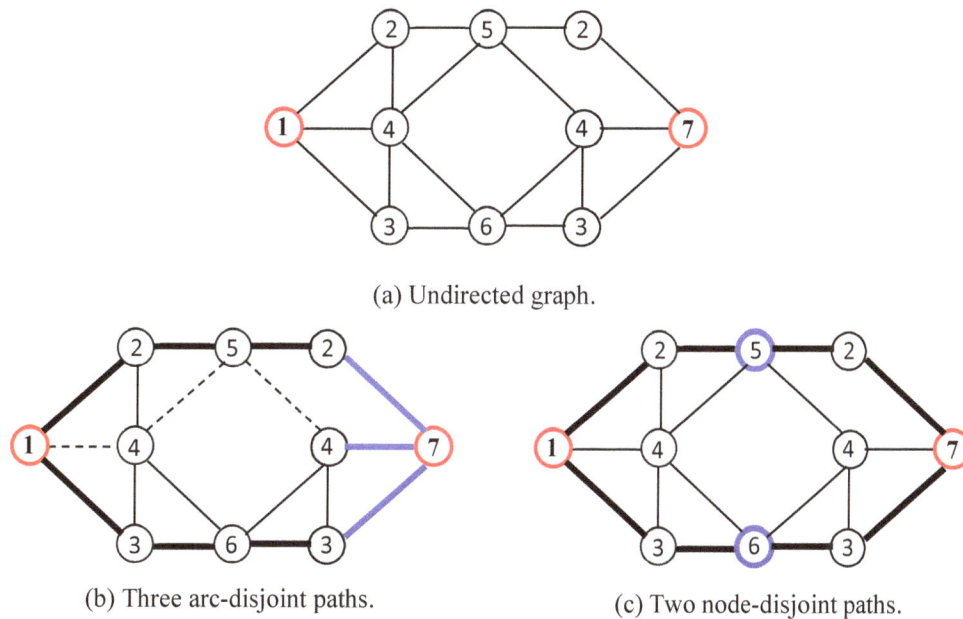

Figure 9.1. Connectivity between nodes 1 and 7.

The following discussions are based upon the work by Frank and Frisch [1, 2] and Kleitman [6] who have extensively studied the connectivity of networks.

9.2.1 Minimum Number of Arcs to Disconnect a Network

First, we consider the problem of *finding the minimum number of arcs* or branches that must be removed in order to disconnect a network. As proved by Frank and Frisch [2], if the capacity of each arc of an undirected network is assigned a value of 1, the *maximal flow* between any two nodes is equal to the minimum number of arcs that must be removed to *break* all paths from one node to the other. The maximal flow is actually equal to the maximal number of *arc disjoint paths* between the nodes. These paths are said to be disjoint because they do not share any arc. If this procedure is repeated for every pair of nodes in the network, the number of branches or arcs to be removed to disconnect the network is equal to the minimum of all maximal flows. This minimum number is a valuable measure of the *vulnerability* of the network.

9.2.2 An Illustrative Example

Figure 9.2 considers a sample network for Problem 1. Figure 9.2(a) shows the undirected 7-node network and Figure 9.2(b) shows the corresponding maximal-flow matrix for all pairs of nodes after setting all arc capacities equal to 1. The matrix was found using NOP for the Gomory-Hu algorithm discussed in Section 2.13.1. From the matrix, we determine that the minimum number of arcs required to disconnect this network is equal to 3. In this example there are several choices to select the three arcs. Figure 9.2(c) shows the disconnected network resulting after the removal of arcs (1,2), (2,4) and (2,5).

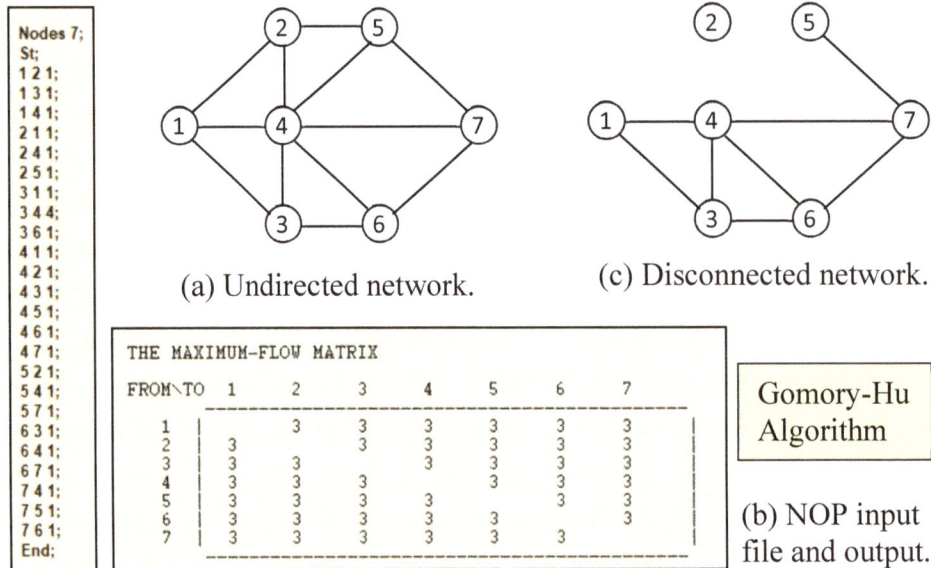

```
Nodes 7;
St;
1 2 1;
1 3 1;
1 4 1;
2 1 1;
2 4 1;
2 5 1;
3 1 1;
3 4 4;
3 6 1;
4 1 1;
4 2 1;
4 3 1;
4 5 1;
4 6 1;
4 7 1;
5 2 1;
5 4 1;
5 7 1;
6 3 1;
6 4 1;
6 7 1;
7 4 1;
7 5 1;
7 6 1;
End;
```

(a) Undirected network.

(c) Disconnected network.

Gomory-Hu Algorithm

THE MAXIMUM-FLOW MATRIX

FROM\TO	1	2	3	4	5	6	7
1		3	3	3	3	3	3
2	3		3	3	3	3	3
3	3	3		3	3	3	3
4	3	3	3		3	3	3
5	3	3	3	3		3	3
6	3	3	3	3	3		3
7	3	3	3	3	3	3	

(b) NOP input file and output.

Figure 9.2. Sample network for Problem 1.

9.2.3 Minimum Number of Nodes to Break all Paths

The second problem we consider is *finding the minimum number of nodes* that need to be removed to *break* all paths between two specified nodes. First, we select any pair of nodes and consider one as source and the other one as terminal. Then we proceed to assign capacities equal to 1 for all nodes, except the source and the terminal. Following the network transformation procedure illustrated in Example 1.7.2 we construct a directed *expanded network* that has capacities for the arcs only. The labeling procedure of Section 2.12 can now be used to find the value of the maximal flow from the source node to the terminal node in the expanded network. It can be proved that this value is equal to the maximum number of paths that do not share any node, except the source and the terminal. These paths are referred to as *node disjoint paths*. The *node connectivity* of a graph is the *minimum number of node disjoint paths* between any pair of nodes.

If the previously described procedure is repeated for all pairs of nodes in the network, we can find the *minimum* of all maximal flows, which is actually equal to the minimum number of nodes that must be removed to disconnect the network. An alternative more efficient procedure was developed by I. T. Frisch [3]. Later on, Steiglitz and Bruno [13] proposed an improved modification of Frisch's algorithm using the Ford and Fulkerson labeling algorithm.

9.2.4 Illustrative Example

As an illustration we will find the node connectivity of the *undirected* graph of Figure 9.3(a). The procedure starts selecting any pair of nodes, such as node 1 (source) and node 5 (terminal). After this choice, the corresponding transformation to convert unit *node capacities* for all nodes, except the source and the terminal, into *arc capacities* is shown in the *directed* graph of the expanded network shown in Figure 9.3(b).

351

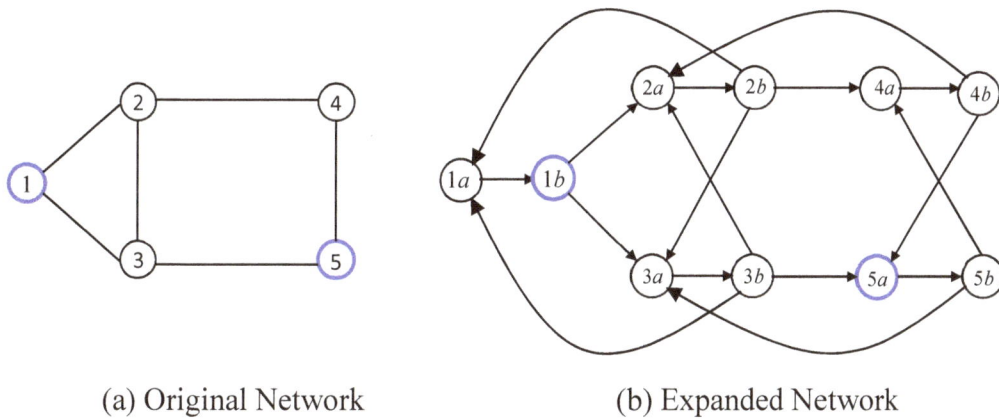

(a) Original Network   (b) Expanded Network

Figure 9.3. Sample networks for Problem 2.

Figure 9.4 shows the expanded network of Figure 9.3 9b) with renumbered nodes. In the expanded network of Figure 9.3 (b) the source is node 1b and the terminal is node 5a. In Figure 9.4 the corresponding source and terminal nodes are node 2 and node 9, respectively. By inspection, we find that the maximal flow from node 2 to node 9 is equal to 2. This means that in the original network the removal of two nodes would break all paths between node 1 to node 5. These two nodes can be node 3 and node 5, or node 8 and node 6. There are other disconnecting pairs.

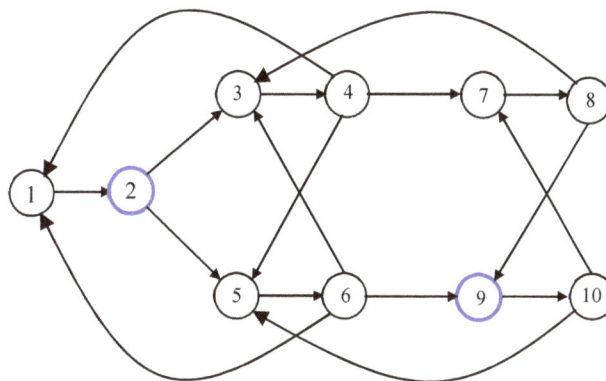

Figure 9.4. Relabeling nodes in sample network for Problem 2.

Proceeding in a similar way, we continue selecting pairs of source and terminal nodes in the expanded network and find the corresponding maximal flows. These flows are arranged in the following matrix.

	1a	2a	3a	4a	5a
1b		2	2	2	2
2b	2		2	2	2
3b	2	2		2	2
4b	2	2	2		2
5b	2	2	2	2	

Note that all maximal flows are equal to 2. Therefore, we conclude that for any pair of source and terminal nodes there are two node-disjoint paths connecting them. Thus, at least two nodes must be removed to break all paths between the chosen pair of nodes. As an illustration to break all paths between node 1 and 5 the nodes to be removed are node 2 and node 3, or node 3 and node 4. As another illustration, all paths between node 3 and node 4 are broken if node 2 and node 5 are removed.

In case that NOP is used to find the maximal flows, for each run of the labeling algorithm it is important to keep in mind that the nodes must be *renumbered* to have node 1 and node 10 being always the source and terminal, respectively.

Although the foregoing procedure is guaranteed to yield the proper solution, the procedure is unwieldy for a large number of nodes. As an illustration, a network with 1000 nodes may require almost 500,000 maximal flow calculations to determine if the removal of as many as four arbitrarily chosen nodes will disconnect the network.

9.2.5 THEOREM [6]

In an undirected network $G = (\mathbf{N}, \mathbf{A})$ the following verifications are sufficient to conclude the existence of M node disjoint paths between each pair of nodes. Start with $j = 1$ and $\mathbf{N}_1 = \mathbf{N}$.

1. Choose any node $n_j \in \mathbf{N}_j$ and verify the existence of M node-disjoint paths from node n_j to all other nodes. Remove node n_j from $\mathbf{N}_j$ and all its connecting branches. Update $j = j + 1$. Update $\mathbf{N}_j = \mathbf{N}_j - \{n_j\}$.

2. Remove node n_j from $\mathbf{N}_j$ and all its attached branches. Verify the existence of $M - j$ node-disjoint paths from node n_j to all other nodes.

3. Stop if $M = 1$. Otherwise, go to Step 1.

Based on the above theorem, Kleitman [6] showed that the following procedure will result in fewer than 4000 maximal flow calculations to determine if the removal of as many as four arbitrarily chosen nodes will disconnect a network with 1000 nodes.

Step 1. Set $M = 4$. Choose any node and check if the number of disjoint paths from this node to every other node is at least M. If true, delete this node and all arcs connected to the node. If not true, stop.

Step 2. Set $M = M-1$. Using the reduced network, choose one of the remaining nodes and check if the number of disjoint paths connecting this node to all others is at least equal to M. If true, delete this node and all arcs connected to the node. Go to Step 3. If not true, stop.

Step 3. If $M = 1$, stop. Otherwise go to Step 2.

If the complete sequence of operations cannot be executed, it is not possible to verify that the required number of paths exists. The procedure is illustrated in Section 9.2.6 on a detailed example where $M = 5$. That is, we want to verify if removing 5 nodes will result in a disconnected network.

9.2.6 An Illustrative Example

Consider the 8-node 22-arc network shown in Figure 9.5. We wish to verify if the removal of as many as five nodes will disconnect the network.

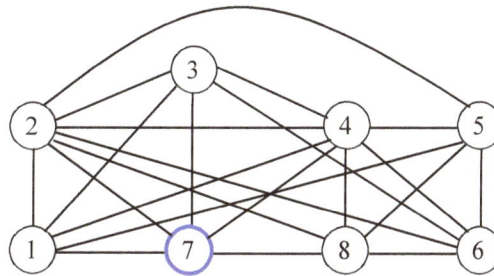

Figure 9.5. Sample network for disconnection analysis.

Step 1. $M = 5$. Choose node 7. One can verify that there are at least five node disjoint paths from node 7 to each of the other nodes. The node disjoint paths from node 7 are listed below:

> For node 1: 7-1, 7-2-1, 7-3-1, 7-4-1, 7-8-5-1
> For node 2: 7-1-2, 7-2, 7-3-2, 7-4-5-2, 7-8-6-2
> For node 3: 7-1-2-3, 7-2-3, 7-3, 7-4-3, 7-8-6-3
> For node 4: 7-1-2-4, 7-2-4, 7-3-4, 7-4, 7-8-5-4
> For node 5: 7-1-5, 7-2-5, 7-3-6-5, 7-4-5, 7-8-5
> For node 6: 7-1-5-6, 7-2-6, 7-3-6, 7-4-5-6, 7-8-6
> For node 8: 7-8, 7-2-8, 7-3-4-8, 7-1-5-8, and 7-4-6-8.

Next, delete node 7 and all arcs connected to it. The reduced network is shown in Figure 9.6.

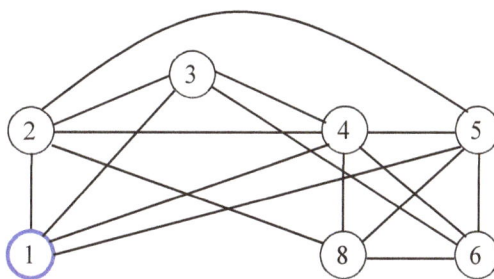

Figure 9.6. First reduced network after removing node 7.

Step 2. Set $M = 4$. Using the reduced network, choose node 1 and check if the number of disjoint paths connecting this node to all others is at least equal to 4. Four disjoint node paths from node 1 are listed below for nodes 2, 3, 4, 5, 6, and 8:

> 1-2, 1-3-2, 1-4-2, 1-5-2
> 1-2-3, 1-3, 1-4-3, 1-5-6-3
> 1-2-4, 1-3-4, 1-4, 1-5-4
> 1-2-5, 1-3-4-5, 1-4-8-5, 1-5

354

1-2-5-6, 1-3-4-6, 1-4-8-6, 1-5-6
1-2-8, 1-4-8, 1-5-8, 1-3-6-8

Next, delete node 1 and all arcs connected to it. The reduced network is shown in Figure 9.7.

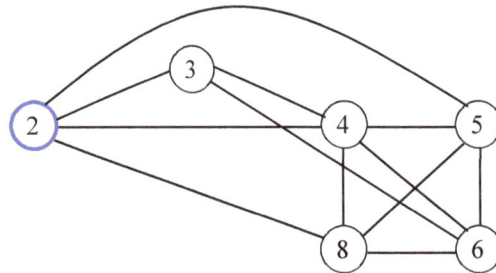

Figure 9.7. Second reduced network after removing node 1.

Set $M = 3$. Using the reduced network, choose node 2 and check if the number of disjoint paths connecting this node to all others is at least equal to 3. Three disjoint node paths from node 2 are listed below for nodes 3, 4, 5, 6, and 8:

2-3, 2-4-3, 2-8-6-3
2-4, 2-3-4, 2-5-4
2-5, 2-3-4-5, 2-8-6-5
2-5-6, 2-3-4-6, 2-8-6
2-8, 2-4-8, 2-5-8

Next, delete node 2 and all arcs connected to it. The reduced network is shown in Figure 9.8.

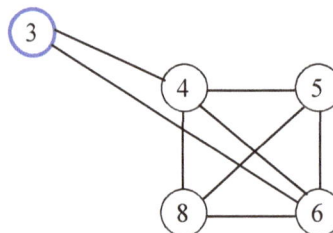

Figure 9.8. Third reduced network after removing node 2.

Set $M = 2$. Using the reduced network, choose node 3 and check if the number of disjoint paths connecting this node to all others is at least equal to 2. Two disjoint node paths from node 3 are listed below for nodes 4, 5, 6, and 8:

3-4, 3-6-4
3-4-5, 3-6-5
3-4-6, 3-6
3-4-8, 3-6-8

Next, delete node 3 and all arcs connected to it. The reduced network is shown in Figure 9.10.

355

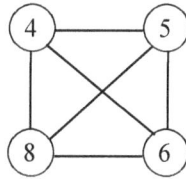

Figure 9.10. Fourth reduced network after removing node 3.

Set $M = 1$. Therefore, the procedure is stopped, and we conclude that the removal of as many as five nodes will disconnect the network.

9.3 DETERMINISTIC NETWORK RELIABILITY

We will consider some typical measures to assess the worst-case operation of a deterministic network $G = (\mathbf{N}, \mathbf{A})$. It is assumed that the working elements of the network may be removed or become inactive as a result of an undesirable incident affecting the connectivity of the network.

First, we will consider the case where the network performs satisfactorily when there is a connecting path from its source node s to its terminal node t. In this case, an appropriate reliability measure is the number of arcs in an *arc disconnecting set* (also referred to as an *arc cutset*) separating node s from node t. We will represent this number by $\lambda(G)$.

Alternatively, if we are considering *node failures* for the same cases considered for satisfactory network operation in terms of arc failures, a relevant measure is the minimum number of nodes in a *node disconnecting set* (or *node cutset*). We will represent this number by $\kappa(G)$. In 1969 Harary [4] proved the result given in Eq. (9-1) regarding bounds on the reliability $\lambda(G)$ of network G:

$$\kappa(G) \leq \lambda(G) \leq \delta(G) \leq \left[\frac{2m}{n}\right] \qquad (9\text{-}1)$$

where $\delta(G)$ is the *minimum number of arcs* incident with a node in network G, m is the number of arcs, n is the number of nodes, and $[2m/n]$ is the integer part of $2m/n$.

9.3.1 Proof

Consider a node $v \in \mathbf{N}$. If all arcs incident with v are removed, an arc cutset is obtained since node v will be isolated. Hence, the cardinality of the minimum arc cutset is equal to or less than the minimum *degree* (number of incident arcs) over all nodes of the graph, $\lambda(G) \leq \delta(G)$.

If $\lambda(G) = 0$ or 1, then $\kappa(G) = \lambda(G)$. If $\lambda(G) = k \geq 2$, let (u_1, v_1) (u_2, v_2), ..., (u_k, v_k) be the arcs of the arc cutset that disconnects G into two components $\mathbf{V}_1$ and $\mathbf{V}_2$. Then, either $\mathbf{V}_1$ contains a node v different from u_1, u_2, ..., u_k, meaning that removing u_1, u_2, ..., u_k, causes v to be disconnected from $\mathbf{V}_2$, or $\mathbf{V}_1 = \{u_1, u_2, ..., u_k\}$, where $|\mathbf{V}_1| \leq k$ (some u_i's might be identical). Now, in this case, u_1 has at most k neighbors: $|\mathbf{V}_1| - 1$ in $\mathbf{V}_1$ and $k - (|\mathbf{V}_1| - 1)$ in $\mathbf{V}_2$. Moreover, $\lambda(G) = k$,

356

thus, the degree of u_1 is equal to k and the removal of the k neighbors of u_1 causes G to be disconnected. Therefore, $\kappa(G) \leq \lambda(G)$.

Finally, note that there are m arcs and so the total degree of the network is $2m$ so the average degree is equal to $\dfrac{2m}{n}$. Hence, $\delta(G) \leq \left[\dfrac{2m}{n}\right]$.

9.3.2 Examples

It is desired to find the value or interval for $k(G)$ for the two given sample undirected networks shown in Figure 9.11.

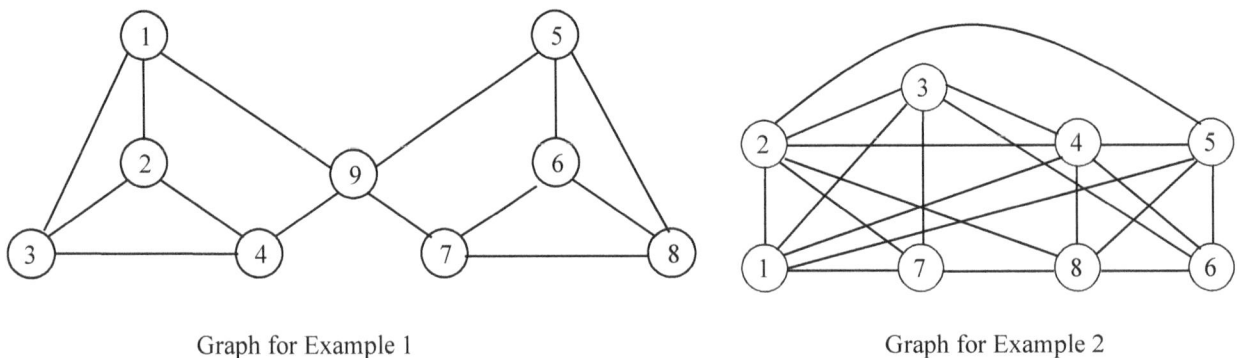

Graph for Example 1   Graph for Example 2

Figure 9.11. Sample Undirected Networks.

The first example, the graph of the network, $m = 14$, $n = 9$, $[2m/n] = [28/9] = 3$. Therefore, $\kappa(G) \leq \lambda(G) \leq \delta(G) \leq 3$. Now, from the graph again, $\delta(G) = 3$ and we conclude that $\kappa(G) \leq \lambda(G) \leq 3$. Note that the network becomes disconnected if we eliminate at least two arcs, such as (1,9) and (4,9). Thus, $\lambda(G) = 2$ and we conclude that $\kappa(G) \leq 2$. From the graph we can verify that if we remove node 9 the network becomes disconnected, and, therefore, $\kappa(G) = 1$.

For the second example, $m = 22$, $n = 8$, $[2m/n] = [5.5] = 5$. Therefore, $\kappa(G) \leq \lambda(G) \leq \delta(G) \leq 5$. From the graph, $\delta(G) = 5$. Thus, $\kappa(G) \leq \lambda(G) \leq 5$. For this network, $\lambda(G) = 5$. Therefore, $\kappa(G) \leq 5$.

9.3.3 Other Reliability Measures

Two networks can have the same connectivity measures but have very different ways in which their elements have been arranged. This could make one network easier to disconnect than the other. So, considering only connectivity measures is not enough. Other measures can be considered, including the number of arcs disconnecting sets of a certain size k or the minimal number of arcs that need to be removed to create a cut node.

Once a failure occurs and the network becomes disconnected, one can measure the magnitude of the disruption by determining the size and number of the disconnected components of the new network. This makes sense as a network disconnected into, say, two components has a higher degree of functionality than one broken into 10 components. One possible way to measure this is the minimum ratio of the size of a node disconnecting set to the resulting number of components. This is known as *node toughness* in the network reliability literature.

Another measure addresses the fact that the removal of different arcs can cause different levels of disruption. If the failure of some arcs and nodes of a network G results in the creation of k separate connected components, then nodes in different components are not able to communicate with one another. The pairs of nodes able to communicate is equal to $\sum_{i=1}^{k} \binom{n_i}{2}$ where n_i is the number of nodes in the *i-th* component of G. An average of the number of nodes that can communicate taken over all possible failures can be a useful measure. This is known as the *pair connectivity*, or *resilience*, of the network. These are just a few ways to measure the reliability of the network; the choice of measures depends significantly on the specific application after considering the nature of the failures.

9.4 PROBABILISTIC NETWORK RELIABILITY

This section provides a survey of several methods used to calculate the reliability of a probabilistic network G. One of the most important reliability measures in this kind of network is the probability that two given nodes a and b are connected by a path. We denote this probability by $R_{ab}(G)$. In particular, we are interested in the case where node a is actually a source node s and node b is a terminal node t. We denote this probability by $R_{st}(G)$ and refer to it as the *source-terminal reliability*. The most basic methodology for calculating this probability uses *state-state enumeration*. This procedure was developed in the mid-1950s, when Moore and Shannon [8], following Von Neumann, studied the topic of network reliability. In addition to this procedure, we will discuss three more: the inclusion-exclusion method, the disjoint product method and the factoring method.

In this introductory chapter on network reliability, we will focus on four conceptually and fundamentally important methods that provide a foundation for the development of many other methods. These are exact reliability evaluation techniques. We will illustrate each procedure with a small sample numerical example based on a common network graph.

1. State-Space Enumeration Method
2. Inclusion-Exclusion Method
3. Disjoint Product Method
4. Factoring Method

9.4.1 State-Space Enumeration Method

This method is the simplest and most direct but computationally impractical approach to calculate the reliability of a network. It is based on the availability of all state vectors for the network, and it analyzes each vector to determine if the network is operational or not. Each arc of a network G assumes two states, either *success* or *failure*. The state of the network can be represented in terms of binary variables taking on the values 0 for failure or 1 for success. If the number of arcs is m, the state of the network can be expressed as a vector $\delta = (\delta_1, \delta_2,\ldots,\delta_m)$. Under the assumption of statistically independent arc failures, the probability of a given state vector $\boldsymbol{\delta}$ is given by Eq. (9-2)

$$P(\boldsymbol{\delta}) = \prod_{k=1}^{m} p_k^{\delta_k} (1 - p_k)^{1-\delta_k} \qquad (9\text{-}2)$$

where p_k and 1-p_k are the probabilities of success and failure, respectively, of arc k.

Now, let us define a 0-1 variable $I_{st}(\boldsymbol{\delta})$ equal to 1 when the subset of arcs having $\delta_k = 1$ connects the source node s to the terminal node t, and equal to 0 otherwise. Then the reliability of the network is given by

$$R_{st}(G) = \sum_{\boldsymbol{\delta} \in \mathcal{D}} I_{st}(\boldsymbol{\delta}) P(\boldsymbol{\delta}) \qquad (9\text{-}3)$$

where $\mathcal{D}$ is the set of all network states. Since the number of members of set $\mathcal{D}$ is 2^m this methodology can be used only in the case of small networks. As the size of the network increases, the method quickly becomes computationally undesirable.

9.4.2   Example of State-Space Enumeration Method

As an illustration, let us consider the network shown in Figure 9.12. The set of nodes is $\mathbf{N} = \{s, 1, 2, t\}$ and the set of arcs is $\mathbf{A} = \{1, 2, 3, 4, 5\}$. Let the probability of success of arc k be p_k and let the probability of failure be $q_k = 1 - p_k$.

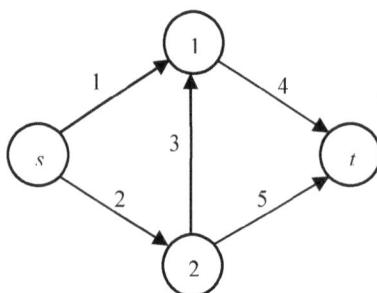

Figure 9.12. Network for Numerical Example.

From Figure 9.12, we observe that the following three paths (described as sequences of arcs) connect the source node to the terminal node: 1-4, 2-3-4, and 2-5. The total number of the five-arc combinations (events) resulting in connections of the source node to the terminal node is equal to 15. The source-terminal reliability can therefore be calculated as shown below:

$$\begin{aligned}
R_{st} = {} & p_1 p_2 p_3 p_4 p_5 + q_1 p_2 p_3 p_4 p_5 + p_1 q_2 p_3 p_4 p_5 + p_1 p_2 q_3 p_4 p_5 + p_1 p_2 p_3 q_4 p_5 \\
& + p_1 p_2 p_3 p_4 q_5 + q_1 p_2 q_3 p_4 p_5 + q_1 p_2 p_3 q_4 p_5 + q_1 p_2 p_3 p_4 q_5 + p_1 q_2 q_3 p_4 p_5 \\
& + p_1 q_2 p_3 p_4 q_5 + p_1 p_2 q_3 q_4 p_5 + p_1 p_2 q_3 p_4 q_5 + q_1 p_2 q_3 q_4 p_5 + p_1 q_2 q_3 p_4 q_5
\end{aligned}$$

After using $q_k = 1 - p_k$, we obtain the following result:

359

$$R_{st} = p_1 p_2 p_3 p_4 p_5 + (1-p_1)p_2 p_3 p_4 p_5 + p_1(1-p_2)p_3 p_4 p_5 + p_1 p_2(1-p_3)p_4 p_5$$
$$+ p_1 p_2 p_3(1-p_4)p_5 + p_1 p_2 p_3 p_4(1-p_5) + (1-p_1)p_2(1-p_3)p_4 p_5$$
$$+ (1-p_1)p_2 p_3(1-p_4)p_5 + (1-p_1)p_2 p_3 p_4(1-p_5) + p_1(1-p_2)(1-p_3)p_4 p_5$$
$$+ p_1(1-p_2)p_3 p_4(1-p_5) + p_1 p_2(1-p_3)(1-p_4)p_5 + p_1 p_2(1-p_3)p_4(1-p_5)$$
$$+ (1-p_1)p_2(1-p_3)(1-p_4)p_5 + p_1(1-p_2)(1-p_3)p_4(1-p_5)$$

This can be further simplified to the following result:

$$R_{st} = p_1 p_4 + p_2 p_5 + p_2 p_3 p_4 - p_1 p_2 p_3 p_4 - p_1 p_2 p_4 p_5 - p_2 p_3 p_4 p_5 + p_1 p_2 p_3 p_4 p_5$$

9.4.3 Inclusion-Exclusion Method

This method is based on the inclusion-exclusion principle to develop an expression for the source-terminal reliability of a network. The method is an alternative and most efficient way to proceed by considering only the events resulting in successful source-terminal paths.

In the mathematical area of combinatorics the *inclusion-exclusion principle* is a counting technique that generalizes the common procedure of finding the number of elements in the union of two sets. If the symbol $|S|$ represents the *cardinality* or number of elements of set A, the number of elements in the union of sets A and B is expressed as

$$|A \cup B| = |A| + |B| - |A \cap B|$$

This formula indicates that adding the elements of both sets results in a total that exceeds the true count since the elements in the intersection are counted twice instead of only once. A generalization of the principle for three sets A, B and C is

$$|A \cup B \cup C| = |A| + |B| + |C| - |A \cap B| - |A \cap C| - |B \cap C| + |A \cap B \cap C|$$

In this case, the number of elements in the mutual intersection of the three sets has been subtracted too often, and thus must be added once to get the correct count.

In the calculation of the source-terminal reliability of a probabilistic network, the inclusion-exclusion principle is used considering events instead of sets and probabilities instead of counts. To illustrate the application of the principle, we consider the simple network shown in Figure 9.13.

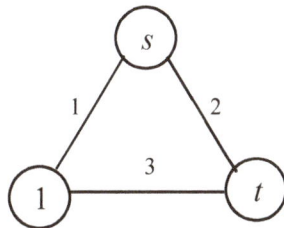

Figure 9.13. Simple network to illustrate the inclusion-exclusion principle.

As can be seen in Figure 9.13, the following two events result in successful paths:

E_1 = event that the path consisting of arcs 1 and 3 operates successfully.
E_2 = event that the path consisting of arc 2 operates successfully.

Therefore, the source-terminal reliability of the network is equal to

$$R_{st} = P(E_1 \cup E_2) = P(E_1) + P(E_2) - P(E_1 \cap E_2) = p_1 p_3 + p_2 - p_1 p_2 p_3$$

To generalize the above formula, let us define E_i as the event that all arcs in path P_i are operating successfully. Then the reliability of the network can be obtained as the probability that at least one path will work satisfactorily, that is, $R_{st} = P(E_1 \cup E_2 \cup \cdots \cup E_n)$. Now we can use the inclusion-exclusion principle to get the result in Eq. (9-5).

$$R_{st}(G) = \sum_i P(E_i) - \sum_{i<j} P(E_i \cap E_j) + \sum_{i<j<l} P(E_i \cap E_j \cap E_l) - \cdots + (-1)^{k+1} P(E_i \cap E_j \cap \cdots \cap E_k) \quad (9\text{-}5)$$

Although this is easier to calculate, compared to the enumeration method, the approach is still computationally inefficient for large networks, since the number of terms is equal $2^n - 1$.

9.4.4 Example of Inclusion-Exclusion Method

In this example, the sample network shown in Figure 9.12 will be considered again. In this network there are three events (k=3) resulting in a successful operational network:

E_1 = path consisting of arcs 1 and 4 operates successfully.
E_2 = path consisting of arcs 2 and 5 operate successfully.
E_3 = path consisting of arcs 2, 3 and 4 operate successfully.

The source-terminal reliability for the network can be calculated as $R_{st} = P(E_1 \cup E_2 \cup E_3)$. Using the inclusion-exclusion principle we get the result in Eq. (9-6).

$$R_{st} = \sum_i P(E_i) - \sum_{ij} P(E_i \cap E_j) + P(E_1 \cap E_2 \cap E_3) \quad (9\text{-}6)$$

From Eq. (9-6) we obtain Eq. (9-7)

$$R_{st}(G) = P(E_1) + P(E_2) + P(E_3) - P(E_1 \cap E_2) - P(E_1 \cap E_3) - P(E_2 \cap E_3) + P(E_1 \cap E_2 \cap E_3) \quad (9\text{-}7)$$

Using Eq. (9-7) we get the following result for the example under consideration.

$$R_{st} = p_1 p_4 + p_2 p_5 + p_2 p_3 p_4 - p_1 p_2 p_3 p_4 - p_1 p_2 p_4 p_5 - p_2 p_3 p_4 p_5 + p_1 p_2 p_3 p_4 p_5$$

9.4.5 Disjoint Product Method

Another approach to computing the reliability of a network is to use *Boolean algebra*, and in particular a *sum of disjoint products* technique. To start our discussion, we consider two events E_1 and E_2 that are not disjoint. The disjoint product method, however, needs to express $E_1 \cup E_2$ as a union of disjoint events. Figure 9.14 illustrates $E_1 \cup E_2 = E_1 \cup \overline{E}_1 E_2$, that $\overline{E}_1$ where is the *complement* of E_1.

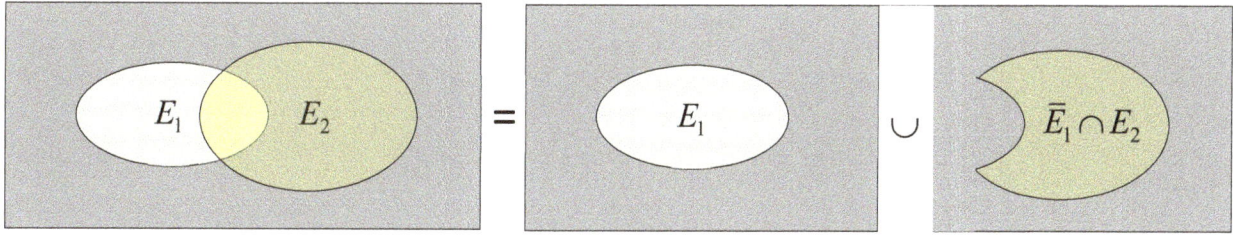

Figure 9.14. Union of two events as union of disjoint events.

The disjoint product method considers $E_1 \cup E_2 \cup \cdots \cup E_n$ as a union of disjoint events. Generalizing the result illustrated in Figure 9.14, the source-terminal reliability is given by

$$R_{st}(G) = P(E_1 \cup E_2 \cup \cdots \cup E_k) = P(E_1 \cup \overline{E_1}E_2 \cup \overline{E_1}\,\overline{E_2}E_3 \cup \cdots \cup \overline{E_1}\,\overline{E_2}\,\overline{E_3} \cdots \overline{E}_{k-1}E_k) \quad (9\text{-}8)$$

Since these events are disjoint, we obtain the result in Eq. (9-9).

$$R_{st}(G) = P(E_1) + P(\overline{E_1}E_2) + P(\overline{E_1}\,\overline{E_2}E_3) + \cdots + P(\overline{E_1}\,\overline{E_2}\,\overline{E_3} \cdots \overline{E}_{k-1}E_k) \quad (9\text{-}9)$$

It is noted that the number of terms has been significantly reduced at the expense of computational complexity. Whereas in the inclusion-exclusion expansion, the terms were numerous but easy to compute, here we have much fewer terms, but each term is more difficult to compute. Another important observation is that the order in which we select the paths to define the events can and does have an impact on the computational performance of this method.

9.4.6 Example of Disjoint Product Method

The disjoint product method is illustrated using the network shown in Figure 9.12 again. From the figure we observe that the following three events result in paths connecting the source node to the terminal node:

E_1 = path consisting of arcs 1 and 4 operates successfully.
E_2 = path consisting of arcs 2 and 5 operates successfully.
E_3 = path consisting of arcs 2, 3 and 4 operates successfully.

Therefore, the source-terminal reliability can be calculated as shown below.

$$R_{st}(G) = P(E_1) + P(\overline{E_1}E_2) + P(\overline{E_1}\,\overline{E_2}E_3)$$

$$P(E_1) = p_1 p_4$$

$$P(\overline{E_1}E_2) = P[(\overline{1} \cup \overline{4}) \cap 2] = p_2 p_5 P[(\overline{1} \cup \overline{4})]$$

$$= p_2 p_5 (q_1 + q_4 - q_1 q_4)$$

$$= p_2 p_5 [1 - p_1 + 1 - p_4 - (1 - p_1)(1 - p_4)]$$

$$= p_2 p_5 [1 - p_1 + 1 - p_4 - 1 + p_4 + p_1 - p_1 p_4]$$

$$= p_2 p_5 [1 - p_1 p_4] = p_2 p_5 - p_1 p_2 p_4 p_5$$

362

$$P(\overline{E}_1\overline{E}_2E_3) = P\left[\left(\overline{1}\cup\overline{4}\right)\cap\left(\overline{2}\cup\overline{5}\right)\cap 234\right]$$

$$= p_2p_3p_4 \, P\left[\overline{15}\right] = p_2p_3p_4(1-p_1)(1-p_5)$$

$$= p_2p_3p_4 - p_1p_2p_3p_4 - p_2p_3p_4p_5 + p_1p_2p_3p_4p_5$$

Thus,

$$R_{st} = p_1p_4 + p_2p_5 + p_2p_3p_4 - p_1p_2p_3p_4 - p_1p_2p_4p_5 - p_2p_3p_4p_5 + p_1p_2p_3p_4p_5$$

It should be noted again that the computational work of this method depends on the order in which we select the paths to define the events E_1, E_2. ..., E_k.

9.4.7 Factoring Method

One way to deal with the inefficiency of generating all states is use the factoring or pivotal decomposition method. If instead of specifying the state of each of the m arcs of a network $G = (\mathbf{N}, \mathbf{A})$, we just consider a particular arc $a\in\mathbf{A}$, then we can decompose the network into two smaller sub-networks, one in which arc a operates successfully ($p_a = 1$) and a second one in which it fails ($p_a = 0$). The first sub-network can be represented by $G|a$ and the second one by $G-a$. Now, if the first sub-network has source-terminal reliability $R_{st}(G|a)$ and the second one has source-terminal Reliability $R_{st}(G-a)$, then the reliability of the original network can be expressed as in Eq. (9-10) in terms of the Maskowitz's *pivotal decomposition formula* [7]:

$$R_{st}(G) = p_a R_{st}(G\,|\,a) + (1-p_a)R_{st}(G-a) \qquad (9\text{-}10)$$

For a directed network G, it is especially useful to do factoring on an arc that exits the source node s or arrives at the terminal node t. When the arc is deleted, the graph is modified by removing it and merging its nodes. It is possible to reduce both $G|a$ and $G-a$ by applying the following three probabilistic rules:

1. **Rule for Parallel Arcs.** Two parallel arcs a and b with probabilities p_a and p_b, respectively, can be replaced with one arc having probability $(1-p_a)(1-p_b) = p_a + p_b - p_ap_b$.

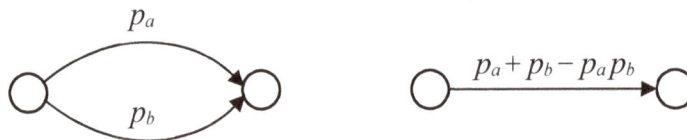

Figure 9.14. Parallel arcs.

2. **Rule for Arcs Connected in Series.** Two arcs a and b connected in series with probabilities p_a and p_b, respectively, can be replaced with one equivalent arc having probability $p_a p_b$.

Figure 9.15. Arcs connected in series.

3. **Rule for single Loops Connected in Series.** Two single *loops* connected in series with probabilities p_a, p_b, p_c and p_d can be replaced by an equivalent single loop with probabilities $p_a p_b$ and $p_c p_d$, respectively.

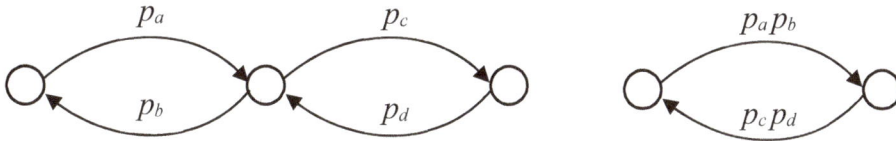

Figure 9.16. Loops connected in series.

9.4.8   Example of Factoring Method

Pivotal decomposition is illustrated on the network shown in Figure 9.12 by conditioning on the state of arc 4, as shown in Figure 9.17.

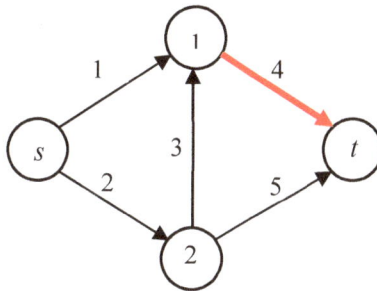

Figure 9.17. Sample network for illustrative example.

If we assume that arc 4 is performing as intended, nodes *t* and 1 can be superimposed and arc 1 is removed, as shown below in Figure 9.18 (a). Using the reduction rules the network can be represented as an equivalent graph with two arcs in parallel, as shown in Figure 9.18 (b).

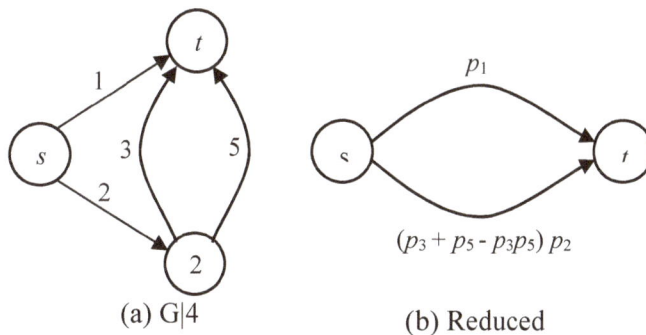

(a) G|4   (b) Reduced

Figure 9.18. Factoring on arc 4.

Note that $R_{st}(G|4) = p_1 + (p_2 p_3 + p_2 p_5 - p_2 p_3 p_5) - p_1 (p_2 p_3 + p_2 p_5 - p_2 p_3 p_5)$. Now assume the arc is not performing as intended. In this case arc 4 must be removed from the graph but its end nodes remain as active components of the network. This is illustrated in the Figure 9.19.

364

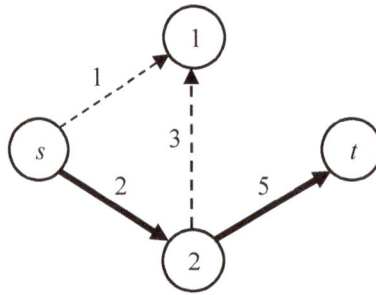

Figure 9.19. *G-4 Network.*

Since arcs 1 and 3 cannot be used to arrive at terminal t from source node s, both arcs can be removed. Therefore, $R_{st}(G-1) = p_2 p_5$. Now we use the pivotal decomposition formula, in Eq. (9-10).

$$R_{st} = p_4 \left[p_1 + (p_2 p_3 + p_2 p_5 - p_2 p_3 p_5) - p_1 (p_2 p_3 + p_2 p_5 - p_2 p_3 p_5) \right] + (1 - p_4) p_2 p_5$$

$$R_{st} = p_1 p_4 + p_2 p_5 + p_2 p_3 p_4 - p_1 p_2 p_3 p_4 - p_1 p_2 p_4 p_5 - p_2 p_3 p_4 p_5 + p_1 p_2 p_3 p_4 p_5$$

Although this is an effective and insightful method for computing the reliability of a network, it may become problematic in the case of larger and more complex networks.

9.5 INTRODUCTION TO ALGEBRAIC METHODS

This section is intended to provide the reader with a brief introduction to algebraic methods for reliability analysis based on the work by Shier [11, 12]. We consider a directed network $G = (\mathbf{N}, \mathbf{A})$ with source node s and terminal node t. It is assumed that only the m arcs in $\mathbf{A}$ can fail. Furthermore, let us associate a variable x_k with each arc $k = 1, 2, \ldots, m$. The main goal of the algebraic methods is to aid in the development of a source-terminal *reliability polynomial* $R_{st}(x_1, x_2, \ldots, x_m)$ having the property that if the reliabilities p_k are substituted for the corresponding variables x_k, for $k = 1, 2, \ldots, m$, then the resulting numerical value of the calculation is equal to the source-terminal reliability R_{st}.

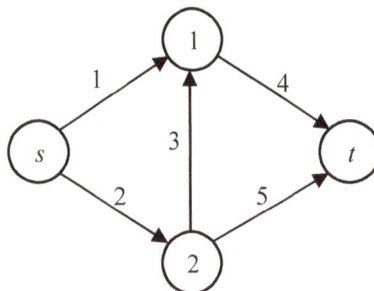

For example, in the case of the network shown in Figure 9.12 (displayed above for convenience), which we have used to illustrate the four methods presented in Section 9.4, it can be proved that the reliability polynomial is formulated as

$$R_{st}(x_1, x_2, x_3, x_4, x_5) = x_1 x_4 + x_2 x_5 + x_2 x_3 x_4 - x_1 x_2 x_4 x_5 - x_1 x_2 x_3 x_4 - x_2 x_3 x_4 x_5 + x_1 x_2 x_3 x_4 x_5$$

365

Now, if we substitute the reliabilities of the arcs for the corresponding variables, we obtain the same result given by the four illustrative examples of Section 9.4.

$$R_{st} = p_1 p_4 + p_2 p_5 + p_2 p_3 p_4 - p_1 p_2 p_3 p_4 - p_1 p_2 p_4 p_5 - p_2 p_3 p_4 p_5 + p_1 p_2 p_3 p_4 p_5$$

From Figure 9.12, we observe that the following three paths (described as sequences of arcs) connect the source node s to the terminal node t:

$$P_1 = \{1, 4\}$$
$$P_2 = \{2, 3, 4\}$$
$$P_3 = \{2, 5\}$$

As previously indicated, the variables x_1, x_2, x_3, x_4, and x_5 are associated with arcs 1, 2, 3, 4 and 5, respectively. Furthermore, each *path* is associated with a *mononomial* defined as the *product* of its *arc variables*. Thus, the three paths we are considering are associated with the following sets S_k and monomials $w^{(k)}$, for $k = 1, 2, 3$:

$$P_1 = \{1, 4\} \Rightarrow w^{(1)} = x_1 x_4, \quad S_1 = \{x_1, x_4\}$$
$$P_2 = \{2, 3, 4\} \Rightarrow w^{(2)} = x_2 x_3 x_4, \quad S_2 = \{x_2, x_3, x_4\}$$
$$P_3 = \{2, 5\} \Rightarrow w^{(3)} = x_2 x_5, \quad S_3 = \{x_2, x_5\}$$

To generalize this illustration, let us define the following multivariable monomials and the associated sets of variables:

$$w^{(i)} = x_{i_1} x_{i_2} \cdots x_{i_r}, \, i = 1, 2$$

$$S_i = \{x_{i_1}, x_{i_2}, \cdots, x_{i_r}\}, \, i = 1, 2$$

The first operation $\otimes$ and the second operation $\oplus$ on $w^{(1)}$ and $w^{(2)}$ are defined as

$$w^{(1)} \otimes w^{(2)} = \prod_{x_k \in S_1 \cup S_2} x_k$$

$$w^{(1)} \oplus w^{(2)} = w^{(1)} + w^{(2)} - w^{(1)} \otimes w^{(2)}$$

These two operations can actually be used in the case of two multivariable *polynomials f* and *g*. In this case we define

$$f \oplus g = f + g - f \otimes g$$

Table 9.1 lists some properties of these two operations.

Table 9.1. PROPERTIES OF THE OPERATIONS

$f \oplus f = f$	$f \otimes f = f$
$f \oplus g = g \oplus f$	$f \otimes g = g \otimes f$
$f \oplus (g \oplus h) = (f \oplus g) \oplus h$	$f \otimes (g \otimes h) = (f \otimes g) \otimes h$
$f \otimes (g \oplus h) = f$	$f \oplus (g \otimes h) = f$
$f \otimes (g \oplus h) = (f \otimes g) \oplus (f \otimes h)$	$f \oplus (g \otimes h) = (f \oplus g) \otimes (f \oplus h)$
$f \oplus 0 = f$	$f \otimes 1 = f$

Now let ϑ_{st} be the set of all simple paths from the source node s to the terminal node t. For any path $P \in \vartheta_{st}$ we define the *value* of the path as the "product", using the $\otimes$ operation, of the variables x_k associated with the corresponding arcs $k \in P$; more specifically,

$$v(P) = \prod_{\otimes} (x_k \mid k \in P)$$

The source-terminal reliability polynomial is equal to the "sum", using the $\oplus$ operation, of path values for all simple paths from the source node to the terminal node of the network; more specifically,

$$R_{st}(\mathbf{x}) = \sum_{\oplus, P \in \vartheta_{st}} P(v)$$

Returning to the illustration based on the network of Figure 9.12, the reliability polynomial is $R_{st}(\mathbf{x}) = x_1x_4 \oplus x_2x_3x_4 \oplus x_2x_5$. To simplify the evaluation of $R_{st}(\mathbf{x})$, we will define $A = x_1x_4 \oplus x_2x_3x_4$ and write $R_{st}(\mathbf{x}) = A \oplus x_2x_5$. Furthermore, $A = x_1x_4 \oplus x_2x_3x_4 = x_1x_4 + x_2x_3x_4 - x_1x_4 \otimes x_2x_3x_4 = x_1x_4 + x_2x_3x_4 - x_1 x_2 x_3x_4$. Therefore, $R_{st}(\mathbf{x}) = A \oplus x_2x_5 = A + x_2x_5 - A \otimes (x_2x_5) = (x_1x_4 + x_2x_3x_4 - x_1 x_2 x_3x_4) + x_2x_5 - (x_1x_4 + x_2x_3x_4 - x_1x_2x_3x_4) \otimes x_2x_5 = x_1x_4 + x_2x_3x_4 - x_1x_2x_3x_4 + x_2x_5 - x_2x_5 \otimes (x_1x_4 + x_2x_3x_4 - x_1x_2x_3x_4)$. This result can be simplified as $R_{st}(\mathbf{x}) = x_1x_4 + x_2x_5 + x_2x_3x_4 - x_1x_2x_3x_4 - x_1x_2x_4x_5 - x_2x_3x_4x_5 + x_1x_2x_3x_4x_5$. Thus, the source-terminal reliability is equal to:

$$R_{st} = p_1p_4 + p_2p_5 + p_2p_3p_4 - p_1p_2p_3p_4 - p_1p_2p_4p_5 - p_2p_3p_4p_5 + p_1p_2p_3p_4p_5$$

9.6 CLOSING PARAGRAPHS

The study of network reliability has evolved considerably over the past few decades. Network models have been formulated by a growing number of authors to study the reliability of several significant engineering systems, especially in the areas of computers and communication, transportation, electric transmission, oil and gas distribution, and, more recently, national security. Estimating the reliability of such networks is a critical undertaking for today's modern societies and nations. The main purpose of this introductory chapter has been to emphasize on the importance of the topic of network reliability and to provide a foundation for network reliability measurement.

The chapter has focused on concepts and metrics to quantify connectivity aspects of deterministic networks and assess the exact reliability of probabilistic networks. Several methodologies have been presented and illustrated with small numerical examples for both types of networks. We have selected topics that are of a foundational nature and illustrated them with small numerical examples. In the case of deterministic networks, the chapter presented two methods to quantify the number of either arcs or nodes to disconnect a network. In the case of probabilistic networks, the methods discussed included state enumeration, inclusion-exclusion, sum of disjoint product, factoring, and a brief introduction to algebraic methods.

There are many other methods that were not included in the chapter; for the reader interested in these methodologies we have selected a few relevant bibliographical references at the end of the chapter. In the current literature of network reliability, the reader can easily find now many articles that present the developments in this domain at different levels of exposition.

EXERCISES

1. Find the minimum number of arcs that need to be removed to disconnect node 2 from node 3.

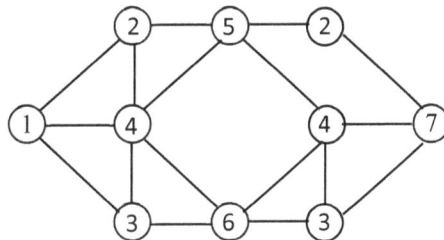

2. In Exercise 1 find (a) the number of node disjoint paths between nodes 3 and 4. (b) the number of arc disjoint paths between nodes 3 and 4. (c) What do these results mean?

3. Obtain the results from the Harary inequality for the network of Exercise 1.

4. In Exercise 1, we wish to determine if the removal of as many as five nodes will disconnect the network.

5. Use the state enumeration method to determine the reliability of the given network. The probabilities of success of arcs 1, 2, and 4 are equal to o.70, 0.90, and 0.75, respectively.

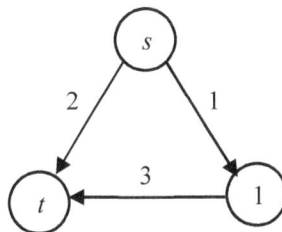

6. Use the state enumeration method to determine the reliability of the given network. Let the probability of success of arc k be p_k and let the probability of failure be $q_k = 1 - p_k$.

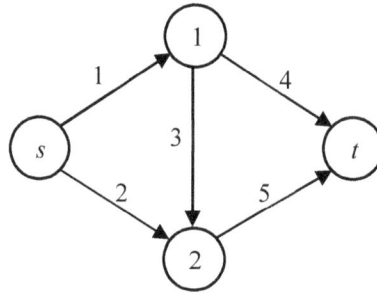

7. In the Example of section 9.4.6 compute the source-terminal reliability assuming the order of events E_1 - E_3 – E_3.

8. For the given graph, use the disjoint product method considering the events in the following order: E_1 = Path 1-5 works; E_2 = Path 1-3-6 works; E_3 = Path 2-6 works; E_4 = Path 2-4-5 works;

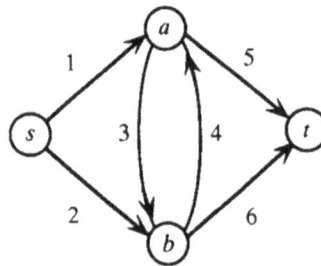

9. Use the factoring method by conditioning on the state of arc 1 in network given for Exercise 6.

10. Use the inclusion-exclusion method for the graph of Exercise 6.

11. Find the reliability polynomial for the graph of the network shown in Exercise 8.

12. Consider two arcs a and b . Suppose that arc a performs successfully if components 1, 2, and 3 work satisfactorily. Similarly, suppose that arc b performs successfully if components 3, 4, 5, and 6 operate satisfactorily. Find the reliability polynomial for this simple system if (a) the arcs are connected in series; (b) if the arcs are connected in parallel.

REFERENCES

[1] FRANK H. AND I. T. FRISCH, "Network Analysis," *Scientific American*, Vol. 223, pp. 94-103, July 1970.

[2] FRANK H. AND I. T. FRISCH, "Communication, Transmission and Transportation Networks," *Addison-Wesley Publishing Co.*, Inc., Reading, Mass, 1971.

[3] FRISCH, I. T., "An Algorithm for Vertex-pair Connectivity," *International Journal of Control*, 6:6, 579-593, (1967).

[4] HARARY, F., "Graph theory," *Addison-Wesley Pub. Co.*, Reading, Mass, 1969.

[5] HWANG, C. L., F. TILLMAN AND M. LEE, "System-Reliability Evaluation Techniques for Complex/Large Systems: A Review," *IEEE Transactions on Reliability*, 1981.

[6] KLEITMAN, D. "Methods for Investigating Connectivity of Large Graphs," *IEEE Transactions on Circuit Theory*, vol. 16, no. 2, pp. 232-233, May 1969.

[7] MOSKOWITZ, F. The Analysis of Redundancy Networks. Transactions of the American Institute of Electrical Engineers, Part I: Communication and Electronics, 77, 627-632, 1958.

[8] MOORE, E.F. AND SHANNON, C.E., "Reliable circuits using less reliable relays," *J. Frankl. Inst.*, **262**(3), 191–208, 1956.

[9] PÉREZ-ROSÉS , H., "Sixty Years of Network Reliability," *Mathematics in Computer Science*, Volume 12, pages275–293, 2018.

[10] REBAIAIA, M.L. AND AIT-KADI, D., "Network Reliability Evaluation and Optimization: Methods, Algorithms and Software Tools," Interuniversity research Centre on Enterprise Networks, *Logistics and Transportation* (Cirrelt), Laval University, Quebec, Canada, December 2013.

[11] SHIER, D. R., "Network Reliability and Algebraic Structures", *Clarendon Press*, Oxford, 1991.

[12] SHIER, D. R. AND D. E. WHITED, "Algebraic Methods Applied to Network Reliability Problems," *SIAM Journal on Algebraic Discrete Methods*, Vol. 8, No. 2, pp. 251-262, 1987.

[13] STEIGLITZ K. AND J. BRUNO, "A New Derivation of Frisch's Algorithm for Calculating Vertex-Pair Connectivity," BIT. *Numerical Mathematics* 11(1): 94-106, February 1971.

BIBLIOGRAPHY

CHAPTER 1

[1] AHUJA R.K., T.L. MAGNANTI, AND J.B. ORLIN, *Network Flows: Theory, Algorithms, and Applications*, Prentice-Hall, Englewood Cliffs, New Jersey, 1993.

[2] BAZARRA, M., AND J. J. JARVIS, *Linear Programming and Network Flows*. John Wiley & Sons, Inc., New York, 1978.

[3] BERTSEKAS, D.P., *Linear Network Optimization*, The MIT Press, Cambridge, Massachusetts, 1991.

[4] BERTSEKAS, D. P., *Network Optimization: Continuous and Discrete Models*. Athena Scientific, 1998.

[5] BIGGS, N.L, E.K. LLOYD, AND R.J. WILSON, *Graph Theory: 1736-1936*, Clarendon Press, Oxford, 1976.

[6] BRADLEY, G. H., "A Survey of Deterministic Networks," *AIIE Transactions, 7*, 222-234 (1975).

[7] BUSACKER, R. G., AND T. SAATY, *Finite Graphs and Networks*. McGraw-Hill Book Company, New York, 1965.

[8] CHARNES, A., AND W. W. COOPER, *Management Models and Industrial Applications of Linear Programming, Vols. 1* and 2. New York: John Wiley & Sons, Inc, 1961.

[9] DANTZIG, G. L., *Linear Programming and Extensions*. Princeton, N.J.: Princeton University Press, 1963.

[10] ELMAGHRABY, *S., Some Network Models in Management Science*. New York: Springer-Verlag, Inc., 1970.

[11] EULER, L., "Solutio Problematis Ad Geometriam Situs Pertinentis [The solution of a problem relating to the geometry of position], Translated into English: BIGGS, N.L, E.K. LLOYD, AND R.J. WILSON, *Graph Theory: 1736-1936*, Clarendon Press, Oxford, 3-11, 1976.

[12] EVANS, J.R. AND E. MINIEKA, *Optimization Algorithms for Networks and Graphs*, 2nd Edition, New York: Marcel Dekker, Inc., 1992.

[13] FORD, L. R., AND D. R. FULKERSON, *Flows in Networks*. Princeton, N.J.: Princeton University Press, 1962.

[14] FRANK, H., AND I. T. FRISCH, *Communication, Transmission, and Transportation Networks*. Reading, Mass.: Addison-Wesley Publishing Co., Inc., 1971.

[15] FULKERSON, D. R., "Flow Networks and Combinatorial Operations Research," *American Mathematical Monthly, 73*, 115-138 (1966).

[16] GLOVER F., KLINGMAN, D., AND N.V. PHILLIPS, *Network Models in Optimization and Their Applications in Practice*, New York: John Wiley & Sons, Inc., 1992

[17] HITCHCOCK, F. L., "The Distribution of a Product from Several Sources to Numerous Localities," *Journal of Mathematics and Physics,* 20, 224-230 (1941).

[18] HU, T. *C., Integer Programming and Network Flows.* Reading, Mass.: Addison Wesley Publishing Co., Inc., 1969.

[19] IRI, M., *Network Flow, Transportation and Scheduling.* New York: Academic Press, Inc., 1969.

[20] JENSEN, P. A., AND W. BARNES, *Network Flow Programming.* New York: John Wiley & Sons, Inc., 1980.

[21] KENNINGTON, J. F., "A Survey of Linear Multicommodity Network Flows," *Operations Research,* 26(2):209-236, 1978.

[22] KOOPMANS, T. C., "Optimum Utilization of the Transportation System," *Proceedings of the International Statistical Conference,* Washington, D.C., 1947.

[23] LAWLER, LENSTRA, RINNOYKAN AND SHMOYS, 1985.

[24] MAGNANTI, T. L., AND B. L. GOLDEN, "Transportation Planning: Network Models and Their Implementation," Working Paper 77-008, University of Maryland, General Research Board, Faculty Research Award, 1977.

[25] MINIEKA, *E., Optimization Algorithms for Networks and Graphs.* New York: Marcel Dekker, Inc., 1978.

[26] NEWMAN, J.R., *The World of Mathematics,* Simon and Schuster, New York, 1956.

[27] PHILLIPS, D.T. AND A. GARCIA-DIAZ, *Fundamentals of Network Analysis,* Englewood Cliffs, New Jersey: Prentice-Hall, Inc., 1981.

[28] PRITSKER, A.A.B., AND W.W. HAPP, "GERT: PART I – Fundamentals," Journal of Industrial Engineering, 17(5), 267 (1966).

[29] SCIENTIFIC AMERICAN, *The Konigsberg Bridges,* vol. 189, pp. 66-70, 1953.

[30] SCHRIJVER, A., "On the History of Combinatorial Optimization (Till 1960)," *Handbooks in Operations Research and Management Science,* Vol. 12, pp. 2-68, ELSEVIER, 2005.

[31] WHITEHOUSE, G.E., *Systems Analysis and Design Using Network Techniques.* Englewood Cliffs, New Jersey: Prentice-Hall, 1973.

CHAPTER 2

[1] BELLMORE, M., G. BENNINGTON, AND S. LUHORE, "A Multivehicle Tanker Scheduling Problem," *Transportation Science,* 5, 36-47 (1971).

[2] BENNINGTON, G. E., "Applying Network Analysis," *Journal of Industrial Engineering,* 6 (1), 17-25 (1974).

[3] BRADLEY, G. H., "Survey of Deterministic Networks," *AIIE Transactions,* 7, 222-234 (1975).

[4] DANTZIG, G. B., W. BLATTNER, AND M. R. RAO, "Finding a Cycle in a Graph with a Minimum Cost to Time Ratio with Applications to a Ship Routing Problem," *Theory of*

Graphs International Symposium, pp. 77-83. Paris: Dunod, and New York: Gordon and Breach, 1966.

[5] DANTZIG, G.B. AND FULKERSON, "Minimizing the Number of Tankers to Meet a Fixed Schedule," *Naval Research Logistics Quarterly*, 1, 217-222 (1954).

[6] DIJKSTRA, E. W., "A Note on Two Problems in Connection with Graphs," *Numerishe Mathematik,* 1, 269-271 (1959).

[7] DREYFUS, S. E., "An Appraisal of Some Shortest Path Algorithms," *Operations Research,* 17, 395-412 (1969).

[8] DREYFUS, S. E., "A Generalized Equipment Replacement Study," *Journal of the Society for Industrial and Applied Mathematics,* 8, 425-435 (1960).

[9] EVANS, J. R., "Network Modeling in Production Planning," *Proceedings of the 1978 AIIE National Systems Conference,* Montreal, Canada, 1978.

[10] FLOYD, R. W., "ALGORITHM 97: Shortest Path," *Communications of the ACM*, 5, 345 (1962).

[11] FORD, L. R., AND D. R. FULKERSON, "Maximal Flow through a Network," *Canadian Journal of Mathematics*, 18, 399-404 (1956).

[12] FORD, L. R., AND D. R. FULKERSON, *Flows in Networks.* Princeton, N.J.: Princeton University Press, 1962.

[13] FRANK, H., AND I. T. FRISCH, *Communication, Transmission and Transportation Networks.* Reading, Mass.: Addison-Wesley, Publishing Co., Inc., 1971.

[14] FREDMAN AND TARJAN, "Fibonacci Heaps and their Use in Improved Network Optimization Algorithms," *Journal of ACM,* 34, 596-615 (1987).

[15] GARCIA-DIAZ, A., "A Network Flow Approach to Airline Fuel Allocation Problems," *The Annals of the Society of Logistics Engineers*, Vol. 2, No. 1, 39-53, 1990.

[16] GARFINKEL, R. S., AND G. L. NEMHAUSER, *Integer Programming.* New York: John Wiley & Sons, Inc., 1972.

[17] GLOVER, F., AND D. KLINGMAN, "Network Applications in Industry and Government," *AIIE Transactions,* 9 (4) (1977).

[18] GOMORY, R. E., AND T. C. HU, "Multi-terminal Network Flows," SIAM *J. Soc. Indust. Appl. Math,* 9, 551-571 (1971).

[19] HELLER, L, AND C. B. TOMPKINS, "An Extension of a Theorem of Dantzig's," in *Linear Inequalities and Related Systems,* ed. H. Kuhn and A. W. Tucker. Princeton, N.J.: Princeton University Press, 1956.

[20] HITCHCOCK, F. L., "The Distribution of a Product from Several Sources to Numerous Localities," *Journal of Mathematics and Physics,* 20, 224-230 (1941).

[21] HOFFMAN, A. J., AND S. WINOGRAD, "Finding All Shortest Distances in a Directed Network," *IBM Journal of Research and Development,* 16, 412-414 (1972).

[22] HU, T. C., *Integer Programming and Network Flows.* Reading, Mass.: Addison Wesley Publishing Co., Inc., 1969.

[23] HU T. C., "The Maximum Capacity Route Problem," *Operations Research,* 9, 898-900 (1961).

[24] JACOBS, W. W., "The Caterer Problem," *Naval Research Logistics Quarterly,* 1, 154-165 (1954).

[25] LAWLER, E. L., "Optimal Cycles in Doubly Weighted Linear Graphs," In *Theory of Graphs: International Symposium*, Paris: Dunod, New York: Gordon and Breach, 209-213 (1966).

[26] MINIEKA, E. T., AND D. R. SHIER, "A Note on an Algebra for the *K* Best Routes in a Network," *Journal of the IMA,* 11, 145-149 (1973).

[27] NEMHAUSER, G.L., AND L.A. WOLSEY, *Integer and Combinatorial Optimization*, Wiley-Interscience, John Wiley & Sons, Inc., New York, 1988.

[28] POLLACK, M., "The Maximum Capacity Route through a Network," *Operations Research,* 8, 733-736 (1960).

[29] SHIER, D. R., "Iterative Methods for Determining the *K* Shortest Paths in a Network," *Networks,* 6, 205-230 (1976).

[30] SHIER, D. R., "Computational Experience with an Algorithm for Finding the K Shortest Paths in a Network," *Journal of Research, National Bureau of Standards,* 78B, 139-165 (July-September 1974).

[31] SMYTHE, W. R., AND L. JOHNSON, *Introduction to Linear Programming with Applications.* Englewood Cliffs, N.J

CHAPTER 3

[1] DURBIN, E. P., "The Out-of-Kilter Algorithm: A Primer," Rand Corporation, Santa Monica, California, December 1967.

[2] FORD, L. R., AND D. R. FULKERSON, "Maximal Flow through a Network," Canadian Journal of Mathematics (August 1956).

[3] FORD, L. R., AND D. R. FULKERSON, Flows in Networks. Princeton, N.J.: Princeton University Press, 1962.

[4] FULKERSON, D. R., "The Out-of-Kilter Method for Minimal Cost Flow Problems," Journal of Applied Mathematics, 9 (1) (March 1961).

[5] JENSEN, P. A., AND W. BARNES, Network Flow Programming. New York: John Wiley & Sons, Inc., 1979.

[6] PHILLIPS, D. T., AND P. A. JENSEN, "Network Flow Optimization with the Out-of-Kilter Algorithm, Part I-Theory," Research Memorandum 71-2, Purdue University, February 1971.

[7] PHILLIPS, D. T., AND P. A. JENSEN, "Network Flow Optimization with the Out-of-Kilter Algorithm, Part II-Applications," Research Memorandum 71-3, Purdue University, February 1971.

[8] PHILLIPS, D. T., AND P. A. JENSEN, "Network Flow Analysis: The Out-of-Kilter Algorithm," Industrial Engineering (February 1974). Portions reproduced by permission of the authors and the American Institute of Industrial Engineers.

[9] PHILLIPS, D. T., A. RAVINDRAN, AND J. J. SOLBERG, Operations Research: Principles and Practice. New York: John Wiley & Sons, Inc., 1977.

[10] SARA, J. L., "An Algorithm for Bus Scheduling Problems," Operational Research Quarterly, 21 (4) (December 1970).

[11] SWANSON, H. S., R. E. D. WOOLSEY, AND H. HILLIS, "Using the Out-of-Kilter Algorithm," Industrial Engineering (March 1974). Portions reproduced by permission of the authors and the American Institute of Industrial Engineers.

[12] SWANSON, H. S., AND R. E. D. WOOLSEY, "An Out-of-Kilter Network Tutorial," *ACM SIGMAP Bulletin*, January 1973.

[13] VAIDA, Mathematical Programming. Reading, Mass.: Addison-Wesley Publishing Co., Inc., 1961.

CHAPTER 4

[1] DANTZIG, G.B. (1951) "Application of the Simplex Method to a Transportation Problem," *Activity Analysis of Production and Allocation*. Koopmans, T.C., Ed., John Wiley and Sons, New York, 359-373, 1951.

[2] JENSEN, P., "Network Flow Programming," Department of Mechanical Engineering, The University of Texas at Austin, 1983.

[3] JENSEN P., "Microsolve Network Flow Programming," Department of Mechanical Engineering, The University of Texas at Austin, 1983.

[4] JOHNSON, E., "Networks and basic solutions", *Operations Research*, 14, 619–623, 1966.

CHAPTER 5

[1] BELLMORE, M. AND G. L. NEMHAUSER, The traveling Salesman Problem: A Survey. *Operations Res.* 16 (1968) 538-558.

[2] BELLMORE, M. AND S. HONG, Transformation of Multi-Salesman Problem to the Standard Traveling Salesman Problem. J. *ACM.* 21 (1974) 500-504.

[3] CHRISTOFIDES, N. AND S. EILON, An Algorithm for The Vehicle Dispatching Problem. *Operational Res. Quart.* 20 (1969) 309-318.

[4] CHRISTOFIDES, N. AND S. EILON, Algorithm for Large-Scale Traveling Salesman Problems. *Operational Res. Quart.* 23 (1972) 511-518.

[5] GARCIA-DIAZ, A., A Heuristic Circulation-Network Approach to Solve the Multi-Traveling Salesman Problem," *Networks*, Vol. 15, pp. 455-467, 1985.

[6] GAREY, M. R. AND D. S. JOHNSON, Computers and Intractability: A Guide to the Theory of NP-Completeness, 1979.

[7] GAVISH, B. AND K. SRIKANTH, An Optimal Solution Method for the Multiple Traveling Salesman Problem. *Operations Research,* 34 (1986), 6, 698-710, 1986.

[8] LAWLER, E. L., J. K. LENSTRA, AH.G. RINNOOY KAN, AND D. B. SHMOYS, The Traveling Salesman Problem, John Wiley & Sons, New York, 1985.

[9] LAWLER, E. L., A Solvable Case of The Traveling Salesman Problem, *Math. Programming* 1, (1971) 267-269.

[10] LITTLE, J.D.C., K. G. MURTY, D. W. SWEENEY, AND C. KAREL, An Algorithm for the Traveling Salesman Problem, *Operations Research,* 11 (1963) 972-978.

CHAPTER 6

[1] DAVIS, E. W., AND G. E. HEIDORN, "An Algorithm for Optimal Project Scheduling under Multiple Resource Constraints," *Management Science* (August 1971).

[2] DAVIS, E. W., "Resource Allocation in Project Network Models-A Survey," *Journal of Industrial Engineering* (April 1966).

[3] ELMAGHRABY, S.E., *Activity Networks: Project Planning and Control by Network Models*, New York: John Wiley & Sons, 1977.

[4] FALK, J.E., AND J.L. HAROWITZ, "Critical Path Problems with Concave Cost-Time Curves," *Management Science*, 19(4), 446-455, 1972.

[5] FULKERSON, D. R., "A Network Flow Computation for Project Cost Curves," *Management Science* (January 1961).

[6] KELLEY, J. E., AND M. R. WALKER, "Scheduling Activities to Satisfy Resource Constraints," in *Industrial Scheduling*, by J. Muth and G. Thompson. Englewood Cliffs, N.J.: Prentice-Hall, Inc., 1963.

[7] KELLEY, J. E., AND M. R. WALKER, "Critical Path Planning and Scheduling," *Proceedings of the Eastern Joint Computer Conference*, 1959.

[8] KELLEY, J. E., AND M. R. WALKER, "Critical Path Planning and Scheduling: Mathematical Basis," *Operations Research* (May-June 1961).

[9] KUYUMCU, A, "A Decomposition Approach to Project Compression in CPM/PERT Networks with Concave Activity Cost Functions," *Master of Science Thesis*, Department of Industrial Engineering, Texas A&M University, August 1991.

[10] KUYUMCU, A. AND A. GARCIA-DIAZ, "A Decomposition Approach to Project Compression with Concave Activity Cost Functions," *IIE Transactions*, Industrial Engineering Research & Development, Volume 26, No. 6, pp.63-73, 1994.

[11] LEVY, F. K., AND J. D. WIEST, A Management Guide to PERT/CPM, Englewood Cliffs, N.J.: Prentice-Hall, Inc., 1969.

[12] MODER, J. J., AND C. R, PHILLIPS, *Project Management with CPM and PERT*. New York: Van Nostrand Reinhold Company, 1964; 2nd ed., 1970.

[13] PANAGIOTAKOPOULOS, D., "Cost-Time Model for Large CPM Project Networks," *Journal of the Construction Division*, ASCE, Vol. 103, No. CO2, pp. 201-211, 1977.

[14] PHILLIPS, S. JR, AND M.I. DESSOUKY, "The Cut Search Algorithm with Arc Capacities and Lower Bounds," *Management Science*, 25(4), 396-404, 1979.

[15] SHAFFER, L.R., J.B. RITTER, AND W.L. MEYER, *The Critical Path Method*. New York: McGraw-Hill Book Company, 1964.

[16] THOMAS, W.H., "Four Float Measures for Critical Path Scheduling," *Industrial Engineering* (October 1969).

[17] TUFEKCI, S., "A Flow Preserving Algorithm for Time-Cost Tradeoff Problem," *AIIE Transactions*, Vol. 12, No. 3, 1982.

CHAPTER 7

[1] MASON, S.J., "Feedback Theory: Some Properties of Signal Flow Graphs," *Proc. Inst. Radio Engrs.*, 41, 1144-1156, September 1953.

[2] MASON, S.J., "Feedback Theory: Further Properties of Signal Flow Graphs," *Proc. Inst. Radio Engrs.*, 44, 920-926, July 1956.

[3] PHILLIPS, D. T. AND D. R. SMITH, "Determination of Probabilistic Time Standards for Tasks Performed Under Uncertainty," Proceedings of the AIIE National Conference, San Francisco, Ca., 1979.

[4] PRITSKER, A. A. B., AND W. HAPP, "GERT: Graphical Evaluation and Review Technique, Part I. Fundamentals," *Journal of Industrial Engineering*, pp. 267-274, May 1966.

[5] PRITSKER, A. A. B., AND G. E. WHITEHOUSE, "GERT: Graphical Evaluation and Review Technique, Part II. Probabilistic and Industrial Engineering Applications," *Journal of Industrial Engineering*, pp. 293-301, June 1966.

[6] PRITSKER, A. A. B., The Production Engineer, pp. 499-506, October 1968.

[7] WHITEHOUSE, GARY E., System Analysis and Design Using Network Techniques. Englewood Cliffs, N.J.: Prentice-Hall, Inc., 1973.

[8] WHITEHOUSE G. E, AND A. B. PRITSKER, "GERT-Generating Functions, Conditional Distributions, Counters, Renewal Times and Correlations," *AIIE Transaction* 1969, 1(1): 45-50.

CHAPTER 8

[1] AHUJA, R., MAGNANTI, T., AND ORLIN, J., *Network Flows-Theory, Algorithms, and Applications*. Prentice-Hall, Inc., 1993.

[2] ASSAD, A. A., "Multicommodity Network Flows-A Survey," *Networks* 8, 37-91 (1978).

[3] BAZARAA, M. S., J. J. JARVIS, AND H. D. SHERALI, *Linear Programming and Network Flows*, John Wiley & Sons, Inc., 1990.

[4] BELLMORE, M., G. BENNINGTON, AND S. LUBORE, "A Multivehicle Tanker Scheduling Problem," *Transportation Sciences, 5,* 35-47 (1971).

[5] DANTZIG G. B. AND P. WOLFE, "Decomposition Principle for Linear Programs," *Operations Research*, 8, 101-111, 1960.

[6] EVANS, J. R., J.J. JARVIS, AND R. A. DUKE, "Graphic Matroids and the multicommodity Transportation Problem," *Mathematical Programming, 13*, 323-328 (1977).

[7] EVANS, J. R., AND J.J. JARVIS, "Network Topology and Integral Multicommodity Flow Problems," *Networks, 8*, 107-119 (1978).

[8] EVANS, J. R., "A Combinational Equivalence between a Class of Multicommodity Flow Problems and the Capacitated Transportation Problem," *Mathematical Programming*, 10, 401-404 (1976).

[9] EVANS, J.R., "A Single Commodity Transformation for Certain Multicommodity Networks," *Operations Research, 26*, 673-680 (1978).

[10] EVANS, J. R., "An Equivalent Formulation of Certain Multicommodity Networks as Single Commodity Flow Problems," *Mathematical Programming*, 15, 92-99 (1978).

[11] EVANS, J. R., "Solving Multicommodity Transportation Problems through Aggregation," ORSA/TIMS National Conference, Los Angeles, November 1978.

[12] EVANS, J. R., "Model Simplification in Multicommodity Distribution Systems though Aggregation," *Proceedings, American Institute of Decision Sciences,* National Meeting, New Orleans, November 1979.

[13] GEOFFRION, A., "Customer Aggregation in Distribution Modeling," Working Paper No. 259, Western Management Science Institute, University of California, Los Angeles, 1976.

[14] HU, T. C., "Multicommodity Network Flows," *Operations Research, 11*, 344-360 (1963).

[15] JARVIS, J. J., AND D. D. Miller, "Maximal Funnel-Node Flow in an Undirected Network," *Operations Research, 11*, 344-360 (1963).

[16] KENNINGTON, J. L., "A Survey of Linear Cost Multicommodity Network Flows," *Operations Research, 26*, 209-236 (1978).

[17] KENNINGTON, J. L. AND R. V. HELGASON, *Algorithms for Network Programming.* John Wiley & Sons, Inc., 1980.

[18] PHILLIPS, D.T. AND A. GARCIA-DIAZ, *Fundamentals of Network Analysis,* Prentice-Hall International Series in Industrial Engineering, 1981.

[19] SAKAROVITCH, M., "The Multicommodity Maximal Flow Problems," ORC 66-25, University of California, Berkeley, 1966.

[20] WANG, I-Lin, "Multicommodity Network Flows: A Survey, Part I: Applications and Formulations," *International Journal of Operations Research,* Vol. 15, No. 4, 145-153 2018).

[21] WANG, I-Lin, "Multicommodity Network Flows: A Survey, Part II: Solution Methods," *International Journal of Operations Research,* Vol. 15, No. 4, 155-173 (2018).

[22] WARDROP, J.G., "Some Theoretical Aspects of Road Traffic Research". *Proceedings of the Institution of Civil Engineers.* **1** (3): 325–362, 1952.

[23] ZIPKIN, P. H., "Aggregation in Linear Programming," Ph.D. dissertation, Yale University, December 1977.

CHAPTER 9

[1] FRANK H. AND I. T. FRISCH, "Network Analysis," *Scientific American*, Vol. 223, pp. 94-103, July 1970.

[2] FRANK H. AND I. T. FRISCH, "Communication, Transmission and Transportation Networks," *Addison-Wesley Publishing Co.*, Inc., Reading, Mass, 1971.

[3] FRISCH, I. T., "An Algorithm for Vertex-pair Connectivity," *International Journal of Control*, 6:6, 579-593, (1967).

[4] *HARARY*, F., *"Graph theory,"* *Addison-Wesley Pub. Co.*, Reading, Mass, 1969.

[5] HWANG, C. L., F. TILLMAN AND M. LEE, "System-Reliability Evaluation Techniques for Complex/Large Systems: A Review," *IEEE Transactions on Reliability*, 1981.

[6] KLEITMAN, D. "Methods for Investigating Connectivity of Large Graphs," *IEEE Transactions on Circuit Theory*, vol. 16, no. 2, pp. 232-233, May 1969.

[7] MOSKOWITZ, F. The Analysis of Redundancy Networks. Transactions of the American Institute of Electrical Engineers, Part I: Communication and Electronics, 77, 627-632, 1958.

[8] MOORE, E.F. AND SHANNON, C.E., "Reliable circuits using less reliable relays," *J. Frankl. Inst.*, **262**(3), 191–208, 1956.

[9] PÉREZ-ROSÉS , H., "Sixty Years of Network Reliability," *Mathematics in Computer Science*, Volume 12, pages275–293, 2018.

[10] REBAIAIA, M.L. AND AIT-KADI, D., "Network Reliability Evaluation and Optimization: Methods, Algorithms and Software Tools," Interuniversity research Centre on Enterprise Networks, *Logistics and Transportation* (Cirrelt), Laval University, Quebec, Canada, December 2013.

[11] SHIER, D. R., "Network Reliability and Algebraic Structures", *Clarendon Press*, Oxford, 1991.

[12] SHIER, D. R. AND D. E. WHITED, "Algebraic Methods Applied to Network Reliability Problems," *SIAM Journal on Algebraic Discrete Methods*, Vol. 8, No. 2, pp. 251-262, 1987.

[13] STEIGLITZ K. AND J. BRUNO, "A New Derivation of Frisch's Algorithm for Calculating Vertex-Pair Connectivity," BIT. *Numerical Mathematics* 11(1): 94-106, February 1971.